D1553980

FLUID MECHANICS AND THERMODYNAMICS OF OUR ENVIRONMENT

S. ESKINAZI

DEPARTMENT OF MECHANICAL AND AEROSPACE ENGINEERING
SYRACUSE UNIVERSITY
SYRACUSE, NEW YORK

ACADEMIC PRESS New York San Francisco London 1975

A Subsidiary of Harcourt Brace Jovanovich, Publishers

ACADEMIC PRESS, INC.
111 Fifth Avenue, New York, New York 10003

United Kingdom Edition published by
ACADEMIC PRESS, INC. (LONDON) LTD.
24/28 Oval Road, London NW1

Library of Congress Cataloging in Publication Data

Eskinazi, Salamon.
Fluid mechanics and thermodynamics of our environment.

Bibliography: p.
1. Geophysics. 2. Fluid mechanics. 3. Atmospheric thermodynamics. 4. Oceanography. I. Title.
QC806.E84 532 73-18993
ISBN 0-12-242540-5

PRINTED IN THE UNITED STATES OF AMERICA

To my mother whose value to me
has been as good as her name

Contents

3 Basic Principles of Heat Transfer—Energy Balance of the Environment

4 Static Equilibrium of the Environment

5 Basic Principles of Surface Tension

6 Kinematics of the Environment

7 Dynamics of the Environment

8 Geostrophic Motion and Applications

Preface

The word "environment" takes a variety of meanings and is used in a number of contexts to describe the physical, economic, political, educational, and many other situations in which man and his work survive. Until the discovery of massive nuclear experiments thirty years ago, man could not conceive that in the span of one generation he was capable of fouling the physical environment of air and water in which he lives and upon which his survival depends. This lack of consciousness was not based on ignorance of the pertinent facts, but rather on a lack of awareness that man can produce pollutants at a rate faster than the recycling rate of nature and conversely, that nature's capabilities in the restoration of our increasingly fouled environment are limited and certainly not as boundless as it was thought.

Man's personal and collective obsession with harnessing the maximum amount of energy to develop goods and machinery to serve him in almost every physical task has led to the mushrooming of industries and products which consume, at an alarming rate, the physical environment of the earth.

Although at first public concern lagged, there has been in the past five years considerable awakening, followed by strong pressures on the government to regulate the quality of our physical environment by setting standards and ordinances for all its users. This era marks the end of the free exploitation of our environment without an equal effort toward its restoration; the earth will no longer be man's waste collector. It is the beginning of a new consciousness factor in man's equation for survival. As in all beginnings, all the facts, processes, and solutions are not known. We now observe the start of man's commitment to comprehend fully, to manage, and to save his physical environment in order to permit a healthy life for himself and for the generations to come.

Education and research are two important means by which knowledge and its application to our environment can be developed, organized, and transmitted from one generation to the next. This book addresses itself to these means. Even if we limit ourselves to physical phenomena, all the problems related to them are not necessarily physical. Almost every discipline and profession can apply itself to problems related to the management of the environment. The engineer and the lawyer, the physician and the economist, the meteorologist and the political scientist, the biologist and the geographer, all claim with equal enthusiasm that the environment is a legitimate concern in their own discipline.

Historically, meteorologists, oceanologists, geologists, and geographers could claim to be the trustees of our physical environment. This was certainly true until the environment ceased to be a science only and became a practical concern. Although most of the credit for organizing the major share of the physical knowledge goes to them, their specialized training and interest is not sufficient today to solve the variety of problems associated with the modification, purification, and overall management of the environment. This is not to say that their continued interest and contributions are not needed. Rather, they must share the great task of monitoring, purification, and management of the environment with a number of other professions.

This book was developed for engineering classes interested in the motion of the environment which is a main carrier of pollutants. Unless the motion is well understood, the underlying concepts of pollution cannot be well understood. The selection of topics and the emphasis make the material primarily suited for engineering work. A colleague meteorologist who attended these lectures attested to this by saying that the development of the material in the book is based on enunciations of one problem after another, presenting methods and means of solution developed from practical information and classes of problems with similar behavior. This, indeed, was the intent of the author and it is hoped that the readers will find it to be so. The material was put into final form in the process of presenting it twice at Syracuse University and once at the University of Poitiers in France.

Traditionally, fluid mechanics and thermodynamics are two independently taught basic courses in the engineering curriculum for most mechanical, aerospace, chemical, and civil engineers. It has been argued a number of times that (a) since fluids are extensively involved in the application of the fundamentals of both of these fields, (b) since pressure, density, and temperature enter as the principal dependent properties in both fields of study, and (c) because there is hardly an interesting applied motion in which energy is not transferred or conversely that energy is transfered without deformable motion, a course of study should be devised to include the harmonized use of the principles of fluid mechanics and thermodynamics. Because the environment, and particularly the atmosphere, constantly undergoes mechanical and thermodynamic changes, this book has been developed to fit that need.

The author made two determined efforts in the development of this material, namely: to present it in a way not requiring formal courses in fluid mechanics and thermodynamics as prerequisites, and secondly, to maintain a unified approach when dealing with the atmosphere and ocean as a deformable environment. The background in college physics and mathematics that science and engineering students receive as undergraduates is sufficient for the study of this material.

Finally, a word about the metric system of units used in this book. Considerable thought has gone into the decision to use metric units, especially since almost all engineering curriculums and applications in the field in the USA today use the English system. Traditionally, justifications for preferring one system over the other have been advanced along two independent lines, namely ease of use and tradition of industrial affiliation. Although the metric system enjoys the incorporation of the decimal system which, to most, seems simpler than the English pseudobinary system, the main reasons that led to the final adoption of the metric system, for this book, are not based on an argument of ease but rather on an argument of future needs.

Fluid Mechanics and Thermodynamics of Our Environment

1

The Nature of Our Physical Environment

1.1 THE EARTH

The earth is perhaps the only planet in our solar system that is able to hold, through its gravitational pull, a life-sustaining atmosphere and the oceans of water upon which depend so many living organisms. To a considerable degree, this ideal climate on earth is due to the fact that the distance from the sun is just right, resulting in a range of terrestrial surface temperatures compatible with the profusion of many types of living systems. The atmosphere and the bodies of water constitute together, from a mechanical point of view, a category of materials called *fluids*. Generally speaking, we refer to the atmosphere and oceans as *geofluids*, meaning fluids at the scale of, and whose motion is influenced by the motion of, the earth, and they constitute the physical environment dealt with in this book.

The concept that the earth is of spherical shape goes back in history to Pythagoras, 2500 years ago. However, it was not until the seventeenth century, with the discovery of sophisticated optics and the establishment of the laws of gravity, that reliable and accurate data about the geometry and physical characteristics of the earth were made possible.

Because of the discovery of centrifugal force and because of the realization that the distance traveled on Earth, northward, to make the azimuthal height of the North Star increase by 1° depended on latitude, early men knew that the earth was not a perfect sphere. Since then, there have been many ways of determining the major and minor radii of the earth. The major radius at the equator is 6,378.4 km and that at the poles is 6356.9 km. Because the distance along a flat surface on the earth from the equator to the poles is approximately 10,000 km, the linear distance corresponding to 1° is approximately 10,000/90 or 110 km.

Although it was not until 1686 that Sir Isaac Newton stated his law of gravity on the basis of Kepler's planetary observations, it was known through an earlier French scientific expedition to the territories in the West Indies that a clock ran more slowly there than in France. This again raised the suspicion that the radius of the earth at small latitudes was larger than that at larger latitudes. This conclusion is based on the fact that for a constant pendulum arm, the period of the pendulum is inversely proportional to the square root of the gravitational pull which, itself, is inversely proportional to the square of the earth's radius. This set of relationships causes the period of the pendulum to be approximately proportional to the radius of the earth.[1] Since the radius of the earth at the equator is now known to be largest, the pendulum has a longer period there than in northern latitudes.

It is of interest to look into the internal composition of the earth in order to have a finer appreciation of the earth's gravitational pull. The inside of the earth, beyond a depth of approximately one-half of its radius, is expected to have a density twice as large as the outer 3000 km. The average density varies from 3.5 g/cm^3 at the surface to 11 g/cm^3 at the center. The average density for the entire planet is 5.52 g/cm^3. The high density inside the earth is due mostly to the extreme pressures there. The pressure inside the earth increases with depth at the rate of 500 atm/km. The pressure at the center of the earth is estimated at 3.7×10^6 atm.

The gravitational pull of the earth and the effects of the earth's rotation must enter into every consideration of equilibrium of the atmosphere and the oceans. In essence, we may conclude that a driving motor force of our geofluid is the earth's movement.

The circumference of the earth is nearly 40,000 km and its mass is approximated at 6×10^{24} kg. The sun, which supplies the heat energy involved in some of these motions, has a mass 330,000 times larger than that of the earth, and the moon, which displays its influence on the geofluid through tidal waves, has a mass 81 times smaller than that of the earth, but is at a considerably shorter average distance from the earth (238,850 km). The surface area of the earth is 510×10^6 km^2; of this one-third is land (including Antarctica). The highest point on the earth's surface is Mt. Everest (8.9 km high) and the lowest point is the Marianas Trench in the Pacific (10.9 km deep).

The angular rotation of the earth plays a very important role in geophysical motions. Its rotation around the sun is complete in 365 solar days, which gives the earth an orbital velocity of 29.8 km/s. The revolution around its own axis

[1] The dependence of the gravitational pull of the earth on the geometry of the earth is discussed in Chapter 2.

takes place in a little less than a solar day. The *sidereal day* is 23 h, 56 min, and 4.09 s (1/365 of a revolution around the sun).

As we shall see later in detail, the earth exchanges heat energy with the atmosphere and the oceans before this energy is released into space. Conversely, the earth also receives energy from the sun and space but this energy must first pass through the geofluid. The average thermal gradient in the outermost layers of the solid earth as well as in the oceans and the *troposphere* (atmospheric layer closest to the earth's surface) is about 10°C/km. Starting with an average earth surface temperature and extrapolating to 3000 km deep in the earth, the temperature in the inner core of the earth would be 4300°C. Most of the heat exchanged between earth and space is through *radiation* which the geofluid allows through special spectral windows which are intimately related to the chemical composition of the geofluid. Because of this, we can understand the concern over the effects of pollution on the chemical composition of the environment and ultimately on the energy balance. The average surface temperature of the earth, which is in the neighborhood of 280°K, will be shown to be the stabilized value when considering the available energies from the sun and the earth, and above all the radiation-filter characteristics of the geofluid. We shall study these phenomena in detail.

Owing to natural causes, the atmosphere and oceans display, to some degree, variations of chemical and thermodynamic properties with the time of the year and with the location on the globe. Owing to man's causes, these compositions that affect the heat balance may vary considerably more with time and location. In the past decade concentrated attempts have been made, at the international level, to standardize and control these radiative properties of our environment. Because of the interest in space flights in the US and the USSR, research and explorations in the atmosphere have yielded faster advances than studies in oceanology and climatology. Because of this lag, there exists today very detailed and precise data on almost all physical characteristics of the atmosphere, particularly in the Northern Hemisphere, whereas the same cannot be said for the oceans, seas, lakes, and rivers.

1.2 THE ATMOSPHERE

From the mechanics point of view, the atmosphere is a thin layer of gaseous mixture surrounding the surface of the earth which remains attached to the earth by the pull of gravity (see Fig. 1.1).

Because the mass of the atmosphere diminishes asymptotically with altitude, it becomes impossible to determine its exact thickness. The density of the atmosphere at 100 km is one-millionth that at the earth's surface. Insofar as mechanics, thermodynamics, and meteorology are concerned, it is safe to assume

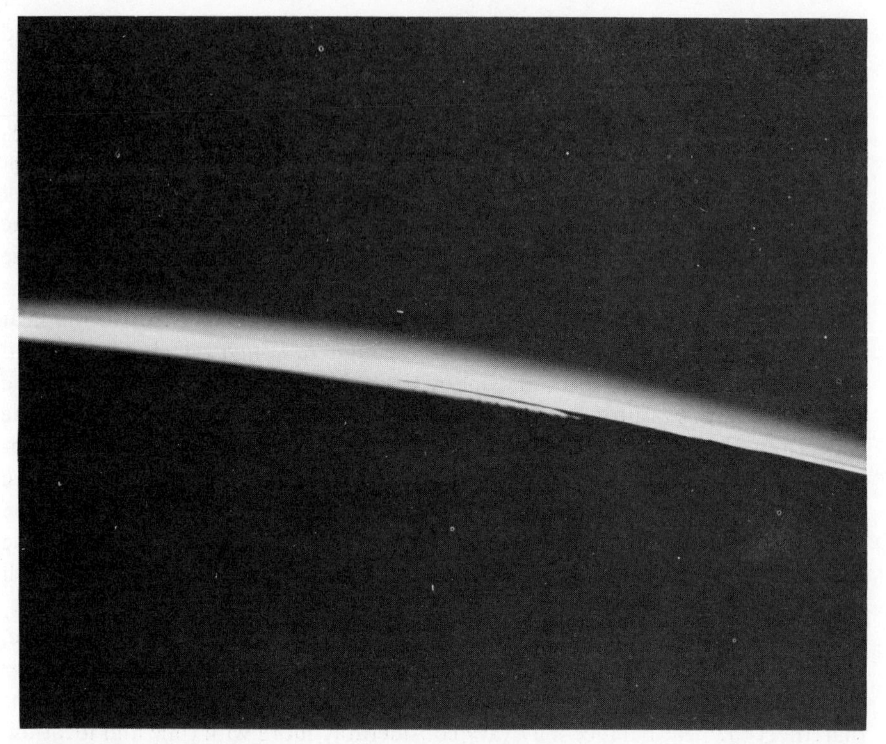

Fig. 1.1 A photograph of the edge of the planet earth seen at twilight. The intense white layer is a silhouette of cloud layers, above which lies an airglow layer. (From Earth's Photographs from Gemini III, IV and V, National Aeronautics and Space Administration SP-129, Washington D.C., 1967.)

that no appreciable effects can take place at higher altitudes and thus we define the thickness of the atmosphere as 100 km. From the point of view of electricity, magnetism, and radiation, the atmosphere may very well start at 100 km.

As we shall see later in detail, the composition of the atmosphere and the characteristics of its energy transfer with space are responsible for the maintenance of its heat balance, which changes little from season to season and latitude to latitude. It is feared that man's pollution may upset this ideal balance of chemistry and energy between earth and space which has been set up by natural causes. The fears are perfectly justified. This delicate balance can conceivably be gradually destroyed over a period of a few generations if waste products are not controlled in a world increasing in population and in the production of consumer goods.

With the exception of water vapor, the chemical composition of the atmosphere is nearly constant up to 100 km. The percent composition by volume is given in Table 1.1. In addition to the gases listed in the table, there are also small traces of krypton, xenon, ozone, and radon. Some dissociation of O_2 into atomic oxygen (O) begins at about 50 km and continues up to 150 km. This dissociation is related to the presence of ozone (O_3) between 10 and 50 km, which has a significant influence on the energy balance of the atmosphere, as we shall see in the course of this book. Also N_2 and CO_2 dissociate at these and higher levels. This dissociation is attributed to the absorption of solar ultraviolet radiation.

TABLE 1.1

Composition of the Atmosphere

Gas	N_2	O_2	Ar	CO_2	Ne	He
Percentage by volume	78.09	20.95	0.93	0.03	0.002	0.0005

The molecular weights of the gases that compose the atmosphere are different and therefore one would suppose that the concentrations of these gases would vary with altitude. For a static atmosphere this would be the case. However, because of the presence of winds and turbulence, there occurs a thorough mixing, and consequently stratification of gases into layers is prevented.

Portions of the atmosphere can be classified on the basis of very marked and specific temperature stratifications. The atmosphere is made up of a number of layers, each characterized by a distinctly different temperature distribution. At mid-latitudes these temperature profiles remain constant with seasonal changes to within 30°K, while the temperature spread in each of these layers is about 100°K. It is because of this remarkably small variation that it is convenient to classify portions of the atmosphere on the basis of certain ranges of temperature distributions. This is shown in Fig. 1.2 where measured values of temperature are compared with the Standard Atmosphere (1966), a standard model. What is significant in this temperature model is that each layer is characterized by a linear temperature change. This makes it simple for identifying regions in the atmosphere and for the determination of other thermodynamic properties from this linear model.

The layer nearest the surface of the earth, characterized by a linear decrease of temperature with altitude, is called the *troposphere* or the *mixed layer* where the average rate of temperature decrease with altitude is approximately 6.5°K/km. We shall see that the temperature gradient in the lowest part of the troposphere varies a great deal, whereas in the upper layers it remains essentially unchanged. Table B.1 in Appendix B, taken from US Standard Atmosphere (1966), shows seasonal variations of this temperature lapse rate. The troposphere contains about

80 percent of the total atmospheric mass. It is the layer in contact with the earth's surface and therefore it is most influenced by energy transfer through radiation, evaporation, condensation, conduction, and convection. As seen from Fig. 1.2 this layer is approximately 15 km in thickness and represents the limit within which conventional air flights take place. (By conventional is meant aerodynamically lifting vehicles.) The troposphere is also the layer in which man-made pollution from industrial wastes is principally confined, and where most cloud formations are found. Over the poles, this layer is thinner, about 8 to 10 km, and in winter it may be entirely absent.

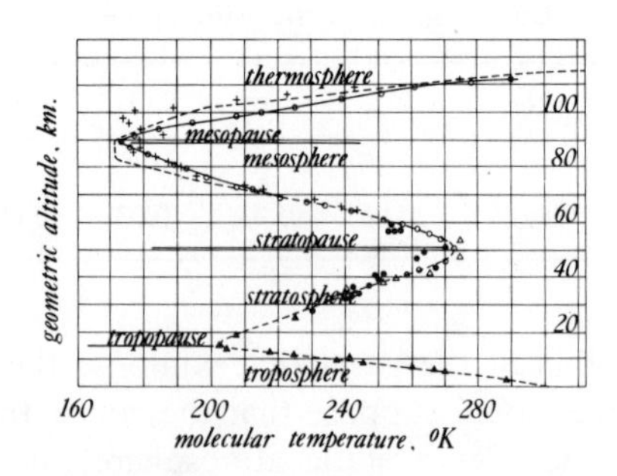

Fig. 1.2 Molecular temperature versus geometric altitude. (After A. C. Faire and K. S. W. Champion, Recent density, temperature and pressure results obtained at White Sands Missile Range compared with IQSY results. *Space Res.* VIII, 1968. Used with permission of North-Holland Publ. Co., Amsterdam.)

Dynamically speaking, the troposphere is stable, but as we shall see in the course of this book those portions of the layer nearest the surface of the earth are often unstable. This explains the presence of updrafts that bring moist air to higher altitudes, which results in the formation of clouds, in water or ice form, depending on the temperature of the environment.

At 15 to 18 km, depending on the season and altitude, the temperature distribution is sharply reversed. This marks the end of the troposphere and the beginning of the *stratosphere*. The region delineating this discontinuity in temperature gradient (viewed at large scale, of course) is called the *tropopause*. In tropical regions this tropopause can be extremely thin, indicating an abrupt reversal in temperature distribution. According to Fig. 1.2, in mid-latitudes the tropopause is also a very thin region. The stratosphere, the layer above the tropopause, is characterized by a positive temperature gradient, hot air above cold air, and is dynamically extremely stable. It extends upward to a height of 50 km where the temperatures at its upper end reach those found on the surface of the earth. We shall see later that this sudden reversal in temperature gradient is due in particular to the presence of large concentrations of ozone and its property of absorbing most ultraviolet solar radiation. Because of the extreme dynamical stability of this layer, there is virtually no vertical motion taking place there.

Whereas the temperature of the troposphere is governed by mixing and convection, the temperature of the stratosphere is controlled by heat radiation. The stratosphere is also largely free from the rainstorms so characteristic of the troposphere. Because the stratosphere is less turbulent than the troposphere, long-distance airplanes are designed for these altitudes. Wind speeds of about 100 m/s are not unusual in this layer and they may reach as high as 150 m/s. The stratosphere has a large-scale circulation that plays an important role in our weather system and in the transport of radioactive fallout material.

Occasionally, clouds resembling "mother-of-pearl" will form in the stratosphere as high as 25 to 30 km. Because of the temperature at that altitude they will be ice clouds with a concentration of ice crystals of a few per cubic centimeter. These clouds are often observed in the North Atlantic regions at dusk or at night when haze is obscured by the earth's shadow. The dryness of the stratosphere is best demonstrated by the absence of contrails of condensed engine exhaust moisture from high-flying aircraft, because the water evaporates instantly. However, these contrails are visible in the upper troposphere where the level of moisture is higher.

The *stratopause* defines the upper edge of the stratosphere and the beginning of the *mesosphere*. This altitude, as shown in Fig. 1.2, is approximately 50 km. The temperature of the stratopause is about equal to that on the surface of the earth. The stratopause is also characterized as a region where clouds cannot form at all, because, as we shall see, the saturation vapor pressure necessary for condensation at the temperature of the stratopause must be equal to or larger than the ambient atmospheric pressure (consult Fig. 4.12, p. 114). The *mesosphere* is confined to a region between 50 and 80 km and because of its negative temperature gradient (temperature decreasing with altitude), vertical motion there is less inhibited. One of the characteristics of this layer is the formation of *noctilucent* clouds, which are observed in the upper northern and southern latitudes. These clouds are believed to occur at the *mesopause*. Cadle (1966) states that the nature of the particles constituting these clouds and the mechanism of the cloud formation have long been a mystery. Three hypotheses have been proposed. The first is that these noctilucent clouds are ice clouds formed by self-nucleation. This is supported by the fact that the mesopause has the lowest temperature in the atmosphere, about 180°K, and that the temperature is the lowest in the summer (consult Table B.2 in Appendix B). The second hypothesis suggests that these clouds consist of nonvolatile solid particles, and is based on correlations that have been observed between meteor showers and the occurrence of the clouds. The third hypothesis is that the clouds consist of particles of ice which have been condensed on nonvolatile solid particles in much the same way that clouds of ice particles are formed in the troposphere. This last hypothesis has received, according to Cadle, considerable support from the results of rocket-borne particle collection. These noctilucent clouds are observed during the summer months, at twilight. They appear to be light blue and resemble an inverted ocean with wave patterns.

The character of the mesosphere differs from that of the lower layers because of ionization of air molecules and atoms, and dissociation of molecular oxygen and other gases. The fact that diffusion is more important there than turbulent mixing, results in relatively high concentrations of light gases. It is believed that above 110 km practically all oxygen is in atomic form. Because of the low densities in this region, and consequently large mean free paths, the heavier gases tend to be less concentrated with increasing height. It is believed that above 500 km helium begins to become the major constituent and that above 3000 km atomic hydrogen becomes the major constituent.

Beyond 500 km, which is outside the scope of this book, in the *exosphere* the neutral atoms diffuse upward and the more energetic ones may escape the earth's gravity. Sometimes this region is referred to as the *magnetosphere* because these ionized particles are bound by the earth's magnetic field.

1.3 THE OCEANS

Because the oceans absorb considerable amounts of solar radiation and are the source for the water vapor in the atmosphere, they are often considered as a subsurface of the atmosphere. Dynamically speaking, water is a Newtonian fluid, as is the atmosphere with the difference that it is an incompressible fluid and in the oceans is confined to two phase changes out of the three possible. In the oceans liquid and solid can occur, with the solid mostly at the interface between the ocean and the atmosphere. The third phase, vapor, is always part of the lower atmosphere.

When one considers the interaction between the atmosphere and the earth's surface, the character of the ocean's surface is found to be different from that of the land. To begin with, the ocean's surface is constantly in motion relative to the ground and possesses different heat transfer characteristics. Also, because of its smoothness compared to land, wind shear forces on the ocean's surface are smaller. The heat absorbed at the ocean's surface is transferred through waves and wind currents to deeper water layers. At one season heat may be absorbed by the ocean from the atmosphere, while at another the ocean provides heat to the atmosphere through convection and particularly through the latent heat of the vapor it provides. Because the oceans provide the greater part of the water vapor in the atmosphere, they are intimately connected with the circulation of the atmosphere and the seasonal climates.

As in the case of gases, the physical properties of pure water depend on pressure and temperature. In the case of seawater, which is not pure, a third variable, *salinity*, must be considered. While some water properties like compressibility, thermal expansion, refraction index, and so on, are hardly affected by the presence of salt, others, like the freezing point, density, and buoyancy, are dependent on the degree of salinity. Properties such as osmotic pressure and electrical conductivity are peculiar to salt water and not to fresh water.

Generally speaking, *salinity* is defined as the ratio of the weight of dissolved materials to that of a sample seawater. Since this turns out to be a very broad definition, an international commission in 1902 defined salinity as the total amount of solid material, in grams, in 1 kg of seawater when all the carbonate has been converted to oxide, the bromine and iodine replaced by chlorine, and all organic matter oxidized. Since chloride ions make up about 55% of the dissolved solids, the following empirical formula is used to determine the salinity $\mathscr{s}$:

$$\mathscr{s} = 0.03 + 1.805 \times \text{chlorinity} \tag{1.1}$$

where both salinity and chlorinity[2] are measured in parts per thousand, or grams per kilograms, and expressed by the symbol ‰. The salinity of the ocean is normally about 35‰. Naturally, in regions of large rainfall or near river discharges it may be considerably less.

We said earlier that most of the earth's surface is covered by water. Seas and oceans cover 70.8% of the globe's surface, which we said was 510×10^6 km^2. One important distinguishing feature of the Southern Hemisphere is that the oceans occupy a larger proportion of the surface than in the Northern Hemisphere. The comparison of the percent area covered by oceans is 81% in the Southern Hemisphere and 61% in the Northern Hemisphere.

Temperatures of the oceans across the globe vary between −1.9 and 30°C; the lower limit is that of ice formation in sea water and the upper limit is that due to the earth's radiation in the tropics. The freezing point, vapor pressure, and maximum density of seawater are affected by the presence of soluble salts. When we speak of freezing point, we mean the first point when an infinitesimal amount of water has formed into ice and is in equilibrium with the rest of the solution. In a sample, when ice forms, the concentration of the dissolved solids increases and the formation of additional ice takes place at lower temperatures. Figure 1.3 shows the variation of freezing point temperature with salinity and the temperature of maximum density. The vapor pressure is also affected by salinity but in most instances the variation of salinity is much smaller than the variation of surface temperature upon which the vapor pressure is more

[2] In 1937, the definition of chlorinity was established as the amount of silver, in grams, necessary to precipitate the halogens contained in 328.52 g of seawater.

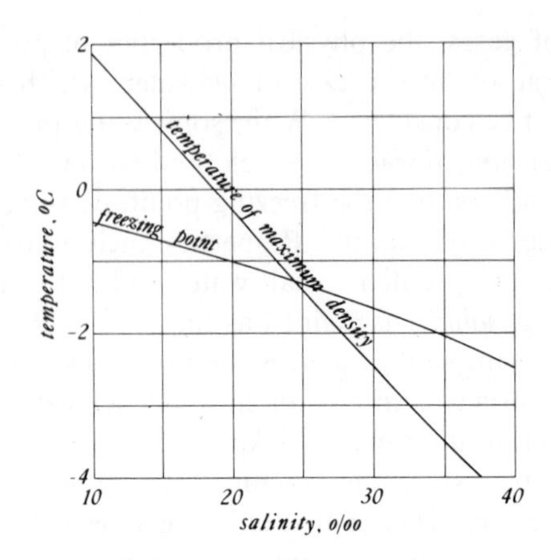

Fig. 1.3 Freezing point and maximum density of salt water.

dependent. It is sufficiently accurate to take the vapor pressure of the seawater as 98 % of that of pure water (Sverdrup, 1942).

The density of seawater depends on three variables, namely salinity $\mathscr{s}$, temperature T, and pressure p. It decreases with temperature rise and increases with salinity and pressure rise, although very little with pressure. Figure 1.4 gives density values of seawater as a function of salinity and temperature drawn from Knudsen's density tables (Knudsen, 1901).

The maximum density of pure water occurs at very nearly 4°C. For seawater the temperature of maximum density decreases with increasing salinity; for salinities larger than 25‰ it is below the freezing point, as can be seen from Fig. 1.3. Stated precisely, at a salinity of 24.7‰ the freezing point is −1.332°C.

The surface tension of seawater is a little greater than that of pure water at the same temperature. In terms of chlorinity it is equal to

$$\sigma = 75.5 + 0.04 \text{ Cl} \tag{1.2}$$

where Cl is the parts per thousand of chlorine.

The specific heat at constant pressure for seawater varies from 0.94 to 0.93 cal/g · °C from 0 to 20°C. The values are not too sensitive to pressure variations. At 4000 db, for instance, it increases only 3 %. The specific heat at constant volume is a little less than that at constant pressure. The ratio of the specific heats $c_p/c_v = 1.0004$ at 0°C and 1.0207 at 30°C.

The latent heat of evaporation for seawater is not different than that for pure water. The specific heat of pure ice is naturally that of pure water, but the specific heat of sea ice depends a great deal on the temperature and the salinity of the ice. For instance, at −2°C the specific heat is 6.7 cal/g · °C, at −4°C it is 1.99 cal/g · °C, at −8°C it is 0.88 cal/g · °C, and at −16°C it is 0.60 cal/g · °C. The latent heat of fusion of pure ice at atmospheric conditions is 79.67 cal/g or 333.1 joules/g. No specific value for the heat of fusion can be assigned to solutions, like seawater, since melting takes place constantly as the temperature rises, regardless of how low the temperature is. Heating values can be assigned to seawater in order to melt all the ice at a given temperature. For instance, it will take 68 cal/g to melt sea ice at $-2°C$ and a salinity of 6‰ and 55 cal/g at $-1°C$ with the same salinity.

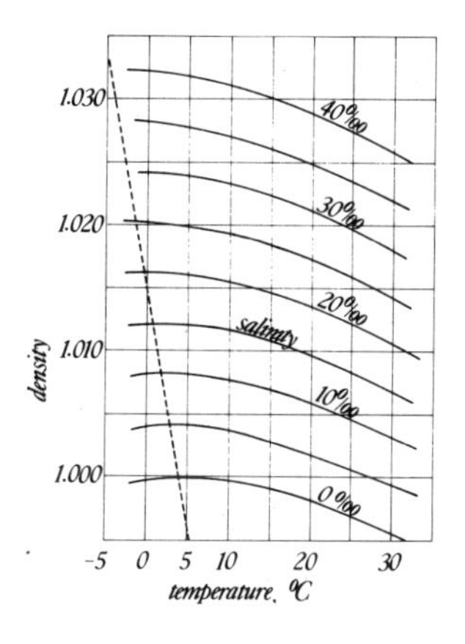

Fig. 1.4 Density of seawater as a function of salinity and temperature. (After Knudsen's density tables; Knudsen, 1901).

The density of seawater increases with depth, and therefore stratification is stable. Unlike the atmosphere, since water is not compressible, the pressure variation with depth affects the density very little. Temperature changes account for most of the density changes. Therefore, as long as the denser or colder water is at the bottom of the stratification, it is stable. With the exception of a thin layer near the surface of the ocean which changes with diurnal changes of temperature, winds, and wind-generated waves, the temperature of the ocean decreases with decreasing depth. A typical temperature and salinity distribution is shown in Fig. 1.5.

For the reasons just given, the distribution of density with depth is closely related to those of temperature and salinity (Fig. 1.5). At the surface of the sea, the density will be affected by the sun's heating, precipitation, meltwater from ice, runoff from land, and cooling due to evaporation or formation of ice. If the density of the surface water is increased beyond that of the layers beneath, vertical convection currents occur which tend to equalize the temperature. In the lower and middle latitudes convection currents, if any, are limited to a thin layer near the surface because cooling, the only influence that would do this, is not pronounced. In the higher latitudes, surface water sinks to lower depths, thus filling the ocean basin with colder water. Because the amount of water that rises toward the surface must exactly equal the amount that sinks, ascending motions must occur in other regions, which turn out to be the western ocean coasts where prevailing winds carry the surface water away from the coast.

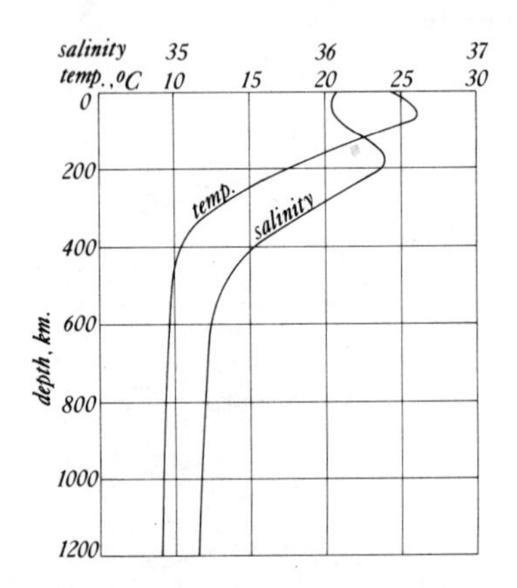

Fig. 1.5 Typical temperature and salinity distribution with depth.

Looking again at Fig. 1.5, the upper layer of warm water is separated from the deep water by a layer in which the temperature decreases rapidly with depth, exhibiting an analogy with the atmosphere. The same terminology of *troposphere* and *stratosphere* can be applied to the first two distinct layers of the ocean. The layer where most mixing occurs, near the surface, is called the troposphere, and is very distinct in middle and lower latitudes. The adjacent layer, the analogous stratosphere, is a nearly uniform mass of cold water.

2

Fundamental Concepts of the Earth and the Geofluid

2.1 INTRODUCTION

With the advent of travel into space and beneath the ocean's surface, there has been a significant amount of contributions made to knowledge concerning the geofluid environment and its behavior. Few new discoveries have been made; for the most part the advance in knowledge has involved the confirmation of theories and the improvement of our quantitative evaluation of the processes of the environment.

Before attempting to derive and gain proficiency in the use of the governing equations of kinematics, dynamics, thermodynamics, and heat transfer of our geofluid, it is essential that we review the origin, definitions, and physical characteristics and relations of concepts affecting the state of the geofluid system itself. This chapter addresses itself to this task. Useful applications of these principles are discussed in subsequent chapters.

Within certain minimum limits, of course, it is not assumed in the development of this book that the reader has a specialized background in fluid mechanics, thermodynamics, heat transfer, or mathematics. However, it is understood that since the environment is of special scientific and engineering interest some basic general scientific facts will be taken for granted. In any case, the initial presentation of concepts will be as self-evident as possible. Since the principles covered in this book deal with subjects that cut across many physical and mathematical disciplines, it is not possible to give a full-scale introduction to each of these subjects without making the book cumbersome for the reader. Thus, it is assumed that the reader will, on occasions, consult elementary works on the basic subjects if he deems necessary. Appendix A is provided for a refresher in vector algebra and calculus.

2.2 SOLID, LIQUID, AND GAS

All matter in nature is found in the form of *solid*, *liquid*, or *gas*, and often a mixture of these forms. These basic forms are identified, thermodynamically, as *phases* which represent a form of matter that is physically and chemically stable. Because of their similarity in dynamical behavior, the liquid and gas phases are designated as *fluids*. A *solid*, in general, is conceived as a substance that offers resistance to change of shape. In contrast, the *fluid*, representing the liquid and gas, does not offer any resistance to change of shape.[1] A simple experiment can demonstrate the difference in behavior between a free volume of solid and a free volume of liquid when they are subjected to similar forces. A glass of water is quickly turned upside down on a flat, level surface. The water trapped between the walls of the glass and the flat surface imposes, owing to its own weight, a force on the glass wall as well as on the flat surface. If the glass is quickly lifted upward, the body of water originally restrained by the glass walls will deform continuously under its own weight. The surface tension finally limits this indefinite deformation to a puddle of water of minimum height. This experiment may be repeated with the same amount of water frozen within the glass. We know from experience that the solid ice will not deform continuously when the glass is removed. Owing to the relief of restraint imposed by the glass walls, the ice will deform slightly under the influence of its own weight. This deformation will be small and will remain the same indefinitely, provided that all conditions remain unchanged.

The principal distinction between a solid and a fluid is, then, the mode of resistance to change of shape. This distinction is definitely dependent on the type of force imposed. For instance, if the glass of water and the glass of ice were not turned upside down, but if a compressive force were imposed in the water and on the ice with the help of a piston fitting snugly into the glass, the water and the ice would compress slightly. We can generalize, then, that *under pure compressive loads, the solid and the fluid will display a finite deformation proportional to the load*. This deformation will generally be accompanied by a change of volume. This proportionality between the load and the deformation is called *Hooke's law*.

What makes the first experiment different from the second is the type of force applied in each case. In the first experiment the sudden relief of restraint from the glass wall induces forces in a direction perpendicular to the glass wall. These forces increase with depth, since the weight above the surface increases with depth. Because of this the layers of water at lower depths will move sideways faster than the upper layers. The difference of velocity between layers will develop frictional forces between them. Since the water deforms continuously, in this

[1] Here, change of shape is conceived as a change of form without a change of volume.

free situation, it cannot resist any change of shape owing to frictional or shear forces. The solid, under the same conditions, will show a finite deformation under the same tangential shear forces. Therefore, *the fluid, unlike the solid, cannot sustain a finite deformation under the action of tangential shear forces. Under shear forces it cannot, like the solid, resist a change of shape.*

Based on experimental evidence, Newton (1686) showed empirically that for most fluids *the time rate of change of the deformation of a fluid element, or parcel, is proportional to the shear force per unit area applied on the surface of the element.* He expressed this behavior of the fluid in his "Principia" in the form of the following hypothesis (Newton, 1946): "The resistance arising from the want of lubricity [slipperiness] in the parts of a fluid, is, other things being equal, proportional to the velocity with which the parts of the fluid are separated from one another."

This difference of behavior between a solid and a fluid now becomes clear. Under a tangential shear force the solid undergoes a *finite deformation* proportional to the force per unit area, and the proportionality factor is the shearing modulus of elasticity. A fluid, under the same force, will deform indefinitely, while *the rate of change of the deformation* will remain proportional to the shearing force per unit area, and this proportionality factor is the *viscosity*. It can be verified easily—with a third experiment—that, for fluids with equal shear forces, the viscosity is what determines the rate of deformation. Two equal volumes of liquid with approximately the same density, originally contained in separate identical glasses, are overturned on a flat surface, as in the previous experiments. At the start of the experiment, the internal force distribution at corresponding levels in both liquids was the same, because of equal weights of fluid at equal depths owing to equal densities. Nevertheless, if the two liquids have different viscosities, it will take the more viscous fluid a longer time to reach its final "puddle" state. This experiment indicates that, for the same shear force, the viscosity determines the time it takes for a given deformation. A more qualitative as well as quantitative description of viscosity is given in Chapter 5.

A *gas* is a fluid and consequently obeys the type of dynamic behavior discussed. It, however, differs from a liquid in its *compressibility*. This implies that a given mass of gas under a given compressive load will change its volume considerably more than does the liquid. Furthermore, unlike a liquid, a gas does not display a free surface; it expands to occupy the entire space.

2.3 SYSTEM, PROPERTY, AND STATE

A *system*, thermomechanically, is an arbitrary volume—within the substance to be analyzed—across whose boundaries no mass is exchanged. A system, like a solid body, may experience a change in momentum and energy but no change in mass, and can be stationary or moving. In a moving system the boundaries will move with the system, and the mass within the system will always remain the same.

In contrast, a *control volume* is an arbitrary volume across whose boundaries mass as well as momentum and energy are transferred. The control volume may be stationary or moving with respect to the motion of the fluid. The fluid enters the boundaries of the control volume on one side and leaves through the other. The gas inside an experimental balloon represents a system, whereas a smoke stack through which mass and energy traverses is a control volume.

The boundaries separate the system and the control volume from the *surroundings*. The surroundings are important in any geofluid consideration since they exchange mass, momentum, and energy with the system or the control volume as the case may be.

A *property*, thermomechanically, is an observable quantity of a system. If measured, its value is always the same when the system is brought to the same state regardless of the way in which it is brought to that state. The properties of the system determine the state of the system. The four fundamental properties from which stem all other dynamic and energetic properties are the mass, the length, the time, and the temperature. Of these four the temperature is the only *intensive* property, meaning that it does not depend on the extent of the system. For instance, two systems at the same temperature, when brought together into one, will still have the same temperature. The same reasoning does not apply to two systems with the same mass or same length. This is because mass and length are *extensive* properties.

Certain properties are characterized by a magnitude alone. For instance, when speaking of volume, temperature, or pressure, a number with appropriate units suffices to characterize that property. These properties are called *scalar properties*. Others, such as velocity, acceleration, and heat flux, cannot be completely described by magnitude alone and accompanying units. The direction of action of these properties is just as important and necessary as the magnitude. These properties are called *vector properties*. They are subject to a more general form of algebra and calculus called vector analysis. Appendix A is a survey of the principles of vector analysis used in this book.

2.4 PROPERTIES IN A CONTINUUM

The general laws of mechanics and thermodynamics establish the basic working scientific principles of the environment. The meaning that properties have in this environment depends on an understanding of a *continuum*. For instance, the meaning of pressure in the ocean and most of the atmosphere is often defined as the total force per unit area imposed on each of the unit areas in this environment and can be thought of as the force per unit area on a solid unit surface immersed at any point in the environment owing to the continuous impingement and bouncing off of molecules at the surface. A given mass of gas in a constant volume and at a constant temperature is always under the same pressure.[2] This law, naturally, begins to lose its meaning when the gas is rare or the number of molecules per unit volume is very small, because pressure then will depend on the probability of a few molecules bouncing off any surface. Consequently under these conditions the property of pressure will not be continuous, constant, or smoothly varying in time.

To illustrate this concept of a continuum let us examine the *density* (mass per unit volume) of a substance with respect to the size of sample considered. Consider an atmospheric column with a cross section of 1 m^2 and reaching to an altitude of 100 km. Let us remove various sizes of samples from this column and determine the density. If we remove samples 1 m in depth beginning with the thermosphere, the first sample will have very few molecules and thus, depending on the time the sample was taken, the density will vary with the probability of the number of molecules in that volume. We should not expect a constant value,

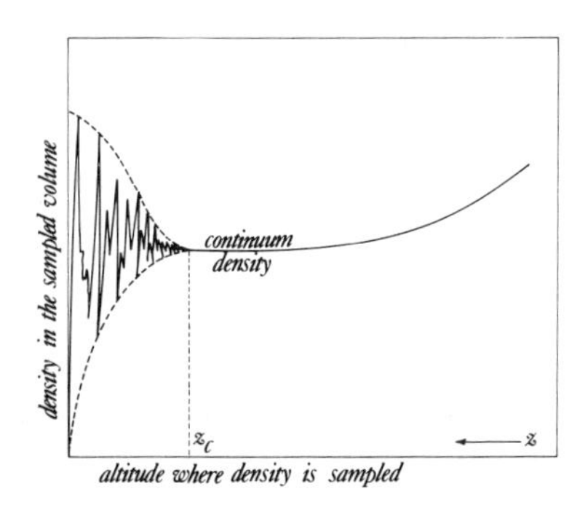

Fig. 2.1 Density at various altitudes.

for the measurements of density in the thermosphere will be erratic. This is shown in Fig. 2.1. The lower the elevation from which we pick our sample, the less erratic will the density be because the molecule population in the same size volume will increase, and the density value will be less and less dependent on the time the sample was taken and more and more constant at a given elevation.

Molecules inside a substance are in constant agitation and collision with one another. Their average motion between collisions, called the *mean free path*, is about 5×10^{-6} cm for sea-level air. This distance becomes larger with altitude

[2] This is true for the thermodynamically *pure substances*. In particular this statement may be verified from Boyle's and Charles's laws pertaining to perfect gases. Boyle's law states that during an isothermal process the ratio of the pressure to the density is constant. Charles's law states that during an isobaric (constant pressure) process the product of the density and the absolute temperature is constant.

as the density decreases. The condition of continuum is reached when the properties of a parcel of air are not probability dependent but have the same values when the state of the substance is the same. We conclude that the smallest parcel to be considered as a continuum depends on size and is related to the mean free path of the molecules. Thus in a continuum a molecule has no significance, making the description of properties and the mathematics much simpler.

From a mathematical point of view, a continuum lets us postulate that the properties at any one point can be expressed in terms of the properties at a neighboring point. This is because the property and its derivatives are continuous in their variations with space. This interrelation between property values at two neighboring points in a continuum comes from calculus and is known as the Taylor's series expansion. This theorem states that if a property P and its derivatives are known at a given point x_0, then that property at a neighboring point x_1 is evaluated from the relation

$$P(x_1) = P(x_0) + (x_1 - x_0)\left.\frac{dP}{dx}\right|_{x_0} + \frac{(x_1 - x_0)^2}{2!}\left.\frac{d^2P}{dx^2}\right|_{x_0} + \cdots \tag{2.1}$$

The use of total differentiation in Eq. (2.1) implies that P is a function of x alone. If it were a function of other variables, then similar terms in accordance with the expansion above should appear with respect to the other variables. For a finite distance $(x_1 - x_0)$ separating the two points where the properties are compared, the series in Eq. (2.1) is infinite. This means that the accuracy with which $P(x_1)$ can be evaluated from $P(x_0)$ and its derivatives at x_0 will depend on the number of terms considered in the expansion. Naturally, the smaller the distance $(x_1 - x_0)$ the smaller will be the contribution of the higher-order terms. It is therefore conceivable that if $(x_1 - x_0)$ approaches an infinitesimally small distance dx, second- and higher-order terms can be neglected compared with the first two terms in the expansion on the right-hand side of the equation. This implies that for very small displacements dx the variation of the property in that distance is linear. Actually, since all differential equations are written for phenomena within a very small volume, this linear relationship between the property value at the two ends of the volume will be used over and over again. These considerations are of immense help in the mathematical analysis of any physical phenomenon treated in a continuum.

2.5 THE EARTH AND ITS GRAVITATIONAL EFFECTS

At the start it must be clear that the dynamical and thermodynamical processes our geofluid environment undergoes are primarily generated by the earth's mechanical and thermodynamical energy storage. To smaller extents, the moon and the sun contribute to the mechanical energy. However, the sun is as large a contributor of thermal energy to the geofluid as the earth is in mechanical and thermal energy.

The earth's gravity has a major influence on the physical makeup and stability of the atmosphere, oceans, and lakes. The total gravity force of the earth is the vector sum of the gravitational attraction of the earth's mass, as stated by Newton's *universal law of gravitation* and the *centrifugal force* caused by the earth's rotation around its axis. This is shown schematically in Fig. 2.2.

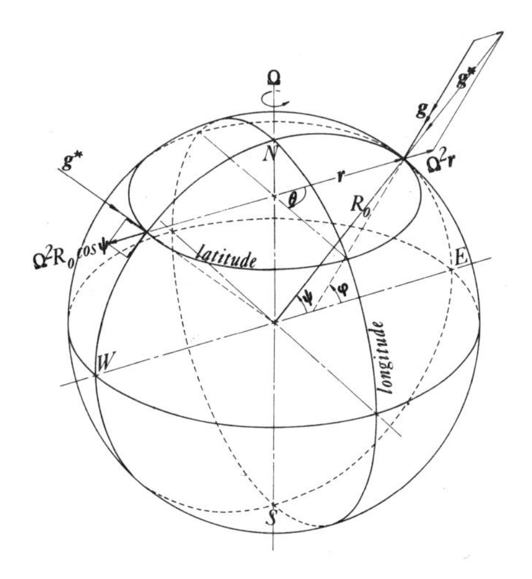

Fig. 2.2 The vector sum of the gravitational and centrifugal forces per unit mass.

Let us consider each of these forces. To begin with, one must recognize *Newton's law of inertia* which relates an inertia-type force to the product of the *mass m* on which the force acts and the *acceleration* $\boldsymbol{a}$ which the force produces. This axiomatic law is expressed as

$$\boldsymbol{F} = m\boldsymbol{a} \tag{2.2}$$

The centrifugal force mentioned above is an inertial force since it is based on motion and an acceleration. The gravitational force may or may not turn into an inertial force, depending on whether or not it produces a motion with acceleration. The physical nature of the gravitational force, owing to the attraction between masses, was also expressed by Newton in the form of a universal law of gravitation. This law, formulated from Kepler's planetary observations, states that the magnitude of the *gravitational force* of attraction between two masses m_1 and m_2 is directly proportional to their product and inversely proportional to the square of the distance R between their mass centers. Thus we can express this law as

$$\boldsymbol{F} = G \frac{m_1 m_2}{R^2} \frac{\boldsymbol{R}}{R}$$

The constant G has been measured in the laboratories and found to be a function of the substances making up the masses. The quantity $\boldsymbol{R}/R$ is the unit radial vector measured from the center of the mass upon which the force of attraction is being considered. Thus $\boldsymbol{R}$ and $\boldsymbol{F}$ are vectors having the same direction.

Applying this to the earth and any other mass, if we let M_e be the mass of the earth acting on another mass m, say of geofluid, if the distance is measured from the earth's center, the force acting on m, from the previous equation, will be

$$\boldsymbol{F} = -G \frac{M_e}{R^2} \frac{\boldsymbol{R}}{R} m \tag{2.3}$$

The minus sign is because the unit vector $\boldsymbol{R}/R$ was defined in the direction of the outer normal to the earth's surface. If the mass m of the geofluid is allowed to fall by being free of any other forces, it will fall with an acceleration

$$\boldsymbol{g}_0{}^* = -G \frac{M_e}{R_0{}^2} \frac{\boldsymbol{R}_0}{R_0} \tag{2.4}$$

called the *gravitational acceleration*, a vector in the direction of $-\boldsymbol{R}_0$, the radius of the earth. The subscript zero will refer to conditions at the surface of the earth.

This law is valid for any altitude z or depth[3] $-z$ from the surface of the earth. Thus $\boldsymbol{g}^*$ is a function of z in the following way:

$$\boldsymbol{g}^* = -G \frac{M_e \boldsymbol{R}_0}{(R_0 + z)^2 R_0} \tag{2.5}$$

and

$$\boldsymbol{g}^* = \boldsymbol{g}_0{}^* \left(1 + \frac{z}{R_0}\right)^{-2} \tag{2.6}$$

[3] In oceanology and limnology the terms in the equations containing the coordinate z have a minus sign.

The internationally accepted value for GM_e is 3.9862216×10^{14} m^3/s^2. The radius of the earth at the *geographic latitude* of $\psi = 45°$ is $R_0 = 6370.1$ km, at the equatorial plane it is 6378.4 km, and at the poles it is 6356.9 km. The gravitational acceleration owing to the pull of the earth, on the surface and at a geographic latitude of $\psi = 45°$, is

$$g_0{}^* = 3.9862216 \times 10^{14}/(6370.1)^2 = 9.82357 \quad \text{m/s}^2$$

From Fig. 2.2 the *centrifugal acceleration* at the same point is $\Omega^2 \boldsymbol{r}$ in the direction of $\boldsymbol{r}$, the radius of the latitude circle. Then to the gravitational pull we should add vectorially the centrifugal force per unit mass and obtain a modified gravitational acceleration $\boldsymbol{g}_0$, such that

$$\boldsymbol{g}_0 = \boldsymbol{g}_0{}^* + \Omega^2 \boldsymbol{r} = -[9.82357(\boldsymbol{R}_0/R_0) - \Omega^2 \boldsymbol{r}] \tag{2.7}$$

The resultant vector is slightly inclined away from $\boldsymbol{R}_0$ as shown in Fig. 2.2 because the order of magnitude of the centrifugal acceleration is smaller.[4] Because of this we often assume that the absolute value of $\boldsymbol{g}$ is in the direction of $\boldsymbol{g}^*$ or $-\boldsymbol{R}_0$. Thus

$$\begin{aligned} |\boldsymbol{g}_0| = g_0 &= g_0{}^* - \Omega^2 r \\ &= 9.82357 - \Omega^2 R_0 \cos^2 \psi \\ &= 9.8066 \end{aligned} \tag{2.8}$$

Thus the contribution of the centrifugal acceleration in the direction $\boldsymbol{R}_0$ is to relieve the earth's pull by $\Omega^2 R_0 \cos^2 \psi = 0.0169$ m/s^2 at $\psi = 45°$.

Although the order of magnitude of the centrifugal force is small, it accounts for the flattening of the earth at the poles and a corresponding bulge at the equator. The earth took this form in order to minimize its internal stresses.

The calculation in Eq. (2.8) applies to the earth's surface and 45° latitude. Owing to the oblateness of the earth, the radius of the earth varies with latitude. Both the gravitational force and the centrifugal force are influenced by the changing radius of the earth with latitude. An exact analysis giving g at any latitude and approved by the International Civil Aviation Organization (US Standard Atmosphere, 1962) is fairly involved and is not reproduced here. Instead, a simplified version is now given that applies equally well to various altitudes and depths.

Consider the radius of the earth at $\psi = 45°$ where $R_0 = 6370.1$ km and s is the increment change of radius from the poles to the equator. Then Eq. (2.8) can be rewritten as

$$g = [GM_e/(R_0 \pm s)^2] - \Omega^2 (R_0 \pm s) \cos^2 \psi$$

[4] The earth revolves about its axis once in every 23 h 56 min and 4 s or a total of 86,164 s. The frequency of rotation or the angular velocity of the earth is

$$\Omega = 2\pi/86{,}164 = 7.292 \times 10^{-5} \quad \text{rad/s}$$

Dividing by $g_0^* = GM_e/R_0^2$ and expanding the first term we have

$$\frac{g}{GM_e/R_0^2} = \frac{R_0^2}{(R \pm s)^2} - \frac{R_0^3\,\Omega^2}{GM_e}\frac{(R_0 \pm s)}{R_0}\cos^2\psi$$

$$= 1 - \frac{2s}{R_0} - 3.44 \times 10^{-3}\left(1 + \frac{s}{R_0}\right)\cos^2\psi \tag{2.9}$$

For the variation of gravitational pull of the earth with oblateness at the surface of the earth the following empirical expression is suggested:

$$g_0^* = g_0^*|_{45^\circ}(1 - 0.000807\cos^2\psi) = 9.82357(1 - 0.000807\cos^2\psi) \tag{2.10}$$

and after introducing it into Eq. (2.9) we obtain the total gravitational acceleration depending on latitude and variation of surface distance s:

$$g = 9.82357(1 - 0.000807\cos^2\psi)\left[1 - \frac{2s}{R_0} - 3.44 \times 10^{-3}\left(1 + \frac{s}{R_0}\right)\cos^2\psi\right] \tag{2.11}$$

For an average value of R_0, the quantity $\Omega^2 R_0^3/GM_e$ is constant and equal to 3.44×10^{-3}. Then, when we expand the first term into a Taylor series, discussed in Section 2.4,

$$g/g_0 = 1.0018[1 - 2s/R_0 - 3.44 \times 10^{-3}(1 + s/R_0)\cos^2\psi] \tag{2.12}$$

The $\pm$ sign has been removed and s is free to carry its own sign. For instance, at the *equator* $s = +8.3$ km and $\psi = 0$. Then we obtain

$$g = 0.9957 g_0 = 9.7648 \quad \text{m/s}^2$$

The accepted measured value at the equator is 9.780 m/s^2 and this approximate method gives an error of 0.15%.

At the *poles*, using the same equation and taking $s = -13.2$ km with $\psi = 90^\circ$, we obtain $g = 9.864$ m/s^2 instead of the accepted measured value of 9.832 m/s^2. The error is 0.3%.

Often Eq. (2.12) can be replaced by an empirical formula for the total surface gravitational acceleration g_0 which gives more accurate values at the equator and the poles:

$$g_0 = 9.8066(1 - 0.00264\cos 2\psi) \tag{2.13}$$

The lines of force (potential) in the atmosphere at various latitudes are shown schematically in Fig. 2.3. Since at the equator the universal gravitational pull and the centrifugal force are in the same direction, there is no angle between these two components and the radial line is the force line or the *plumb line*. At the poles, also, the plumb line is a radial line because the centrifugal force is zero.

At any other latitude, the centrifugal force makes an angle with g^* and thus the resultant $\boldsymbol{g}$ is inclined away from the radial line toward the south, in the Northern Hemisphere, as seen in Fig. 2.2. As the altitude increases in the atmosphere, $\boldsymbol{g}^*$ reduces according to Eq. (2.6) and the centrifugal acceleration increases because of the increase of r. Thus the angle between the radial line and the line of force increases. Figure 2.3 shows the *plumb line* or direction of the line of force at mid-latitude, the pole, and the equator. It also shows the difference between the geocentric and geometric latitudes.

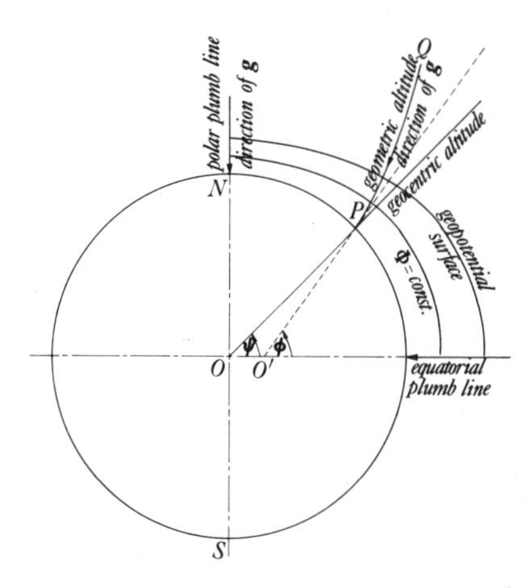

Fig. 2.3 Lines of force.

The variation of gravitational acceleration with altitude is given by Eq. (2.12) simply by replacing s by z. Again for $z > 0$ we imply elevation above sea level and for $z < 0$ we will find values for depth. Thus,

$$g = 1.0018 g_0 [1 - 2z/R_0 - 3.44 \times 10^{-3}(1 + z/R_0) \cos^2 \psi] \qquad (2.14)$$

where

$$g_0 = g_0^*/1.0018 \qquad \text{and} \qquad g_0^* = g^*(1 + z/R_0)^2$$

At $z = 10$ km above sea level and $\psi = 45°$, Eq. (2.14) yields a value for $g = 9.7765$ m/s^2, while the International US Standard Atmosphere tables give 9.7759 m/s^2. At $z = 100$ km and the same latitude the value of g from Eq. (2.14) is 9.498 m/s^2 compared to the accepted value of 9.505 m/s^2, a good agreement. Figure 2.4 gives a plot of the gravitational acceleration with the geometric altitude.

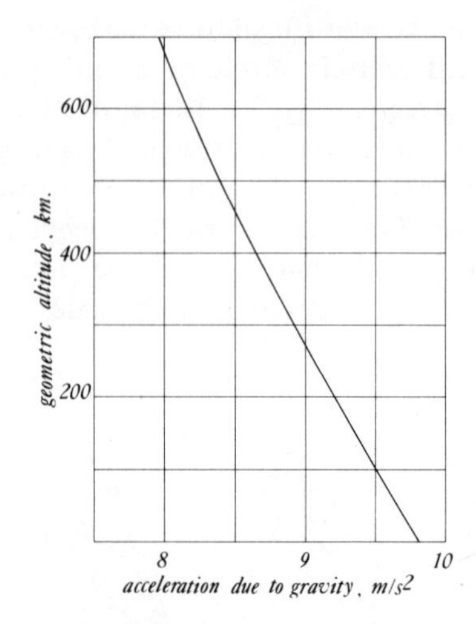

Fig. 2.4 Gravitational acceleration as a function of geometric altitude z.

2.6 THE GEOPOTENTIAL

Because the gravitational force field is conservative, it can be represented as the gradient of a scalar function. This is ultimately tied to the line integral which is discussed in Appendix A and in Section 6.11. It is easy to guess that this scalar function Φ must have the units of energy per unit mass since dimensionally the gradient is the reciprocal of distance and the energy over distance has the unit of force. The energy, then, must be the sum of the gravitational and centrifugal potential energies. For convenience, we shall express the force field per unit mass or $\boldsymbol{g}$ in terms of the gradient of the sum of the *gravitational potential* Φ_g and the *centrifugal potential* Φ_c

$$\boldsymbol{g} = -\nabla(\Phi_g + \Phi_c) = -\nabla\Phi_E \tag{2.15}$$

Multiplying both sides of the equation by $d\boldsymbol{R}$ we have

$$\begin{aligned} \boldsymbol{g} \cdot d\boldsymbol{R} &= -\nabla(\Phi_g + \Phi_c) \cdot d\boldsymbol{R} = (\boldsymbol{g}^* + \Omega^2 \boldsymbol{r}) \cdot d\boldsymbol{R} \\ &= -d(\Phi_g + \Phi_c) = -d\Phi_E \end{aligned} \tag{2.16}$$

as in Eq. (A.44) in Appendix A. The quantity $\nabla\Phi_E$ is called the *gradient of the geopotential* and Φ_E is the *geopotential function* or the *gravitational potential.* Then Φ_E = constant are surfaces which increase in value with altitude along the line of force. The gradient is positive in the direction of the elevation. Arbitrarily, $\Phi_E = 0$ can be assigned on the surface of the earth. The altitude along which the geopotential varies the fastest is in the direction of the line of force shown in Fig. 2.3. For this reason, the radial line could be called the *geocentric altitude* and the distance along the line of force the *geometric altitude*. This slight difference, accentuated in Fig. 2.3 for illustrative purposes, is negligible for most practical problems. Thus,

$$\Phi_E = -\int_A^B (\mathbf{g}^* + \Omega^2 \mathbf{r}) \cdot d\mathbf{R} \tag{2.17}$$

Integrating along a plumb line or line of force, and using Eq. (2.14) we have

$$\begin{aligned} \Phi_E &= \int_0^z \left[\frac{g_0^*}{(1 + z/R_0)^2} - \Omega^2 (R_0 + z) \cos^2 \psi \right] dz \\ &= \frac{g_0^* z}{1 + z/R_0} - \Omega^2 z \left(R_0 + \frac{z}{2} \right) \cos^2 \psi \end{aligned} \tag{2.18}$$

This equation is algebraically complicated to use. For values of z of concern in this book, i.e., for the atmosphere and the oceans, $|z/R_0| \ll 1$ and the centrifugal part may in most cases be ignored. Then the geopotential function is just $g_0^* z$.

2.7 THE GEOPOTENTIAL HEIGHT

We have discussed the two kinds of altitudes, namely the geocentric and the geometric. We can define a third altitude called the *geopotential height*. It is a kind of normalized altitude that is a measure of the work in raising a unit mass from sea level to the geometric altitude z. This is defined as

$$H = \Phi_E / g_0 = \int_0^z (g/g_0)\, dz \tag{2.19}$$

Using Eq. (2.18) and the approximation $g_0 = g_0^*/1.0018$, at $\psi = 45°$ we have

$$H = 1.0018 \left(\frac{z}{1 + z/R_0} - 3.44 \times 10^{-3} z \cos^2 \psi \right) \tag{2.20}$$

Of course, at the poles $g_0 = g_0^*$ and at the equator the ratio is 0.9975.

The geopotential height of Eq. (2.19) is measured along z or along the line of force PQ as shown in Fig. 2.3. The values of z and H differ by Eq. (2.20). As we said, H can be considered as the work in raising a unit mass from sea level to the geometric altitude z. Table 2.1 gives a comparison of z and H at two mid-latitudes.

TABLE 2.1

Relationship of Geometric (z) and Geopotential (H) Altitudes[a]

	H (km)	
z (km)	45°N latitude	30°N latitude
0	0	0
10	9.984	9.971
20	19.937	19.910
40	39.750	39.695
60	59.439	59.357
80	79.006	79.896
90	88.743	88.620
100	98.451	98.314

[a] From US Standard Atmosphere (1962).

2.8 THE TEMPERATURE

We discussed continuum in Section 2.4, and noted that in the entire geofluid system continuum would be a problem in the upper atmosphere. The oceans do not pose this problem. Since temperature, and particularly the way it is measured, depends on the existence of a continuum, it is best to speak of *molecular scale temperature* which, according to the definition that follows, takes care of the degree of rarity of the atmosphere:

$$T_m = (M_0/M)T \tag{2.21}$$

where M_0 and M are the molecular weights at sea level and at a given altitude, and T is the *absolute kinetic temperature* in degrees Kelvin. Measurements actually yield T/M. Figure 2.5 shows the departure of T_m from T above 100 km.

For easy calculation of the molecular temperature with altitude, the US Committee on Extension to the Standard Atmosphere (COESA) has approximated the temperature variations in Fig. 2.5 into a set of linear segments of H between 0 and 88.743 km and of z between 90 and 700 km. Tables 2.2 and 2.3 give the origins and ends of the linear segments as well as the lapse rates of the molecular scale temperature. Thus with the help of these tables we can find T_m at any altitude $0 \leq H \leq 88.743$ km from

$$T_m = T_{m,0} + \frac{dT_m}{dH}(H - H_1) \tag{2.22}$$

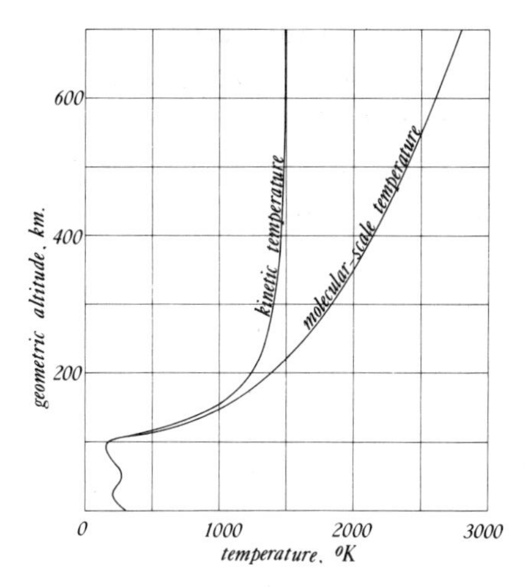

Fig. 2.5 Variations of T and T_m with altitude. (From US Standard Atmosphere, 1962.)

TABLE 2.2

Linear Temperature Segments of the Atmosphere up to $H = 88.743$ km for Mid-latitudes[a]

Altitude, H (km)	Molecular scale temperature, $T_{m,o}$ (°K)	Temperature lapse rate, $dT_m/dH = (-\alpha)$ (°K/km)
0	288.15	
		−6.5
11	216.65	
		−0.0
20	216.65	
		+1.0
32	228.65	
		+2.8
47	270.65	
		0
52	270.65	
		−2.0
61	252.65	
		−3.4
79	190.65	
		0
88.743	190.65	

[a] Spring–Fall, US Standard Atmosphere (1962). Constant molecular weight $M = 28.9844$; also $T = T_m$.

TABLE 2.3

Linear Temperature Segments of the Atmosphere from $z = 90$ to 700 km[a]

Altitude, z (km)	Molecular temperature, $T_{m,0}$ (°K)	Temperature lapse rate $dT_m/dz = -\alpha$	Molecular weight M	Kinetic temperature, T (°K)
90	180.65		28.9644	180.65
		+3.0		
100	210.65		28.88	210.02
		+5.0		
110	260.65		28.56	257.00
		+10.0		
120	360.65		28.07	349.49
		+20.0		
150	960.65		26.92	892.79
		+15.0		
160	1 110.65		26.66	1 022.2
		+10.0		
170	1 210.65		26.40	1 103.4
		+7.0		
190	1 350.65		25.85	1 205.4
		+5.0		
230	1 550.65		24.70	1 322.3
		+4.0		
300	1 830.65		22.66	1 432.1
		+3.3		
400	2 160.65		19.94	1 487.4
		+2.6		
500	2 420.65		17.94	1 499.2
		+1.7		
600	2 590.65		16.84	1 506.1
		+1.1		
700	2 700.65		16.17	1 507.6

[a] US Standard Atmosphere (1962).

and for $90 \le z \le 700$ km from

$$T_m = T_{m,0} + \frac{dT_m}{dz}(z - z_1) \tag{2.23}$$

where $T_{m,0}$ is the temperature at the beginning of the linear segment and H_1 and z_1 are the altitudes at the origin of each segment.

As far as the environmental properties of concern to this book, we need to focus attention only up to the atmosphere's 100-km geometric altitude. The reasons were already discussed in the first chapter, namely that this layer controls most of what occurs on the earth's surface and is of frequent interest to man.

In this range there has been much accurate experimentation with rockets in the US and abroad. At any altitude the temperature may vary from day to day and from season to season. However, within a year the temperature on the surface of the earth and the atmospheric layer nearest it varies the most, about 50°C. What is surprising, at first thought, is that the shape of the temperature distribution with altitude varies little with seasonal changes at a given latitude. Figure 2.6, taken from the US Standard Atmosphere (1962, 1966), is compiled from experimental values obtained at White Sands Missile Range.[5] These values show a sharper identification of the tropopause, the stratopause, and the mesopause, which were previously assumed to be constant temperature zones at all seasons. The data above the stratopause were obtained from the density derived in the measured aerodynamic drag of falling spheres. We will see later that breaking the temperature variations in the atmosphere in a set of linear segments provides a convenient way of evaluating the pressure, density, and other temperature-dependent properties.

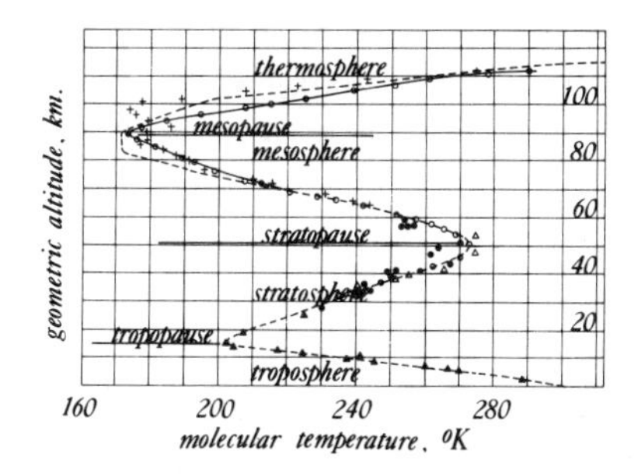

Fig. 2.6 Molecular temperature versus geometric altitude. (After A. C. Faire and K. S. W. Champion, Recent Density, Temperature, and Pressure Results Obtained at White Sands Missile Range Compared with IQSY Results, *Space Res.* VIII, 1968.) Dashed line US Atmosphere Supplement (1966) (30°N, July). + Sphere data, 1966. ○ Sphere data, 1964. △ Rocketsonde data, 1964. ▲ Radiosonde data, 1964. ● Arcasonde data, 1966. Used with permission of North-Holland Publ. Co., Amsterdam.

2.9 DENSITY, SPECIFIC WEIGHT, SPECIFIC GRAVITY, AND SPECIFIC VOLUME

Density is the property of the substance which denotes its mass per unit volume. Consequently, if Newton's inertial law is divided by the volume we have

$$F/\mathscr{V} = ma/\mathscr{V} \tag{2.24}$$

[5] The authors of this study have published additional data for the atmosphere, in April, over Florida for 1967. This is presented in Faire and Champion (1969).

If the acceleration is replaced by the gravitational acceleration g, then F is the weight of the mass m, $F/\mathcal{V}$ is the *specific weight*, and $m/\mathcal{V}$ is the density. Thus the specific weight is the density multiplied by the gravitational acceleration. Even though the density is invariant with geographic location, the specific weight is not, because it depends on the gravitational acceleration, which is a function of geography.

The specific gravity of a substance is the ratio of its density to that of pure water at 4°C and 76 cm of mercury. The *specific volume* is the reciprocal of the density.

The units of density can be expressed in grams per cubic centimeter or kilograms per cubic meter or other decimal derivatives. The units of specific weight are dynes per cubic centimeters or newtons (10^5 dynes) per cubic meter. Often, the units of mass, grams or kilograms, are *misused* for units of force or specific weight. This is unfortunate because it creates confusion and sometimes errors in interpretation of scientific and technical work. The supporters of this unfortunate use explain that one could call a gram-force the force acting on a gram-mass at the standard gravitational acceleration, which is taken to be 980.66 cm/s^2 or actually identify a gram-force as 980.66 dynes.

For the units of force we can use the *dyne*, which is the amount of force necessary to accelerate 1 g of mass 1 cm/s^2. The *newton*, which is the force capable of accelerating 1 kg of mass 1 m/s^2 is the internationally accepted unit of force in the SI unit system.

2.10 THE EQUATION OF STATE

The geofluid in our environment is not only acted on by motor forces arising from the earth's rotation and the earth's and moon's gravity, but it is also heated, cooled, and driven by thermal energies coming from the sun and the earth. Actually, in the whole environment we cannot neglect the thermal effects, as they are often equal in importance to the dynamic effects. Thus thermodynamics must play a key role in the environment.

Thermodynamics enters into the study of the geofluid in two essential ways. First, it enters for the evaluation of property changes caused by other property changes, and second, in doing so, it permits the maintenance of a balance of energy changes the same way dynamics permits the maintenance of a balance of forces. When conversion of energy takes place in the geofluid from one form into another, it is accompanied by changes of properties and thus changes of state which finally affect the motion itself.

The basic equation that describes the relationship between thermodynamic properties at *all* states of the geofluid is the *equation of state*. Depending on the substance, fluid as well as solid, this equation can be simple or complicated in form. Fortunately, for most substances of interest in engineering, including our geofluid, the equation of state requires only two independent thermodynamic properties to express any third one. For instance, if the pressure and temperature are two independent variables, then the equation of state

$$\rho = f(p, T) \tag{2.25}$$

gives the density ρ in terms of the pressure p and the temperature T. The atmosphere and the oceans have equations of state[6] such as Eq. (2.25) but their forms, represented by the function f, are different. This type of an equation of state, with only two independent variables, is called an equation of state for *pure and simple homogeneous substances.*[7] The simplicity of Eq. (2.25) stems from the fact that it is in terms of two independent variables only. As we said, the form of the function, however, may be complicated as in the case of liquids.

Experimenting with air at ordinary temperatures, Boyle and Mariotte arrived separately at the following conclusion: During a constant-temperature (*isothermal*) process, the ratio of the pressure to the density is constant. The same experiments show that the scales of pressure and temperature have to be absolute, that is, measured from vacuum for pressure, and in degrees Kelvin for temperature. Obviously, to keep a gas at constant temperature while varying the pressure and density, we must add or subtract heat. For instance, if an expansion is planned to lower the pressure in a container, the gas will experience a cooling at the same time, and to maintain it at a constant temperature as in Boyle's experiment we must add heat to restore the temperature at this new state with lower pressure and density.

Charles and Gay-Lussac performed independent observations with gases at constant-pressure (*isobaric*) processes. Their results show that the product of the density and the absolute temperature remains constant in this process.

As shown in Fig. 2.7, any two points 1 and 3 on the p–ρ coordinates can be connected with the two processes just mentioned, namely, the *isothermal process* and the *isobaric process*. For the gases studied by Boyle, which include air, the relationship of the properties of points 1 and 2 is

$$p_1/\rho_1 = p_2/\rho_2 = p_3/\rho_2$$

The third expression is true because $p_2 = p_3$. Also during the isobaric process

$$\rho_2 T_2 = \rho_3 T_3 = \rho_2 T_1$$

[6] During a change of phase, as we shall see in Section 2.14, pressure and temperature are dependent, consequently one of these must be changed for a different property, for instance, entropy. As we shall see in Chapter 4, the salinity variation with depth correlates well with the temperature variation with depth so that a linear relationship can be written between T and $\mathscr{S}$ and the equation of state for the ocean is reduced to Eq. (2.25).

[7] The word "pure" need not have anything to do with chemical purity.

Here also, the third expression is true because $T_1 = T_2$. Dividing these two sets of relations, the equation of state for a *perfect gas* is obtained

$$p_1/\rho_1 T_1 = p_2/\rho_2 T_2 = p_3/\rho_3 T_3 = \text{constant}$$

If the constant is denoted by R, its units will be $N \cdot m/kg \cdot {}^\circ K$ or $J/kg \cdot {}^\circ K$. Thus

$$\rho = p/RT \tag{2.26}$$

It is also observed that the constant R is a function of the gas. This equation is a special case of the pure-substance equation (2.25), and it is called the *perfect gas law*. Since the endpoints 1 and 3 on the p–ρ diagram were chosen arbitrarily, the perfect gas law can be written for any point (state) such as 2′ and 4 for this pure substance. Consequently, Eq. (2.26) is not a process equation but an equation describing the thermodynamic equilibrium states for all processes of perfect gases.

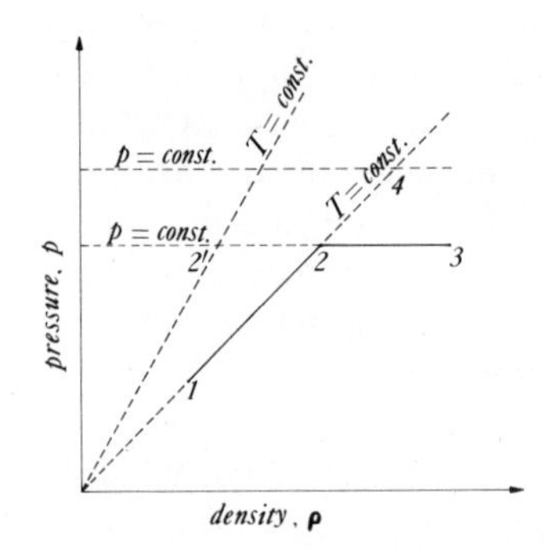

Fig. 2.7 Isothermal and isobaric processes on the p–ρ diagram.

We said that the constant R is dependent on the gas used. We can see this from Table 4.2. Since gases have different masses in 1 mole, the equation of state is often expressed in terms of the molecular weight M by defining a new gas constant R^* such that $R = R^*/M$. Then

$$p = (R^*/M)\rho T \tag{2.27}$$

In this representation, the new gas constant R^*, which has been normalized with the molecular weight, so to speak, turns out to be a *universal constant* for *all* perfect gases and equal to 8.3143 J/g mole · °K. Either of the two equations (2.26) or (2.27) can be used. If the first one is used, we must remember R for the gas used; and if the second one is used, we must remember the universal constant and the molecular weights of the gases.

2.11 COEFFICIENTS OF COMPRESSIBILITY, OF THERMAL EXPANSION, AND OF TENSION

Since density is the reciprocal of the specific volume v, Eq. (2.25) can be written as

$$p = \phi(v, T) \tag{2.28}$$

According to the rules of differentiation, the change of pressure of any system can be evaluated in terms of the corresponding changes in temperature and volume for that pure substance:

$$dp = \left(\frac{\partial p}{\partial v}\right)_T dv + \left(\frac{\partial p}{\partial T}\right)_v dT \tag{2.29}$$

Dividing by dT and assuming that there was no change of pressure (after all, $dp = 0$), we obtain an important relation among the partial derivatives:

$$\left(\frac{\partial v}{\partial T}\right)_p = -\left(\frac{\partial p}{\partial T}\right)_v \Big/ \left(\frac{\partial p}{\partial v}\right)_T \tag{2.30}$$

The subscripts indicate that the process of constant property represented by the subscript is maintained during the evaluation of the partial derivative. Actually, Eq. (2.30) would be true for any mathematical relationship among three variables but it has a special significance in the dynamic and thermodynamic behavior of fluids. Equation (2.30) can be considered as the differential form of the equation of state for a pure substance. The three partial derivatives have the following physical significance. By definition, the quantity

$$\beta = -\frac{1}{v}\left(\frac{\partial v}{\partial p}\right)_T = \frac{1}{\rho}\left(\frac{\partial \rho}{\partial p}\right)_T \tag{2.31}$$

is called the *coefficient of compressibility* which is a measure of the percent change in volume per change of pressure when this measurement is done at constant temperature. Obviously, this is a very important yardstick that will separate the fluids (and the domain of fluids) that are elastic from those that are not. According to what we said in Section 2.2 if we applied Eq. (2.31) to a liquid, like ocean water, the coefficient β will be practically zero, while for the atmosphere it will have a very definite value.

In a similar fashion, the *coefficient of thermal expansion* is defined as

$$\beta_1 = \frac{1}{v}\left(\frac{\partial v}{\partial T}\right)_p = -\frac{1}{\rho}\left(\frac{\partial \rho}{\partial T}\right)_p \tag{2.32}$$

This coefficient describes the change of volume produced by the change of temperature at a constant-pressure process. We know from elementary physics that this volume coefficient of thermal expansion is equal to three times the value of the linear coefficient of thermal expansion, provided the substance is *isotropic* (has no preferred direction) in its expansion characteristics.

Finally, the third derivative in Eq. (2.30) is the *coefficient of tension*:

$$\beta_2 = \frac{1}{p}\left(\frac{\partial p}{\partial T}\right)_v \tag{2.33}$$

When the volume[8] is maintained constant, this coefficient indicates the sensitivity of pressure to change in temperature. Substituting these coefficients into Eq. (2.30) we have

$$p\beta = \beta_1/\beta_2 \tag{2.34}$$

Illustrative Example 2.1

Calculate the three coefficients β, β_1, and β_2 for a perfect gas, such as the atmosphere, at standard pressure and temperature of 760 mm Hg and 20°C.

If the density ρ is substituted for the mass m per unit volume $\mathscr{V}$ in Eq. (2.26), we have

$$p\mathscr{V} = mRT$$

Since the compressibility coefficient is obtained at constant temperature, for a constant-mass system

$$p\,d\mathscr{V} + \mathscr{V}\,dp = 0$$

Consequently,

$$\beta = -\frac{1}{\mathscr{V}}\left(\frac{d\mathscr{V}}{dp}\right)_T = \frac{1}{p}$$

The atmospheric pressure was given as 760 mm Hg which corresponds approximately to 1013 mb, or 1.013×10^6 dynes/cm^2, or 1.013×10^5 newtons/m^2. Then $\beta = 1/1.013 \times 10^5$ m^2/newton.

The other two coefficients are the same:

$$\beta_1 = \beta_2 = 1/T = 1/(273 + 20) \quad {}^\circ\mathrm{K}^{-1}$$

It is easy to verify that Eq. (2.34) is satisfied by these values.

Illustrative Example 2.2 treats liquids for a comparison.

[8] In all these coefficients the specific volume can be replaced by the volume $\mathscr{V}$ for a given mass, and the partials can be replaced by total derivatives since one of the properties is always maintained constant.

2.12 COMPRESSIBLE AND INCOMPRESSIBLE SUBSTANCES

The compressibility coefficient β introduced in the preceding section was said to be a measure of the compressibility of a substance. *Compressibility* actually implies that the volume of a substance is a sensitive function of the pressure level. It must be emphasized, also, that certain states of the same substance are more susceptible to compressibility than others. This term is often erroneously identified as the ability of the substance to change shape. All fluids change shape under the influence of an external force, but what is important here is whether or not a change of volume has taken place. Conversely, *incompressibility* is the inability to change the volume of a given mass even though there has been a change of pressure. This term, too, is often erroneously used to indicate that the volume or the density remains constant. We should not forget that although a substance can be incompressible, its volume or density is free to change with temperature. There is a proper name for a constant-density process, i.e., *isochoric*. This naturally implies that the volume or density is not at all sensitive to changes of pressure as well as temperature.

Liquids are generally considered to be incompressible substances since their density is insensitive to changes of pressure. *Hydrodynamics* is the general study of the dynamics of liquids as well as gases in a domain of states where the variations of pressure are very small. This actually occurs in the atmosphere on horizontal surfaces where the pressure variations are small enough to ignore density variations on these surfaces.

Often, as in the case of solids, the *bulk modulus of elasticity* is used as a measure of compressibility. This is the reciprocal of the coefficient of compressibility β:

$$E = \frac{1}{\beta} = -\left(\frac{dp}{d\mathscr{V}/\mathscr{V}}\right)_T \tag{2.35}$$

Since for an increase of pressure the volume always decreases, even when very small, the derivative $dp/d\mathscr{V}$ is always negative; this makes the compressibility coefficient and the modulus of elasticity always positive. Thus for an incompressible fluid, $\beta = 0$ and $E = \infty$.

In general, the compressibility coefficient is a function of pressure and temperature. In Illustrative Example 2.1 it was found that for a perfect gas the compressibility coefficient is a function of pressure only. Thus for a perfect gas the modulus of elasticity is the pressure itself.

At ordinary pressures and temperatures water has a modulus of elasticity $E = 21 \times 10^6$ mb. At the same temperature and at atmospheric sea-level pressure the modulus of elasticity for air is 1000 mb. We can see then that water is 2.1 $\times\ 10^4$ times less compressible than air.

For seawater the compressibility coefficient has been found to vary from 4.6×10^{-5} to 4.0×10^{-5}. An approximate expression for the specific volume of seawater is

$$v = v_0(1 - \beta p) \tag{2.36}$$

where p is in bars.

Equation (2.35) is made more general when the compressibility coefficient and the modulus of elasticity are defined for any process, including isothermal. The subscript T can be removed and

$$E = \frac{1}{\beta} = -\frac{dp}{d\mathscr{V}/\mathscr{V}} \tag{2.37}$$

This will imply that these coefficients will depend on the process employed to evaluate $dp/d\mathscr{V}$. For an *adiabatic* (no heat exchange) process it can be shown that the special relationship between p and $\mathscr{V}$ is

$$p\mathscr{V}^n = \text{constant} \tag{2.38}$$

and, after differentiating this equation, the modulus of elasticity for an adiabatic process is

$$E = -\frac{dp}{d\mathscr{V}/\mathscr{V}} = np \tag{2.39}$$

This result is in accordance with that of Illustrative Example 2.1 for $n = 1$, an isothermal process. In fact, it is easy to see from Fig. 2.7 that any process can be represented fully or approximated by Eq. (2.38) by choosing the proper exponent n.

Illustrative Example 2.2

Find the increase in pressure necessary to produce a 1% reduction in the volume of seawater at ordinary pressure and temperature. Compare the findings with those of air compressed adiabatically.

For seawater at about 10°C, the value of E is approximately 21×10^6 mb as stated in this section. Then, for a finite pressure rise of seawater

$$\Delta p = -E(\Delta\mathscr{V}/\mathscr{V}) = -21 \times 10^6(-0.01) = 210 \times 10^3 \text{ mb}$$

Since air can be considered a perfect gas, for an adiabatic compression, from Eq. (2.39) the modulus of elasticity is $1.4 \times 1013 = 1418.2$ mb. Then the pressure rise necessary to cause the percent change in volume for air is

$$\Delta p = -1418.2(-0.01) = 14.18 \quad \text{mb}$$

Thus we conclude that a 1% change in volume is very likely to occur in the atmosphere owing to zonal pressure changes, while in the ocean a 1% volume change requires about a 210-atm pressure or equivalently, a depth of over 2 km. This problem illustrates well the comparative order of magnitude of compressibility of air and water. This question of compressibility is generalized in Chapter 7.

2.13 THE PERFECT GAS LAW, THE ATMOSPHERE AND ITS CONSTITUENTS

The dry and clean air near sea level is composed of several gases, as we saw in Table 1.1. Nitrogen and oxygen make up 99% of the total volume. In order of decreasing volume, the following gases are also present: argon, carbon dioxide, neon, helium, krypton, xenon, hydrogen, methane, nitrous oxide, ozone, sulfur dioxide, nitrogen dioxide, ammonia, carbon monoxide, and iodine. Table 2.4 lists the accepted standard clean dry air composition at sea level. Although water vapor is plentiful in the atmosphere, it is never included in standard tables because its content varies considerably from day to day and from season to season. We study this question separately in future sections.

TABLE 2.4

Standard Composition of Clean Dry Air at Sea Level

Gas	Percent by volume	Molecular weight
Nitrogen, N_2	78.084	28.0134
Oxygen, O_2	20.9476	31.9988
Argon, Ar	0.934	39.948
Carbon dioxide, CO_2	0.0314	44.01

Thus the atmosphere is a gas mixture, specifically a perfect gas mixture since experiments show that each constituent as well as water vapor obeys the perfect gas law. In combining Eqs. (2.21) and (2.27) we have a modified gas law in terms of the molecular temperature:

$$p = (R^*/M_0)\rho T_m \tag{2.40}$$

We said that R^* was the universal gas constant equal to 8.3143 J/g mol · °K, and M is the local molecular weight, while M_0 is the molecular weight at sea level. The difference between T and T_m was explained in Section 2.8. It is important to find a way to use Eq. (2.40) for the mixture. The short treatment given here can be found in many thermodynamics books.

When a number of gases, each of mass m_i, are mixed in a fixed volume $\mathscr{V}$, the density of each of the constituents is given by

$$\rho_i = m_i/\mathscr{V} \tag{2.41}$$

where the subscript i refers to the particular constituent. According to Eq. (2.40) each of the constituents having the same volume and the same temperature[9] will have different partial pressures p_i:

$$p_i = \frac{R^*}{M_i}\frac{m_i}{\mathscr{V}}T \tag{2.42}$$

According to *Dalton's law*, which in essence allows the whole mixture to be a perfect gas as well, the pressure of the mixture is the sum of the partial pressures of the constituents. Thus

$$p = \frac{R^*T}{\mathscr{V}}\sum_i \frac{m_i}{M_i} \tag{2.43}$$

If we define the molecular weight M of the mixture as the weighted sum of the constituents' molecular weights,

$$M = \sum_i m_i / \sum_i \frac{m_i}{M_i} = m / \sum_i \frac{m_i}{M_i} \tag{2.44}$$

then Eq. (2.43) becomes

$$p = \frac{R^*T}{\mathscr{V}}\frac{m}{M} = \frac{R^*}{M}\rho T \tag{2.45}$$

where, now, ρ, M, p, and T are properties of the mixture.

The molecular weight of dry air at sea level is $M = 28.9644$. The gas constant for air $R = R^*/M = 0.291$ J/g · °K and the perfect gas equation for the air mixture is

$$p = \rho R T \tag{2.46}$$

2.14 WATER VAPOR

We have not yet considered the influence of water vapor on the state of the atmosphere. However, we mentioned that vapor played a major role since its presence, although variable, had a considerable influence on all aspects of the thermodynamics of the atmosphere. As long as there is no condensation, vapor

[9] Each constituent in the mixture must occupy the whole volume and be at the same equilibrium temperature. In what follows, the subscript m is dropped from T_m.

may be treated as a perfect gas. Experimental data verify this. In the range of conditions in which we consider vapor in the atmosphere the perfect gas law applies very well even at states very near saturation. Since the saturation vapor pressure is the highest vapor pressure possible at the temperature of the vapor, as can be seen from Fig. 2.8, we would like to compare the measured density of the saturated vapor at a given temperature to that calculated from the perfect gas law. The ratio of these two densities should be a good measure of the applicability of the perfect gas law to vapor, even when near saturation.

Table 2.5 is given for this comparison. It shows the temperature of air, and naturally vapor, the partial pressure p_s of vapor at the saturated state, and the ratio of the densities of the real vapor to those calculated from the gas law.

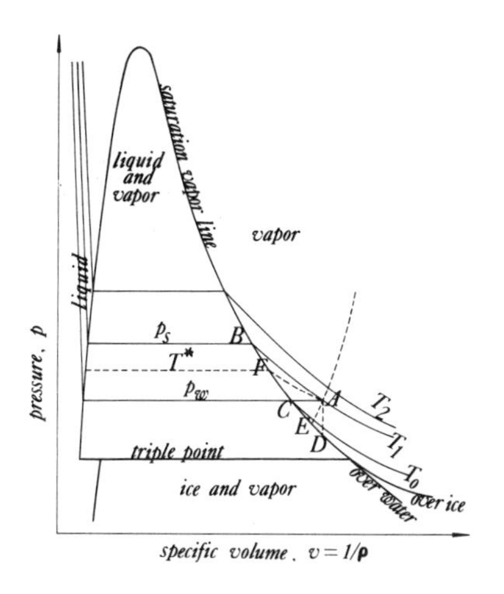

Fig. 2.8 Phase diagram for water, ice, and vapor.

TABLE 2.5

Comparison of Densities of Vapors
(Table Values over Perfect Gas Values)

Temperature (°C)	Pressure p_s (mb)	Density ratio
−10	2.86	1.0003
0	6.108	1.0005
10	12.27	1.0008
20	23.37	1.0012
30	42.41	1.0018
40	73.75	1.0027

The molecular weight of vapor is $M_w = 18$. Since vapor must be in equilibrium at its partial pressure, like any other constituent, the gas law applied to it reads:

$$p_w = \rho_w R^* T / M_w \tag{2.47}$$

Introducing $R = R^*/M$ for the mixture and since the ratio of the molecular weights $M/M_w = 1.6091$, Eq. (2.47) becomes, after letting $R^*/M_w = RM/M_w$,

$$p_w = 1.6091 \rho_w RT \tag{2.48}$$

When we combine water vapor with dry air in a unit volume, the density of the *moist air* is the sum of the densities of the vapor and the dry air. For an atmospheric pressure p the partial pressure of the dry air is $(p - p_w)$. Thus,

$$\rho = \rho_w + \rho_a = \frac{1}{1.6091} \frac{p_w}{RT} + \frac{p - p_w}{RT}$$

$$= \frac{p}{RT}\left(1 - 0.3785 \frac{p_w}{p}\right) \tag{2.49}$$

This result indicates that since p_w/p is always positive, *dry air is always heavier than moist air because the weight of the vapor is lighter than that of the air it replaces.*

Let us examine p_w a little closer. Since vapor in the air does not remain constant in the atmosphere, the implication is that condensation and evaporation take place often. In turn this implies that the thermodynamic states of vapor in the air are normally close to the vapor saturation conditions.

The partial pressure p_w of water vapor has an upper limit for a given temperature. This limit is the saturation pressure p_s. Thus, if $p_w < p_s$, more evaporation can take place. If $p_w = p_s$, the air is said to be saturated and will remain in this state unless there is a change of temperature or a wind bringing new air at a different state. If the temperature drops and thus p_s drops, tending to make $p_w > p_s$, condensation occurs. The magnitude of p_w is less than the ambient atmospheric pressure. A temperature of 100°C is necessary for the vapor pressure to reach the value of 1 atm. This is not likely to occur in our environment.

The change of phase of water is shown in Fig. 2.8 in a p–v–T diagram. In this diagram the saturation line and three isotherms are shown, such that $T_0 < T_1 < T_2$. In Fig. 2.9 we see a three-dimensional representation of the three phases of water and their mixtures.

Consider the vapor side of these diagrams and let us apply it to a moist atmosphere at the temperature T_1. If the partial pressure of vapor in the air is p_w at the temperature T_1, it places the state of this vapor in the atmosphere at the point A. There are a number of ways to saturate the atmosphere so that the point A moves to a point on the *saturation vapor line* in Fig. 2.8. We can cool the air at constant pressure so that the vapor with it will move from A to C. When we reach the temperature T_0, at C, condensation will begin. The point C is called the *dew point* and the *dew point temperature* of a gas–vapor mixture is the temperature at which the vapor condenses or solidifies when it is cooled at constant pressure. In other words, the dew point temperature is the saturation temperature corresponding to a given partial pressure of the vapor in the atmosphere. Thus the dew point temperature of the vapor in equilibrium at A is T_0, while its equilibrium temperature is T_1.

Another way of saturating the air is to cool it at constant volume or constant density. In the representation of Fig. 2.8 this cooling takes place along a vertical line from A to D. Since the temperature at D is lower than that at C, we conclude that we must remove more heat energy from the atmosphere to perform a constant-volume precipitation than a constant-pressure precipitation.

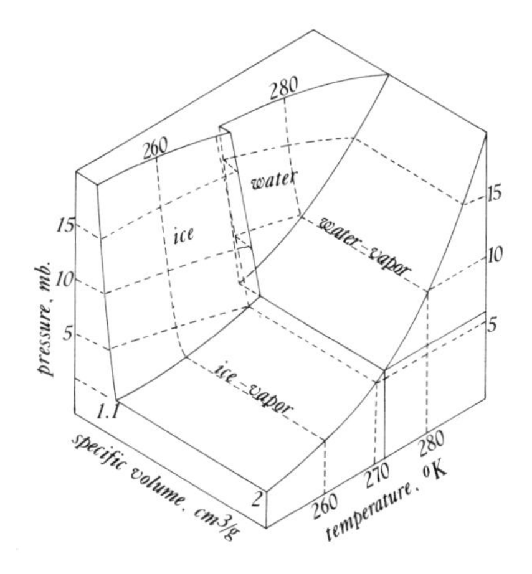

Fig. 2.9 Three-dimensional representation of the phase changes of water.

A third way of reaching the saturation line is by an adiabatic process (no net heat exchange with the surroundings). This means that the internal energy of the vapor and the air will drop in compensation for the work due to the change of ρ that takes place. This process is shown in Fig. 2.8 as AF. The temperature at F is called the *temperature of adiabatic saturation* or the *wet bulb temperature.* This temperature and its use in determining the humidity in the atmosphere are discussed in following sections. We can conclude that if we knew the temperature of the moist air T_1 and any of the other temperatures $T_0 = T_C$ or $T_F = T^*$ or T_D, we could locate the state of the vapor in the air by finding the intersection of BA with CA or FA or DA. Actually, the constant-volume process AD is never used. The dew point temperature or the wet bulb temperature together with the ambient temperature is commonly used to determine the state of vapor in the air.

2.15 METASTABLE THERMODYNAMIC EQUILIBRIUM

We have indicated in the preceding section, and on the basis of Fig. 2.8, that condensation of vapor in the air begins as soon as its state begins to cross to the left of the saturation line *BFCED*, even though the saturated state and the condensing states are at the same pressure and temperature during this change of phase.

Practical observations in the atmosphere and in machinery carrying vapor indicate that if the process by which vapor reaches the saturation vapor line is somewhat rapid, then this vapor could reach a state to the left of the saturation line without condensation occurring. This vapor could remain there without condensation for a considerable period of time, and for this reason these states are called *metastable states*, which implies that this momentary equilibrium could be upset in abrupt condensation, sometimes referred to as *condensation shock*.

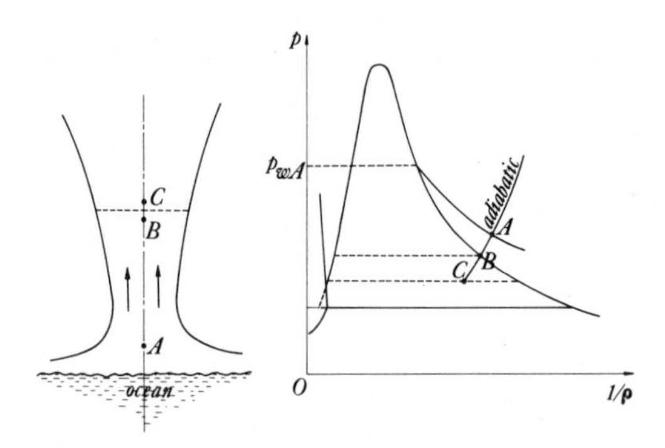

Fig. 2.10 Thermodynamic metastable states.

Let us illustrate this phenomenon with an atmospheric example. On a clear day, the heat from the sun warms the earth's surface which in turn warms the air layer next to it. As the temperature of the air near the earth's surface increases, its ability to hold larger amounts of vapor increases. If this air layer is on the ocean, evaporation begins to take place. Because the air layer near the ocean surface has heated more than the layers above it, and because moist air is lighter than dryer air, the moist air near the surface will tend to rise because of an upset in the mechanical equilibrium. This air will rise, with its moisture, in a vertical column until it seeks another mechanical equilibrium. As we shall see later, the ambient pressure of the atmosphere decreases with altitude. We saw already that temperature also decreases with altitude. The thermodynamic process that this rising moist air column will undergo is similar to an expansion (decrease of pressure) in a diverging nozzle. Figure 2.10 shows this schematically.

The water evaporated just at the surface of the ocean; at the point A it has a vapor state at the temperature of the ambient air T_A and the partial pressure p_{wA}. Let us suppose that air rises along ABC without exchanging heat with the surroundings, an adiabatic thermodynamic process. The actual process will be close to that. In the accompanying p–$(1/\rho)$ diagram the thermodynamic changes the vapor will undergo will be along the adiabatic line ABC. In this expansion one expects that when the vapor pressure reaches p_{wB} on the saturation line, condensation will begin followed by precipitation. In reality, this may not happen. The column continues to rise and expand to a pressure p_C where the partial pressure of the vapor is p_{wC}, which is below the saturation pressure corresponding to the ambient temperature T_B. Between these two neighboring points B and C the water remains in the vapor state but in a *supercooled* or *supersaturated state*. This is known as a *metastable state*.

This phenomenon could have occurred easily around the triple point in Fig. 2.8 since air actually rises to elevations where the ambient temperature is around freezing. Thus, around the triple point it is possible to have metastable states of vapor states in an ice state domain, or an ice state in a vapor state domain. Table 2.6 shows two values of p and ρ for the same temperature, over water and over ice. Insofar as the temperature and pressure are concerned, these metastable states are located very close to the saturation values so no significant errors are made in misjudging p and T. However, since the release of latent heat depends on condensation taking place, any serious misjudgment will be in the correct value of the enthalpy of the air.

2.16 HUMIDITY

The amount of vapor in air is identified by two widely used concepts, namely the *relative humidity* and the *humidity ratio* or the *absolute humidity*. Relative humidity is defined as the ratio of the partial pressure of the vapor in air to that of the saturated vapor at the same temperature. In other words, this measure is relative because it compares the actual vapor present in air to the maximum amount air could absorb at the same temperature. If we designate the relative humidity by ϕ, we have

$$\phi = p_w/p_s = \rho_w/\rho_s \tag{2.50}$$

For the same temperature, the perfect gas law permits us to equate the ratio of pressures and densities.

The relative humidity does not directly give the amount of vapor contained in air. For this we define the *humidity ratio*, sometimes called the *absolute humidity* or the *mixing ratio* ω, as the ratio of the mass of water vapor to the mass of dry air in the mixture. Thus

$$\omega = m_w/m_a \tag{2.51}$$

TABLE 2.6

Properties of Saturated Water Vapor over Water and Ice[a]

Temperature (°C)	Saturation pressure, p_s (mb)		Density ρ (g/m^3)		Enthalpy, h (J/g)	Latent heat, L (J/g)
	Over water	Over ice	Over water	Over ice		
−40	—	0.13	—	0.12	2 427	2 839
−30	—	0.38	—	0.34	2 446	2 840
−25	—	0.64	—	0.56	2 454	2 840
−20	1.24	1.04	1.07	0.89	2 464	2 839
−15	1.91	1.66	1.63	1.39	2 473	2 838
−10	2.86	2.60	2.36	2.14	2 482	2 837
−8	3.34	3.10	2.74	2.54	2 486	2 837
−6	3.90	3.69	3.17	3.00	2 490	2 837
−4	4.54	4.37	3.66	3.53	2 493	2 836
−2	5.27	5.17	4.22	4.14	2 497	2 835
0	6.108		4.847		2 501.6	2 501.6
1	6.566		5.192		2 503.4	2 499.2
2	7.055		5.558		2 505.2	2 496.8
4	8.129		6.358		2 508.9	2 492.1
6	9.345		7.258		2 512.6	2 487.4
8	10.720		8.267		2 516.2	2 482.6
10	12.270		9.396		2 519.9	2 477.9
12	14.014		10.66		2 523.6	2 473.2
14	15.973		12.06		2 527.2	2 468.5
16	18.168		13.63		2 530.9	2 463.8
18	20.62		15.36		2 534.5	2 459.0
20	23.37		17.29		2 538.2	2 454.3
22	26.42		19.42		2 541.8	2 449.6
24	29.82		21.77		2 545.5	2 444.9
26	33.60		24.37		2 549.1	2 440.2
28	37.78		27.23		2 552.7	2 435.4
30	42.41		30.37		2 556.4	2 430.7
32	47.53		33.82		2 560.0	2 425.9
34	53.18		37.59		2 563.6	2 421.2
36	59.40		41.72		2 567.2	2 416.4
38	66.24		46.24		2 570.8	2 411.7
40	73.75		51.16		2 574.4	2 406.9

[a] Values below triple point are converted values from Keenan and Keyes (1936). Values above triple point are from Schmidt (1969). Enthalpy based on 0°C liquid.

The conventional units for the humidity ratio are grams of vapor per kilogram of dry air. Using the perfect gas law

$$\omega = \frac{m_w}{m_a} = \frac{\rho_w}{\rho_a} = \frac{p_w}{p_a}\frac{R_a}{R_w} = 0.6215\,\frac{p_w}{p - p_w} \tag{2.52}$$

Since vapor and air are in thermal equilibrium, their temperatures are the same, the ratio of the gas constants is 0.6215, and p is the atmospheric pressure, p_w the partial pressure of the vapor, and p_a the partial pressure of the dry air.

Often in the literature a third definition of humidity is used, called the *specific humidity* ω^0, giving the ratio of the mass of vapor to the mass of the *moist* air. Then[10]

$$\omega^0 = \frac{m_w}{m_a + m_w} = \frac{\omega}{1 + \omega} \tag{2.53}$$

and for very low partial pressures the two relations become approximately equal.

A relationship can be established between the relative humidity and the humidity or mixing ratio as follows:

$$\omega = \frac{\rho_w}{\rho_a} = \phi\,\frac{p_s}{p_a}\frac{R_a}{R_w} = \phi\,\frac{\rho_s}{\rho_a} \tag{2.54}$$

The properties of the liquid and vapor states of water have been tabulated extensively and in detail in many references. In the vapor state, water at a given ambient temperature can exist at many partial pressures, as certified by the perfect gas law. However, at the saturated states, there is only one saturation partial pressure for an ambient temperature. It is not possible to measure directly the partial pressure of vapor in the atmosphere. It is easier to measure the temperature and the humidity, and thus determine p_w. Tables 2.6 and 2.7 give the properties of saturated liquid and vapor for pressure and temperature ranges of interest in the environment. Table 2.6 is tabulated beginning with rounded increments of ambient temperatures, and Table 2.7 gives the same information but beginning with rounded increments of saturation pressures.

Although the data given in Tables 2.6 and 2.7 are useful for calculations of humidity and properties of air before and after precipitation, they do not adapt to analytical treatment. For mathematical analysis we often require an analytical relationship between the pressure and temperature along the saturation line. This is obtained through what is called the *Clapeyron equation*.

[10] Substituting for ω the expression of Eq. (2.52) we have

$$\omega^0 = \frac{0.6215\,p_w}{p_a + 0.6215 p_w} \simeq 0.6215\,\frac{p_w}{p_a + p_w} = 0.6215\,\frac{p_w}{p}$$

TABLE 2.7

Properties of Saturated Water Vapor[a]

Saturation pressure, p_s (mb)	Temperature, T (°C)	Specific volume (m³/kg)	Density, ρ (kg/m³)	Enthalpy, h (J/g)	Latent heat, L (J/g)
10	6.9828	129.20	0.007739	2 514.4	2 485.0
15	13.036	87.98	0.01137	2 525.5	2 470.7
20	17.513	67.01	0.01492	2 533.6	2 460.2
25	21.096	54.26	0.01843	2 540.2	2 451.7
30	24.100	45.67	0.02190	2 545.6	2 444.6
35	26.694	39.48	0.02533	2 550.4	2 438.5
40	28.983	34.80	0.02873	2 554.5	2 433.1
45	31.035	31.14	0.03211	2 558.2	2 428.2
50	32.898	28.19	0.03547	2 561.6	2 423.8
55	34.605	25.77	0.03880	2 564.7	2 419.8
60	36.183	23.74	0.04212	2 567.5	2 416.0
65	37.651	22.02	0.04542	2 570.2	2 412.5
70	39.025	20.53	0.04871	2 572.6	2 409.2

[a] After Schmidt (1969).

We know that the change of phase takes place at constant pressure and temperature. If we differentiate the *Gibbs function*, defined as

$$G = h - Ts \tag{2.55}$$

where h is the enthalpy and s the entropy.

During the change of phase, we have

$$dG = dh - T\,ds - s\,dT \tag{2.56}$$

However, from the combination of the first and second laws in thermodynamics we have

$$dh = T\,ds + \frac{1}{\rho}\,dp \tag{2.57}$$

During the change of phase, it turns out, then, with the use of these equations and $dp = dT = 0$, that

$$dG = 0 \tag{2.58}$$

Thus, along the saturation line $dG = 0$ or $G =$ constant and from Eq. (2.56) we have

$$dh = T\,ds + s\,dT \tag{2.59}$$

Using Eq. (2.57) again we have finally

$$dp/dT = \rho s \tag{2.60}$$

where s is the *entropy* of the vapor. Since entropy, as a property, is cumbersome to work with, primarily because it cannot be measured, we relate it to the *latent heat* L given in Tables 2.6 and 2.7. Thus, $L = T(s_v - s_l)$ where s_v is the entropy of the vapor and s_l is the entropy of the liquid at the same temperature. But since s_l is so much smaller than s_v, especially for the temperatures of the atmosphere, we can write that $L = sT$. We are ready to substitute this as well as the perfect gas law for ρ into Eq. (2.60). It follows that

$$\frac{dp}{dT} = \frac{Lp}{R_w} \frac{1}{T^2} \tag{2.61}$$

This is the *Clausius–Clapeyron equation* or simply the Clapeyron equation. Then along the saturation line if we replace p by p_s, we have a differential expression for p_s in terms of T which should correspond to the values tabulated in Tables 2.6 and 2.7.

Integrating Eq. (2.61) we have

$$\ln p_s = -\frac{L}{R_w} \frac{1}{T} + \text{constant} \tag{2.62}$$

When we use an approximate average of L in the ranges of atmospheric temperatures, $L = 2500$ J/g, and for $R_w = 0.464$, and if we evaluate the constant in Eq. (2.62) by picking $p_s = 6.1$ at $T = 273°\text{K}$ as in Table 2.6 we have

$$\ln p_s = 21.548 - (5388/T) \tag{2.63}$$

The units in this equation are millibars for pressure and degrees Kelvin for temperature. Equation (2.63) determines the pressure–temperature relationship along the saturation vapor line *above the triple point*. The following Illustrative Example develops relationship below the triple point for vapor temperature below 0°C.

Illustrative Example 2.3

Find an expression similar to Eq. (2.63) for the saturation line below the triple point between ice and vapor.

The only difference between Eq. (2.63) and the expression that is sought is that the value of the latent heat of solidification must replace the latent heat of condensation in Eq. (2.62). Below the triple point the vapor becomes ice without passing through the liquid phase. The latent heat of solidification is 2835 J/g (see Table 2.6). Using again $p_s = 6.1$ mb at 273°K to determine the constant of Eq. (2.62) we obtain the relation sought:

$$\ln p_s = 24.203 - (6114/T)$$

This is not in bad agreement with values in Table 2.6 for vapor over ice.

2.17 WET BULB TEMPERATURE AND THE DETERMINATION OF HUMIDITY

We mentioned earlier that the partial pressure of water vapor is not readily measurable in the atmosphere. The state of water vapor in the atmosphere cannot be determined from a *dry bulb temperature* alone. One way of determining the state of the vapor is to have it undergo a known thermodynamic process until it reaches the saturation vapor line as shown on Fig. 2.8. With one known state on the saturation vapor line, the process, and the temperature of the air it is possible to locate or calculate the original state of the vapor.

The process that is commonly used to bring vapor in the air to saturation is *adiabatic saturation.* This process consists of passing moist air, whose vapor properties are to be determined, over a sponge or cotton containing sufficient amounts of water and inside of which is a thermometer, *wet bulb*, which measures the temperature of the air that became saturated in this way. The evaporation of this water will be accompanied by a reduction of the temperature of the water in the sponge or cotton and of the air up to the point of complete saturation. The final equilibrium temperature T^* is called the *wet bulb temperature* or the *temperature of adiabatic saturation.* Actually the cooling process thus described is more an isenthalpic process than an adiabatic process as we shall see. The water in the sponge has evaporated and in doing so has taken heat from the remaining water for its latent heat of evaporation, and this reduces the temperature of the remaining water in the sponge to the temperature T^*.

Obviously the level of T^* depends on the humidity already present in the air blowing over the sponge. If this air was already saturated, it cannot evaporate any more water around the sponge and T^* will be T, the temperature of the dry bulb. Conversely, the dryer the air blowing over the sponge, the larger the temperature difference between the dry and wet bulbs.

We can conceive a *wet and dry bulb thermometer* which has two separate thermometer units over which the air to be analyzed for humidity is passed. This is shown in Fig. 2.11. The thermometer to the left is dry. The second thermometer has a wick around its bulb immersed in a large supply of water. The same air is passed over the two bulbs.

When we take into account the energy balance in this process (see Wood, 1969), we can show that it is nearly *isenthalpic* (constant enthalpy), i.e., that the enthalpy of the moist air at the ambient temperature T_1 is equal to the enthalpy of the saturated air at T^*. Figure 2.8 also shows this process AF to saturation.

When we consider this isenthalpic process

$$h_{a1} + \omega_1 h_{w1} = h_a^* + \omega^* h_s^* \tag{2.64}$$

where h_{a1} is the enthalpy of dry air at the dry bulb temperature T_1, ω_1 is the humidity ratio of air, which we are seeking, h_{w1} is the saturation enthalpy of

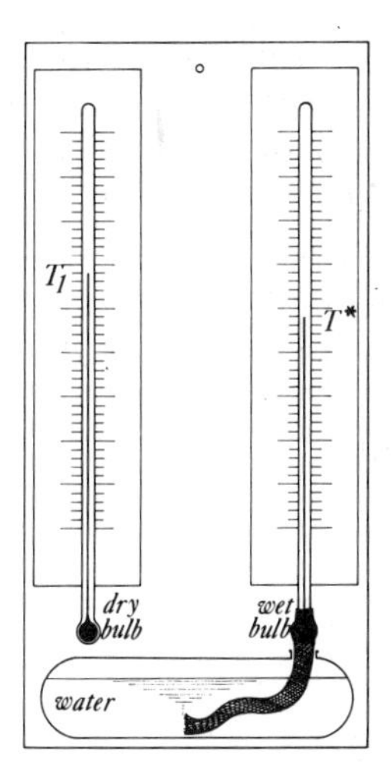

Fig. 2.11 Wet and dry bulb thermometer.

vapor in the air at the temperature T_1 as taken from Table 2.6, $h_a{}^*$ is the enthalpy of dry air at T^*, ω^* is the humidity ratio of air at T^*, and $h_s{}^*$ is the saturation enthalpy of vapor at T^* taken from Table 2.6 also. The enthalpy of the liquid that was evaporated, which is equal to $(\omega^* - \omega_1)h_l$, has been neglected because it is small. The humidity ratio ω^* can be calculated from ρ_w/ρ_a at T^*. The enthalpies of dry air $h_{a1} = c_p T_1$ and $h_a{}^* = c_p T^*$ where the quantity c_p is the specific heat of air, which is known. Finally we can solve for ω_1, the humidity ratio of air:

$$\omega_1 = \frac{\omega^* h_s{}^* + (h_a{}^* - h_{a1})}{h_{w1}} \tag{2.65}$$

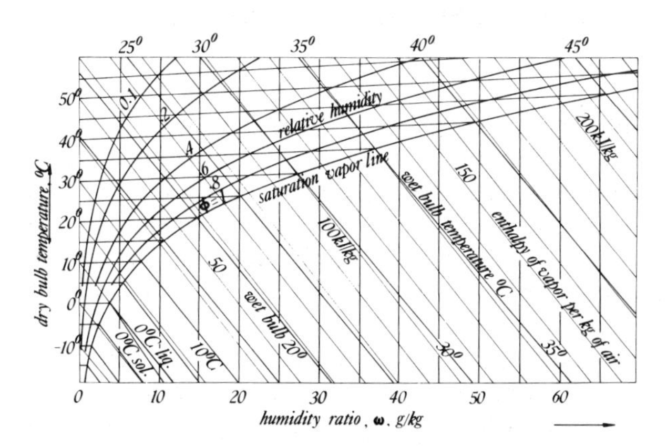

Fig. 2.12 Psychometric chart.

Graphical determination of the humidity ratio can be obtained from psychometric charts such as the one shown in Fig. 2.12. Notice that this chart is made for a standard ambient atmospheric pressure, and that corrections or different charts must be provided for different altitudes.

Figure 2.13 represents a typical humidity distribution with altitude, over a period of one day, at Washington, D.C. Unlike the smooth distribution of pressure and density, the humidity ratio is irregular. The question is raised as to whether there exists a cutoff altitude where humidity becomes negligible. Other soundings suggest that perhaps above 10 km or 300 mb such a cutoff could exist (see also Table 3.2).

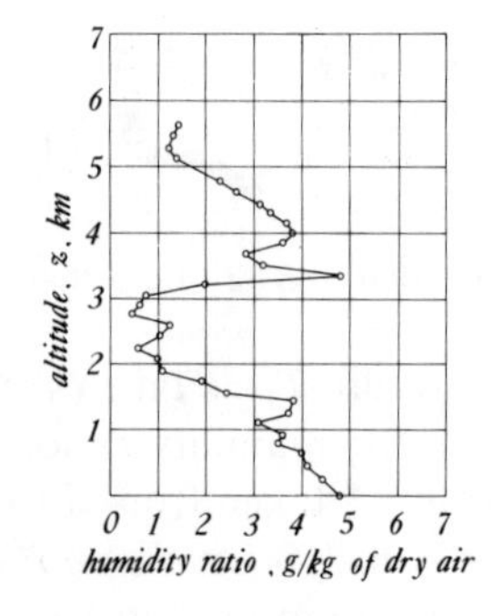

Fig. 2.13 Variation of humidity ratio with altitude. (After F. F. Fischback, M. E. Graves, and L. M. Jones, Microwave refractions as a technique for satellite meteorology, *Space Research*, IX, 1969.) Used with permission of North-Holland Publ. Co., Amsterdam.

2.18 THE GENERAL BEHAVIOR OF WINDS AND CURRENTS

We shall see later in the dynamical considerations of the environment that in order to impart inertia to the atmosphere and oceans, internal and external forces must develop on them. The external forces, such as gravity. Coriolis, centrifugal, and to a negligible degree electric and magnetic, are extremely predictable, whereas the internal forces of pressure and friction are generated by local changes. Winds and currents are produced by the sum total of these external and internal forces.

The conventional direction of winds and currents is taken as that from which the wind or current comes from. Thus a *westerly* is a motion coming from the west. The wind or current velocity is a *vector* quantity. This means that this velocity has a magnitude and a direction as well. Furthermore, we know that the velocity varies not only from locality to locality but also in time. Thus we can say that the totality of this information is a *vector field*. Atmospheric winds near the earth's surface are generally very turbulent and gusty. The average magnitude of the velocity varies with altitude, and the level of turbulence or gustiness depends on the local stability, and on the roughness of the terrain. For these reasons it is best to define the velocity vector in terms of coordinate location and time. We must realize, also, that there are a number of factors that cause time dependence of the velocity, such as molecular agitation, turbulence, zonal perturbations, seasonal perturbations, and so on. Each of these influences has a different time scale. This will become important when we attempt to define time averages.

Let us designate by $\tilde{\boldsymbol{u}}$ the instantaneous wind velocity as a function of position $\boldsymbol{r}$ and time t as follows:

$$\tilde{\boldsymbol{u}} = f(\boldsymbol{r}, t) \tag{2.66}$$

Now, the time average velocity $\boldsymbol{U}$ is defined[11] for a characteristic time τ, i.e., a τ representative of the type of perturbation, with respect to which we wish to find an average. Thus

$$\boldsymbol{U} = \lim_{T \to \tau} (1/T) \int_0^T \tilde{\boldsymbol{u}}(\boldsymbol{r}, t)\, dt \tag{2.67}$$

When we speak of wind or current *speed*, we mean the absolute value of $\boldsymbol{U}$ or its magnitude. Often it is customary to separate the average part of $\tilde{\boldsymbol{u}}$ from its time-dependent part such that

$$\tilde{\boldsymbol{u}}(\boldsymbol{r}, t) = \boldsymbol{U}(\boldsymbol{r}) + \boldsymbol{u}(\boldsymbol{r}, t) \tag{2.68}$$

Figure 2.14 shows this representation of the instantaneous, average, and fluctuating parts. For illustrative purposes we show two different time averages: one with respect to a scale a little larger than the turbulent scale, which normally is of the order of minutes, and the other an average for a day.

Wind velocities vary with altitude, and one way to measure their average value is by seeding sodium clouds that drift with the wind and whose velocity is measured from the ground. Another method of measurement is with balloons made to stay at a fixed altitude. Figure 2.15 shows typical wind velocity

[11] Each of the perturbations discussed at the beginning of this section has a time scale $T = \tau$. Therefore, depending on whether we wish to average with respect to turbulence, zonal changes, or seasonal changes, we must make a choice of the limiting τ. This is discussed in Section 7.12.

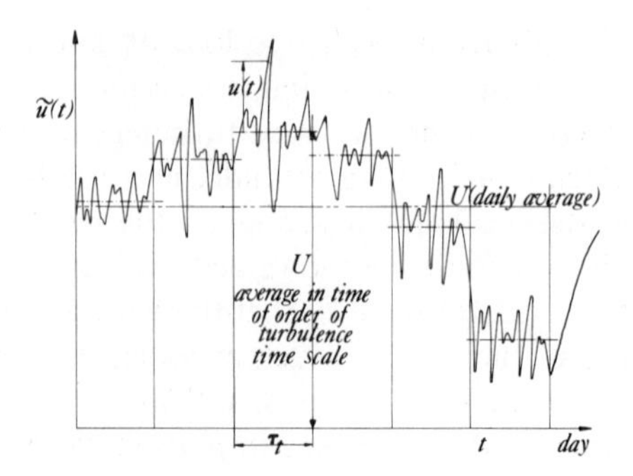

Fig. 2.14 Schematic representation of wind velocity over a period of a day, at one location.

components extending into the thermosphere. In this figure only the dominant easterly and westerly components are shown. Up to the mesopause the north–south components are small and of an order less than 10 m/sec. These components are not shown in the figure, whose sole purpose at this point is to acquaint the reader with the ranges of wind velocities in the atmosphere. The vertical component of the wind increases with altitude as a consequence of the conservation of mass as shown in Illustrative Example 6.2. This increase takes place up to about 100 km at the rate of 0.17 m/s per km and up to a value of 30 m/s. Then it decreases with altitude in the thermosphere.

The altitude coordinate in Fig. 2.15 starts at 20 km. Figure 2.16 shows a typical wind velocity distribution of a westerly in the troposphere. This figure shows the strong upper tropospheric winds that take place at 10 km altitude where most of our long-distance flights occur.

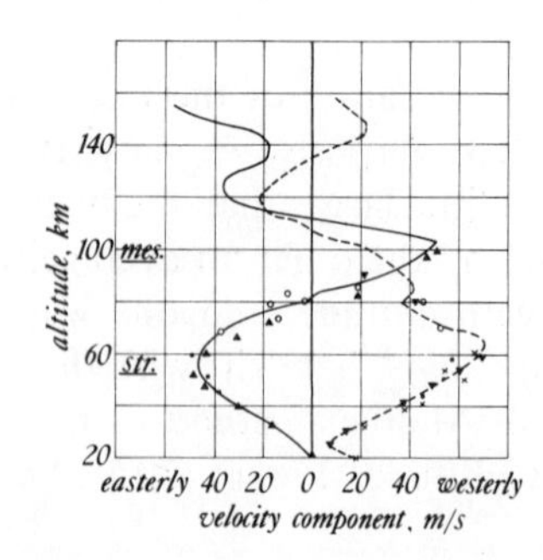

Fig. 2.15 Steady components of prevailing atmospheric wind velocity near 38°N latitude in the US. (After Kochanski, 1964.) ○ Tonopah 38°N and White Sands 34°N. ▲▼ 35°N and S (Batten, 1961). ■ US 28–40°N (Appleman, 1963). × Wallops Island (Newell, 1963).

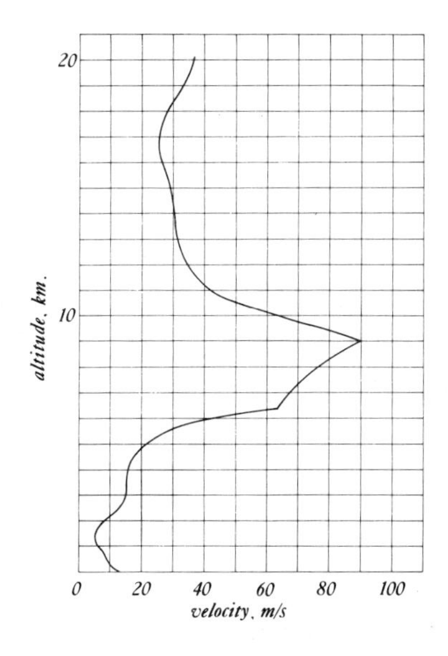

Fig. 2.16 Vertical distribution of velocity of westerly winds. (After Defant.)

3

Basic Principles of Heat Transfer—Energy Balance of the Environment

3.1 MODES OF HEAT TRANSFER

Basically, there are three different ways for heat energy to transfer from one mass point to another, namely, *conduction*, *convection*, and *radiation*. These three modes of heat transfer display different transfer mechanisms but they depend on the temperature levels involved, as we would expect.

Heat conduction is the mechanism of heat flow *through matter* by molecular contact or molecular diffusion. The presence of mass is necessary for this mechanism of heat exchange to take place. The *heat flux* of heat conduction is a vector directed opposite to the temperature gradient in the mass, while its magnitude is proportional to the temperature gradient and the proportionality coefficient is dependent on the molecular structure of the substance. This statement is the Fourier law of conduction which can be expressed, mathematically, in the form

$$\boldsymbol{q}_c = -k\,\nabla T \tag{3.1}$$

where $\boldsymbol{q}_c$ is the heat flux of conduction in calories per unit area and time, and k is the *thermal conductivity* of the substance in which this transfer of heat is taking place. As long as the temperature gradients are not very large, k is constant for a substance. The temperature gradient $\nabla T = \boldsymbol{n}\, dT/dn$ is in the direction $\boldsymbol{n}$ of maximum temperature difference and the units of k must be homogeneous with Eq. (3.1), that is, cal/s · cm · °C.

Since there is very little conduction at high altitudes where the mass points are far apart because of low densities, and because convection and radiation are the principal modes of heat transfer at low altitudes, conduction is a negligible mode of heat transfer in most of the large-scale environment. On the ocean surface, convection and radiation play the most important role. However, at greater depths where current velocities are very small and where radiation from the sun cannot penetrate, conduction is the only mode of heat transfer.

Heat convection is a mode of transporting heat energy from one mass point to another using the macroscopic motion of the matter as a carrier of this energy. Thus this energy transfer is intimately coupled with the motion of the substance. The motion of the environmental substance will be developed in the course of this book.

Heat convection is an important mode of heat transfer in the troposphere and on the surface of the ocean. In considering problems of thermal pollution from stacks and from hot-water discharges in oceans and lakes, convection and conduction (turbulent) become the significant modes of heat transfer.

Heat radiation between two mass points propagates as electromagnetic waves do, and therefore does not require the presence of matter between the two mass points. Thus it could propagate across a vacuum, where conduction and convection cannot take place. The space between the earth and the sun is near vacuum and thus heat radiation is the only means of receiving heat energy from the sun.

The radiation received by a material point from another material point is not dependent on its own temperature but on the temperature of the emitting source. Since a receiving point can also emit its own heat through radiation, the energy emitted is a function of its temperature.

Before dealing with the quantitative aspects of heat radiation, it is important to study the structure and mechanism of its propagation. We said earlier that heat radiation propagates in the form of waves and consequently its mechanism should be looked on as composed of waveforms propagating in a specific range of wavelengths λ. Starting from a unified theory of wave propagation, we must conceive of heat radiation as energy that begins as heat in an emitting source, converts itself into electromagnetic energy for propagation, and then on encountering matter reconverts itself back into heat energy. There are many types of energy radiated in our environmental space in the form of electromagnetic waves. These energies are classified and named on the basis of the range of wavelengths in which they propagate. Table 3.1 identifies these radiating phenomena. Judging from this table, radiated heat energy of concern to the geofluid is confined to the ultraviolet, visible, and infrared regions. The *spectrum* of wavelengths involved in this discussion is from 20 nm to about 100 μm. As we shall see, this depends on the temperatures of radiation.

Materials subjected to heat radiation will not necessarily absorb all the energy that falls on them. The fraction of the incoming energy they absorb will be identified as the *absorptivity*. Part of the incoming energy may be reflected from the material as it reaches the surface. This portion of the energy is the *reflectivity*. A third portion will be transmitted, without necessarily being first transformed into heat. This part of the energy is the *transmissivity*. In the final balance, the sum of the absorptivity, reflectivity, and transmissivity must equal unity or the full amount of the incoming energy.

Analogously, in the case of light falling on materials, some of the light is reflected, some is transmitted, and some is absorbed. Opaque materials transmit less energy than transparent materials; they reflect and absorb larger amounts. With respect to heat radiation, the earth and most minerals are *not* transparent, meaning that they absorb and reflect most of the radiant energy falling on them. However, it must be remembered that all materials do not have the same trans-

TABLE 3.1

Classification of Electromagnetic Waves

Type of radiation	Wavelength range,[a] λ
Cosmic rays	Up to 0.5 pm
Gamma rays	0.05–10 pm
X rays	1.0 pm–10 nm
Ultraviolet rays	20 nm–0.4 μm
Visible light	0.4–0.8 μm
Infrared or heat energy	0.8 μm–0.8 mm
Radio waves	Over 0.2 mm

[a] The wavelengths are given in microns ($\mu m = 10^{-6}$ m), in nanometers ($nm = 10^{-9}$ m, or millimicrons), and in picometers ($pm = 10^{-12}$ m, or micromicrons).

parent or opaque qualities in the entire range of wavelengths given in Table 3.1. For instance, we know that minerals are more transparent to X rays than they are to visible light. The same will be true for gases and liquids. We conclude that each material must have its own characteristics toward absorption, transmission, and reflection with range of wavelength radiation. This concept is extremely important in understanding the heat balance of the atmosphere, oceans, and the earth.

Figure 3.1 is the absorption spectrum of the gases constituting the atmosphere and the sum total spectrum of the air. On this same figure we have also represented, in a normalized amplitude scale, the ranges of wavelengths through which the sun's energy comes to us, as well as the ranges of wavelengths through which the earth radiates its own energy.

First, the energy from the sun at wavelengths centered around 0.5 μm must try to pass through an atmosphere that has strong absorption or is opaque to ultraviolet wavelengths less than 0.3 μm because of oxygen and ozone absorptivities. Insofar as the sun's radiation the atmosphere has a very definite *window* in the range 0.3 $\mu\text{m} < \lambda < 1.0\ \mu\text{m}$. The earth which radiates its own heat energy in wavelengths centered around 10 μm, also finds that atmospheric gases, especially water vapor and CO_2, play an important role in what is absorbed by them and what ultimately goes back into space. The dynamics of this interplay of energy between the sun, the earth, and space is discussed in more detail in later sections.

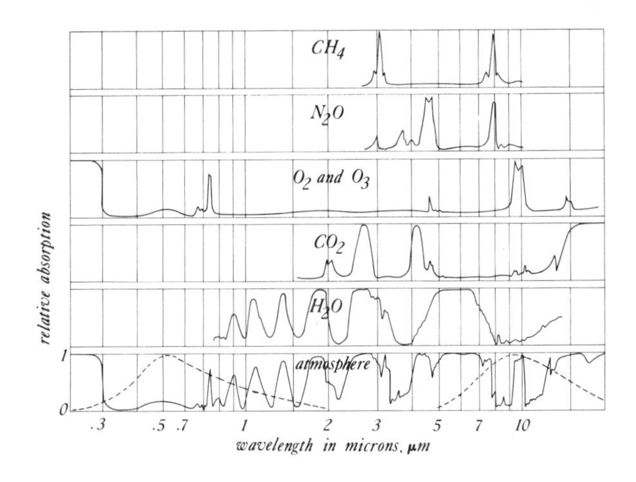

Fig. 3.1 Absorption spectra. (After Fleagle and Businger, 1963.)

Insofar as incoming radiation is concerned we divided it into three parts: absorption, reflection, and transmission. However, a mass that absorbs energy from another source can in turn *emit* its own energy but not necessarily at the same wavelength as the energy it received. We shall see that emission depends on the temperature of the emitting source, and absorption on the substance of the absorbing material. Thus, this explains why the sun and the earth do not emit at the same range of wavelengths, as shown in Fig. 3.1.

Since all substances absorb, reflect, and transmit varied amounts, we define a standard *blackbody* as that which absorbs and emits the maximum energy at a given temperature.

In 1860, G. Kirchhoff maintained that *of two surfaces of equal size and temperature, the surface that absorbed the most emitted the most.* Quantitatively stated, a surface that absorbs n times more radiation than another surface of equal area, emits n times more of the same kind of radiation at the same temperature. However, unlike absorption, monochromatic radiation from blackbodies depends on the temperature of the body that radiates. According to

Planck this radiated energy is distributed over a *continuous* range of wavelengths λ according to

$$I_{\lambda,T} = c_1\lambda^{-5}/[\exp(c_2/\lambda T) - 1] \tag{3.2}$$

$I_{\lambda,T}$ is the intensity of the energy radiated in cal/cm^2 · s in a narrow range of wavelengths $d\lambda$ between λ and $\lambda + d\lambda$. (In essence this intensity is cal/cm^2 · s/cm.) The *universal* constants are $c_1 = 0.321 \times 10^{-15}$ kcal · m^2/h or 3.74×10^{-12} W · cm^2, and $c_2 = 1.438$ cm · °K. It is obvious from this equation that the intensity of emitted radiation $I_{\lambda,T}$ is a function of the wavelength λ and of the absolute temperature T. One can plot this relation in Eq. (3.2) and obtain the

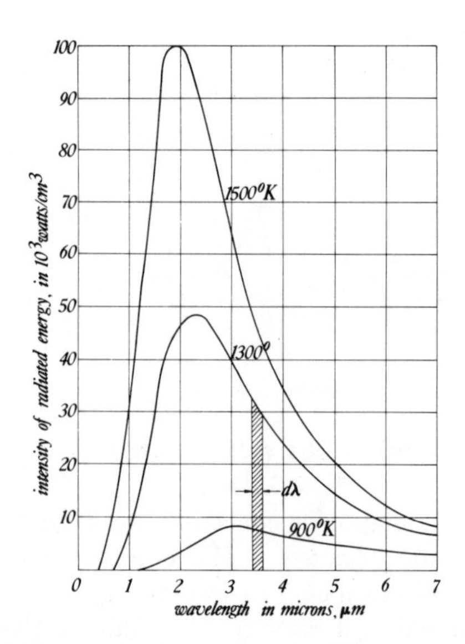

Fig. 3.2 Monochromatic intensity of radiation from a blackbody.

graph shown in Fig. 3.2. Each curve in this figure is an isotherm and from the ranges of temperatures shown, it is seen that it is difficult to represent the isotherms of the sun and the earth on the same linear ordinate. For this reason, we consider the amount of energy per unit area and time contained in the range of wavelengths λ and $\lambda + d\lambda$ and plot these values in a logarithmic scale in Fig. 3.3. This gives a larger range of temperatures in the same plot.

If we represent by $E_{\lambda,T}$ the radiated energy contained in $d\lambda$, it will be related to the intensity $I_{\lambda,T}$ as follows:

$$dE_{\lambda,T} = I_{\lambda,T}\, d\lambda \tag{3.3}$$

The value of $dE_{\lambda,T}$ is the ordinate in Fig. 3.3. For small values of λT, Eq. (3.2) can be approximated and is known as *Wien's law*:

$$I_{\lambda,T} = c_1 \lambda^{-5} \exp(-c_2/\lambda T) \tag{3.4}$$

Integrating with respect to λ, and with the use of Eq. (3.2) we have

$$E_T = (6.49 c_1/c_2{}^4) T^4 = \sigma T^4 \tag{3.5}$$

which is known as the *Stefan–Boltzmann law of blackbody radiation*. The radiated energy is seen to be proportional to the fourth power of the temperature of the emitting body and the proportionality constant σ is called the Stefan–Boltzmann constant which is equal to 0.817×10^{-10} cal/cm^2 · min · °K^4, often the *langley* (ly) is used instead of calories per square centimeter. The location of the position

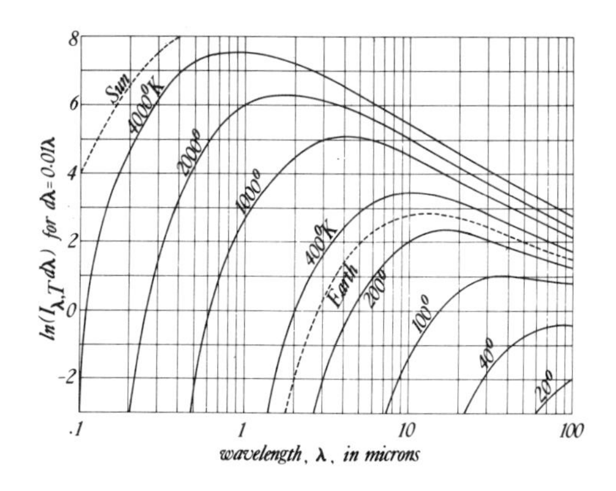

Fig. 3.3 Logarithmic representation of Planck's law.

of the maximum energy in the curves of Fig. 3.2 can be obtained by differentiating Eq. (3.2) and setting it equal to zero. This gives *Wien's displacement law* which locates the value of λ for maximum intensity at a given temperature of radiation. Thus

$$\lambda T|_{\max} = 0.2896 \quad \text{cm} \cdot {}^\circ\text{K} \tag{3.6}$$

This equation permits the evaluation of the temperature of stellar bodies on the basis of spectrographic data giving $\lambda_{\max}$. Substituting this equation into Eq. (3.4) we obtain

$$I_{\lambda,T}|_{\max} = c_3 T^5 \tag{3.7}$$

where $c_3 = 1.126 \times 10^{-5}$ kcal/m^3 · h · °K^5. For example, according to Eq. (3.6) the wavelength at which the maximum intensity from the sun's radiation occurs is 0.5 μm, which corresponds to a temperature of 5800°K. It is also clear that the earth's maximum intensity occurs at 10 μm if we take the earth's average temperature to be 290°K.

3.2 RADIATION ABSORPTION

According to Kirchhoff's law stated in the preceding section, the absorption of a substance is linearly related to the emission from the same substance, which we have just discussed for blackbodies. Let us look into the absorptive properties of substances before we relate absorption rates to emission rates as defined in Eq. (3.2).

Let us consider a gas layer of thickness dx and upon whose surface falls a monochromatic radiation of intensity J_λ. Let us also assume that this layer is a perfect absorber such that its emission is negligible. It is the property of this gas to absorb a small part of this incident radiation, say an amount dJ_λ, and transmit the rest. The differential law of absorption is then

$$dJ_\lambda = -a_\lambda J_\lambda \tag{3.8}$$

We immediately notice that absorption is not a function of temperature independently, as in the case of emission [Eq. (3.2)]; it is a function of wavelength only. Experience shows that the coefficient a_λ is not dependent on temperature either; it depends on the density of the gas, the thickness of the layer, and a gas absorption coefficient k_λ which is dependent on the state of the gas and the wavelength of the incident radiation. Thus Eq. (3.8) becomes

$$dJ_\lambda = -\rho k_\lambda J_\lambda \, dx \tag{3.9}$$

Integrating this equation over a finite distance x yields

$$J_{\lambda x} = J_{\lambda 0} \exp\left(-\int_0^x \rho k_\lambda \, dx\right) \tag{3.10}$$

where $J_{\lambda x}$ is the transmitted intensity after the incident intensity $J_{\lambda 0}$ has crossed a distance x of this gas or substance. We call the integral $\Lambda = \int_0^x \rho \, dx$ the *optical thickness* and then

$$J_{\lambda x} = J_{\lambda 0} \exp(-\Lambda k_\lambda) \tag{3.11}$$

This relation is called *Beer's law* for pure absorption.

In considering the sun's rays falling on an atmospheric layer, for instance, it is important to note that we are often interested not in the total path crossed by the sun's ray but in a vertical height of a gas layer which may have a zenith angle β to the sun. This is shown in Fig. 3.4. Then the relationship of dx to the vertical is given by $dx = -\sec \beta \, dz$. Equation (3.11) is modified by introducing this trigonometric expression, and

$$J_{\lambda x} = J_{\lambda 0} \exp\left(\int_0^z \rho k_\lambda \sec \beta \, dz\right) \tag{3.12}$$

According to Eq. (3.2) the argument of small emissions is only true for very small wavelengths. However, in the infrared region, gases do emit to other parts of the atmosphere, to space, and to the surface of the earth. In that case, the original equation, (3.8), should be altered to include emission $I_{\lambda,T}$ which, as we saw, is a function of wavelength and temperature. According to Kirchhoff's law, the absorption is only proportional to emission and therefore the modified absorption relation is

$$dJ_\lambda = \rho k_\lambda (I_{\lambda T} - J_\lambda)\, dx \tag{3.13}$$

where $I_{\lambda\tau}$ and J_λ are the emitted and incident intensities, respectively, dJ_λ is the net decrease of incident intensity due to absorption, and ρ is the medium density.

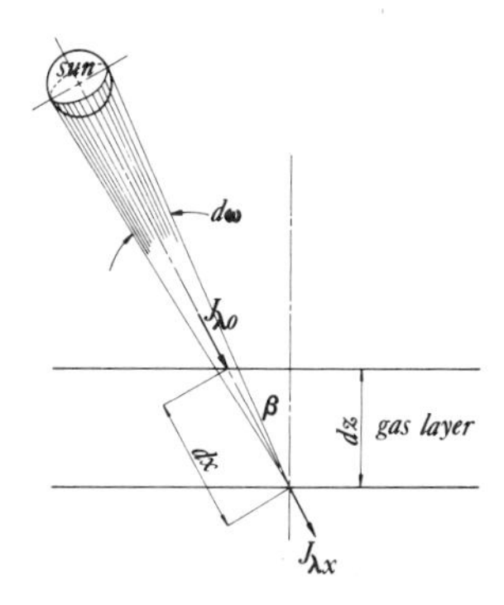

Fig. 3.4 Absorption from the sun with zenith angle β.

Realizing that the absorption spectrum of gases is an irregular function of λ, as we saw in Fig. 3.1, Elsasser (1942) replaced the absorption coefficient k_λ and the use of Eq. (3.13) by a simpler concept. First, he defined a *transmission function* τ, based on the average over a range of wavelengths, by the ratio of transmitted to incident intensities:

$$\tau = (J/J_0)_{\mathrm{av}} \tag{3.14}$$

which according to Beer's law [Eq. (3.11)] is

$$\tau(\Lambda, \lambda) = [\exp(-k_\lambda \Lambda)]_{\mathrm{av}} \tag{3.15}$$

where again Λ is the optical depth and is equal to $\int_0^x \rho\, dx = -\int_0^z \rho \sec \beta\, dz$. Since in the absorption band the spacing between the lines is large compared to the half-width of the line, the average was carried out mathematically and found to be

$$\tau = 1 - \frac{2}{\sqrt{\pi}} \int_0^{k_\lambda' \Lambda/2} \exp(-\eta^2)\, d\eta \tag{3.16}$$

The probability integral is well known and tabulated in many handbooks. For $\Lambda = 0$, Eq. (3.16) gives the correct answer: $\tau = 1$, meaning that all the incident energy is transmitted. If $\Lambda \to \infty$, then $\tau \to 0$. The value of k_λ' is an average or smoothed absorption coefficient in which detailed line structure has been averaged out.

Furthermore in vertical transmission we have variations of T and p with altitude and this absorption coefficient k_λ' is found to depend on both of these properties. Based on experimental data it is proposed that

$$k_\lambda' = (p/p_0)^{1/2} k_{\lambda 0}'$$

where the subscript zero refers to a reference altitude, conveniently taken at the surface of the earth, or a $p_0 = 1000$-mb level, or a saturated state of the air. Since k_λ' is a function of pressure, it is also a function of z.

In Eq. (3.16) the quantity

$$k_\lambda' \Lambda = -\int_0^z \rho k_\lambda' \sec \beta \, dz = -k_{\lambda 0}' \int_0^z \rho (p/p_0)^{1/2} \sec \beta \, dz$$

If we define $\Lambda_0 k_{\lambda 0}' = k_\lambda' \Lambda$, we obtain

$$\Lambda_0 = \int_0^x \rho (p/p_0)^{1/2} \, dx \tag{3.17}$$

For the case of water vapor only, the mass of water in the column of air of density ρ is $\rho\omega^0$ where ω^0 is the specific humidity ratio and ρ the density of the air. Let us define a *modified optical thickness*, or *depth*, Λ' such that

$$\Lambda' = \int_0^x \rho \omega^0 (p/p_0)^{1/2} \, dx \tag{3.18}$$

which represents the amount of water vapor contained in a column of unit cross section. Now we can use this equation in conjunction with Eq. (3.16) to find the transmission function and also the net absorption. Therefore Λ' is the average amount of water vapor (in grams per square centimeter) contained in a column of unit cross section from ground level to a height x.

The transmission is then

$$\tau = f(\Lambda' k_{\lambda 0}') \tag{3.19}$$

Equation (3.18) could have been written in terms of dz, and $\rho\, dz$ can be replaced by $-dp/g$ in the hydrostatic equation which we discuss in Chapter 4. Then a modified optical depth reads

$$\Lambda' = \int_0^z \frac{\omega^0}{g} \left(\frac{p}{p_0}\right)^{1/2} dp \tag{3.20}$$

Through measurements of ω^0, p and T with altitude, we can run a finite difference method to calculate Λ'. Table 3.2 gives typical measured data in the first three columns and calculations to obtain the optical depth.

In Fig. 3.4, the flux of incoming radiation can be obtained by integrating the intensity over the solid angle $d\omega$ of the sun. Thus

$$F = \int J \cos \beta \, d\omega \tag{3.21}$$

TABLE 3.2

Optical Depths for the Atmosphere[a]

p (mb)[b]	T (°C)	ω^0 (g/kg)	ω_i^0 (g/kg)[c]	dp	$\left(\frac{p}{1000}\right)^{1/2}$	Λ' ref. 974 mb
974	16	5.6				0
			5.9	14	0.98	
960	19	6.2				0.08
			5.9	38	0.97	
922	19	5.5				0.30
			4.5	80	0.93	
842	11	3.5				0.64
			3.5	52	0.91	
790	8	3.5				0.81
			3.5	48	0.88	
742	4	3.5				0.96
			2.9	62	0.83	
680	2	2.3				1.11
			1.9	75	0.80	
605	−3	1.4				1.22
482	−15	0.7				1.31
322	−37	0.1				1.35
279	−44	0.1				1.35

[a] On the basis of radiosonde soundings from "Handbook of Meteorology" by F. A. Berry and E. Bollay. Copyright 1945 by McGraw-Hill. Used with permission of McGraw-Hill Book Co.

[b] The values of p, T, and ω^0 are measured through atmospheric soundings. We calculate $\Delta\Lambda' = (\omega^0/g)(p/1000)^{1/2}\, \Delta p$ and add it to the previous value of Λ'.

[c] The value of ω_i^0 is the center value of the specific humidity ratio in the layer represented by dp.

This solid angle can be represented in terms of β and the azimuthal angle α. Then the incoming radiation flux is

$$dF_\lambda = \int_0^{2\pi} d\alpha \int_0^{\pi/2} dJ_\lambda \cos\beta \sin\beta \, d\beta \tag{3.22}$$

Notice that this expression still represents the differential flux over a narrow wavelength band $d\lambda$. Ultimately to obtain F_λ we must integrate with respect to λ. According to Eq. (3.15) if dJ_λ is the differential intensity of incoming radiation, the amount transmitted is $dJ_\lambda \exp(-k_\lambda \Lambda' \sec\beta)$, and according to Kirchhoff's law if we replace $dJ_\lambda = I_\lambda k_\lambda \sec\beta \, d\Lambda'$, we have

$$dF_\lambda = 2\pi I_\lambda \, d\Lambda' \int_0^{\pi/2} k_\lambda \exp(-k_\lambda \Lambda' \sec\beta) \sin\beta \cos\beta \, d\beta \tag{3.23}$$

Denoting by τ_f the integral $2\int_0^{\pi/2} \exp(-k_\lambda \Lambda' \sec\beta) \sin\beta \cos\beta \, d\beta$, we have

$$dF_\lambda = \pi I_\lambda \frac{d\tau_f}{d\Lambda'} d\Lambda' \tag{3.24}$$

Integrating first in Λ' and then in λ for all wavelengths, we have the total net flux at the base of the layer

$$F = \int_0^{\infty} \int_0^{\Lambda_1'} \pi I_\lambda \frac{d\tau_f}{d\Lambda'} d\Lambda' \, d\lambda \tag{3.25}$$

where Λ_1' is the optical depth of the whole layer and πI_λ is the monochromatic blackbody flux. The net radiative flux by the atmosphere is expressed by Eq. (3.25), which must be integrated.

We know that absorption does not depend on temperature but rather on the optical depth Λ' and the standard absorption coefficient $k_{\lambda 0}$. However, emission depends on T as well as on Λ' and $k_{\lambda 0}$. Thus the net transfer of radiation expressed by Eq. (3.25) is a function of T, Λ', and $k_{\lambda 0}$. After having integrated for all wavelengths the result of Eq. (3.25) will be a function of T and Λ'. Elsasser (1942) has provided special charts for the graphical evaluation of this equation.

In order to obtain the absorption and emission characteristics of an atmospheric layer let us consult Elsasser's chart which is provided in Fig. 3.5. This is essentially a plot of Λ' versus T, where the isotherms are vertical and the lines of constant Λ' (*isopleths*) are slightly curved lines converging at $T = -273°C$ on the right-hand side of the graph.

The diagram is constructed in such a way that the area to the right of an isotherm, from $\Lambda' = 0$ to $\Lambda' = \infty$ (abscissa), is proportional to the energy emitted by blackbody radiation at that temperature. The temperature range between

-80 and $-273°C$ is outside the meteorological range and therefore is not included in these charts. The chart is used in conjunction with atmospheric soundings giving Λ' as indicated in Table 3.2, and corresponding temperatures T which must be plotted on this chart to obtain the area represented by Eq. (3.25) for H_2O in the atmosphere. Since CO_2 in the 15-μm band acts differently than water in the sense that it reemits *all* the energy absorbed from the layer on one side to the layer on the other side, only the last layer near the surface of the earth

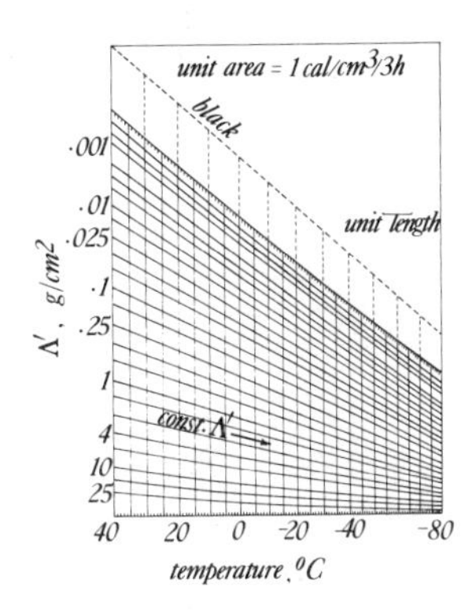

Fig. 3.5 Elsasser's radiation chart. The unit area equivalence is given on the chart.

enters into the energy consideration for CO_2. This final emission of the last layer is done at the temperature of the earth. At a given temperature, this CO_2 radiation is represented by the area where the dashed lines are to the right of the isotherms. The amount of energy radiated upward by the earth is represented by the whole triangular area to the right of the isotherm. The difference is the net exchange between space and earth. The following illustrative example puts this figure to use.

Illustrative Example 3.1

Calculate the flux arriving at the earth's surface with an atmosphere above it having the properties given in Table 3.2, where at the surface of the earth $T = 16°C$ and $\Lambda' = 0$.

The first point plotted on Fig. 3.6 is A, the conditions on the surface of the earth. Then we try to plot all the other values of T and Λ' from Table 3.2 onto this chart, to obtain a state curve ABC. From A to B the curve makes an area to the left of the isotherm AD, and from B to C it is to the right of AD. It is clear from the values of the temperature in Table 3.2 that AB represents an *inversion* with $dT/dz > 0$. Notice also that Λ' does not vary beyond 300 mb because the moisture above that level is negligible. The area *ABCEA* represents the downward flux of radiation from H_2O in the entire column. The area *AB* to the left of *AD* must be negative, or upward flux, because of the inversion. The net flux of radiation to the earth's surface due to H_2O is the area *ABCEA* less area *AB*. The contribution due to CO_2 is the area *AFEA*.

With clear skies, since the earth radiates upward at the temperature of *A*, the energy is given by the whole area *FDEF*. Thus the net radiation transfer from the earth is the difference between these areas, which is *DBED*. Notice that the amount transferred by CO_2 cancels out in the difference.

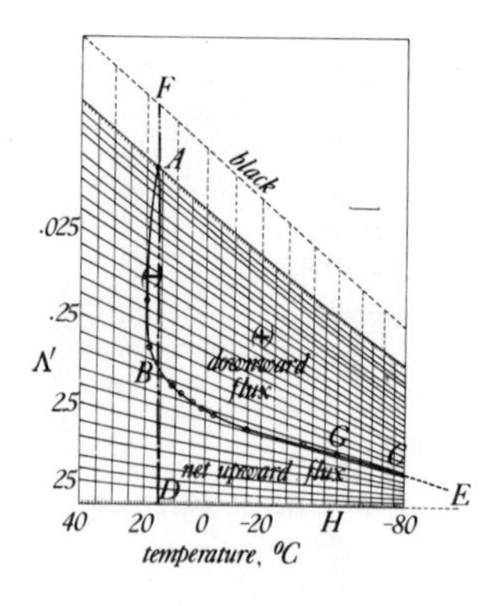

Fig. 3.6 Energy process diagram.

If, for instance, there are clouds at an altitude corresponding to $T = -50°C$, or the point *G* on Fig. 3.6, the net amount of radiation reaching the ground from the clouds will be limited to the area *FABGHEF*. As before, the radiation traveling upward from the ground is the area *FADEF*, and the difference between these two areas, which is *ABGHDB*, is the net loss from the ground.

We have been talking of long-wave radiation all along, corresponding to temperatures of the order of that of the earth's surface. The earth receives short-wave radiation also, to compensate for this loss.

Empirical relations have been used for emitted energy from gases in the atmosphere. In a theoretical analysis it becomes necessary to express emission in terms of an analytical expression. One simple expression based on empiricism and applicable to any constituent in the atmosphere is

$$F = K(pL)^m(T/100)^n \tag{3.26}$$

where *K* is a constant that depends on the gas, *p* is the partial pressure of the gas (in atmospheres), *L* is the length of the column traversed by radiation (in meters), and T is the absolute temperature of the gas. The exponents m and n are also gas dependent. For carbon dioxide, for instance, $K = 0.027$, $m = 0.4$, and $n = 3.2$.

It must be remembered that pollution gases have their own absorption properties, for which new charts are necessary. Suspensions in the form of solid particles have scattering and absorbing properties which should be added to the known effects of H_2O and CO_2. The charts provided in Fig. 3.7 combine calculations of the sun's downward radiation reaching the ground, taking into account the moisture in the air ω_s (in grams per kilogram) to obtain a percent transmission from it, and a chart at the right for dust and haze concentration d (in particles per cubic centimeter). The two transmission factors are combined in a center scale to give the product of the two.

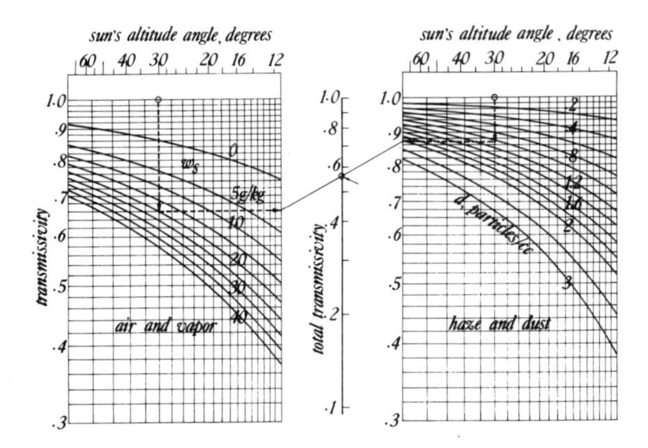

Fig. 3.7 Calculation chart sunshine intensity. (After F. A. Brooks, 1959).

3.3 HEAT BALANCE OF EARTH AND THE ATMOSPHERE

We have seen in the preceding section that the earth, the oceans, and the atmosphere receive almost all of the sun's radiant heat on the average wavelength of 0.5 μm, corresponding to a blackbody temperature of approximately 5800°K. In turn the earth radiates part of this converted energy to outer space at the lower temperature of approximately 290°K at around 10 μm.

In 1957 the International Radiation Committee adopted a value of 1.98 cal/min $\cdot$ cm^2 or 0.1382 W/cm^2 as the radiated solar energy on a unit surface normal to the solar beam located just outside the earth's atmosphere. Because of the uncertainty in the accuracy of this value, the value 2.0 cal/min $\cdot$ cm^2 has been adopted as the *solar constant*. The total surface of the earth is four times the area of its cross section normal to and exposed to the sun. Thus the average solar constant for the entire surface of the earth is one-quarter of that falling on a normal surface, or 0.5 cal/min $\cdot$ cm^2 for the entire surface of the earth.

Naturally, the distribution of the solar radiation varies from latitude to latitude and from month to month, depending on the orientation of the earth's axis to its orbital plane. The *flux density* q is the energy per unit time and unit surface normal to the sun beam. If β is the azimuthal angle, which is a function of time of the day and season, then

$$q = q_0 \cos \beta \tag{3.27}$$

The quantity q_0 is the maximum value of the flux at $\beta = 0$, which is the solar constant. The angle β is a function of time, and the maximum heat flux is inversely proportional to the square of the distance from the sun. This distance

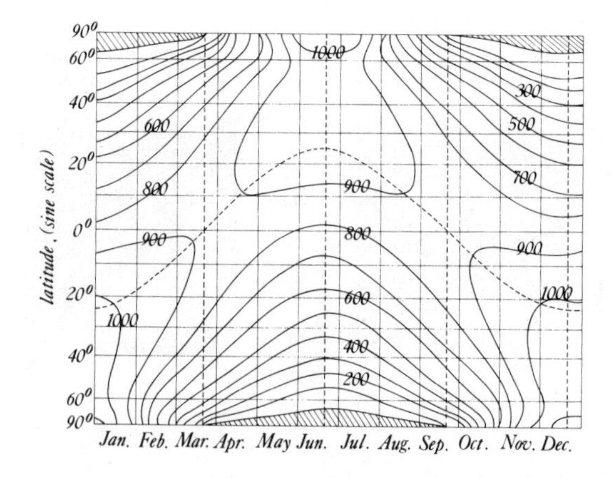

Fig. 3.8 Solar radiation in calories per square centimeter during daylight hours at the top of the atmosphere as a function of latitude and month of the year. (" Meteorological Tables," Smithsonian Institute, 1961.)

varies seasonally, from $L = 147 \times 10^6$ km at the perihelion on January 3, to 152×10^6 km on July 5 at the aphelion. The average distance can be taken as $\bar{L} = 149.5 \times 10^6$ km. Thus the daily radiation from the sun on the earth's upper atmosphere is

$$q = q_0 \left(\frac{\bar{L}}{L}\right)^2 \int_{\text{sunrise}}^{\text{sunset}} \cos \beta(t)\, dt \tag{3.28}$$

Figure 3.4 must be consulted for the trigonometric relationship. Knowing $\beta(t)$ and the hours from sunrise to sunset, this equation can be computed for every latitude and every day of the year. In this equation it is assumed that the solid angle $d\omega$ is constant with β. Figure 3.8 shows the result of this integration. The figure gives the flux of heat from the sun during daylight hours which varies with season, distributed in latitude. The lines are constant heat flux in calories per square centimeter or *langleys* (ly) in a daylight day. For a 12-h day if an area on the earth remained normal to the sun for 720 min, it would receive a maximum of 1440 cal/cm^2.

First we notice from Fig. 3.8 that the radiated energy is not symmetric with respect to the equator. This is because the earth is closer to the sun during the southern summer; the daily solar radiation for the summer at any latitude in the Southern Hemisphere is greater than the solar radiation in the equivalent northern latitudes in the northern summer. Furthermore, it may look deceiving to notice that the maximum radiation per day occurs at the poles, particularly at the south pole. This is true not because the flux there is largest, but because the daylight hours are longer.

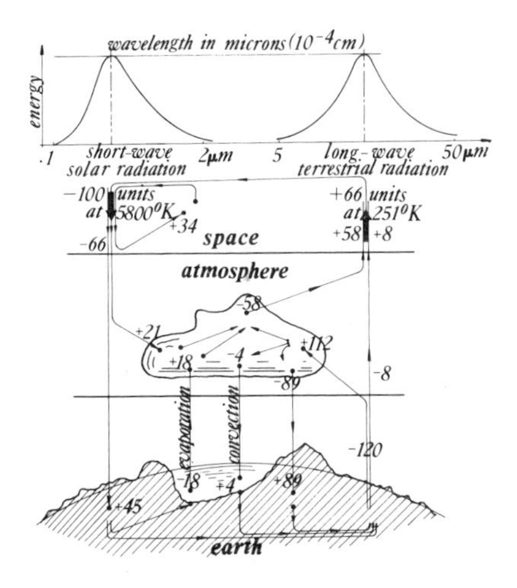

Fig. 3.9 Energy balance between earth and space.

In comparison with conventional fuels, the heat radiating from the sun is infinitesimally less, say, than that from coal. The immense intensity of the sun's heat represents only 3.8×10^{-6} cal/cm$^3 \cdot$ min on the basis of the sun's unit volume, while a cubic centimeter of coal has a heating value approximately 10^9 times larger. Yet it is difficult to conceive that with such a relatively smaller energy density the earth has maintained a nearly constant temperature level for thousands of years. Therefore what matters is not the amount of energy from the sun but the net heat balance necessary to keep the earth at a steady temperature.

To illustrate this heat balance we should look at Fig. 3.9. Not all the solar radiation reaches the surface of the earth. Let us consider that 100 units of solar radiation represent solar radiation's spatial average of 0.5 cal/cm$^2 \cdot$ min or 0.5 ly/min reaching the upper atmosphere. Out of these 100 units, 34 are reflected by the earth's atmosphere and consequently do not enter into the heat balance of the earth and the atmosphere.[1] This reflected radiation is called the *earth's albedo*. The brightness of the earth from out of space is determined by the intensity of the earth's albedo. The remaining 66 units of solar radiation are absorbed by the atmosphere and the earth's surface to be distributed, converted and reemitted. So we can say that there are 66 units of usable radiation coming to the earth from the sun. These 66 units of usable energy correspond then to $0.5 \times 0.66 = 0.33$ cal/cm^2 · min, and as shown in Fig. 3.9 the wavelengths are those at the sun's temperature of 5800°K.

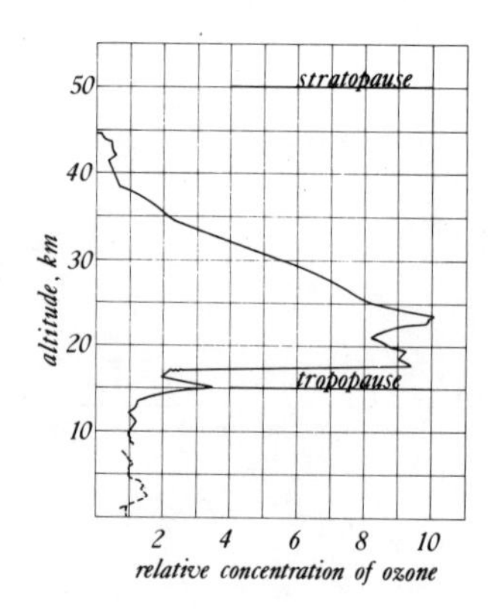

Fig. 3.10 Ozone distribution at White Sands Missile Range (May 1, 1968). (After Randhaowa, 1969.)

Twenty-one of the 66 units entering the atmosphere are absorbed by the atmosphere. These 21 units of solar radiation are distributed as follows: One unit is absorbed in the thermosphere, which accounts for the filtering of approximately 95% of the ultraviolet rays and probably for the increase in temperature in this layer as shown in Fig. 2.6. The ozone content in the stratosphere absorbs approximately 2 units which also explains the rise in temperature there. The ozone distribution in the statospheric region shown in Fig. 3.10 and the absorption properties of ozone shown in Fig. 3.1 are responsible for the increase of temperature in the stratosphere. It can be calculated from the amount of ozone in Fig. 3.10 and its absorption characteristics that the total absorption is sufficient to increase the temperature of the stratosphere by 10°C per day. Of course, this does not happen because that layer also emits an equal amount to remain in steady state. Finally the water vapor in the troposphere absorbs the

[1] The heat balance figures given here are averages. It is conceivable that these figures may vary a great deal from one reference to another. Qualitatively, these comparisons are valid.

remaining 18 units. This amount of heat absorption by water vapor is about constant throughout the troposphere. This is because at lower layers where there is more water vapor, the air is also denser and the energy partition seems to equalize. The heat absorbed in the troposphere divided by the entire mass of air turns out to be equivalent to a rise in temperature at the rate of 1°C per day. Of course air does not keep or store this energy. It is concluded that this energy has little influence on the temperature distribution of the troposphere.

The remaining 45 units of solar radiation that pass through the atmosphere are absorbed by the oceans and land. Since water absorbs 95% of the incident radiation and land absorbs 85–90%, and since there is twice as much water surface, approximately 15 of these 45 units are absorbed by land and 30 units by water. These distributions are also tabulated in Tables 3.3–3.5.

TABLE 3.3

Absorption of Solar Radiation

	Percent	Intensity ($cal/cm^2 \cdot min$ or ly/min)
Radiation scattered and reflected by		
Upper atmosphere	34	0.170
Atmosphere		
Absorption in the thermosphere (short ultraviolet)	1	0.005
Absorption by ozone in the stratosphere (ultraviolet)	2	0.010
Water vapor in the troposphere (infrared)	18	0.090
Earth		
Absorption by land	15	0.075
Oceans, lakes, and rivers	30	0.150
Total solar radiation to earth	100	0.500

So far we have considered the distribution of the incoming solar energy. Let us look at the earth as a source of radiation. On the same unit basis the earth radiates 120 units out into the atmosphere and space, in the range of temperatures between 210 and 300°K, which is contained in the wavelength range 5 to 50 μm. For comparison the amplitude-modulated spectra of the sun and earth are represented in Figs. 3.9 and 3.1. At first thought, the 120 units of radiation from earth may seem difficult to comprehend since the amount is about twice that penetrating the atmosphere from the sun's radiation, and under three times the solar radiation reaching the earth. The answer to this is quickly realized when

TABLE 3.4

The Heat Balance of the Earth

Source	Gain		Loss	
	Units	$cal/cm^2 \cdot min$ or ly/min	Units	$cal/cm^2 \cdot min$ or ly/min
Direct solar radiation	45	0.225		
Direct from atmosphere	89	0.445		
Conduction and convection by winds	4	0.020		
Total gain	138	0.690		
Radiation to atmosphere			112	0.560
Radiation to space			8	0.040
Loss due to evaporation			18	0.090
Total loss			138	0.690

we look at Fig. 3.9 and see that the atmosphere sends down to earth a net amount of 75 units, which it collects from the sun, evaporation, and the earth itself. A large amount of the earth's radiation is absorbed by the atmosphere because of vapor and CO_2, in the range of 8.5 to 11 μm.

So of the 120 units the earth radiates, 112 are absorbed by the atmosphere and 8 units are transmitted into space directly. From turbulent heat transfer through convection, which we mentioned in Section 3.1, the earth gains 4 units from the atmosphere but loses 18 units by evaporation of water at the surface.

TABLE 3.5

The Heat Balance of the Atmosphere

Source	Gain		Loss	
	Units	$cal/cm^2 \cdot min$ or ly/min	Units	$cal/cm^2 \cdot min$ or ly/min
Direct from space	21	0.105		
Direct from earth	112	0.560		
From condensation of vapor	18	0.090		
Total gain	151	0.755		
Radiation to earth			89	0.445
Radiation to space			58	0.290
Convection by winds			4	0.020
Total loss			151	0.755

This heat of evaporation is the latent heat. When the vapor condenses in the form of rain it stays in the atmospheric air. This is discussed in Section 3.5.

The atmospheric heat balance is given in Table 3.5. We have discussed all the units involved in the atmosphere in connection with the sun's or earth's radiation, except the 58 units of energy that are radiated by the atmosphere back to space. Space is also in balance because it lost 66 units to earth and the atmosphere and received (58 + 8) units from the atmosphere and earth, respectively.

In considering these figures we must realize that considerably more heat is received in the low latitudes, and that the amount of cloud formation will affect these figures. It has been estimated, for instance, that earth and its atmosphere may receive as much as 80 units of solar radiation at the equator and as little as 20 units at the poles. This is plotted in Fig. 3.11, along with a curve showing the radiation from earth, and the deficit and surplus regions.

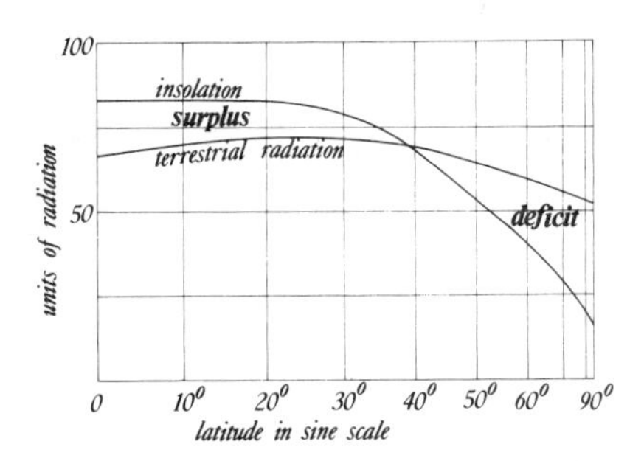

Fig. 3.11 Annual insolation and outgoing radiation in the tropopause.

3.4 THE GREENHOUSE EFFECT OF THE ATMOSPHERE

The atmosphere is composed of a number of gases that absorb and emit the radiation they receive from the sun and the earth. We have seen that all gases do not have the same absorption properties with respect to wavelength. The energy that enters the atmosphere from space was estimated to be 66 units or $0.5 \times 0.66 = 0.33$ ly/min. As we look at Fig. 3.1, we see that this energy enters in the wavelength range $0.3\ \mu m < \lambda < 0.8\ \mu m$. The earth, in turn, emits in longer waves ($5\ \mu < \lambda < 50\ \mu m$) at the rate of 120 units or 0.6 ly/min. Considering the earth to be a blackbody emitter, Eq. (3.5) can give an estimate of the earth's temperature averaged over its entire surface. Thus

$$\begin{aligned} q &= 0.6 \quad \text{cal/cm}^2 \cdot \text{min} \\ &= 0.817 \times 10^{-10} T^4 \end{aligned}$$

where the value of the Stephan–Boltzmann constant σ has been substituted. A solution of this equation yields an average equilibrium temperature of the earth's surface of 290°K with maximum intensity at 10 μm according to Wien's displacement law. In contrast, the moon, which is approximately at the same distance from the sun as the earth, receives the same solar constant of 0.5 ly/min; it reflects 10% of this energy, and since it has no atmosphere, it must emit all the energy it receives (90 units) to keep at a steady state. The same calculation as above for $0.5 \times 0.9 = 0.45$ ly/min yields a moon surface temperature of 265°K. If the earth did not have an atmosphere, it would have emitted about 90 units instead of 120 units and have the same equilibrium temperature of the moon. Thus let us look once more at the contribution of the atmosphere.

A first-hand conclusion is that the earth is warmer than the moon because the atmosphere manages to reflect back to earth a net amount of 30 units, which the earth must dispose of at a higher temperature (about 290°K). This blanketing effect is due to the near opacity of the water vapor and carbon dioxide in the troposphere with respect to the long wavelengths from terrestrial radiation. We can see this clearly from Fig. 3.9. Of the 120 units emitted by the earth, 112 units are absorbed by the atmosphere or nearly 93% absorption or opacity to terrestrial radiation, whereas the same atmosphere is less opaque (32%) to solar radiation since it only absorbs 21 units of the 66 that enter it. This difference in opacity with respect to incoming and outgoing radiation is called the *greenhouse effect*.

While the troposphere absorbs energy mostly from terrestrial radiation, it emits both to earth and space. Thus, again according to Fig. 3.9, the fact that the troposphere emits more units to space than it receives, accounts for the lower temperatures there than on the earth's surface.

All the figures given in Fig. 3.9 are approximate and vary with season, latitude, and time of day. We know that, locally, at night terrestrial radiation is always present while insolation is absent. It is of interest to set up an instantaneous heat balance in order to estimate the relative role of each term in the balance. If T is the equilibrium temperature on the earth's surface, we know that the terrestrial radiation is always $\sigma T^4 = q_t$. This is a loss for the earth, indicated as 120 units on Fig. 3.9. There is another loss as shown on the figure and in Table 3.4, that of evaporation. This, of course, will vary locally and in time. If we let u_t be the actual terrestrial radiation units represented by q_t and u_e be the actual units due to evaporation, then the loss in langleys per minute due to evaporation will be $(u_e/u_t)\sigma T^4$. Actually, the gain claimed from convection may or may not be a gain locally. If we represent the number of units by u_c, then the number of langleys per minute due to convection will be $(u_c/u_t)\sigma T^4$. There are two definite gains in energy by the earth's surface; these are the radiation from the atmosphere to earth given by $(u_a/u_t)\sigma T^4$, and the insolation, which will be taken as 43 units instead of 45 because according to Table 3.3, 2 units absorbed by the stratosphere never get to the earth. Thus the insolation is $0.5 \times 0.43 = 0.215$ ly/min. Actually if one wishes to be more exact locally, one must take the local insolation rate, q_i. At any rate, since the net heat balance must be momentarily in equilibrium we can write

$$\text{Net loss or gain} = \sigma T^4\left(1 + \frac{u_e}{u_t} \pm \frac{u_c}{u_t} - \frac{u_a}{u_t}\right) - q_i \tag{3.29}$$

When the value of this equation is positive it represents a loss, and when negative a gain. At any rate, the equilibrium temperature of the earth's surface is a solution of this equation. For the entire earth, the net result of Eq. (3.29) must be zero, and locally or temporally it will be the amount of heat transported from one latitude to another.

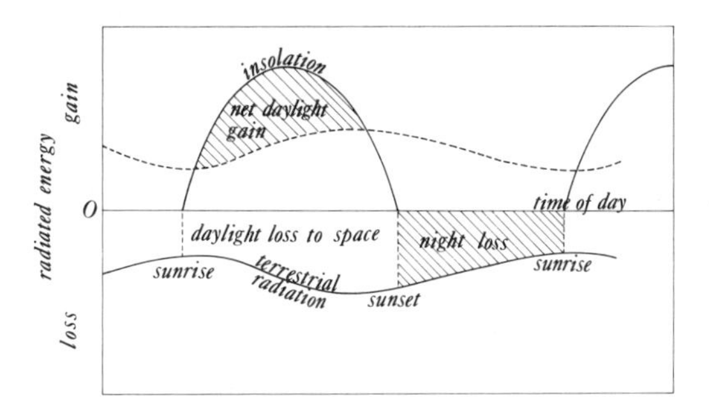

Fig. 3.12 Diurnal energy transfer between earth and space. (After Scorer, 1958.)

According to Fig. 3.11 the terrestrial loss of heat is nearly constant with latitude; this is represented by the terms in parentheses in Eq. (3.29). It is the insolation q_i that varies more with latitude, as seen in Figs. 3.8 and 3.11. Since at the global level there is no net heat lost or gained, there are larger heat concentrations at lower latitudes. We shall see later that these energy gradients will be the source of large atmospheric circulations tending to homogenize the temperatures around the earth. We shall see also that extreme energy concentrations can create hurricanes, tornados and strong winds.

It is obvious that pollution plays a very important role in this heat balance. All the arguments we presented so far are based on a standard atmosphere where H_2O and CO_2 are the two *natural* gas by-products of life on earth and have a major role in the heat balance of the troposphere. The increased release of other gases from industrial effluents, depending on their absorption characteristics, can alter q_i in Eq. (3.29) and all of the units of radiation in Table 3.5, and consequently u_a, thus allowing the establishment of a different equilibrium temperature T on the surface of the earth and in the troposphere.

Figure 3.12 is a graphical representation of Eq. (3.29) during a 24-h period. The intensity of energy radiation (ordinate) is shown as a function of the time of day (abscissa). The periodic wave below the origin O (which is continuous in time) represents the terrestrial radiation into space and for this reason is marked as a loss. Its fluctuation in amplitude during one cycle (day and night) is not very large because emission depends on temperature, and the temperature of the earth does not vary much between one night and the next. However, the insolation curve above the origin is not continuous in time. It begins at sunrise and goes to zero intensity at sunset. During the night it is zero. The areas under either the insolation or terrestrial radiation curves represent the total energy flux (in langleys) for a daily cycle. If we assume that locally the mean temperature of the earth has not changed, then the area above the origin has to equal the area below it. By reflecting these curves on the same side of the abscissa we can clearly identify the daylight gain, the daylight loss, and the *net* daylight gain as opposed to the night loss on the surface of the earth. The net daylight gain should equal the night loss for a steady-state condition in the daily cycle.

Illustrative Example 3.2

Let us consider a simple infrared absorption spectrum for an atmosphere containing a fixed amount of water vapor and CO_2. Let this spectrum be represented by the simplified model[2] shown in Fig. 3.13. Compute the energy lost to space (stratosphere) as well as the energy absorbed by the atmosphere at the temperature of 210°K if the earth's temperature is at 280°K. This will be a simplified approach to the analysis given in Section 3.2.

We start by drawing blackbody radiation spectra for the two temperatures from Eq. (3.2), similar to the graph shown in Fig. 3.2. These two curves are shown

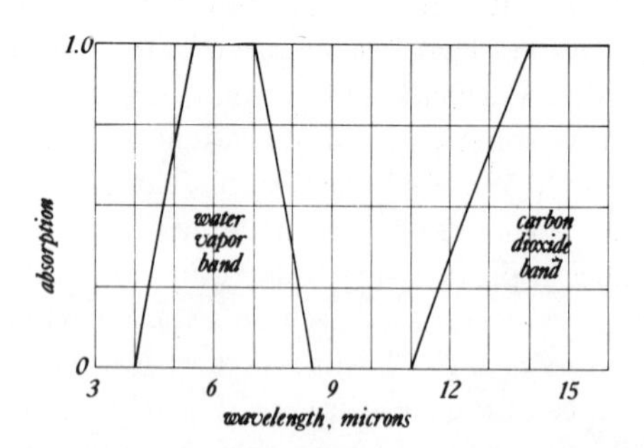

Fig. 3.13 Simplified absorption spectrum for a layer of moist air.

[2] This simplified model was first proposed by Simpson (1928). It appears in a number of meteorology books; see Belinskii (1948).

in Fig. 3.14. At the temperature of the earth, the energy emitted by it is labeled E in this figure. Part of this radiated energy is absorbed by the first moist atmospheric layer and part is transmitted to the upper layer at its own temperature, which is normally a little lower than the surface temperature. Assuming that the second layer has the same moisture content, it will radiate half its energy to the lower layer, compensating for what it received from it, and half to the upper layer. This radiation passes upward in the atmosphere until it reaches a layer that no longer has the absorbing gases which we assumed to be H_2O and CO_2. The last layer will emit to space at its final temperature which was taken as 210°K.

On Fig. 3.14 we can draw another blackbody radiation spectrum at 210°K labeled S. Obviously, the entire area under the S curve will be lost to space since we have assumed that beyond the altitude corresponding to 210°K there are no absorbing gases. However, not all of the area under E is lost to space. Between E and S we have an atmosphere with absorption characteristics as shown in Fig. 3.13 and thus we must take these characteristics into account to find the portion of terrestrial radiation that will pass through into space.

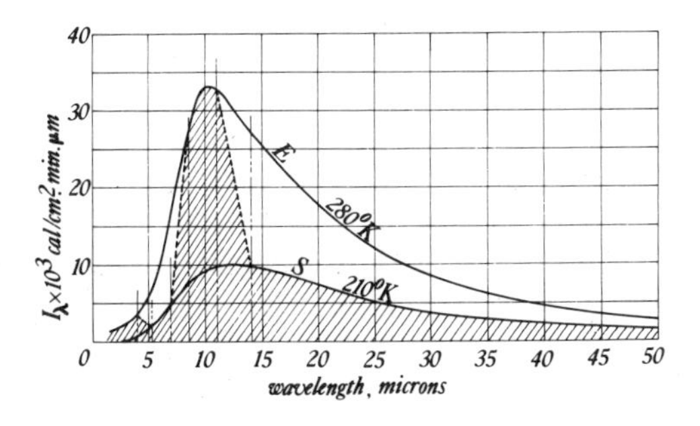

Fig. 3.14 Radiated energy balance.

From 8.5 to 11 μm all the infrared energy radiated by the earth passes into space through the window shown in Fig. 3.13. Between 7 and 8.5 μm we shall suppose that the atmosphere is opaque and lets through one-half of this energy, because of the triangular shape of the spectrum. The same is true between 4 and 5.5 μm and also 11 and 14 μm. Thus the shaded areas shown in Fig. 3.14 are the radiated energy passing from the earth's surface to space. The clear areas represent the energy absorbed by the atmosphere and ultimately returned back to earth. Comparing this with Fig. 3.9 we can see that of the 120 units emitted by the earth, ultimately 66 units go back to space and the remainder stays in the atmosphere to be radiated back to earth again. The ratio of the areas in Fig. 3.14 is in the same proportion.

This simplified model neglects the presence of a mass of clouds which, if present, could absorb and emit as blackbodies at intermediate temperatures. In that case, there will be another spectrum between E and S in Fig. 3.14 and the exchange will be the same but at three levels instead of two.

3.5 HEAT BALANCE OF THE OCEANS

On dry earth, the principal heat transfer processes are radiation and smaller amounts of conduction-convection. The atmosphere itself exchanges heat by radiation, condensation, and convection. In the heat transfer of the oceans we must consider radiation, conduction-convection with the atmosphere, evaporation, conduction-convection with the ocean floor, chemical processes in organic and inorganic reactions, and ultimately frictional heat from wind stresses and tides.

Since the surface of the ocean is twice as large as that of dry land, almost two-thirds of the heat balance in Table 3.4 takes place on the ocean surface. This is not exactly true since the 18 units of evaporation must take place, for the most part, on the ocean surface. Conduction-convection from the ocean floor is estimated at 50 to 80 ly/yr; this corresponds to only 0.15×10^{-3} ly/min or 3/100 of a unit. It is too small to make a difference in the overall heat balance. The frictional energy due to wind stresses and tides is again a very small amount, estimated at 10^{-4} of the total units exchanged. The chemical processes are estimated to add 0.8% of the total heat exchanged. Figure 3.15 shows this heat balance.

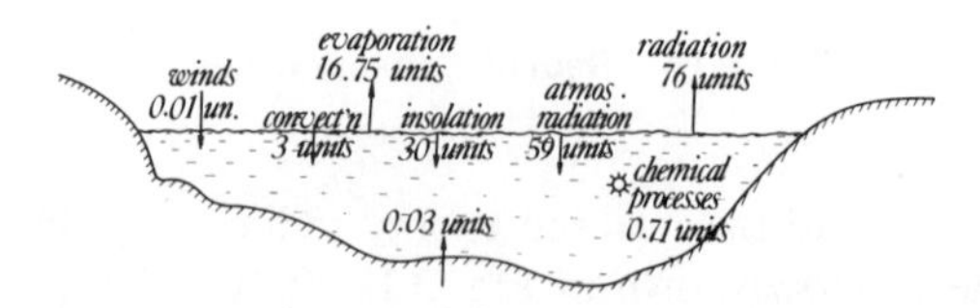

Fig. 3.15 Heat balance of the oceans. A unit is equivalent to one-hundredth of the solar constant.

The ocean water has a relatively large absorption coefficient and consequently very poor transmission into deeper layers. Most of the incoming radiation on the surface of the ocean is absorbed in the first 100 m. Tables 3.6 and 3.7 give the absorption properties of water with incident wavelength (Lacombe, 1965) and the percent absorption with depth (Sverdrup, 1942).

The convection between the atmosphere and the ocean which begins as conduction in a small thermal layer at the interface above sea level can be estimated with a modified Fourier law as

$$q_c = -k\left(\frac{dT}{dz} - \alpha_{ad}\right) \tag{3.30}$$

similar to Eq. (3.1) where k is an *eddy thermal diffusivity*, appropriate for turbulent transfer of heat, dT/dz is the temperature variation with altitude, and α_{ad} is the adiabatic lapse rate (constant) or the temperature variation with altitude appropriate for an adiabatic atmosphere.[3] Comparison with the adiabatic lapse rate determines whether or not an actual temperature lapse rate will produce a vertical motion which will be responsible for the convection. In essence the difference $(dT/dz - \alpha_{ad})$ determines the degree of stability of the atmosphere, and can be written also as $d(T - T_{ad})/dz$. This vertical motion is discussed in Section 4.12.

TABLE 3.6

Absorption and Transmission Properties of Water

Wavelength (μm)		Absorption coefficient per meter	Transmission coefficient per meter
Ultraviolet	0.32–0.4	0.58–0.072	56×10^{-2}–93×10^{-2}
Violet	0.42–0.44	0.041–0.023	98×10^{-2}
Blue	0.46–0.5	0.0155	98.5×10^{-2}
Yellow/green	0.54–0.58	0.24–0.055	97×10^{-2}
Orange	0.60–0.62	0.125–0.178	88×10^{-2}–83×10^{-2}
Red	0.64–0.70	0.21–2.40	81×10^{-2}–9.1×10^{-2}
Infrared	0.85–2.5	4.12–8500	1.6×10^{-2}–2×10^{-8}

The evaporation from the ocean depends also, as the Fourier law of diffusion would predict, on the gradient of *specific humidity*, defined in Eq. (2.53). Thus the amount of water evaporated from the surface of the ocean per unit time and area is (in g/cm^2·s)

$$W = -\rho D_v \frac{d\omega^\circ}{dz} \tag{3.31}$$

[3] This is discussed in detail beginning with Sections 4.10 and 4.11.

where D_v is the *eddy diffusivity* for vapor in air in units of square centimeters per second, and ρ is the density of moist air. Replacing the definition of $\omega°$ in the equation yields

$$W = -0.6215 D_v \rho \frac{dp_{ws}}{p\,dz} \tag{3.32}$$

We have said in Section 1.3 that the partial pressure of ocean water p_{ws} is slightly less (98%) than that of fresh water at the same temperature. The salinity $\mathscr{s}$ is responsible for this difference. An empirical relation has been proposed relating the dependence of the partial pressure of seawater to its salinity:

$$p_{ws} = p_w(1 - 0.0053\mathscr{s}) \tag{3.33}$$

where p_{ws} is the partial pressure of salt water, p_w is the partial pressure of fresh water at the same temperature, and $\mathscr{s}$ is the salinity in ‰.

TABLE 3.7

Percent of Total Incident Energy at Various Depths

Depth (m)	Pure water	Ocean water	Coastal water
0	100	100	100
1	38.9	35.2	26.7
2	33.7	28.0	17.0
5	28.0	17.3	5.95
10	22.0	9.5	1.21
20	15.8	3.72	0.064
50	7.64	0.311	
100	3.04	0.0057	

Knowing the amount of evaporation W through Eq. (3.32), the amount of heat removed from the ocean surface due to this evaporation is determined when multiplying Eq. (3.32) with the latent heat L of seawater, which in Section 1.3 was found to be essentially the same as that of pure water. This heat of evaporation q_e in the same units as q_c in Eq. (3.30) (cal/cm^2 · s) is

$$q_e = -0.6215 \rho D_v L \frac{dp_{ws}}{p\,dz} \tag{3.34}$$

The ratio of the heat convected[4] to the heat evaporated, called the *Bowen number B*, is

$$B = 0.6215 \frac{kp}{\rho D_v L} \frac{dT}{dp_{ws}} \tag{3.35}$$

In multiplying this equation, numerator and denominator, by the specific heat c_p of the moist air, we can immediately recognize a dimensionless number, the *Prandtl number* $\rho c_p D_v/k$, which is $\mathscr{P}r \simeq 0.75$ for gases. This is discussed thoroughly in Chapter 6.) Then Eq. (3.35) becomes

$$B = 0.6215\, \mathscr{P}r^{-1} \frac{pc_p}{L} \frac{dT}{dp_{ws}} \tag{3.36}$$

In Eq. (3.36) we recognize the rate of change of temperature with respect to the partial pressure of seawater. During evaporation, the likelihood of the vapor

TABLE 3.8

Rates of Evaporation

Ocean	Amount (km^3/yr)	Land	Amount (km^3/yr)
Evaporation	334,000	Evaporation	62,000
Precipitation	297,000	Precipitation	99,000
Runoff from land	37,000	Runoff from ocean	37,000

state being near saturation is very high. Thus we can use the Clapeyron equation [Eq. (2.61)] in terms of this rate of change along the saturation line. Equation (3.36) becomes

$$B = 0.6215\, \mathscr{P}r^{-1} \frac{c_p p R_w T^2}{L^2 p_{ws}} \tag{3.37}$$

Here p and p_{ws} must be distinguished as different values. The first is the atmospheric pressure and the second is the partial pressure of the water. Knowing the ambient pressure and temperature, these values can be determined.

The total evaporation from the oceans and land has been estimated. Table 3.8 gives these annual figures. Considering that the mass of 1 km^3 of water is one million tons, these quantities are indeed very large. Furthermore, since each gram of water requires 600 cal for evaporation, each ton would require 6×10^8 cal.

[4] Since the adiabatic temperature T_{ad} is not dependent on the vapor pressure, it does not appear in Eq. (3.35).

Illustrative Example 3.3

Calculate the heat involved in the evaporation rates of Table 3.8 and compare the value (in langleys per minute) to the value given in Table 3.4.

The total amount of water evaporated from the ocean and land is

$$334{,}000 + 62{,}000 = 396{,}000 \quad \text{km}^3/\text{yr}$$

The kilometer has 10^5 cm and one year has 365 days, each 23 h, 56 min, and 4.09 s, giving a total of 524,164 min. When this amount of water is spread out over the globe with an area of $4\pi\,(6370 \times 10^5)^2$ cm^2, since each cubic centimeter of water weighs 1 g and requires 600 cal, the heat involved in evaporation in cal/min $\cdot$ cm^2 of earth surface is 0.089 ly/min. When this is compared to the value of Table 3.4, giving 0.090 ly/min, we realize that the agreement is excellent.

4

Static Equilibrium of the Environment

4.1 INTRODUCTION

In the considerations of the atmosphere and the ocean, sooner or later we must look into the conditions that enable these masses to be in *mechanical* and *thermal equilibrium*. The word *static* is derived from the Greek word *statikos*, meaning causing to stand still. The Latin word *equilibrium* implies a state of balance among the forces acting on a system. *Thermodynamic equilibrium* implies mechanical, thermal, and chemical equilibria.

On the basis of Newtonian mechanics, the sum of all *external forces* and moments on a given mass system must balance out to zero for mechanical equilibrium. In case this fails to be true, an accelerating or decelerating motion will develop in which the net resulting force will be the product of the mass of the system and the acceleration, and the direction of the force will be in the direction of acceleration, according to Newton's law of inertia [Eq. (2.2)]. This is Newton's second law. The third law deals with the balancing of *internal forces*, and the first law states that when the resultant external force is zero, the system is either at rest or moves at a constant velocity (constant direction and magnitude).

Since external and internal forces play an important role in the state of the environment, it is important that we describe them here. External forces are imposed on a mass system from its environment. These forces can be imposed by contact with the surface of the system, or from afar, as with gravitational force (see Fig. 4.1). *Frictional* and *pressure forces* act through contact on the surface, while *body* or *volumic forces* such as gravity and Coriolis act on the whole of the volume.

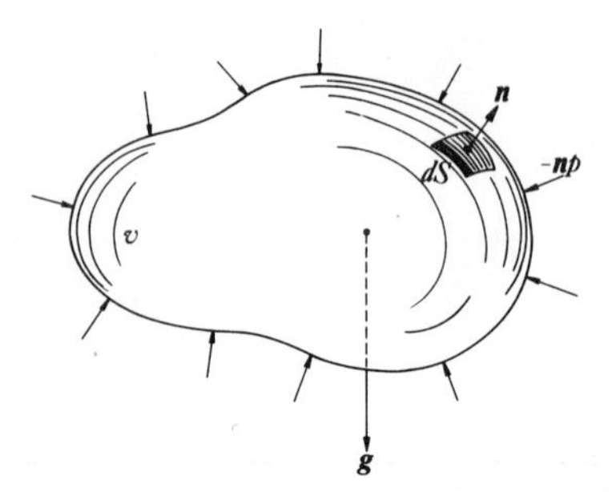

Fig. 4.1 Fluid parcel under surface and body forces.

4.2 BODY AND SURFACE FORCES

Body forces act on the extent of the mass. The gravitational force is a body force, and according to Eq. (2.7)

$$\boldsymbol{g} = -G \frac{M_e}{R^2} \frac{\boldsymbol{R}}{R} + \Omega^2 \boldsymbol{r}$$

is the body force per unit mass. When the system has a total volume $\mathscr{V}$, the total body force on the system is given by

$$\boldsymbol{F}_b = \int_{\mathscr{V}} \rho \boldsymbol{g} \, d\mathscr{V} \tag{4.1}$$

where $\rho \, d\mathscr{V}$ is the unit mass, and the body force $\boldsymbol{F}_b$ is a vector in the direction of $\boldsymbol{g}$.

If we are to take a similar approach to surface forces, we realize that the unit area, because of its changing orientation, is a vector quantity and the surface force per unit area is a vector field. Since the final result of this summed product must be a vector, it is best to examine this in closer detail.

Consider Fig. 4.2 which represents a volume of geofluid from which a small tetrahedron-like portion has been removed with three normal surfaces in (b) with the exterior surface defined by the unit normal vector $\boldsymbol{n}$ as shown in (c). Since this elemental tetrahedron-like parcel is very small, let the surface forces on the exterior surface ΔS in (c) be represented by $\Delta \boldsymbol{F}_n$. The *surface stress vector* is defined as

$$\boldsymbol{P}_n = \lim_{\Delta S \to 0} \frac{\Delta \boldsymbol{F}_n}{\Delta S} \tag{4.2}$$

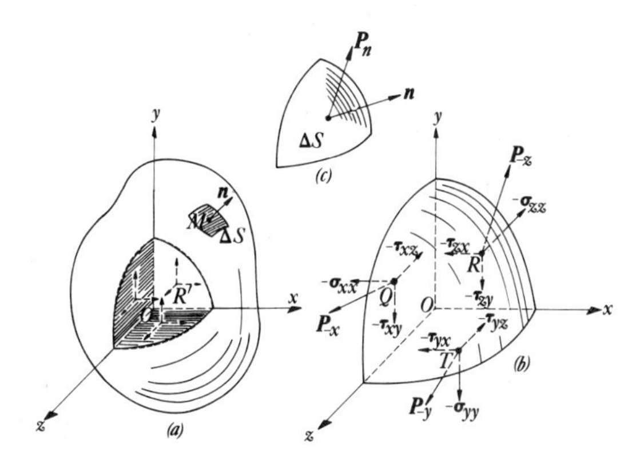

Fig. 4.2 Surface forces.

In the same way, on the other three surfaces, $\boldsymbol{P}_{-x}$, $\boldsymbol{P}_{-y}$, and $\boldsymbol{P}_{-z}$ are stress vectors on planes normal to $-x$, $-y$, and $-z$, respectively, but not necessarily perpendicular to these planes. Throughout this book, the positive unit normal is taken as the *outer* normal, which is the reason the directions of the three internal surfaces are $-x$, $-y$, and $-z$.

Each of these surface stress vectors can be decomposed into three components called *surface stresses*. This makes a total of 12 in the tetrahedron, but not all are independent of each other as we shall see. The symbol σ is used for normal stress and τ for tangential stress. Since there are so many surface and normal stresses, we need two indices for each stress to distinguish one from the other. The first subscript denotes the surface on which the stress is located, and the second identifies the direction on that surface. For instance τ_{xy} indicates a tangential stress on a plane normal to x and pointing in the y direction on that plane, and so on. Newton's third law pertaining to the balance of internal forces requires that $\boldsymbol{P}_{-x} = -\boldsymbol{P}_x$. Also in component form

$$\begin{aligned} -\boldsymbol{P}_x &= -\boldsymbol{i}\sigma_{xx} - \boldsymbol{j}\tau_{xy} - \boldsymbol{k}\tau_{xz} \\ -\boldsymbol{P}_y &= -\boldsymbol{i}\tau_{yx} - \boldsymbol{j}\sigma_{yy} - \boldsymbol{k}\tau_{yz} \\ -\boldsymbol{P}_z &= -\boldsymbol{i}\tau_{zx} - \boldsymbol{j}\tau_{zy} - \boldsymbol{k}\sigma_{zz} \end{aligned} \tag{4.3}$$

where $\boldsymbol{i}, \boldsymbol{j}$, and $\boldsymbol{k}$ are unit vectors in the x, y, and z directions, respectively. Let us examine how we can relate $\boldsymbol{P}_x$, $\boldsymbol{P}_y$, and $\boldsymbol{P}_z$ to $\boldsymbol{P}_n$ in the whole tetrahedron. For equilibrium (mechanical) of this differential element, the sum of forces must be zero. For the moment, let us assume that by equilibrium we mean static equilibrium. We shall see that motion will not affect these results. Since ΔS is the outer surface, the quantity $\Delta S \cos(\boldsymbol{n}, x)$ will denote the projection of the elemental surface ΔS on the y–z plane, also $\Delta S \cos(\boldsymbol{n}, y)$ would denote the projection on the x–z plane, and so on, where, for instance, $(\boldsymbol{n}, x)$ is the angle made by $\boldsymbol{n}$ and x. Summing surface forces and the weight of the tetrahedron for equilibrium, we have

$$-\boldsymbol{P}_x \, \Delta S \cos(\boldsymbol{n}, x) - \boldsymbol{P}_y \, \Delta S \cos(\boldsymbol{n}, y) - \boldsymbol{P}_z \, \Delta S \cos(\boldsymbol{n}, z) + \boldsymbol{P}_n \, \Delta S + \tfrac{1}{3}\rho \boldsymbol{g} \, \Delta S \, \Delta h = 0 \tag{4.4}$$

where Δh is the height of the tetrahedron from the apex O to the surface ΔS.

Since all terms contain ΔS, this unit area can be divided through to obtain

$$\boldsymbol{P}_x \cos(\boldsymbol{n}, x) + \boldsymbol{P}_y \cos(\boldsymbol{n}, y) + \boldsymbol{P}_z \cos(\boldsymbol{n}, z) - \boldsymbol{P}_n = \tfrac{1}{3}\rho \boldsymbol{g} \, \Delta h \tag{4.5}$$

Now in the limit, if we make the elemental volume approach zero size, but not quite, the left-hand side of Eq. (4.5) is not at all affected by the limit because the size of the element does not appear there. The right-hand side, which contains Δh, must go to zero. If we keep in mind the vector geometry:

$$\cos(\boldsymbol{n}, x) = \boldsymbol{n} \cdot \boldsymbol{i}, \qquad \cos(\boldsymbol{n}, y) = \boldsymbol{n} \cdot \boldsymbol{j}, \qquad \cos(\boldsymbol{n}, z) = \boldsymbol{n} \cdot \boldsymbol{k}$$

then in the limit Eq. (4.5) becomes

$$\boldsymbol{P}_n = \boldsymbol{n} \cdot \boldsymbol{i}\boldsymbol{P}_x + \boldsymbol{n} \cdot \boldsymbol{j}\boldsymbol{P}_y + \boldsymbol{n} \cdot \boldsymbol{k}\boldsymbol{P}_z = \boldsymbol{n} \cdot (\boldsymbol{i}\boldsymbol{P}_x + \boldsymbol{j}\boldsymbol{P}_y + \boldsymbol{k}\boldsymbol{P}_z) \tag{4.6}$$

The quantity in parentheses is the sum of three *dyads*[1] called the *stress tensor* $\mathscr{S}$ whose nine components are given in Eq. (4.3). Thus, when we generalize Eq. (4.6) we have a very fundamental relation from equilibrium

$$\boldsymbol{P}_n = \boldsymbol{n} \cdot \mathscr{S} \tag{4.7}$$

We must point out here that had there been motion of the tetrahedron-like elemental volume, Eq. (4.4) would have included an inertial term very similar to the last body force, and when the limit had taken place, it would have been ignored as the body force. So Eq. (4.7) must be valid for static equilibrium and for motion. It is clear that if $\mathscr{S}$ is known everywhere in the environment, given any surface defined by $\boldsymbol{n}$ passing through any point, then the surface stress vector $\boldsymbol{P}_n$ on that surface will be determined by a simple dot product of $\boldsymbol{n}$ and the stress tensor.

4.3 THE CONCEPT OF PRESSURE

When all the tangential (shear) stresses τ are zero, a case of frictionless motion, the remaining components of the tensor are the diagonal components of Eq. (4.3): σ_{xx}, σ_{yy}, and σ_{zz}. This could also be the case of no motion. Since the concept of *pressure in thermodynamics* implies a force per unit area which when increased, decreases the volume, this force must have a direction opposite to

[1] See Appendix A.

those of σ as defined in Fig. 4.2. Let us then define the pressure p as the average of all three normal stresses. This mechanical definition of pressure is in total agreement with the kinetic theory definition of pressure in the range of continuum of the atmosphere and certainly of the ocean.[2] Thus

$$p = \tfrac{1}{3}(\sigma_{xx} + \sigma_{yy} + \sigma_{zz}) \tag{4.8}$$

In the absence of friction or motion the three normal components are equal and

$$p = -\sigma_{xx} = -\sigma_{yy} = -\sigma_{zz} \tag{4.9}$$

This is called *Pascal's principle*. Thus for static equilibrium[3]

$$\boldsymbol{n} \cdot \mathcal{S} = -\boldsymbol{n}p \tag{4.10}$$

and then it follows that this is a special tensor

$$\mathcal{S} = -\mathcal{I}p \tag{4.11}$$

where $\mathcal{I}$ is the idem or unit tensor with unit diagonal components only, such that $\mathcal{I} = \boldsymbol{ii} + \boldsymbol{jj} + \boldsymbol{kk}$.

4.4 THE HYDROSTATIC EQUATION

In the absence of motion, the sum of the body forces and surface forces must be zero for a static equilibrium. Considering the arbitrary volume of Fig. 4.1 with an exterior surface S defined in direction at every point by the external unit normal $\boldsymbol{n}$, the mechanical equilibrium requires that

$$\int_{\mathcal{V}} \rho \boldsymbol{g}\, d\mathcal{V} + \oint_S \boldsymbol{n} \cdot \mathcal{S}\, dS = 0 \tag{4.12}$$

The circled integral sign implies that it is taken over a closed surface bounding the volume $\mathcal{V}$, as in Fig. 4.1. In hydrostatics, since there is no motion, the only stress components in the stress tensor are the normal stresses and the stress tensor is the special one of Eq. (4.11), $\mathcal{S} = -\mathcal{I}p$. In order to obtain a differential equation from this integral equation it is first necessary to change the surface

[2] The only cases where disagreement may occur between the two definitions are for very fast rates of divergence on the order of large explosions.

[3] For equilibrium, the summation of moments must be zero as well. It can be shown that for most deformable substances, including water and air, this condition is automatically satisfied for a symmetrical stress tensor. In statics since all τ's are zero symmetry is clearly satisfied.

integral into a volume integral through Gauss's theorem.[4] It turns out that the divergence of the tensor $-\mathscr{I}p$ is the gradient[5] of p. Thus Eq. (4.12) becomes

$$\int_{\mathscr{V}} (\rho \boldsymbol{g} - \nabla p)\, d\mathscr{V} = 0 \tag{4.13}$$

Because the volume $\mathscr{V}$ was completely arbitrary, meaning that neither its shape nor extent came into consideration in the last equation, the result of Eq. (4.13) must be true for any volume. Thus we conclude that it must be the integrand in the integral that is zero. This gives the *hydrostatic equation* which will be the basic equation governing the vertical stability of the entire environment:

$$\nabla p = \rho \boldsymbol{g} \tag{4.14}$$

One of the properties of this vector equation is that when it is multiplied by an elemental length $\boldsymbol{ds}$ on both sides of the equation,

$$\nabla p \cdot \boldsymbol{ds} = \rho \boldsymbol{g} \cdot \boldsymbol{ds} \tag{4.15}$$

the left-hand side gives the total change of p in that distance, which equals, according to the right-hand side and to Eq. (2.16), $-\rho\, d\Phi$.

If instead of any $\boldsymbol{ds}$ we choose the geometric altitude dz which is in the same sense but opposite in direction to $\boldsymbol{g}$, then two important corollaries follow:

(a) On a surface normal to $\boldsymbol{g}$ (geopotential surface), p is constant. This can be derived from the geometry of Eq. (4.14). Thus, if for altitudes and depths of concern to us we define our coordinate system in the environmental space as the altitude z, the latitude ψ, and the longitude θ, then under static equilibrium

$$\partial p/\partial \psi = \partial p/\partial \theta = 0$$

This implies that on geopotential surfaces (almost parallel to the surface of the earth, especially near the surface of the earth) the pressure is constant. Since Eq. (4.14) is not valid for horizontal motion the corollary is not true for that case. However, as we shall see later, even in the case of motion the variations of pressure along ψ and θ are orders smaller than the dominant vertical pressure variation.

(b) The second corollary gives a quantitative value for the change of pressure with altitude or depth. If the displacement ds is taken as the elevation dz (opposite in direction to $\boldsymbol{g}$), then the dot product on the right-hand side of Eq. (4.15) is negative, and

$$\frac{dp}{dz} = -\rho g \tag{4.16}$$

[4] This theorem is presented in Appendix A.

[5] This can be easily verified.

When the ocean is considered, the positive coordinate is always depth ($-z$) and thus the sign of Eq. (4.16) becomes positive. This hydrostatic equation implies that pressure *must* decrease with altitude, since ρ and g are both positive scalars, for equilibrium to exist. The exact function that gives this decrease of pressure cannot be obtained until both ρ and g are specified as functions of z. We have determined the variation of g with z in Eq. (2.14). Judging from Fig. 2.4, if we consider the troposphere (20 km high), the error introduced by considering g constant is $(1 + 20/6000)^2$, or about 0.6%, which might be tolerable. However, for higher altitudes it will become more and more pressing to keep g variable with z. In the considerations that follow, g is not considered constant. Instead the geopotential altitude is used as defined in Section 2.7 such that $g\,dz$ is replaced by $g_0\,dH$.

4.5 VERTICAL PRESSURE VARIATION IN THE ATMOSPHERE AND OCEANS

We are ready to integrate Eq. (4.16) but we need information on ρ and g. Since $g\,dz$ can be replaced by $g_0\,dH$ and g_0 is not a function of z, the difficulty is surmounted with the choice of a new distance H as tabulated in Tables 2.1 and 2.3.

Thus Eq. (4.16) can be written as

$$\frac{dp}{dH} = -\rho g_0 \tag{4.17}$$

where g_0 is constant.

Thermodynamics must enter here in order for us to be able to express ρ in terms of p and T for the atmosphere, and in terms of p, $\mathscr{s}$, and T for the ocean. Let us consider the atmosphere first since its equation of state is that of a perfect gas: $\rho = p/RT$.

A Pressure Variation in the Atmosphere

Equation (4.17) becomes

$$\frac{dp}{p} = -\frac{g_0\,dH}{RT} \tag{4.18}$$

and upon integration we have

$$p = p_0 \exp\left(-\frac{g_0}{R}\int_{H_0}^{H}\frac{dH}{T}\right) \tag{4.19}$$

where p_0 corresponds to the initial pressure at the altitude $H = H_0$. However, with the use of Tables 2.2 and 2.3 we can integrate piecemeal into the atmosphere, for each linear temperature segment of Eq. (2.22), by considering H_0 and p_0 to be the beginning of one linear segment and consequently the end of the linear segment below.

In the first 100 km, R is constant and the temperature can be expressed in linear segments

$$T = T_0 - \alpha H \tag{4.20}$$

where α is the actual *lapse rate of temperature*[6] $-dT/dH$, given also in Table 2.2. Thus for each linear segment the integral of Eq. (4.19) becomes

$$\int_{H_0}^{H} \frac{dH}{T_0 - \alpha H} = -\frac{1}{\alpha} \ln \frac{T}{T_0}$$

Then for each segment of *linear temperature*

$$p = p_0 \left(\frac{T}{T_0}\right)^{g_0/\alpha R} = p_0 \left(\frac{T_0 - \alpha H}{T_0}\right)^{g_0/\alpha R} \tag{4.21}$$

From Table 2.2 we can see that certain segments are *isothermal*, and for such segments Eq. (4.19) integrates to

$$p = p_0 \exp\left[\frac{-g_0(H - H_0)}{RT_0}\right] \tag{4.22}$$

With the help of Table 2.2 and Eqs. (4.21) and (4.22) we can compute the pressures at any geopotential altitude. Figure 4.3 shows the pressure ratio p/p_0 for spring and fall in the mid-latitudes of the Northern Hemisphere. Table 4.1 shows the calculated values from these equations and the corresponding values accepted as the standard atmosphere.

B Pressure Variation in the Ocean

We have seen in Chapter 1 that seawater density was a function of p, $\mathscr{s}$, and T. However, measurements such as those given in Fig. 1.5 show that in the areas of interest to the oceanographer we can also break the distribution of $\mathscr{s}$ and T in linear segments such as

$$T = T_0 - \alpha H$$

and

$$\mathscr{s} = \mathscr{s}_0 - \varepsilon H \tag{4.23}$$

[6] The lapse rate is defined as the negative value of the actual temperature gradient. In the troposphere the value of the lapse rate is positive.

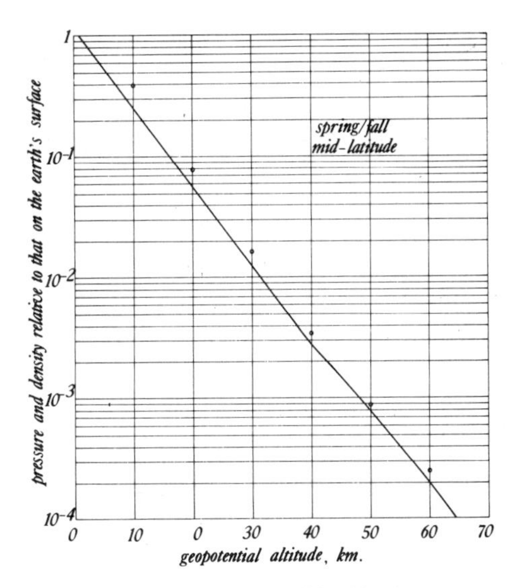

Fig. 4.3 Atmospheric pressure variation with altitude: — pressure; ○ density.

where α and ε are the temperature and salinity lapse rates. Furthermore, it is observed that on a T–s diagram the saltwater states line up nearly on straight lines. This indicates that within the accuracy of this statement (Fig. 1.5 seems to confirm this) if the salinity states vary linearly with the temperature states, then the equation of state can always be reduced to that of a pure substance, as Eq. (2.25). Since the density of seawater is less sensitive to p than to s and T, it is also possible to write as an approximation that

$$\rho = f(s, T) \tag{4.24}$$

If the thermal expansion coefficient for water is β and the coefficient of salinity is γ such that from Eq. (4.24)

$$d\rho = \left(\frac{\partial \rho}{\partial s}\right)_T ds + \left(\frac{\partial \rho}{\partial T}\right)_s dT$$

then the first partial derivative is γ and is always positive while the second is $-\beta$ and is always negative. Thus the density of seawater can be expressed for small changes of density as

$$\rho = \rho_0[1 + \gamma(s - s_0) - \beta(T - T_0)] \tag{4.25}$$

Substitution for $(s - s_0)$ and $(T - T_0)$ in terms of the geopotential depth H for linear segments as in Eq. (4.23) yields

$$\rho = \rho_0[1 + (\alpha\beta - \gamma\varepsilon)H] \tag{4.26}$$

TABLE 4.1

Pressure Calculations and Comparison with US Standard Atmosphere (1966) Mid-Latitude Spring–Fall[a]

Geopotential altitude H (km)	Lapse rate, α (°K/km)	T (°K)	T_0 (°K)	T/T_0	$g_0/\alpha R$	g_0/RT_0	p/p_0 [Eq. (4.21) or (4.22)] $\left(\frac{T}{T_0}\right)^{g_0/\alpha R}$	$\exp\left[-\frac{g_0}{RT_0}(H-H_0)\right]$	p/p_0 US Standard Atmosphere (1966)
0	6.5	—	288.15	—	—	—	1.0	—	1.0
5	6.5	255.65	288.15	0.887	+5.256	—	0.535	—	0.533
11	6.5	216.65	288.15	0.752	+5.256	—	0.221	—	0.223
15	0	216.65	—	—	—	0.157	—	0.1185	0.1188
20	0	216.65	—	—	—	0.157	—	0.0539	0.0540
26	−1.0	222.65	216.65	1.028	−34.163	—	0.02105	—	0.02125
32	−1.0	228.65	216.65	1.055	−34.163	—	0.00869	—	0.00857
40	−2.8	251.05	228.65	1.098	−12.2	—	0.00275	—	0.00274
47	−2.8	270.65	228.65	1.184	−12.2	—	0.001105	—	0.001095
52	0	270.65	—	—	—	0.126	—	0.00059	0.000582
61	+2.0	252.65	270.65	0.933	+17.08	—	0.000183	—	0.000180
70	+3.4	217.65	252.65	0.861	−10.05	—	0.0000414	—	0.0000484
79	+3.4	190.65	252.65	0.755	+10.05	—	0.0000108	—	0.00001071
88.74	0	190.65	—	—	—	0.179	—	0.00000189	—

[a] $g_0/R = 34.163$.

Then the hydrostatic equation for the ocean, upon integrating $dp/dH = \rho g_0$, becomes

$$p = p_0 + \rho_0 g_0 \left[H + \frac{(\alpha\beta - \gamma\varepsilon)}{2} H^2 \right] \tag{4.27}$$

Starting from Eq. (4.25) we could have also replaced $(\sigma - \sigma_0)$ by $(\kappa + \varepsilon T)$ and have a temperature relationship for ρ, which when introduced in the hydrostatic equation gives $dp = \rho_0 g_0 f(T)\, dH$. Actual measurements of T with H can give the pressure distribution as in the case of the atmosphere.

4.6 IMPORTANCE OF HYDROSTATIC EQUILIBRIUM IN A MOVING ENVIRONMENT

There is an important characteristic of the environment (atmosphere and ocean) such that, even when it is moving, the vertical component of the equation of motion is predominantly governed by the hydrostatic terms. So our interest in the hydrostatic equation is not just for a *static environment*, which is merely a textbook example, but also for an actively moving environment. This implies four important things: (a) in the vertical component of the equations of motion the Coriolis, acceleration, and viscous terms are negligible compared with the gravitational acceleration and the pressure variation with altitude; (b) the horizontal motion has little dynamical influence on the vertical equilibrium; (c) static stability is sufficient for vertical stability with or without motion; and (d) the horizontal components of the motion must satisfy themselves nearly independently of the vertical equilibrium. (This is thoroughly discussed in Chapter 6.)

This can be shown with the following order of magnitude analysis. In the z component of the equation of motion, as we shall see in the dynamical considerations, there will be terms of acceleration, Coriolis, pressure, gravitational, and friction. Using Table 7.1 for order of magnitude considerations, the largest part of the acceleration will be represented typically by $W\,\partial W/\partial z$, which has an order of magnitude between 10^{-7} and 10^{-3}; the vertical component of the Coriolis represented by $2\Omega_y U$ is of the order of magnitude of 10^{-4} to 10^{-3}; and the viscous terms represented by $\mu\,\partial^2 W/\partial z^2$ would be of the order of 10^{-11} to 10^{-9}. However, the only two sizable quantities would be $(1/\rho)\, dp/dz$ which is of the order of 10, and g which is also of the order of 10. These two are at least four orders of magnitude larger than the largest other term in the vertical equilibrium.

Thus vertical static equilibrium holds even when there is horizontal motion.

4.7 UNITS OF PRESSURE

Pressure was defined as the normal surface force per unit area. Therefore the units of pressure can be expressed as newtons per square meter (pascals) which is equivalent to 10 dynes/cm^2. In meteorology and oceanology a *bar* is defined as the pressure of 10^6 cgs units or 10^6 dynes/cm^2. Atmospheric pressure at sea level varies from day to day, but is nearly 1 bar. The bar is 0.987 of a standard atmosphere which has also been defined in terms of a column of mercury. By definition the standard atmosphere is 1.01325×10^6 dynes/cm^2 or 101325 N/m^2. This pressure is obtained by taking 76 cm of mercury at a density of 13.5951 g/cm^3 and a gravitational acceleration of 980.665 cm/s^2, giving the standard atmospheric pressure in dynes per square centimeter.

For pressure and pressure differences it is convenient to speak of *millibars*, which are 10^{-3} bar, or 10^3 dynes/cm^2, or 100 N/m^2. The abbreviation for millibar is mb.

4.8 DENSITY VARIATION WITH ALTITUDE

Equations (4.21) and (4.22) expressed the pressure variations in atmospheric segments with linear temperature change and with constant temperature. Applying the perfect gas law to the same segments, for a *linear temperature* segment

$$\frac{\rho}{\rho_0} = \frac{p}{p_0}\left(\frac{T}{T_0}\right)^{-1} = \left(\frac{T}{T_0}\right)^{(g_0/\alpha R)-1} \tag{4.28}$$

and for an *isothermal* segment $\rho/\rho_0 = p/p_0$, and

$$\frac{\rho}{\rho_0} = \exp\left[-\frac{g_0(H - H_0)}{RT_0}\right] \tag{4.29}$$

Calculations similar to those performed for the pressure and listed in Table 4.1 can be performed for the density ratio starting from standard atmospheric temperature profiles. These calculations are plotted in Fig. 4.3 along with the pressure ratios. Equation (4.26) gives the variation of density with depth for the ocean and for linear segments.

4.9 MECHANICAL EQUILIBRIUM

We know from experience that neither the atmosphere nor the ocean is in a state of static equilibrium. At this point it becomes important to examine the mechanical as well as the thermodynamical conditions that will permit or upset a stable static equilibrium. First, let us look into the mechanical factors involved which are responsible for convective currents.

Let us define the concept of *stability*. A fluid element is said to be in a *stable mechanical equilibrium* such that when the fluid is displaced slightly from its position, the forces created by the displacement are such that they return the fluid right back to its original position. Conversely, an *unstable mechanical equilibrium* exists when during the same displacement of the fluid parcel the forces created are such that they take the parcel further and further away from where it started. A *neutral equilibrium* exists when the displaced parcel is in a state of equilibrium at the new position as well.

We have seen in the preceding section that the density of our environment increases with depth. Consider Fig. 4.4 and let a fluid parcel at an elevation z_2

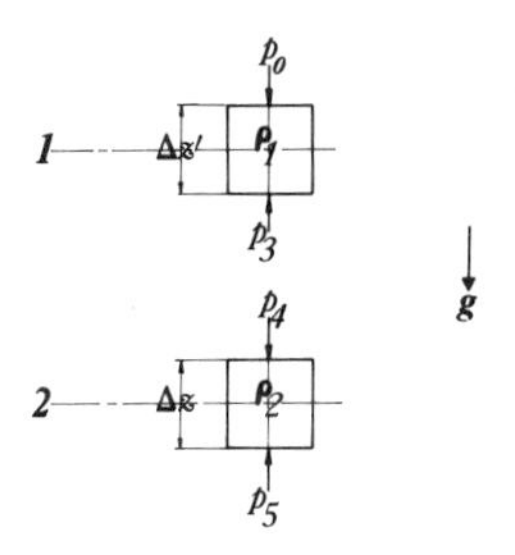

Fig. 4.4 Isothermal stability of equilibrium.

and density ρ_2 be raised to a level z_1 where the density of the surrounding fluid is ρ_1. For the case we are discussing $\rho_1 < \rho_2$. Since the displaced parcel originally at position 2 was in equilibrium at 2, then

$$p_5 - p_4 = \rho_2 g\, \Delta z$$

by the principle of the hydrostatic equation (4.16). In other words, the sum of the surface forces balanced the weight. When this parcel is at its new position 1, because of the new ambient pressure, the dimensions of the parcel will expand to a new dimension $\Delta z'$ such that its mass remains the same with a corresponding change in density to ρ_1. The volume of the element must increase by the ratio ρ_2/ρ_1 and $\Delta z' = \Delta z(\rho_2/\rho_1)^{1/3}$. The resultant ambient pressure difference at this new location is $p_3 - p_0 = \rho_1 g\, \Delta z'$ and in general, it may not be able to support the weight of the displaced element 2 which remains $\rho_2 g\, \Delta z$. Consequently, this displaced element, at its new location, will be acted upon with a resultant force:

$$\begin{aligned} \varepsilon &= \rho_2 g\, \Delta z - (p_3 - p_0) = g(\rho_2\, \Delta z - \rho_1\, \Delta z') \\ &= \rho_2 g\, \Delta z[1 - (\rho_1/\rho_2)^{2/3}] \end{aligned}$$

This new resultant force may be zero, positive, or negative. The quantity ε is defined so that if positive it will oppose the infinitesimal displacement. Consequently, from the equation above we can see that, if $\partial\rho/\partial z < 0$, or ρ is decreasing with altitude, $\rho_1 < \rho_2$, then $\varepsilon > 0$, and the parcel 2 is pushed back in the direction from which it came. This will be *stable*. If, however, $\rho_1 = \rho_2$ or $\partial\rho/\partial z = 0$, the restoring force $\varepsilon = 0$ and the element will find itself in *neutral equilibrium*. Finally, when $\rho_1 > \rho_2$ or $\partial\rho/\partial z > 0$, the restoring force $\varepsilon < 0$ and the parcel will continue to move in the direction from which it was originally displaced. The fluid is then mechanically *unstable*.

Illustrative Example 4.1

A good example of an unstable equilibrium is the flow of gases in a chimney. The equilibrium is upset when the weight of a parcel of gas in the chimney is less than the pressure forces acting on it because of the buoyancy. The resultant upward force on the parcel of gas will give it, according to Newton's law of inertia, an upward acceleration. This acceleration can be calculated simply, following the sketch in Fig. 4.5.

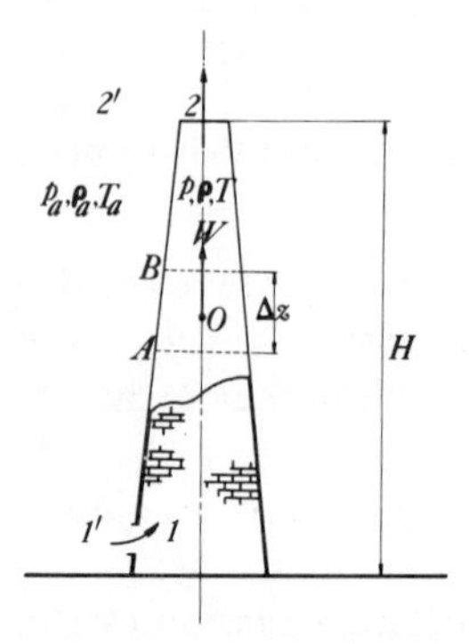

Fig. 4.5 Unstable equilibrium in a chimney.

Let us consider a small section of chimney such that at the center of the section AB the gas velocity is W. At the cross section A, $\Delta z/2$ lower than O, the velocity must be, according to Taylor's expansion, $W - (dW/dz)(\Delta z/2)$. At section B it will be faster and equal to $W + (dW/dz)(\Delta z/2)$. The acceleration at O is then the difference of velocity between A and B divided by the time it takes the gas element to go from A to B. This time is $\Delta z/U$. Thus, in performing this division, we realize that the acceleration of the gas parcel is

$$W\frac{dW}{dz} = -\frac{1}{\rho}\frac{dp}{dz} - g$$

and equal to the unbalanced hydrostatic equation which represents the sum of external forces.

It is logical to conceive that since the entrance and exit of the chimney are connected with the outside atmosphere, the pressure drop inside and outside the chimney would be the same. However, the density and the temperature are certainly not the same. So dp/dz inside the chimney, in the equation we have developed, can be replaced by dp_a/dz, which is $P_a g$. Thus the preceding equation becomes

$$W\frac{dW}{dz} = \left(\frac{\rho_a}{\rho} - 1\right) g = \frac{1}{2}\frac{dW^2}{dz}$$

Again, integration of this exact differential equation becomes possible when the densities of the gas and the outside atmosphere are known as a function of elevation. The first simple idea that comes to mind is a well-insulated chimney with a height H where the densities vary little in that height. Thus, with this approximation, the integration follows and

$$W(z) = \left[2\left(\frac{\rho_a}{\rho} - 1\right) gz\right]^{1/2}$$

The lower limit for the velocity at the bottom of the chimney was taken arbitrarily as zero.

In conclusion, the chimney draft, represented by the velocity, increases as the square root of the height of the chimney. The draft also increases with increase of temperature of the gas which produces a decrease in its density. Obviously, if the cross section of the chimney varied, the conservation of mass for the gas would have required that the same flux of gas pass through each cross section $A = f(z)$. This would have imposed a condition on ρ before integration.

4.10 THERMODYNAMIC EQUILIBRIUM

We indicated in Section 4.1 that *thermodynamic equilibrium* included *mechanical* and *thermal* equilibria. Through Maxwell's thermodynamic relations, J. Willard Gibbs recognized a thermodynamic property, called the *Gibbs function* G, or the *thermodynamic potential* (Gibbs, 1948; see also Zemansky and Van Ness, 1966), which is an important criterion for thermodynamic equilibrium. Combining Eqs. (2.56) and (2.57) this function is defined as

$$dG = \frac{1}{\rho} dp - s\, dT \tag{4.30}$$

where s is the entropy. We have seen in Section 2.16 that since changes of phases occur at constant p and T, this function G is constant during the change of phase.

For thermodynamic equilibrium to exist thermal equilibrium must also exist. Then $dT = 0$ and the differential form of the Gibbs function must be

$$dG = \frac{1}{\rho}\, dp \tag{4.31}$$

or in vector notation

$$\nabla G = \frac{1}{\rho}\, \nabla p \tag{4.32}$$

Using the hydrostatic equilibrium of Eq. (4.14) we conclude that

$$\nabla G = \boldsymbol{g}$$

Thus this relation indicates that along a geopotential surface, the Gibbs function is also constant. Or if we introduce the geopotential function as in Eq. (2.15), then

$$\nabla(G + \Phi_{\mathrm{E}}) = 0 \tag{4.33}$$

or, upon integration

$$G + \Phi_{\mathrm{E}} = \text{constant} \tag{4.34}$$

This constitutes the thermodynamic equilibrium equation for a system with zero acceleration and under gravitational force.

4.11 STABILITY OF THE ENVIRONMENT

We have seen all along that the atmosphere and the ocean are not in thermal equilibrium since they display large temperature variations. The question now arises as to whether there can be some form of stability (mechanical) without thermal equilibrium. Would all the lapse rates set up convective currents that will destroy the static equilibrium and produce a mixing of the geofluid to equalize the temperature, thus attempting to achieve thermal equilibrium by upsetting mechanical equilibrium? Let us investigate the mechanical and thermodynamical implications of a mechanical equilibrium.

Consider in Fig. 4.6 a parcel or differential element of geofluid in equilibrium at a position 2 with pressure and entropy p_2 and s_2. Let us move this parcel a small distance dz to a new position of the environment where the stable state of the fluid around it is s_1 and p_1. During the process of displacement we shall say that the parcel 2 would expand to the new ambient pressure p_1 but is likely to maintain its entropy if the time involved in the displacement is not too long since air is a poor conductor of heat and the displacement process can be considered adiabatic.

Thus the density of the parcel 2 in the new environment 1 is $\rho_3(s_2, p_1)$ while the density of the environment at 1 is $\rho_1(s_1, p_1)$. In order for equilibrium to exist

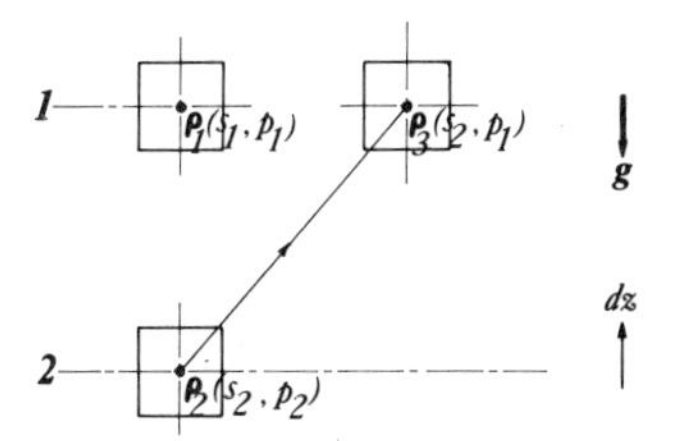

Fig. 4.6 General stability considerations.

in the environment the density ρ_3 must be larger than ρ_1 so that the parcel can return to its original position. Thus

$$\rho_3(s_2, p_1) - \rho_1(s_1, p_1) > 0$$

Since we have only moved a small displacement dz these two states can be expressed in terms of a Taylor series expansion for s. Thus

$$\rho_1(s_1, p_1) = \rho_3(s_2, p_1) + \left(\frac{\partial \rho}{\partial s}\right)_p ds$$

Substituting this expansion into the condition for mechanical equilibrium and since the variation of density occurs in z, we can conclude that

$$\frac{d\rho}{dz} = \left(\frac{\partial \rho}{\partial s}\right)_p \frac{ds}{dz} < 0$$

We have said that the displaced parcel and the environment were at the same pressure p_1 and thus if an energy transfer is to take place after the displacement, it must be at constant pressure. The first and second law combination in thermodynamics [Eq. (2.57)] for a constant pressure process becomes

$$T\,ds = c_p\,dT$$

and thus

$$\left(\frac{dT}{ds}\right)_p = \frac{T}{c_p}$$

where c_p is the specific heat at constant pressure. Multiplying both sides by $(d\rho/dT)_p$, we have

$$\left(\frac{d\rho}{dT}\right)_p \left(\frac{dT}{ds}\right)_p = \frac{T}{c_p}\left(\frac{d\rho}{dT}\right)_p = \left(\frac{d\rho}{ds}\right)_p$$

Then from the equilibrium inequality it follows that[7]

$$\frac{T}{c_p}\left(\frac{d\rho}{dT}\right)_p \frac{ds}{dz} < 0$$

[7] Since the environment is a pure substance and only two properties are sufficient to define the state, $(\partial\rho/\partial s)_p = (d\rho/ds)_p$.

For most substances, excluding changes of phase, the density decreases with increasing temperature, at the same pressure. Thus $(d\rho/dT)_p < 0$ and since T and c_p are always positive, the last inequality yields

$$ds/dz > 0 \tag{4.35}$$

This is the basic *thermodynamic condition* on the environment for it to be in mechanical equilibrium only. This implies that for a parcel to be in mechanical equilibrium with its environment, *the entropy of the environment must increase with altitude. The entropy lapse rate must be positive.*

In order to express this inequality in terms of the temperature lapse rate, we proceed as follows: Again for a pure substance, $s = f(p,T)$. Then the *entropy lapse rate* can be expressed as

$$\frac{ds}{dz} = \left(\frac{ds}{dT}\right)_p \frac{dT}{dz} + \left(\frac{ds}{dp}\right)_T \frac{dp}{dz} > 0$$

It has already been established here that $(ds/dT)_p = c_p/T$. Also from the hydrostatic equation $dp/dz = -\rho g$ and when substituted[8] into the last expression we conclude that

$$\frac{dT}{dz} > \frac{T\rho g}{c_p}\left(\frac{ds}{dp}\right)_T \tag{4.36}$$

In order to evaluate $(ds/dp)_T$ in Eq. (4.36) we start with Eq. (2.57. As the derivative implies we are concerned with an isothermal process. Since this is the last step in the anslysis, we shall consider three types of environment: A, an atmospheric environment that exhibits no change of phase, in other words no precipitation or evaporation; B, the ocean; and C, an atmospheric environment where there is a change of phase, with precipitation or evaporation.

A Mechanical Equilibrium of the Atmosphere without Change of Phase

For gases[9] at a constant temperature, $dh = 0$ because $dh = c_p\, dT$ with no change of phase, and from Eq. (2.57)

$$\left(\frac{ds}{dp}\right)_T = -\frac{1}{\rho T}$$

[8] We should remember to change the sign for the ocean where the direction of z is reversed.

[9] Remembering that $h = c_p T$ for all processes of ideal gases and only for constant-pressure processes in liquids. Thus $dh = 0$ for an isotherm is only valid for the atmosphere. We shall see the difference in Section 4.11.B.

Substitution into Eq. (4.36) yields the final result desired for the *dry atmosphere*:

$$\frac{dT}{dz} > -\frac{g}{c_p} \tag{4.37}$$

Since α was defined as the negative value of dT/dz in Table 2.2 and Section 4.5, the condition for this environment to be mechanically stable is that its temperature lapse rate α be less than the *adiabatic lapse rate* $\alpha_{ad} = g/c_p$ or that $(dT/dz)_{act} < (dT/dz)_{ad} = -\alpha_{ad}$. So this is the thermodynamic condition imposed on mechanical stability. Table 4.2 gives values of c_p for various gases and for water.

TABLE 4.2

Specific Heat of Atmospheric Gases and Seawater

	J/g · °K			
	c_p	c_v	R	$\gamma = c_p/c_v$
Monoatomic gases				
Helium (He)	5.226	3.206	2.020	1.630
Argon (Ar)	0.531	0.322	0.209	1.648
Diatomic gases				
Hydrogen (H_2)	14.058	9.991	4.067	1.407
Nitrogen (N_2)	1.004	0.717	0.287	1.401
Oxygen (O_2)	1.05	0.752	0.298	1.396
Triatomic gases				
Water vapor (H_2O)	1.912	1.448	0.464	1.32
Carbon dioxide (CO_2)	0.822	0.636	0.186	1.293
Ozone (O_3)			0.1732	
Dry air	1.015	0.724	0.291	1.402
Seawater	3.92		—	1.0004–1.0207

For dry air $c_p = 1.015$ J/g·K and $g = 9.8066$ m/s^2; then[10] the adiabatic lapse rate $\alpha_{ad} = 9.66$°K/km.

B Mechanical Equilibrium in Liquids

All the steps up to Eq. (4.36) are valid for gases and liquids. The only difference for the ocean or lakes as compared to the analysis in Section 4.11.A is the evaluation of $(ds/dp)_T$. Although it is true that for liquids $dh = c_p\,dT$ is valid for a constant-pressure process, we need an isothermal process to evaluate $(ds/dp)_T$. For this we should return to combining Eqs. (2.56) and (2.57), giving

$$dG = \frac{1}{\rho}\,dp - s\,dT$$

[10] The ratio of the units of g (m/s^2) and those of c_p (J/g·K) convert into degrees Kelvin per kilometer. In the stratosphere since dT/dz is positive in its entire range, Eq. (4.37) is automatically satisfied.

Applying the mathematical condition for dG to be an exact differential we obtain what is called *Maxwell's fourth equation*:

$$\left(\frac{ds}{dp}\right)_T = -\left[\frac{d(1/\rho)}{dT}\right]_p$$

According to Eq. (2.32) the right-hand side[11] is $-\beta_1/\rho$ and the *stability of the ocean* is

$$\frac{dT}{dz} > \frac{gT\beta_1}{c_p} \tag{4.38}$$

where β_1 is the coefficient of thermal expansion as defined as in Eq. (2.32). Thus unlike the dry atmosphere the *adiabatic stability lapse rate*[12] for the ocean and lakes is a function of the temperature. Furthermore, the thermal expansion coefficient in the vicinity of 0°C shows anomalies because seawater reaches a maximum density around this temperature. Around the maximum density point, when β_1 changes sign it will create vertical motions.

We have indicated in Chapter 1 that the specific heat of seawater at constant pressure is from 0.94 to 0.93 cal/g · °K or about 3.92 J/g · °K. The coefficient of thermal expansion for seawater is 1.57×10^{-4}/°K. Then at 10°C or 283°K the vertical temperature gradient dT/dz is 0.111°K/km. This is so small that it is not difficult to imagine that Eq. (4.38) will be satisfied most of the time.

C Stability of the Atmosphere with Change of Phase

The question that arises here is what is special during the change of phase that was not treated in Section 4.11.A? During a change of phase, evaporation or condensation takes place, and the latent heat of evaporation (or condensation or sublimation) L must be considered in the energy balance of the first and second law combination, which we have been using in the analysis all along. Actually, the relationship between T and p and T and s, and the values of R and c_p do not change that much because of the presence of vapor.

Taking into account the latent heat, the combined first and second law becomes

$$T\,ds - L\,d\omega_s + \frac{1}{\rho}\,dp = dh \tag{4.39}$$

[11] To check our result with Section 4.11.A, for a perfect gas $-\beta_1/\rho = -1/\rho T$.

[12] When the hydrostatic equation was used for the liquid, z was taken positive in the direction of depth and the sign was changed.

where ω_s is the saturation humidity ratio defined as in Eq. (2.52) with $p_w = p_s$. When condensation takes place $L > 0$ and $d\omega_s < 0$, while the reverse is true when evaporation takes place. Thus the product of the two always has the same sign.

Starting again with Eq. (4.36) we must evaluate $(ds/dp)_T$ for the case of change of phase. From Eq. (4.39) we have

$$\left(\frac{ds}{dp}\right)_T = \frac{L}{T}\frac{d\omega_s}{dp} - \frac{1}{\rho T}$$

Using the hydrostatic equation we have

$$\left(\frac{ds}{dp}\right)_T = -\frac{L}{T\rho g}\frac{d\omega_s}{dz} - \frac{1}{\rho T}$$

Substituting this into Eq. (4.36) we have

$$\frac{dT}{dz} > -\frac{L}{c_p}\frac{d\omega_s}{dz} - \frac{g}{c_p} \tag{4.40}$$

Defining the *saturation adiabatic lapse rate*[13] for the atmosphere as $\alpha_{ads} = -(dT/dz)_{sat}$ its relation to the adiabatic lapse rate for dry air is

$$-\left(\frac{dT}{dz}\right)_{sat} = \alpha_{ads} = \alpha_{ad} + \frac{L}{c_p}\frac{d\omega_s}{dz} \tag{4.41}$$

As in the case of Eqs. (4.37) and (4.38), for stability $(dT/dz)_{act} > -\alpha_{ads} = (dT/dz)_{sat}$. Furthermore, $d\omega_s/dz$ can be expanded as

$$\frac{d\omega_s}{dz} = \frac{d\omega_s}{dT} \cdot \frac{dT}{dz}$$

Finally, Eq. (4.41) becomes

$$\alpha_{ads} = -\left(\frac{dT}{dz}\right)_{sat} = \frac{\alpha_{ad}}{1 + (L/c_p)\, d\omega_s/dT} \tag{4.42}$$

This relationship, although accurate, may be a bit cumbersome to use because $d\omega_s/dT$ may not always be known. Because of this we may begin the analysis from a different point. The definition of ω_s, from Eq. (2.52) can be differentiated and we have

$$\frac{d\omega_s}{\omega_s} = \frac{dp_s}{p_s} - \frac{dp}{p}$$

[13] The process is still called adiabatic even though the latent heat is taken away or given to the environment. In the case of the stratosphere, since $(dT/dz)_{act}$ is positive in its entire range, Eqs. (4.37) and (4.40) are automatically satisfied.

Using the hydrostatic equation and the gas law for the atmosphere we obtain

$$d\omega_s = \omega_s\left(\frac{dp_s}{p_s} + \frac{g}{RT}\,dz\right)$$

Substitution of this into Eq. (4.39) gives

$$L\omega_s\left(\frac{dp_s}{p_s} + \frac{g}{RT}\,dz\right) = c_p\,dT + g\,dz$$

where for the mixture dh has been replaced by $c_p\,dT$. Dividing by dz and arranging, we obtain

$$-L\omega_s\left(\frac{1}{p_s}\frac{dp_s}{dT}\frac{dT}{dz} + \frac{g}{RT}\right) = c_p\frac{dT}{dz} + g$$

Factoring out dT/dz

$$\alpha_{\text{ads}} = -\frac{dT}{dz} = g\,\frac{1 + (L\omega_s/RT)}{(L\omega_s/p_s)(dp_s/dT) - c_p}$$

Finally using the Clapeyron equation (2.61) we have

$$\alpha_{\text{ads}} = \alpha_{\text{ad}}\,\frac{c_p T(RT + L\omega_s)}{c_p RT^2 + 0.6215L^2\omega_s} \tag{4.43}$$

Saturation adiabatic lapse rates have been computed from Eq. (4.42) and are shown in Table 4.3 and Fig. 4.7 with $L = 2501$ J/g, $c_p = 1.015$ J/g · °K, and the

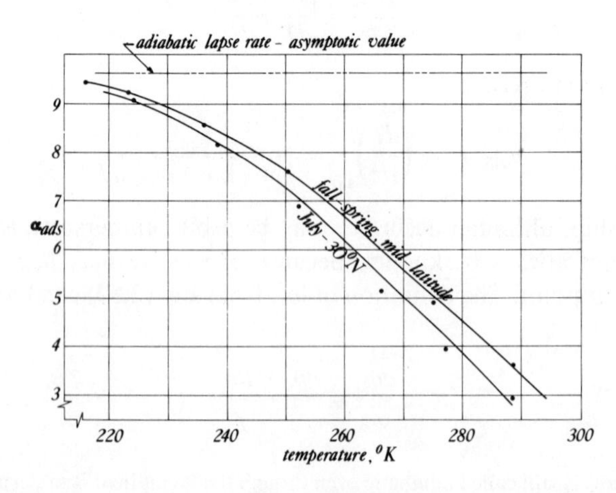

Fig. 4.7 Saturation adiabatic lapse rate for mid-latitudes (spring–fall) and 30°N (in July).

TABLE 4.3

Saturated Pressures, Humidity ratios, and Saturated Adiabatic Lapse Rates for Northern Mid-Latitudes and 30°N

Geopotential altitude	Temperatures (°K)		Atmospheric pressure (mb)		Saturation pressure (mb)		Saturation humidity ratio, ω_s		Saturation adiabatic lapse rate, α_{ads}	
H (km)	Mid-latitude fall–spring	30°N July	Mid-latitude	30°N	Mid-latitude	30°N	Mid-latitude	30°N	Mid-latitude	30°N
0	288.15	304.58	1.013×10^3	1.013×10^3	17.30	47.5	0.0107	0.030	3.58	1.87
2	275.15	289.54	7.950×10^2	8.084×10^2	7.145	18.9	0.0056	0.014	4.88	2.93
4	262.15	277.82	6.164×10^2	6.325×10^2	2.705	8.62	0.0027	0.0086	6.33	3.93
6	249.15	266.44	4.718×10^2	4.920×10^2	0.929	3.76	0.00123	0.0048	7.66	5.11
8	236.15	252.27	3.560×10^2	3.781×10^2	0.282	1.21	0.00049	0.0019	8.64	6.85
10	223.15	238.18	2.644×10^2	2.861×10^2	0.0746	0.343	0.000175	0.00075	9.23	8.22
12	216.65	224.15	1.933×10^2	2.129×10^2	0.0362	0.083	0.000116	0.00024	9.36	9.07
14	216.65	210.15	1.410×10^2	1.554×10^2	0.0362	0.017	0.00016	0.000067	9.25	9.47
16	216.65	203.15	1.029×10^2	1.113×10^2	0.0362	0.007	0.00022	0.00004	9.09	9.54
18	216.65	207.55	7.505×10	7.981×10	0.0362	0.012	0.0003	0.00009	8.91	9.38
20	216.65	211.95	5.475×10	5.762×10	0.0362	0.021	0.0004	0.00023	8.66	9.05
25	221.65	222.15	2.511×10	2.623×10	0.633	0.066	0.0015	0.0016	6.79	6.80
30	226.65	232.15	1.172×10	1.237×10	0.108	0.190	0.0063	0.011	3.89	2.88

US standard atmosphere for mid-latitude spring–fall and for July at 30°N latitude as if saturation conditions existed. The values of $d\omega_s/dT$ are estimated either from

$$\frac{d\omega_s}{dT} = \frac{d\omega_s}{dp_s} \cdot \frac{dp_s}{dT}, \qquad \frac{dp_s}{dT} = \frac{5388 p_s}{T^2}$$

and

$$\frac{d\omega_s}{dp_s} \simeq \frac{0.6215}{p}, \qquad \frac{d\omega_s}{dT} \simeq \frac{3348.6}{T^2}\frac{p_s}{p}$$

with Eq. (4.42), or from Eq. (4.43) and knowing T, p_s, and ω_s we obtain α_{ads} directly.

The ratio of the saturated adiabatic to adiabatic lapse rates in terms of the altitude (in millibars) and temperature is also given in Table 4.4. These figures

TABLE 4.4

Ratio of Lapse Rates: α_{ads}/α_{ad}

Pressure	Temperature (°K)								
(mb)	220	230	240	250	260	270	280	290	300
700	0.99	0.97	0.92	0.84	0.72	0.56	0.40	0.27	0.17
900	0.99	0.97	0.94	0.87	0.76	0.62	0.46	0.32	0.21
1000	0.99	0.97	0.94	0.88	0.78	0.64	0.48	0.34	0.22

are for the standard atmosphere and are rounded off. Equation (4.43) shows that as $\omega_s \to 0$, $\alpha_{ads} \to \alpha_{ad}$. At high altitudes where $\omega_s \to 0$ the adiabatic saturated lapse rate will become the adiabatic lapse rate, as shown in Fig. 4.7.

The specific value for moist air will be different from that for dry air because of the higher values of c_p and c_v for vapor. These values will depend on the amount of water vapor in the air, in other words, on the specific humidity ratio ω^0. Thus

$$c_p = (1 + 0.90\omega^0)c_{pa}, \qquad c_v = (1 + 1.02\omega^0)c_{va}$$

where c_{pa} and c_{va} are values for dry air.

Illustrative Example 4.2

Determine the signs of $d\omega_s/dT$ and $d\omega_s/dz$ as they enter into Eq. (4.42).

We have established from the Clasius–Clapeyron equation (2.62) that along the saturation line the saturated pressure p_s and T have the relationship

$$\ln p_s = \text{constant} - \frac{L}{R_w T}$$

When we consider ω_s we must consider that the process at that state is a phase change. Although the relationship $Tp^{-\kappa} = \text{constant}$ is for a non-phase-changing a diabatic process, we can say that

$$\ln p_{ad} = (1/\kappa) \ln T - c_1$$

and that at the same temperature $p_{ad} > p_{ads}$. Thus, in combining first the adiabatic and saturation conditions we have

$$\left(\frac{p_s}{p}\right)_{ad} = c_2 \exp\left[-\left(\frac{L}{R_w T} + \ln T\right)\right]$$

and

$$\left(\frac{p_s}{p}\right)_{sat} > c_2 \exp\left[-\left(\frac{L}{R_w T} + \ln T\right)\right]$$

and finally

$$\omega_s > c_3 \exp\left[-\left(\frac{L}{R_w T} + \ln T\right)\right] = \frac{c_3}{T \exp(L/R_w T)}$$

where $c_3 = 0.6215 c_2$.

Since L, R_w, and T are positive values, and $\omega_s > 0$, we conclude from the last expression that when T decreases ω_s also decreases; thus $d\omega_s/dT > 0$, and since T decreases with increasing z, $d\omega_s/dz < 0$.

4.12 VERTICAL DISPLACEMENT DUE TO BUOYANCY

We have established in the preceding section that the thermodynamic condition for mechanical stability is that the actual lapse rate α_{act}, defined as the negative value of $(dT/dz)_{act}$, be less than the adiabatic lapse rate in the case of no change of phase, and less than the saturation adiabatic lapse rate in the case of change of phase. The same applies to the ocean using Eq. (4.38). Since we know from experience that the atmosphere and the oceans do not always have stable conditions, vertical motion sets in when the static stability is upset. Figure 4.8a shows the ranges of stability for the atmosphere and the oceans.

At a given altitude z_0 shown in Fig. 4.8b let us consider an air parcel at the same pressure as its environment and in thermal equilibrium with it, implying that the temperature of the parcel T is equal to the temperature of the environment T_a at that elevation. In order to find the kind of mechanical equilibrium to which the parcel is subjected, we must displace it up or down from this initial level z_0 and study the balance of forces. As shown in Fig. 4.8b T_a is the temperature profile with altitude of the ambient environment, and T is the temperature

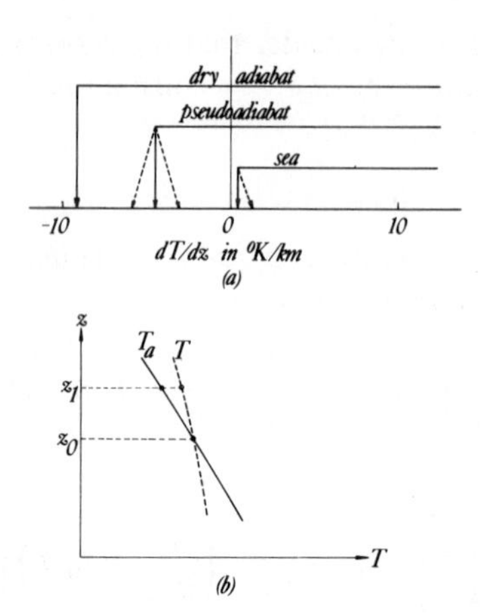

Fig. 4.8 Comparative ranges of stability.

profile with altitude that the displaced parcel will follow when its thermodynamic properties change adiabatically. When the parcel is displaced at an elevation z_1, there will be two forces per unit volume acting on the parcel and their summation will determine if there is a net resultant force and in which direction it is acting. These forces are the weight of the parcel ρg, directed downward, and the *hydrostatic flotation* or pressure difference imposed by the ambient environment, directed upwards. The flotation is controlled by the hydrostatic equilibrium of the environment and is equal to $\rho_a g$, where ρ and ρ_a are the densities of the displaced parcel and the ambient environment, respectively. The resultant of these two forces $g(\rho_a - \rho)$ is the buoyant force and is directed upward if $\rho_a > \rho$ and vice versa. This buoyancy force, if nonzero, will impart upon the parcel an inertial force equal to ρa and the acceleration will be

$$a = \frac{d^2 z}{dt^2} = g\,\frac{\rho_a - \rho}{\rho} \tag{4.44}$$

Since the pressure will equalize at the new elevation we can use the equation of state in the case of the atmosphere and replace the density ratio $\rho_a/\rho = T/T_a$ to obtain

$$a = g\left(\frac{T - T_a}{T_a}\right)$$

The temperature can be expanded in terms of the lapse rates

$$T_a = T_0 - \alpha_a\,\Delta z, \qquad T = T_0 - \alpha\,\Delta z$$

and the expression for the acceleration takes the form[14]

$$-a = \frac{g\,\Delta z}{T_a}(\alpha - \alpha_a) \tag{4.45}$$

When $\alpha > \alpha_a$, which implies that the adiabatic lapse rate α is less than the ambient lapse rate, the acceleration will be negative and the parcel will return to z_0, implying stable equilibrium.

Let us define a *static stability parameter* such that

$$\Pi = -\frac{a}{g\,\Delta z} = \frac{a}{\Delta\Phi} = \frac{1}{T_a}(\alpha_a - \alpha) \tag{4.46}$$

In conclusion, we can establish that (see also Fig. 4.8)

$$\alpha_{ads} < \alpha < \alpha_{ad}$$

i.e., the atmosphere is *conditionally stable*, because it is stable for unsaturated air and unstable for saturated air. In the range

$$\alpha < \alpha_{ads} < \alpha_{ad}$$

the state of the atmosphere is *absolutely stable* since being stable for unsaturated air implies stability for saturated air. Furthermore we can summarize that

$$\alpha = \alpha_{ads}$$

implies *neutral stability for saturated air*,

$$\alpha = \alpha_{ad}$$

implies *neutral stability for unsaturated air*, and finally

$$\alpha > \alpha_{ad}$$

implies an *absolute instability*.

4.13 ADIABATIC CONDITIONS

Although adiabatic conditions may not always exist in the atmosphere, they have become a major index, from the vertical stability point of view. The true sense of adiabatic condition, for stability, requires that no heat is exchanged between fluid parcels during any displacement. If we idealize this adiabatic condition by saying that it is also *isentropic*, then the change of entropy ds is zero.

[14] For the case of the moist atmosphere, since α_{ads} is the relevant stability lapse rate it should be used instead of α. The symbol α_{ad} has not been used for α because the thermodynamic process of the displaced parcel may not always be adiabatic.

These two terms adiabatic and isentropic are often used interchangeably even though isentropic implies adiabatic, but adiabatic does not imply isentropic. For an isentropic process in the atmosphere the first law reads

$$c_v\,dT + p\,d\frac{1}{\rho} = \frac{c_v}{R}\frac{dT}{T} + \rho\,d\frac{1}{\rho} = 0 \tag{4.47}$$

where the internal energy has been replaced by $c_v\,dT$ and the perfect gas law has been used. Integrating, we obtain

$$\begin{aligned} T\rho^{-R/c_v} &= T\rho^{-\gamma\kappa} = \text{constant} \\ Tp^{-R/c_p} &= Tp^{-\kappa} = \text{constant} \\ p\rho^{-c_p/c_v} &= p\rho^{-\gamma} = \text{constant} \end{aligned} \tag{4.48}$$

where $\gamma = c_p/c_v$, and $\kappa = R/c_p$. From these expressions, since we defined $\alpha_{\text{ad}} = -(dT/dz)_{\text{ad}} = g/c_p$, for the isentropic atmosphere we can conclude that the pressure, density, and temperature vary with altitude as

$$\begin{aligned} T &= T_0 - \frac{g}{c_p}(z - z_0) \\ p &= p_0\left[1 - \frac{g}{c_p T_0}(z - z_0)\right]^{1/\kappa} \\ \rho &= \rho_0\left[1 - \frac{g}{c_p T_0}(z - z_0)\right]^{1/\gamma\kappa} \end{aligned} \tag{4.49}$$

This is done by taking the derivative of Eqs. (4.48) with respect to z and integrating. It is easy to verify that for a linear temperature change with altitude, the exponents in Eqs. (4.48) need not be the adiabatic exponents. We have seen this in Eq. (4.21).

4.14 COMPARATIVE STABILITY POSTURES

From the information derived in the previous sections, it is of interest to compare stable and unstable conditions on the same diagram. Figures 4.9 and 4.10 have been drawn to achieve this comparison. According to Eqs. (4.49), plotting in (T, z) coordinates is the same as plotting in (T, p) coordinates. Figure 4.9 shows a temperature altitude distribution for a US standard atmosphere for spring–fall at mid-latitudes together with a hypothetical model. The lines for adiabatic lapse rates (9.66°K/km) are also plotted. In Fig. 4.10 the saturated adiabatic or the *pseudoadiabatic* lapse rate is also shown for comparison. Its values are computed in Table 4.3.

Looking at the standard atmosphere, it is evident that with the exception of

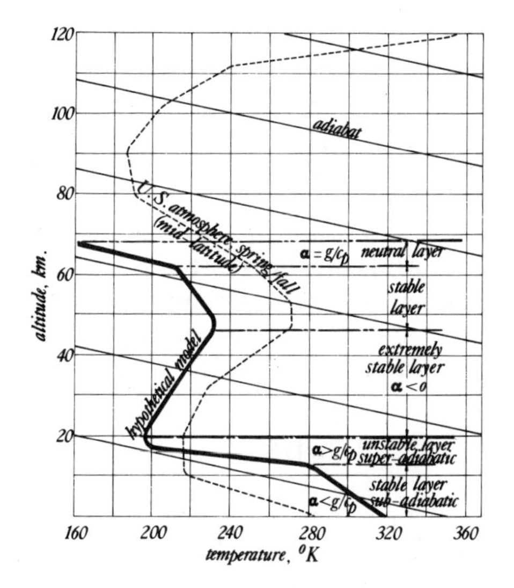

Fig. 4.9 Temperature profile and static stability criteria for a US model and a hypothetical atmosphere.

the troposphere, all portions of the atmosphere are stable for dry and saturated air since the actual temperature lapse rates are less than both the adiabatic and the saturated adiabatic lapse rates. This is explained in Section 4.11. In the lower troposphere, the dry air is normally stable but saturated air may not be if

$$\alpha_{\text{ads}} < \left|\frac{dT}{dz}\right|_{\text{act}} < \alpha_{\text{ad}}$$

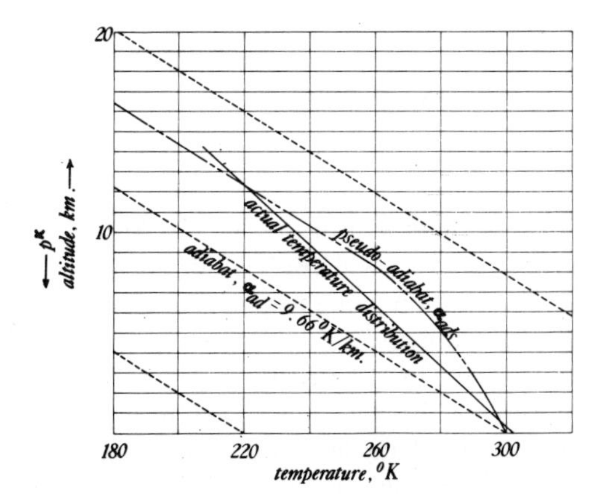

Fig. 4.10 Comparison of temperature lapse rates: adiabatic, saturated adiabatic, and actual.

As long as $\alpha < \alpha_{ad}$, a dry fluid element displaced higher in altitude will experience a downward acceleration. The saturated fluid element will experience, according to the results of Section 4.11, an upward acceleration. During this climb, the moist fluid element will carry moisture and higher enthalpy to higher altitudes. The hypothetical model of atmosphere is shown with segments ranging from extreme stability to instability.

The first segment adjacent to the earth's surface is stable because the lapse rate is less than the adiabatic lapse rate. For this reason this layer is often called *subadiabatic*, while the layer next to it is unstable and called *superadiabatic*. The third layer has a positive temperature gradient and therefore is extremely stable. The fourth layer is similar to the first and the last layer is *adiabatic* or neutral.

4.15 FORMATION OF CLOUDS

The upper movement of moisture carrying hotter air will form clouds when it reaches certain stable altitudes that have the colder temperatures necessary for condensation or sublimation to occur. Let us consider the stable unsaturated layer of the troposphere AE in Fig. 4.11, established during a spell of prolonged

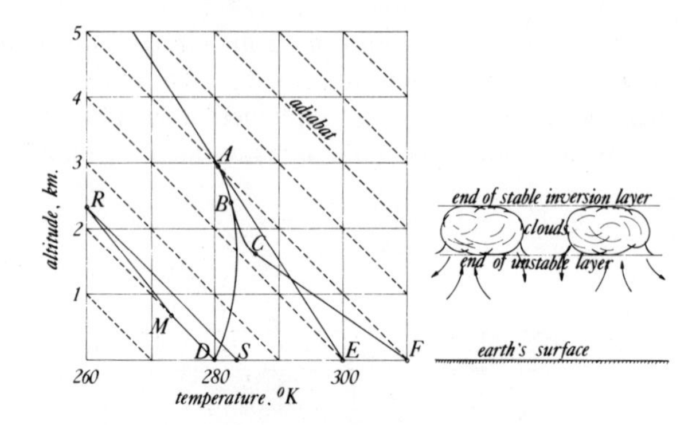

Fig. 4.11 Cloud formation.

mild weather. The daylight temperature is 300°K. At night the earth's surface and the lower tropospheric layer will cool rapidly to about 280°K because the earth continues to emit radiation while not receiving any from the sun. Cooling will take place as indicated along ABD. This layer is perfectly stable. If the following day is sunny and warmer, say 310°K, the layer closest to the earth's surface may take the temperature profile $ABCF$ as shown in Fig. 4.11. The layer CF has a lapse rate larger than the adiabatic lapse rate and therefore is unstable. According to Section 4.12 warm layers near the surface will rise to upper layers, carrying humidity with them. There will be corresponding downcurrents of cooler air.

At a given altitude C the temperature and the humidity ratio will be such that moisture will start condensing in the form of liquid clouds or subliming in the form of ice clouds. Since the drier air at this altitude is slightly stable and since it is heavier than the uprising damp air, it will descend to lower layers, creating a continual growth of cloud formation up to a point B where there is an *inversion* marking the beginning of a very stable layer. Vertical motion also changes the lapse rate, by expanding in ascent to lower pressures and by compressing in descent to larger pressures.

4.16 CLOUDS AND CLOUD EXCLUSION

In addition to all the factors discussed so far, condensation and precipitation have a significant influence on the local dynamical and thermodynamical states of the atmosphere. Near the earth's surface, air is rarely saturated and consequently, its temperature is higher than the dew point. Because of this, evaporation is always taking place above the sea, lakes, rivers, and vegetation. In the diurnal variations of temperature some of this evaporated vapor condenses as dew when the surface temperature drops sufficiently in the early hours of the day, while the remaining vapor is convected by upward currents.

The temperature of the atmosphere drops with altitude in the first 10 km and the vapor undergoes a quasi-adiabatic process of cooling as it rises. Condensation takes place in the form of large cloud masses which ultimately precipitate in the form of rain, snow, or hail. During the cycle of evaporation and condensation, a considerable amount of heat is transported into the troposphere where condensation takes place.

Because of the hotter temperatures in the tropical regions, the greatest amounts of evaporation take place there over the oceans and a considerable portion of this evaporation does not precipitate at the same location, but is convected toward the poles because of the prevailing circulation system, which we discuss in Section 8.10. So at the tropics evaporation is greater than precipitation while at the northern latitudes it is the opposite.

The rates of evaporation are controlled by the vapor pressures at the surface of the earth and above it. We have seen this in Section 3.5. Since vapor pressures are functions of temperature at a given ambient pressure, temperature becomes a dominant factor in these rates of evaporation.

On the basis of the temperature distribution in the atmosphere and on the basis of the pressure–temperature relationship of the saturated states (Fig. 2.8) it is possible to combine these diagrams and determine the regions in which condensation can occur and regions in which clouds cannot form at all. In other words, the altitude axis of the atmosphere becomes a pressure axis in the p–$(1/\rho)$ diagram.

Consider Fig. 4.12 where we have plotted a standard atmosphere and a p–T saturation and sublimation line for water and other substances that change phase. The saturation and sublimation line for water is the same as that in Fig. 2.8 except that in Fig. 4.12 pressure decreases with increasing ordinate. Above the saturation line DE and the sublimation line FD water can exist *only* in vapor form. Below DE it will exist in liquid form and below DF in ice form. This diagram is a side view of the three-dimensional schematic shown in Fig. 2.9, viewed from a constant-density plane. The change of phase takes place at constant temperature in a direction perpendicular to the plane of the paper.

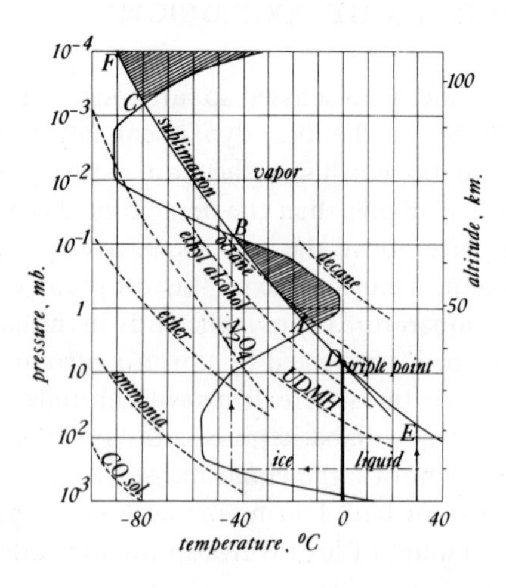

Fig. 4.12 Atmospheric equilibrium and saturation equilibrium. (After McDonald 1964.)

We should recall that if the saturation vapor pressure is equal to the atmospheric pressure at a given altitude, then for this situation to be possible, the atmospheric sample at that altitude must be all water—an impossible situation. Thus the saturation–sublimation line $EDABC$ represents a limiting condition for the saturation pressure and the standard atmosphere line represents the actual ambient pressure. We know from Chapter 2 that p_w is always less than or equal to p_s at the same temperature. We also know that the ambient atmospheric pressure p is larger than the partial pressure. Thus when

$$p_w < p_s < p$$

no condensation occurs. However, if

$$p_s = p_w < p$$

condensation or freezing takes place. Ultimately if

$$p_w < p \leq p_s$$

condensation or freezing can never occur, and there can be no cloud formations in that region. The shaded areas in Fig. 4.12 show such regions where $p \leq p_s$ and where it is impossible to have clouds form. The same arguments are valid for other substances such as the one shown in Fig. 4.12 whose sublimation lines cross the standard atmosphere pressure distribution.

Illustrative Example 4.3

If a mass of moist air at relative humidity $\phi = 50\%$ on the earth's surface, moves up to an altitude of 10 km, will there be cloud formation?

The temperature on the earth's surface is taken as 27°C at a pressure of 1000 mb. From Fig. 4.12 the saturation vapor pressure at E is 36 mb (see also Table 2.6). The vapor pressure that exists on the earth's surface is, from Eq. (2.50),

$$0.5 = p_{w0}/p_{s0}$$

where the subscript zero stands for the earth's surface. Then $p_{w0} = 18$ mb. The humidity ratio, from Eq. (2.52) is

$$\omega_0 = 0.6215 \frac{18}{1000 - 18} = 0.01135 \text{ g/g of dry air}$$

As air rises to 10 km, the atmospheric pressure will drop from 1000 to 280 mb, and the temperature from 27 to -35°C. Using the perfect gas law, the density of air will drop by a ratio of 8.5, and the mass of dry air will decrease by 8.5. If the mass of moisture raised from the ground remains the same, the ratio of the mass of water vapor to the mass of dry air, at 10 km, will be approximately 8.5 times larger or $\omega_{10} = 0.0968$. The ambient pressure, as given above, is 280 mb and the partial pressure will become

$$p_{w10} = 37.7 \quad \text{mb}$$

The saturation pressure corresponding to -35°C is 0.230 mb; thus vapor would have condensed long before reaching the altitude of 10 km. Looking at Fig. 4.12 one might be interested in knowing the probability of cloud formation at various altitudes. The line segments AB and CF in this figure represent zero probability.

In the definition of the humidity ratio ω, if we arbitrarily replace p_w by p_s, we will obtain a saturation humidity ratio ω_s at saturation or $\phi = 1.0$:

$$\omega_s = 0.6215 \frac{p_s}{p - p_s} \tag{4.50}$$

Looking at Fig. 4.12 since the standard atmosphere pressure plot diverges furthest to the left of the saturation–sublimation line at the tropopause, p_s/p will be at its lowest value there and ω_s its smallest. Therefore the tropopause has a high probability of cloud formation since it requires the smallest amount of vapor mass per mass of air, provided of course that vertical motion extends that far to bring the vapor from the earth's surface, which is not always the case. Examining the shape of the saturation–sublimation line, the value of ω_s increases to infinity at A, B, and C, implying that all the atmosphere at these points must be vapor which is not probable at all. At the earth's surface

$$\omega_s = 0.6215 \frac{36}{1000 - 36} = 0.023$$

With the use of Eq. (4.50) and the data in Fig. 4.12 we can compute at every altitude the maximum value of the humidity ratio, or ω_s, necessary for condensation or freezing. These computed values are plotted in Fig. 4.13 for summer

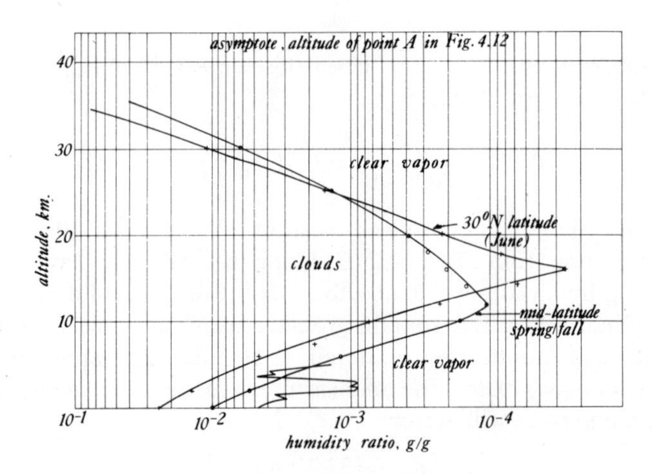

Fig. 4.13 Saturation humidity ratio in the troposphere and the stratosphere.

at 30°N latitude. In the same figure is plotted the measured values of the humidity ratio obtained in Washington, D.C., as shown in Fig. 2.13. This comparison shows that up to 5 km the sky was clear on that day with the possibility of some clouds at 3 to 4 km.

This analysis could be repeated for every latitude and every season. Schilling (1964) has calculated, with the data available at that time, the variation of the saturation humidity ratio ω_s with latitude in the winter and summer. One must remember that although the pressure–temperature distribution of the saturation–

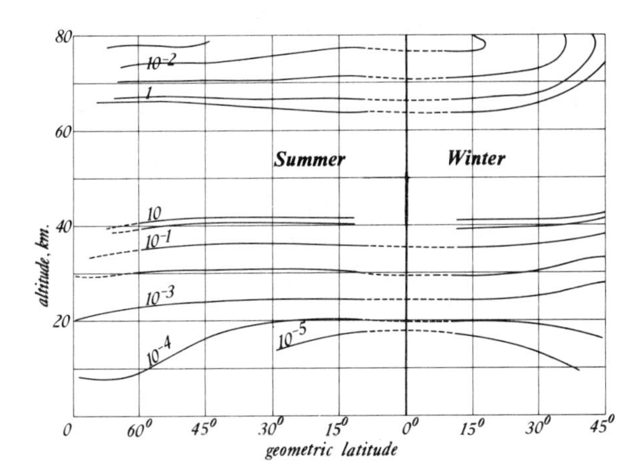

Fig. 4.14 Calculated humidity ratios in grams of vapor per grams of air based on saturation pressure that would give vapor condensation or solidification. (After Schilling, 1964.)

sublimation line depends only on substance and not on time, the pressure–temperature distribution of the atmosphere is latitude and season dependent. Schilling's results are shown in Fig. 4.14.

Meteorologists have classified the types of clouds as shown in Table 4.5.

TABLE 4.5

Types of Clouds and Their Basic Characteristics

Name	Appearance	Height (km)	Precipitation
Fog	Stratiform	Earth surface	Fine drizzle
Stratus (ST)	Stratiform	0–3	Drizzle, fine snow
Nimbostratus (Ns)	Stratiform	0–3	Light, moderate or heavy rain or snow
Altostratus (As)	Stratiform	3–7	Light or moderate rain or snow
Cirrostratus (Ca)	Stratiform	Higher than 7	
Cumulus (Cu)	Heap-convective	0.3–3	Heavy rain, hail, snow
Cumulonimbus (Cb)	Heap-convective	0.3–3	Thunderstorms
Altocumulus (Ac)	Heap-convective	3–7	Light precipitation aloft
Cirrus (Ci)	Streak	Higher than 7	Ice particles aloft
Stratocumulus (Sc)	Stratiform-heap (uncommon)	3–7	Light showers or snow
Cirrocumulus (Cc)	Stratiform-heap (uncommon)	Higher than 7	

4.17 THE POTENTIAL TEMPERATURE

Since we have found that the adiabatic process is a limiting process insofar as vertical stability is concerned, a gas particle moving vertically with constant entropy experiences a change of temperature according to Eq. (4.48). The constant of integration in that equation can be found by knowing *one* state in that process. Then the *potential temperature* θ is defined as that temperature which would result if the gas parcel were brought from its actual pressure to the standard pressure of 1000 mb isentropically. Thus

$$\theta(1000)^{-\kappa} = Tp^{-\kappa} = \text{constant}$$

or

$$\theta = T(1000/p)^{\kappa} = Tp^{-\kappa}(1000)^{\kappa} = \text{constant}(1000)^{\kappa} \tag{4.51}$$

In oceans, seas, and lakes, the potential temperature is similarly defined as the temperature that a water sample would attain if raised adiabatically to the surface. Of course, Eq. (4.51) cannot be used since water is not a perfect gas and also because salinity plays an important role.

When dry air undergoes adiabatic changes, the potential temperature remains constant. Thus according to the adiabatic process T and p^{κ} are linearly related to each other. The vertical ordinates of Figs. 4.9–4.11 could have been labeled p^{κ} with adiabatic lines having a slope of $\theta/(1000)^{\kappa}$. On these figures, the potential temperature at any altitude is the temperature value the adiabat passing through the point has at 1000 mb.

Differentiating Eq. (4.51) we obtain

$$\frac{d\theta}{dz} = \frac{\theta}{T}\left(\frac{dT}{dz} + \frac{\kappa g}{R}\right)$$

The hydrostatic equation and the perfect gas law have been used to eliminate the pressure. In terms of the lapse rates,

$$\frac{d\theta}{dz} = \frac{\theta}{T}(\alpha_{\text{ad}} - \alpha) \tag{4.52}$$

It is clear that when the actual temperature lapse rate α is equal to the adiabatic lapse rate, θ remains constant with elevation. Then along an adiabat, $\theta = \text{constant}$ and this constant is the temperature on the ground at $p = 1000$ mb. Taking the derivative of Eq. (4.51) and arranging differently than for Eq. (4.52) by dividing through by T/θ and replacing $\kappa = R/c_p$,

$$c_p \frac{d\theta}{\theta} = c_p \frac{dT}{T} - R\frac{dp}{p} \tag{4.53}$$

From the first and second law combination for an adiabatic process with change of phase, $dh = c_p\, dT$ and $\rho T = p/R$ in Eq. (4.39) we have

$$-\frac{L}{T} d\omega_s = c_p \frac{dT}{T} - R\frac{dp}{p} \tag{4.54}$$

Comparing Eqs. (4.53) and (4.54) we may write

$$-L\frac{d\omega_s}{T} = c_p \frac{d\theta}{\theta} \tag{4.55}$$

To integrate this expression we note that $(d\omega_s/T) \simeq d(\omega_s/T)$. This is because in the expansion

$$d\left(\frac{\omega_s}{T}\right) = \frac{d\omega_s}{T} - \omega_s \frac{dT}{T^2}$$

it is evident that in the second term ω_s is a small number and is divided by T^2, a very large number. So using this approximation

$$\frac{d\omega_s}{T} \simeq d\left(\frac{\omega_s}{T}\right)$$

Eq. (4.55) becomes an exact differential equation

$$-L\, d\left(\frac{\omega_s}{T}\right) = c_p \frac{d\theta}{\theta}$$

which when integrated yields

$$-\frac{L\omega_s}{T} = c_p \ln \theta + A \tag{4.56}$$

where A is the constant of integration. We can relate this constant to the logarithm of a potential temperature θ_e where the parcel of air has lost all the vapor and all its latent heat has returned to the dry air. So actually, A is the value of $c_p \ln \theta_e$ when $\omega_s = 0$. Then after transposing

$$\theta = \theta_e \exp(-L\omega_s/c_p T) \tag{4.57}$$

Physically speaking, if we have a moist air parcel at the conditions p and T corresponding to a potential temperature θ, we expand this parcel adiabatically, until all the vapor condenses. Then we take the dry air and compress it back to the base pressure of 1000 mb which gives a new temperature T_e which will correspond to θ_e. The quantities T_e and θ_e are called the *equivalent temperatures*. Since T_e and θ_e are related through Eq. (4.51), we conclude that the same relationship holds for

$$T_e = T \exp(L\omega_s/c_p T) \tag{4.58}$$

This indicates that T_e must always be larger than T. This is not surprising because the released latent heat has been absorbed in the process.

Looking again at the definition of the potential temperature θ, if we devise a coordinate system such that on the ordinate we place values of pressure plotted as p^{κ} decreasing upward (this is what takes place in the atmosphere) and on the abscissa we place linear values of T, then $\theta = \text{constant}$ lines are straight lines, all beginning at $p = 0$ and $T = 0$, and the values of the potential temperatures on each of these lines correspond to the temperature at 1000 mb. Figure 4.15

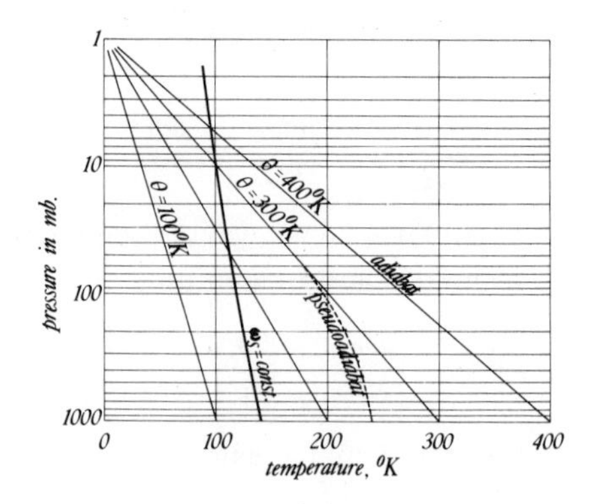

Fig. 4.15 The potential temperature, adiabat, pseudoadiabat, and constant ω_s on a p^{κ}–T plot.

shows this convenient plot. The slopes of the $\theta = \text{constant}$ lines are, from Eq. (4.51), $-(1000)^{\kappa}/\theta$. This diagram representation is called a *Stüve diagram* and is discussed further in Section 4.20. We have already used a comparable diagram representation in Fig. 4.10, and in Fig. 4.18 it is also identified as a pseudo-adiabatic chart.

4.18 EFFECTS OF LARGE-SCALE VERTICAL MOTION ON THE LAPSE RATE AND ON STABILITY

Referring to Fig. 4.10 again, if an element of dry air is lifted and cooled adiabatically, the temperature of the surroundings, judging from the actual temperature distribution, decreases at a rate that is slower than the adiabatic rate. The displacement of this parcel attains, at its new elevated position, a temperature lower than that of the ambient air. The pressure of the parcel will equalize rapidly with that of the environment. However, the density and weight of the displaced element will be greater than those of the surroundings. We saw that this will tend to return the parcel to the position from which it originated. This explanation assumes that the motion of the displaced parcel did not influence the state of the ambient surroundings.

When large-scale vertical motion takes place, the displaced masses will eventually exchange some heat with the environment, by mixing or radiation, and they will follow a path on the T–z or the T–p^{κ} coordinates intermediate between the adiabatic and the actual lapse rate. Continued up-and-down motion will equalize this difference and have no further effect on the lapse rate. Consequently for an environment originally subadiabatic, the lower levels are heated and the upper levels are cooled in the process.

In the case of the atmosphere, the situation just described was for dry air. If moist air is forced upward by winds in the vicinity of a mountain or by saturated atmospheric conditions, it will be cooled pseudoadiabatically until the vapor pressure equals the saturation pressure. When the air reaches the top of the mountain, as shown in Fig. 4.16, it is cooled and a cloud is formed. When

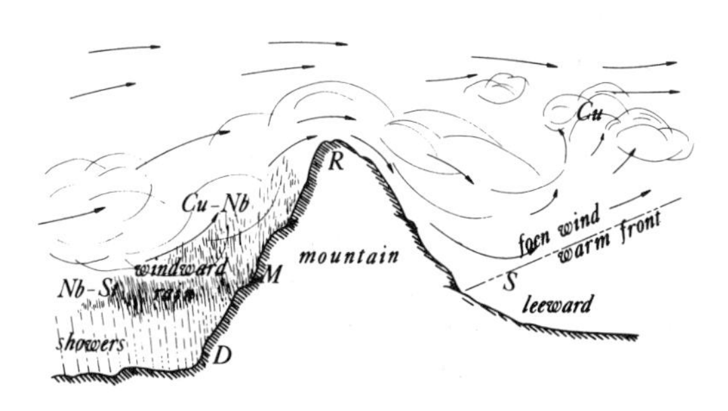

Fig. 4.16 Foehn or chinook wind over a mountain.

precipitation begins, the mass of water or the specific humidity reduces and warms up by absorbing the latent heat of condensation released by the vapor during precipitation, and when the air descends on the lee side of the mountain it warms up pseudoadiabatically according to the saturated conditions until it is no longer saturated. Then, further descent warms the air adiabatically, making the temperature on the leeward side of the mountain considerably warmer and much drier because of the loss of moisture due to some precipitation during the entire thermodynamic process from one side of the mountain to the other. This temperature difference could be as much as 25°C. The warmth and dryness of these *foehn* or *chinook* winds on the eastern slope of the Rockies are noted for being able to melt as much as 50 cm of snow in 24 h. The same phenomenon is observed north of the Alps and in the Himalayas.

In conclusion we can say that air rising adiabatically in the troposphere will bring cooler temperatures if the actual lapse rate is subadiabatic. The reverse will take place if cooling is done along the pseudoadiabat for saturated conditions. The direction of the temperature changes will be opposite in the descending motions.

4.19 VARIATION OF LAPSE RATE AND TEMPERATURE DUE TO VERTICAL MOTION

The results of the discussion in the preceding section are independent of the scale and speed of the vertical motion. Let us consider large-scale vertical motions, through an adiabatic process, and let us consider their effect on the temperature and pressure fields. We have seen that rising air will be cooler than a subadiabatic environment and warmer than a superadiabatic environment. Normally since the troposphere is subadiabatic, the rates of cooling of temperatures can be estimated from the acceleration equation (4.45). Since g/a in that equation will be constant under a given situation, taking the derivative of the equation with respect to time yields

$$\frac{dT}{dt} \sim -\frac{dz}{dt}(\alpha_{ad} - \alpha) = -W(\alpha_{ad} - \alpha) \tag{4.59}$$

where W is the vertical velocity defined as positive upward. When the conditions are those of saturated air α_{ad} is replaced by α_{ads}. We have established that $\alpha_{ad} = 9.66°\text{K/km}$ and that the standard US atmosphere for mid-latitude, spring–fall, has a lapse rate of $\alpha = 6.5°\text{K/km}$. If we assume $W = 0.2$ km/h and g/a of the order of unity, then

$$\frac{dT}{dt} \quad \text{is of the order of} \quad -0.6°\text{K/h}$$

As seen from Fig. 3.9 this heat convection accounts approximately for 18 units, nearly the amount the atmosphere receives directly from the sun's radiation.

The following is a method of calculating the change in the lapse rate of the atmosphere due to vertical motions. During unstable conditions air rises with a resulting acceleration, or an increase in velocity, and the cross section of the air column must decrease with altitude.[15] The ascending air is under expansion; the geometry and properties of the column are shown in Fig. 4.17.

[15] We mentioned in Section 2.18 that these vertical components of wind may increase at the rate of 0.17 m/s · km, sometimes up to the thermosphere. See also Illustrative Example 6.2.

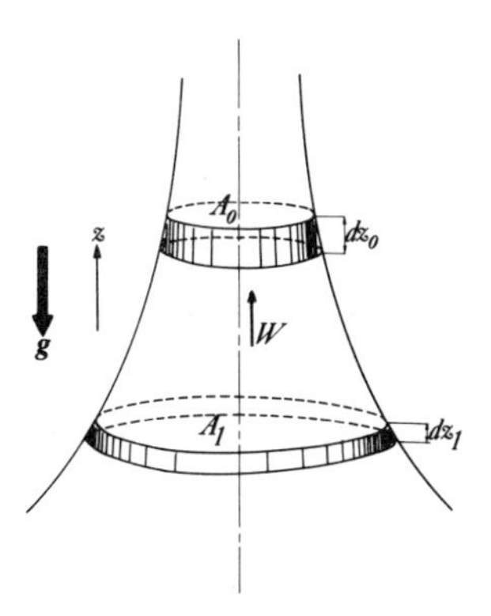

Fig. 4.17 Ascending air column.

Conservation of mass requires that the rate of mass flow crossing section 1 must equal that crossing at 0. This mass rate is easily seen to be represented by

$$\frac{d(\rho_0 A_0\, dz_0}{dt} = \frac{d(\rho_1 A_1\, dz_1)}{dt}$$

and since z is the only quantity varying with time in a steady motion,

$$\rho_0 A_0 W_0 = \rho_1 A_1 W_1 \tag{4.60}$$

We have seen from Eq. (4.51) that the potential temperature is not explicitly dependent on p or T; for this reason we shall choose it as our variable. We can write

$$\left(\frac{d\theta}{dz}\right)_1 = \left(\frac{d\theta}{dz}\right)_0 \frac{dz_0}{dz_1}$$

This assumes that the distance between 1 and 0 is infinitesimal. In the same lapse of time $dz_0/dz_1 = W_0/W_1$. Thus

$$\left(\frac{d\theta}{dz}\right)_1 = \left(\frac{d\theta}{dz}\right)_0 \frac{W_0}{W_1} = \left(\frac{d\theta}{dz}\right)_0 \frac{A_1 p_1 T_0}{A_0 p_0 T_1}$$

The conservation of mass and the gas law equations have been used. From Eq. (4.52) we can also write

$$\frac{\theta}{T_1}(\alpha_{\text{ad}} - \alpha_1) = \frac{\theta}{T_0}(\alpha_{\text{ad}} - \alpha_0)\frac{A_1 p_1 T_0}{A_0 p_0 T_1}$$

Simplifying,

$$\alpha_1 = \alpha_{\text{ad}} - \frac{A_1 p_1}{A_0 p_0}(\alpha_{\text{ad}} - \alpha_0)$$

If the change of cross section can be ignored in comparison with the change of pressure, then

$$\alpha_1 = \alpha_0 - (\alpha_{ad} - \alpha_0)\frac{\Delta p}{p} \tag{4.61}$$

where $\Delta p = p_1 - p_0$. Let us consider the following cases, where in each case the state 0 refers to the *former state* in the displacement.

A Originally Stable Atmosphere: $\alpha_{ad} > \alpha_0$ and $\Delta p > 0$

According to the shape of the column given in Fig. 4.17 and Bernoulli's equation[16] the motion is downward since $\Delta p > 0$ or $p_1 > p_0$ and $A_1 > A_0$. According to Eq. (4.61), $\alpha_1 < \alpha_0$ or the lapse rate becomes smaller and further away from the adiabat. We conclude that *vertical descent increases the departure from the adiabatic lapse rate*. The value of Δp could be sufficiently large to make α_1 zero or negative if this mixing spreads out to encompass a large mass of air. An *inversion* will form by the sinking and spreading of the air. *Thus a descending motion of air that was originally stable makes that column even more stable.*

B Ascending Air $\Delta p < 0$ of Originally Stable Air: $\alpha_{ad} < \alpha_0$

Again from Eq. (4.61), in the latter state 1, $\alpha_1 > \alpha_0$ or the lapse rate increases and gets closer to the adiabatic lapse rate.

C An Unstable Situation

When $\alpha > \alpha_{ad}$, which might be the case under saturated conditions, the sign of Δp produces effects opposite to those described under B above.

D Adiabatic State

When $\alpha_0 = \alpha_{ad}$ it is obvious that the pressure differential does not affect the lapse rate.

Now going back to Fig. 4.16 let us follow the thermodynamic path and process the air follows over the mountain. Let us assume that the mountain is 2.25 km high and that the state of the air on the windward side is approximately 280°K

[16] This equation is not derived until Section 8.2. However, all we need to know at this time is that when the pressure drops the velocity must increase, and vice versa.

and 75% relative humidity, or 0.0047 g/g absolute humidity. This complies with the conditions in B. An ascent will make the temperature lapse rate larger and therefore closer to the adiabatic. The process is shown along DM in Figs. 4.11 and 4.18. At the elevation M, precipitation is likely to occur and at that altitude the saturation humidity ratio is approximately 0.0047 g/g as seen from Fig. 4.18. Since each gram of vapor releases a latent heat of 2500 J, the ambient air will be

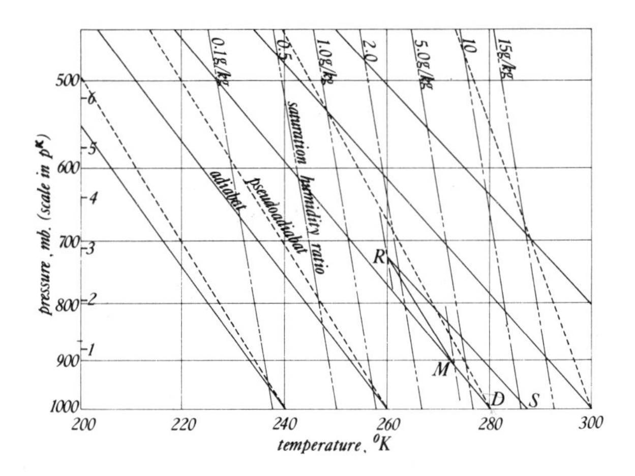

Fig. 4.18 The pseudoadiabatic chart.

heated. Thus from M the rise must proceed at higher temperatures than adiabatic. This process is shown by MR. At R, the peak of the mountain is reached, and the drier air begins to descend. This condition will be equivalent to that of A. When the air reaches the plain on the leeward side at S, it will have a temperature 7°C higher than that on the windward side and be considerably drier, about 20%. Section 4.21 treats this problem in more numerical detail.

4.20 TYPES OF COORDINATES, DIAGRAMS AND CHARTS REPRESENTING THERMODYNAMIC PROCESSES OF THE ATMOSPHERE

In the previous sections we have shown many different diagrams representing the thermodynamic processes of the atmosphere. In meteorological literature there are a number of convenient coordinate systems that are used. It is not possible, within the scope of this book, to give a thorough familiarization with each and every diagram that is used. Instead, we choose to discuss the general theory behind these charts and diagrams.

Whichever the choice of coordinate representation, we know that from a pure substance point of view, the state of air and water must be given in terms of two independent thermodynamic properties. Including the dependent variable, the total number of variables involved is three and on two-dimensional diagrams the third variable must be represented as a process line (isopleth) on which the process or one variable remains constant. Since there are many different processes and many thermodynamic properties, we may conclude that there could be many ways of representing the same state.

In choosing a specific kind of coordinate system, for diagrams that have a practical use, we prefer to select this coordinate representation in such a way that the main processes of interest become straight lines or nearly straight lines for ease of interpolation and computation.

Obviously, another argument in this choice of coordinates is that areas which represent energy in transfer are conserved from one diagram to the other. For instance, in a $(p, 1/\rho)$ representation or any other general (ξ, η) representation, the areas under the same closed process must be conserved. One obvious choice for another coordinate system is $(T\text{–}s)$, which according to the first law of thermodynamics is $\oint pd\,(1/\rho) = \oint T\,ds$. Nevertheless, in general, we must have

$$\oint p\,d(1/\rho) = \oint \xi\,d\eta$$

and the relationship between these variables is obtained by rewriting

$$\oint [p\,d(1/\rho) - \xi\,d\eta] = 0$$

which implies that the integrand must be an exact differential of another property Π as a consequence of the line integral theorem. Thus

$$p\,d(1/\rho) - \xi\,d\eta = d\Pi \tag{4.62}$$

In this equation, if ξ, for instance, is replaced by T and η by the entropy s, then Π can only be the negative value of the internal energy, according to the first law. In general we write

$$\Pi = f(v, \eta)$$

where $v = 1/\rho$, the specific volume defined in Section 2.9. By differentiation of Π we also write

$$d\Pi = \left(\frac{\partial \Pi}{\partial v}\right) dv + \left(\frac{\partial \Pi}{\partial \eta}\right) d\eta$$

and comparing with Eq. (4.62) we have

$$p = \left(\frac{\partial \Pi}{\partial v}\right) \qquad \text{and} \qquad \xi = -\left(\frac{\partial \Pi}{\partial \eta}\right) \tag{4.63}$$

The conditions for an exact differential are

$$\left(\frac{\partial p}{\partial \eta}\right)_v = -\left(\frac{\partial \xi}{\partial v}\right)_\eta \tag{4.64}$$

As long as this relationship is maintained the areas will be conserved in the two coordinate systems (p, v) and (ξ, η).

Let us proceed at applying these relations to reinvent, so to speak, other coordinate representations. Choosing $\eta = T$ from the left-hand side of Eq. (4.64)

$$\left(\frac{\partial p}{\partial T}\right)_v = \frac{R}{v}$$

for a perfect gas and it follows also that

$$-\left(\frac{\partial \xi}{\partial v}\right)_T = \frac{R}{v}$$

Before integrating for ξ we must remember that it is a partial differential equation and that we shall have *arbitrary functions* of T instead of an arbitrary constant. Thus

$$\xi = -R \ln v + f(T) \tag{4.65}$$

If we take $f(T) = R \ln R + R \ln T$, for instance, Eq. (4.65), using the equation of state, becomes

$$\xi = R \ln p \tag{4.66}$$

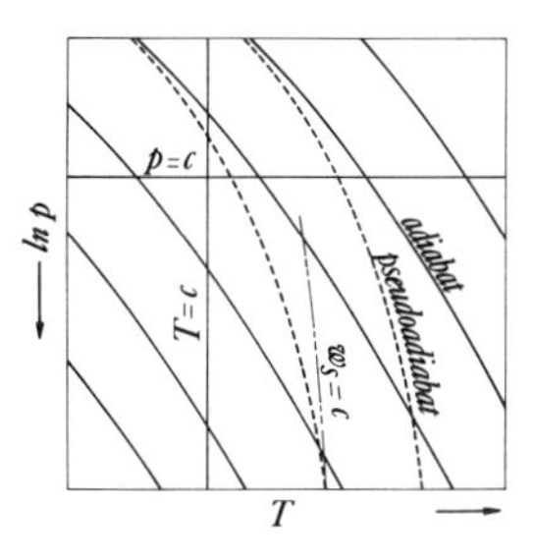

Fig. 4.19 The emagram ($\ln p$, T).

Therefore $\ln p$ and T are two compatible new coordinates. We did not use these coordinates in any of the previous representations, but we used (p^κ, T) and (z, T) which obviously do not conserve the areas.

This new system ($\ln p$, T) is called an *emagram*. Figure 4.19 shows this coordinate system in which adiabats, pseudoadiabats, and a saturated humidity ratio are shown. Since we already know that an adiabat in a (T, p^κ) system is linear, it cannot be linear in the emagram. However, the slopes of the adiabats in the emagram do not change very much, which is an advantage for interpolation.

Another kind of diagram can be obtained from Eq. (4.65) if we substitute the equation of state for υ and use the definition of the potential temperature given in Section 4.16. Thus

$$\frac{\theta}{T} = \left(\frac{1000}{p}\right)^\kappa = \left(\frac{1000\upsilon}{RT}\right)^\kappa$$

Taking the logarithm and solving for $\ln \upsilon$ we have

$$\begin{aligned} R \ln \upsilon &= c_p \ln \theta + R \ln\left(\frac{R}{1000}\right) - \ln T^{(R+c_p)} \\ &= c_p \ln \theta + f(T) \end{aligned}$$

Substitution into Eq. (4.65) yields

$$\xi = -c_p \ln \theta \tag{4.67}$$

and

$$\eta = T \tag{4.68}$$

Since constant θ implies constant entropy, this system is called a *tephigram*.[17] In this diagram $(\ln \theta, T)$ are the two orthogonal coordinates. Again, from the potential temperature definition for an isobar $\ln \theta = \ln T +$ constant. Therefore, in this diagram, isobars are logarithmic curves with small change of slope. Figure 4.20 shows the same information as the emagram.

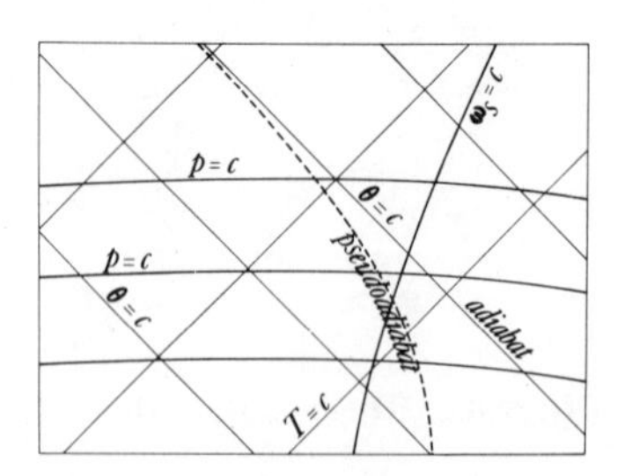

Fig. 4.20 The tephigram $(\ln \theta, T)$.

[17] Sir Napier Shaw, who introduced this system, identified the entropy with the symbol ϕ and called it the T–ϕ diagram or the tephigram. This is the same as the T–s diagram.

A third type of area-conserving system is obtained when $\eta = -R \ln p$. Then from Eq. (4.64)

$$-\left(\frac{\partial \xi}{\partial v}\right) = -\frac{\partial p}{R\,\partial(\ln p)}, \qquad \frac{\partial \xi}{\partial v} = \frac{p}{R}$$

Integration of this partial differential equation yields

$$\xi = \frac{pv}{R} + f(p) = T + \ln p \tag{4.69}$$

We chose the arbitrary function $f(p)$ as $\ln p$ because we took it as one of the coordinates. Thus this coordinate system is composed of ($\ln p$, $T + \ln p$). This is called the *skew T–log p* diagram. The isobars are straight lines; the isotherms, according to Eq. (4.69), are represented by the equation $\xi = T - \eta/R$, which are also straight lines. The adiabats are obtained by taking the logarithm of the definition of the potential temperature

$$\ln T = (R/c_p) \ln p + \text{constant}$$

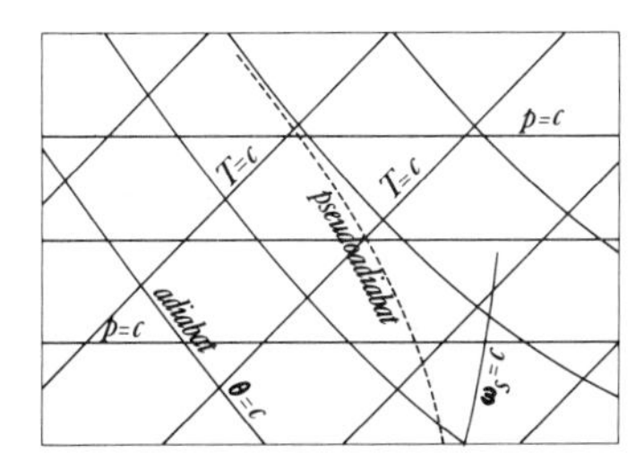

Fig. 4.21 The skew T–log p diagram ($\ln p$, $T + \ln p$).

These are not straight lines. Figure 4.21 diagrammatically shows the makeup of this coordinate system with the same information shown as in the other diagrams. Meteorologists use this diagram extensively. Appendix B gives a standard skew T–log p diagram with numerical values, with a standard atmosphere plotted on it.

4.21 THE PSEUDOADIABATIC CHART

Figure 4.18 is the *pseudoadiabatic chart* with coordinates (T, p^κ). Although, as we have just seen, this system does not conserve the areas, it has the significant advantage that adiabats, isotherms, and isobars are straight lines. From the definition of Eq. (4.51), $T = Cp^\kappa$, the inverse of the constant C is the slope of the adiabats and varies from one adiabat to the other. This constant can be computed at the 1000-mb abscissa by substituting the values of T through which an adiabat passes. These constants are given in Table 4.6.

TABLE 4.6

T (°K)	240	260	280	300	320
C	34.45	37.39	40.19	43.06	45.93

The saturated adiabatic lines or the *pseudoadiabats* can be calculated according to Eq. (4.41) beginning with

$$-dT = (g/c_p)\, dz + (L/c_p)\, d\omega_s$$

From the hydrostatic equation and the perfect gas law, dz can be replaced by the pressure change. Thus

$$dT + \frac{L}{c_p}\, d\omega_s = \frac{RT}{c_p}\frac{dp}{p}$$

Integrating, we obtain

$$\ln\left[\left(\frac{p}{p_0}\right)^{\kappa}\left(\frac{T_0}{T}\right)\right] = \int \frac{L}{Tc_p}\, d\omega_s \tag{4.70}$$

which is another form of Eq. (4.57). For dry adiabats the integral is zero because of Eq. (4.48). The subscript zero refers to an initial state. For drawing the pseudoadiabats in Fig. 4.18 these initial states are taken where the dry adiabats intersect the abscissa, that is, at 1000 mb and 240, 260, 280, and 300°K. The saturation humidities ω_{s0} and ω_s are functions of the saturation pressure and the atmospheric pressure, as we have seen:

$$\omega_s = 0.6215\, \frac{p_s}{p - p_s}$$

and

$$\ln p_s = 21.548 - \frac{5388}{T}$$

Thus ω_{s0} values are computed from these two equations. With $\omega_s = f(p, T)$ and Eq. (4.70) a numerical calculation can be carried out for p and T values along the pseudoadiabats which are plotted in Fig. 4.18.

These pseudoadiabats have steeper slopes than the dry adiabats because the heat of condensation gives higher temperatures at the same pressure. This departure from the dry adiabat is larger the higher the humidity ratio. This is the reason they have a slight curvature since they cross different values of ω_s. On the same chart there are lines of constant specific humidity. To obtain these lines we use the last two relationships for values of p and T for constant humidity.

Returning to the problem of the chinook wind discussed in Section 4.18, let us follow the state of the air as it rises over the 2.25-km mountain. We assume that the air on the windward side, at ground level, has a pressure of 1000 mb, a temperature of 280°K, and a relative humidity of 75%. Point D in Figs. 4.11 and 4.18 describes the state of the air at ground level.

The partial pressure of the vapor can be computed from the definition of the relative humidity to be $p_w = 7.54$ mb and the corresponding humidity ratio is $\omega = 4.7$ g/kg. The saturation mixing ratio $\omega_s = 6.3$ g/kg. When air rises, as it is not completely saturated, it follows an adiabat from D and in the process maintains the humidity ratio of 4.7 g/kg until it reaches an elevation corresponding to a pressure where it becomes saturated. This is indicated as the point M which corresponds to a pressure of 910 mb and an elevation of nearly 1 km. Condensation begins at this point which is the intersection of the adiabat beginning at 280°K and the line of $\omega_s = 4.7$ g/kg. At this point we can compute p_s, T, and p.

As air continues to climb and condensation occurs, it will be in a state of saturation and from point M will follow a pseudoadiabat (parallel to the dashed lines in Fig. 4.18). If this continues to the top of the mountain, which is at 720 mb and shown as the point R, the temperature of the air will be at 260°K and there will be a saturation humidity ratio of 1.9 g/kg. The difference of the mixing ratios of the air at point D and that at R ($4.7 - 1.9 = 2.8$ g/kg) is the amount that has precipitated.

When air begins to descend on the other side of the mountain it will do so along the dry adiabat passing through the point R, because in the downward direction the humidity ratio will be smaller than what is required for saturation at any of the lower levels. The final point S at 1000 mb on the lee side will be at approximately 287°K, a gain of 7°C, and the moisture it contained at the top point R will remain at the lower humidity ratio of 1.9 g/kg, while the saturation humidity ratio at the point S is 10 g/kg. The relative humidity will be approximately 20%. Thus air left the windward side at 75% humidity and came down on the lee side at 10%.

5

Basic Principles of Surface Tension

5.1 INTRODUCTION

The phenomenon of surface tension displays itself when dealing with problems involving the interface between two different fluids or a fluid and a solid surface. The interface behaves in a way similar to a thin stressed membrane under tension. From the molecular point of view, this difference of behavior of the interface compared to that of the rest of the fluid can be explained by considering the balance of cohesive forces between molecules on either side of the interface which are different because of the difference in molecular structure. Thus the surface tension properties of an interface can be changed by dissolving other fluids at the interface.

The surface a liquid displays at the top level in a container is called a *free surface*. This free surface is one of the characteristics that differentiates a gas from a liquid. The slight rise or depression of this free surface near the wall of the container is a consequence of surface tension. The formation of droplets and soap bubbles and the state of equilibrium of a drop of liquid on a solid surface are due to surface tension. Lenses of fat on the surface of a good broth are maintained in equilibrium because of surface tension.

The fact that the surface tension force will act to stretch the surface of a liquid film can be shown through the following example. A thin loop of thread is tied to a rigid loop of wire, as shown in Fig. 5.1. In Fig. 5.1a, the soap film in the wire loop is shown to have formed all the way through the thread loop. If the soap film in the thread loop is pierced, the remaining film will pull the string all the way around, until the soap film attains its smallest possible surface, as shown

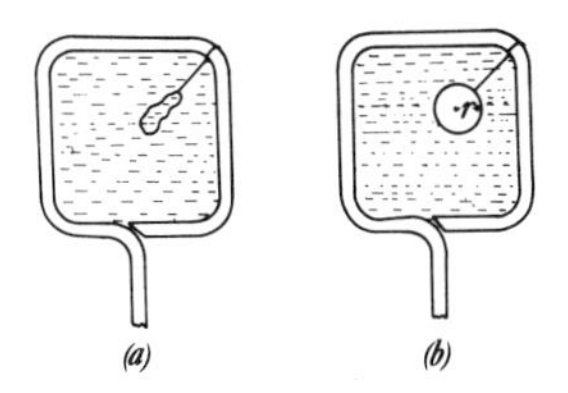

Fig. 5.1 Thin film on a loop.

in Fig. 5.1b. The tension on the thread loop is maintained by the surface tension. Similarly, in the absence of other external forces, a liquid drop will take a spherical shape which is the volume that has the minimum area. A drop resting on a solid dry surface is slightly out of round because of its own weight; the bigger the drop the more the drop is flattened. A falling drop is deformed from its spherical shape as a result of the drag resistance during the fall.

The formation and destruction of droplets of liquid, principally water, in the atmosphere is a very frequent event. The surface of these free droplets behaves somewhat like a stretched membrane, in such a manner that work must be done in order to alter the shape of the free surface. Since drops seek their smallest surface, the fluid must experience a force toward its center in order to prevent it from taking any other form. This force is called the *surface tension force*, and as we said, it is associated with the cohesive force structure of molecules on both sides of the interface. The attraction between molecules decreases as the distance separating them increases, and consequently it can be neglected in the case of gases.

Inside the liquid the cohesive forces compensate each other, but those on the surface, or interface, are acted upon by their own kind of molecules on one side to prevent them from escaping while the other side of the interface will contain different kinds of molecules and different cohesive forces. The surface will find a final equilibrium shape, which for a free surface is a sphere. This surface tension force is what keeps the two halves of a bubble together or what maintains a liquid in a capillary column, as shown in Fig. 5.2.

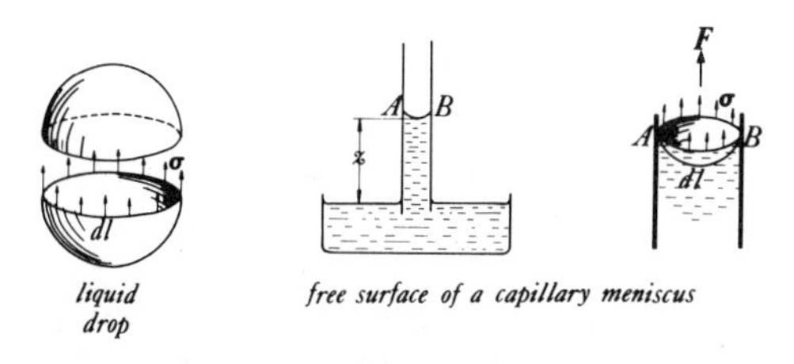

Fig. 5.2 Free liquid surfaces.

The *surface tension coefficient* is defined as the force per unit length of line, dl, on the free surface of a liquid necessary to hold that surface in equilibrium. Thus

$$F = \int \sigma \, dl \tag{5.1}$$

The symbol σ is the surface tension coefficient, which depends on the properties of the free surface, the liquid surface, and the surroundings. Since the spacing between molecules depends on the temperature, the surface tension coefficient is a function of temperature. Table 5.1 gives some typical values of surface tension coefficients for various liquids in different surroundings.

TABLE 5.1

Surface Tension Coefficients

Substance	Temperature (°C)	surface tension coefficient (dynes cm)	Surrounding fluid
Acetone	20	18.6	Air
NaCl in water (g/g)			
0.10	20	69.5	Air
0.35	20	63.0	Air
Water	0	75.6	Air
Water	20	72.8	Air
Water	100	58.9	Air
Mercury	15	487	Air
Mercury	20	375	Water
Silver	970	800	Air
Lead	350	450	Air
Alcohol	0	25	Air
Glycarol	20	63.4	Air
Octane	20	50.8	Water
Benzene	20	35	Water
Ether	20	17.0	Air
Benzene	20	28.9	Air

5.2 MECHANICAL EQUILIBRIUM OF THE FREE SURFACE

Consider a completely closed surface, such as a bubble or a raindrop, and let us consider an infinitesimal portion of this surface, as shown in Fig. 5.3. In general, the curvature of the free surface may not be homogeneous, so that A_1B_1 may have a different curvature than A_2B_2. The forces keeping the film in the equilibrium condition shown are the surface tension forces on all its sides and of these forces. Let p be this internal pressure on the film and p_0 the outside

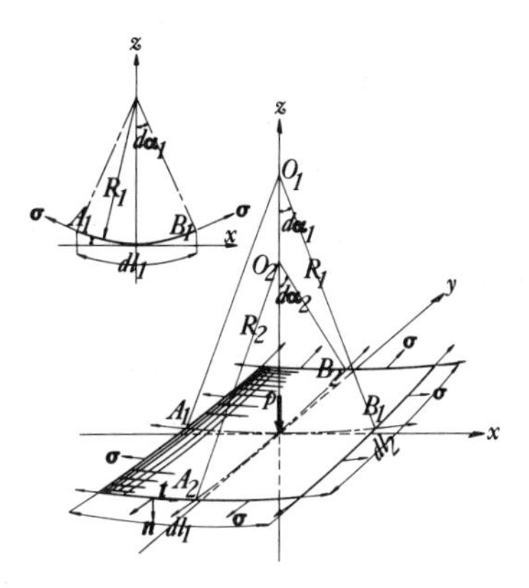

Fig. 5.3 Forces on a curved fluid film.

the internal pressure force which eventually must balance the radial components ambient pressure. Owing to equilibrium conditions on a plane normal to the radius, the components of the surface tension forces in the x and y directions must cancel. Thus the equilibrium in the radial direction is

$$(p - p_0)\, dl_1 dl_2 = 2\sigma\, dl_2 \sin d\alpha_1 + 2\sigma\, dl_1 \sin d\alpha_2$$

For an infinitesimal geometry such as the one in Fig. 5.3, $\sin d\alpha_1 \simeq d\alpha_1 \simeq dl_1/2R_1$. The same is true for the other direction. When these values are substituted into the equilibrium equation, the internal pressure of the film related to the radii of curvature is found to be

$$p - p_0 = \sigma\left(\frac{1}{R_1} + \frac{1}{R_2}\right) \tag{5.2}$$

This expression indicates that the pressure difference is inversely proportional to the radius of curvature. Ideally and, as we shall see, actually, when the radius of curvature is zero the pressure inside the drop is infinite. This explains why gas bubbles (in boiling) never start in the midst of the liquid but seek nucleation sites that have a finite roughness or rugosity upon which to form the bubble, and that in the formation of clouds it is necessary to have dust particles or other droplets to serve as nucleation sites. Otherwise it would be impossible to develop an infinite pressure in order to start a bubble from zero radius.

The principal radii of curvature R_1 and R_2 are positive if the center of curvature lies inside the bubble or drop. In the case of spherical drops $R_1 = R_2$ and the pressure inside the drop is

$$p - p_0 = 2\sigma/R \tag{5.3}$$

Surface tension plays a major role in condensation and evaporation. The surface energy on a curved fluid film plays an important part in the equilibrium vapor pressure and on the rate of evaporation from water droplets.

5.3 MECHANICAL EQUILIBRIUM OF AN INTERFACE BETWEEN TWO PHASES

As long as we do not cross the interface between two phases, the equilibrium is controlled by the hydrostatic equation on both sides of the interface. For equilibrium crossing the interface we must apply the surface tension equilibrium derived in Eq. (5.2).

For an application of these principles consider first the two phases of Fig. 5.4 near a solid boundary, and a point P on the interface. In the heavier phase with

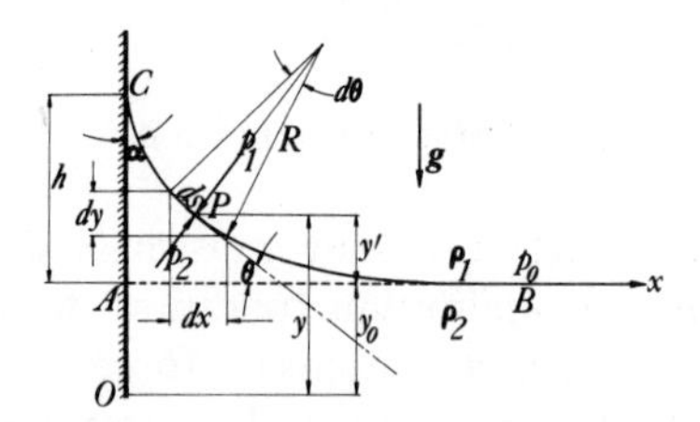

Fig. 5.4 The geometry of the interface.

density ρ_2 the pressure p_2 at the point P can be written from hydrostatics as $p_2 = p_0 - \rho_2 g(y - y_0)$. As long as we do not cross the interface, at the same point in the other phase $p_1 = p_0 - \rho_1 g(y - y_0)$. Then without having to cross the interface $p_1 - p_2 = g(\rho_2 - \rho_1)y'$. In the environment it is likely that the phase with density ρ_2 will be water and the phase with density ρ_1 will be air or vapor. Thus if we let $\rho_2 = \rho_w$, the density of water, and if the density of air is too small compared to ρ_w then

$$p_1 - p_2 = \rho_w g y' \tag{5.4}$$

Without having to cross the interface, the hydrostatic equation already states that $p_1 > p_2$. Now crossing the interface, Eq. (5.2) also gives

$$p_1 - p_2 = \sigma\left(\frac{1}{R_1} + \frac{1}{R_2}\right) \tag{5.5}$$

where R_1 and R_2 are the two radii of curvature at the point P. We can equate the two independent results, so that if we let $a^2 = \sigma/\rho_w g$, called *Laplace's capillary constant*, we have

$$\sigma\left(\frac{1}{R_1} + \frac{1}{R_2}\right) - \rho_w g y' = 0 \tag{5.6}$$

or if we introduce any other reference level O from which the point P is a distance y from O, and since y_0 on the figure is constant, Eq. (5.6) becomes

$$\frac{1}{R_1} + \frac{1}{R_2} - \frac{y}{a^2} = \text{constant} \tag{5.7}$$

In particular if one of the radii of curvature is very large compared with the other, then taking a single radius of curvature R, Eq. (5.7) becomes

$$1/R = y/a^2 \tag{5.8}$$

if the reference level is moved to the line AB on the figure.

This equation can be resolved in two ways. First from the geometry of the figure, a small element of interface $ds = R\,d\theta$ and $1/R = d\theta/ds$. Then

$$d\theta/ds = y/a^2 \tag{5.9}$$

and since $ds \sin\theta = dy$, we have a simple differential equation to integrate:

$$\sin\theta\,d\theta = y\,dy/a^2$$

which yields

$$y^2 + 2a^2 \cos\theta = \text{constant}$$

The boundary condition at $y = 0$ requires that $\theta = 0$ and $\cos\theta = 1$ which gives the value of $2a^2$ for the constant of integration. The equation of the contour is then

$$y^2 = 2a^2(1 - \cos\theta) \tag{5.10}$$

In particular at the point C the contact angle with the three phases is $\alpha = 90 - \theta$. Thus the height of the interface h at the solid surface is

$$h^2 = 2a^2(1 - \sin\alpha) \tag{5.11}$$

Beginning with Eq. (5.8) we could have written an expression for $1/R$ in terms of the first and second derivatives of the interface equation $y = f(x)$. Instead of the solution (5.10), this will give us a contour equation in x and y. From differential geometry

$$\frac{1}{R} = \frac{d^2y/dx^2}{[1 + (dy/dx)^2]^{3/2}}$$

and then

$$\frac{y}{a^2} - \frac{y''}{(1 + y'^2)^{3/2}} = 0 \tag{5.12}$$

Integrating once gives the result

$$C - \frac{2y^2}{a^2} = \frac{1}{(1 + y'^2)^{1/2}}$$

The constant of integration $C = 1$, since $y = y' = 0$ at $x = \infty$, where the primes indicate derivatives with respect to x. Another integration of Eq. (5.12) gives the final solution sought:

$$x = a\left(4 - \frac{y^2}{a^2}\right)^{1/2} - a \cosh^{-1} \frac{2a}{y} + x_0 \tag{5.13}$$

The constant of integration x_0 can be evaluated from the condition that at $x = 0$, $y^2 = h^2 = 2a^2(1 - \sin \alpha)$, as found in Eq. (5.11).

5.4 THE HANGING DROP

Consider the drop shown in Fig. 5.5, hanging by being supported either by electrostatic forces, as is the case in some cloud formations, or by surface tension forces on a rigid surface. Again as in the case of the preceding section, we can write equilibrium conditions within one phase through the hydrostatic equation, and when crossing the interface, we make use of surface tension equilibrium considerations.

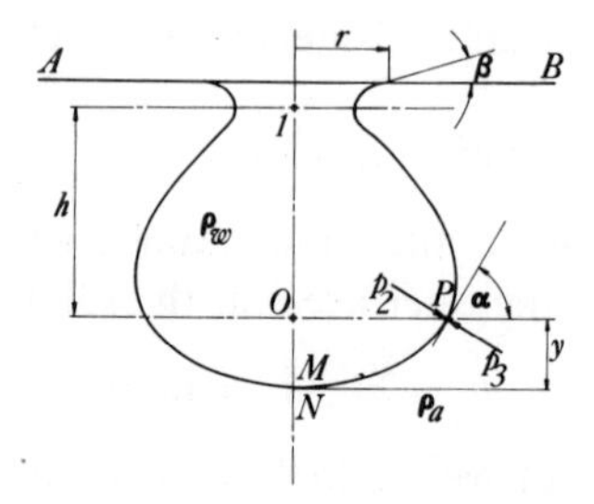

Fig. 5.5 The geometry of the hanging drop.

It is clear from the figure that within the drop $p_2 = p_M - \rho_w g y$ and outside the drop $p_3 = p_N - \rho_a g y$. The reference point has been chosen as the bottom of the drop. This is not necessary. The point 1 at the neck of the hanging drop could have been chosen equally well, as we shall soon see when we need to cross the interface for an additional relation.

The advantage in choosing M as the reference level is the likelihood of having one radius of curvature at the point M due to the axisymmetry of the drop. Crossing the interface at P we have

$$p_2 - p_3 = \sigma\left(\frac{1}{R_1} + \frac{1}{R_2}\right) \tag{5.14}$$

Also crossing the interface at N and M we have

$$p_M - p_N = \sigma \frac{2}{R_M} \tag{5.15}$$

If we had chosen the point 1 as a reference, when we crossed the interface at the neck, we might have needed two radii of curvature if both were of the same order of magnitude.

Now we can combine the four equations we have developed to obtain

$$\sigma\left(\frac{1}{R_1} + \frac{1}{R_2}\right) - \frac{2\sigma}{R_M} + \rho_w g y = 0 \tag{5.16}$$

by taking into account that $\rho_a \ll \rho_w$. Another advantage in choosing the point M as a reference is that it is likely that the radius of curvature R_2 is very near in value to R_M at all points P. This is the radius on the plane normal to the axis OM. In that case Eq. (5.16) simplifies to

$$\frac{\sigma}{R} - \frac{\sigma}{R_M} + \rho_w g y = 0 \tag{5.17}$$

recalling that R_M is the limit of R as y tends to zero, and is constant for a given drop configuration.

The procedure used to obtain Eq. (5.12) can be applied here for a solution. We have additional information available, if necessary, at the upper limit AB, where we know that the weight of the entire drop is supported by the surface tension force (vertical component) at the contact perimeter $2\pi r$.

The equilibrium condition of the entire drop is obtained by equating the weight of the drop $\rho_w g \mathscr{V}$ to the vertical component of the surface tension force $2\pi r\sigma \sin\beta$. The volume of the drop $\mathscr{V}$ is also dependent on $y = f(x)$, the contour of the drop, which is to be obtained from Eq. (5.17).

5.5 ENERGY CONSIDERATIONS

The surface energy released by a circular drop that is changing radius from r to $r + dr$ can be calculated by finding the work necessary to increase or decrease its volume due to evaporation or condensation. This change in volume is $d\mathscr{V} = 4\pi r^2\, dr$, and the work necessary for this change of volume to occur in the presence of a pressure p is

$$p\, d\mathscr{V} = (2\sigma/r)4\pi r^2\, dr = 8\pi\sigma r\, dr$$

This is comparable to an elastic force on a balloon. This energy is stored in pressure energy when the volume is decreased and can be recovered when the volume increases again.

If this change of volume is caused by evaporation or condensation, the work per unit mass can be found by dividing by the mass of the droplet $4\pi r^2 \rho_w \, dr$. This work is then $(2\sigma)/\rho_w r$, where ρ_w is the density of the liquid drop. In any drop or bubble there are two phases to consider, that of the liquid phase, for which we have just calculated the work against p due to a change of volume resulting from evaporation or condensation, and the gas phase inside the bubble, which may undergo work of the form $p \, d(1/\rho)$. In fact if we wrote the first law for the entire bubble, we will have

$$T \, ds = du + p \, d(1/\rho) + d(2\sigma/r\rho_w)$$

where p and ρ should be given the subscript "v" referring to the vapor state around the drop. The Gibbs function was used in Eq. (4.30) and in the presence of evaporation or condensation, the work due to the loss or gain of liquid mass must be added; thus

$$dG = -s \, dT + (1/\rho_v) \, dp_v - d(2\sigma/r\rho_w)$$

In the change of phase the Gibbs function remains constant, or $dG = 0$, and since this change of phase occurs at constant temperature also, we conclude that

$$\frac{1}{\rho_v} \, dp_v = -\frac{2\sigma}{\rho_w r^2} \, dr$$

Since p_v is the vapor pressure around the drop, we use the gas law $\rho_v = p_v/R_v T$, and integrate from r to r_∞, where r_∞ corresponds to a radius when saturation is reached and p_v becomes p_s. Thus

$$\ln \frac{p_v}{p_s} = \frac{2\sigma}{R_v T \rho_w r}$$

or

$$p_v = p_s \exp\left(\frac{2\sigma}{\rho_w R_v T r}\right) \tag{5.18}$$

This relationship is attributed to Lord Kelvin. From this expression we see that the partial pressure of the vapor is larger than the saturation vapor pressure at the same temperature in order to have equilibrium of the drop and to preserve the energy balance. This is because all the values in the exponent of Eq. (5.18) are positive. In thermodynamically stable states, it should have been the other way around, that the saturation vapor pressure be larger than the limiting value of the vapor pressure. Thus, we conclude that in a water cloud, owing to surface tension, the droplets are in a supersaturated state or a metastable state as discussed in Section 2.15. It should be pointed out that the value of the exponent in Eq. (5.18) is a very small number and that p_v is not very far in value from p_s. Thus the metastable states inside the saturation line are not very far from the saturation line.

If water contains an electrolyte in solution, the equilibrium vapor pressure over the surface of the solution reduces according to Raoult's law. This change is given by

$$p_v/p_s = m/(m + M)$$

where $m/(m + M)$ is the molar concentration of water in the total mass of the electrolyte. For a circular drop, since $(m + M)$ is proportional to r^3, this ratio can be written in terms of the radius of the drop and the mass of the water. It can be shown that water drops carry an electric charge q in space, and that Eq. (5.18) can be modified to read

$$p_v = p_s \exp\left[\left(2\sigma - \frac{q^2}{8\pi r^3}\right)\Big/ \rho_w R_v Tr\right] \qquad (5.19)$$

Since these minute droplets seldom carry more than one electron charge, the influence of the electric charge on p_v/p_s is insignificant.

6

Kinematics of the Environment

6.1 INTRODUCTION

Kinematics is the part of mechanics which deals with quantities derivable from units of space and time. By units of space we imply units that measure position relative to a frame of reference and displacement. For instance, velocity and acceleration are two kinematic quantities. Although forces are the causes for accelerated or decelerated motion, kinematics is solely concerned with the description of motion in terms of displacement, velocity, acceleration, or other quantities derivable from length and time.

The displacement, velocity, or acceleration of a fluid parcel must be described relative to a frame of reference, fixed or moving, which we call the coordinate system. An observer on a bridge will describe the motion of a log on a river with respect to a coordinate system fixed to the earth, but a fisherman drifting in a boat will describe the same motion with respect to his moving frame of reference. The two descriptions will not be the same in details, since at the same time, each observer sees different positions of the log relative to his frame of reference. It is understood that by taking into account the motion of the boat relative to the bridge, the two descriptions can be made to coincide. This is fine for kinematics, but we shall see later that the laws of dynamics are based on an inertial or absolute frame of reference, which we must specify.

We shall define a *Lagrangian coordinate system* as one that is fixed to a moving parcel of the environment and describes the properties of the motion in time. A *Eulerian coordinate system* is one located in a fixed position from which the displacement, velocity, etc., of the parcel are observed at various locations

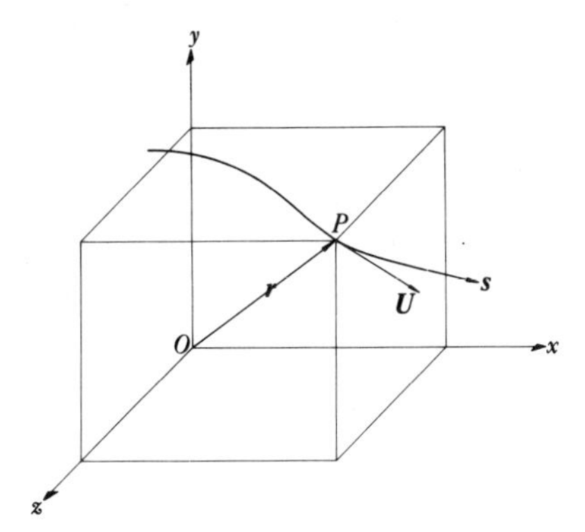

Fig. 6.1 Motion at P with respect to O.

$\boldsymbol{r}$, with time. Following Fig. 6.1, if O is chosen as the origin of this Eulerian system, any point P on the trajectory of a fluid parcel having the coordinates x, y, and z from O can be represented by a single vector distance $\boldsymbol{r}$ which is the vector sum of the coordinate components x, y, and z. The velocity and acceleration of P are also vector quantities which can be expressed in terms of $\boldsymbol{r}$ and t, the time. Reference will be made to these coordinate systems in future developments.

6.2 CLASSIFICATION OF TYPES OF MOTION

At the beginning of Chapter 4 we discussed the types of forces that produce motion of the geofluid. In Chapter 7 we study the types of motion generated by an unbalance of these forces. It is important at this time that we categorize a number of these types of motion.

For instance, there will be many space domains of the environment where viscous forces will affect the motion very little. This is not because the viscosity of the fluid vanishes, but because in such space domains the relative motion of a layer of fluid with respect to its neighboring layer will be so small that the motion can be considered *frictionless*. We discussed compressibility in Section 2.12 and we established in Illustrative Example 2.2 that for a 1% change in volume, a sea parcel requires a pressure increase of 210 bars. For the atmosphere the same percentage in change of volume would necessitate a 14-mb pressure rise. There is no question that we could consider the motion of the ocean as an *incompressible motion* because pressure forces of this magnitude are never behind the motion. But what about the motion of the atmosphere? The principal motion of the atmosphere is confined to the horizontal direction, and as such we must find out if the resultant accelerations can be large enough to generate pressure changes in a given volume of a level to make significant changes in the density or the specific volume of the gas.

For a given velocity U of the atmosphere, the acceleration can be determined from the variation of the kinetic energy from one point to another or $d(\frac{1}{2}U^2)/dx$. This acceleration would have been created by a pressure difference per unit mass $(1/\rho)\, dp/dx$. To tell if this pressure difference is large enough to make ρ a variable in the problem we introduce the compressibility coefficient β of Eq. (2.31) as a yardstick. If

$$\frac{d}{dx}(\tfrac{1}{2}U^2) = -\frac{1}{\rho}\frac{dp}{dx} = -\frac{1}{\rho^2\beta}\frac{d\rho}{dx}$$

where the minus sign indicates that if U^2 increases with x, p must decrease with x to produce the acceleration. The definition of β has been introduced in Chapter 2. If β does not vary much in a given change of velocity and density, we can integrate the equation just given to obtain

$$\beta = 2/\rho U^2$$

Since for a gas we showed in Eq. (2.39) that β must also equal $1/np$, where for an adiabatic compression $n = \gamma$ [as shown in Eq. (4.48)], then if the dynamically produced $\beta = 2/\rho U^2$ approaches the thermodynamical $\beta = 1/\gamma p$, we can certainly say that compressibility may be taken into account since 14 mb is all it takes to produce a 1% change in volume. Thus for the two values of β to be of the same order we must conclude that their ratio must be unity. Thus

$$\frac{\rho U^2}{2\gamma p} = \mathcal{O}(1)$$

We know that the velocity of sound in the gas is given by the relation $c^2 = (\gamma RT)^{1/2} = (\gamma p/\rho)^{1/2}$. Thus the ratio can be seen to be controlled by the square of the *Mach number*, which is defined as U/c (see also Section 7.8). It turns out that unless the Mach number is larger than 0.5, the dynamic effects on compressibility can be neglected. This corresponds to wind speeds not realizable in the troposphere. We conclude that insofar as the motion of the geofluid in the troposphere is concerned, we can treat it as an *incompressible* fluid, also. This does not imply, however, that thermodynamic changes with altitude will not generate considerable changes in the density. We have attested to this in Chapter 4.

Regardless of the level of elastic pressure forces, flows are also classified as *laminar* or *turbulent*, depending on the ratio of the inertia to viscous forces. This ratio is called the *Reynolds number* and as the Mach number controls the level of elasticity, the Reynolds number controls the state of smoothness and turbulence in the flow. Applications in the environment are almost always in a truly turbulent state of motion. There will be many times, however, when we can neglect the effects of turbulence, without having to admit that the flow is void of internal irregularities with respect to time and space.

In the Eulerian sense, a *steady motion* is one in which the properties of the fluid do not vary in time at a given point in space with reference to the Eulerian frame. We shall see that the variation of properties in time and space is partial in the Eulerian sense. An *unsteady or transient motion* does exhibit variations with respect to time.

Since we have said that turbulent flows are irregular in space and in time, we conclude that they cannot be considered steady in the strict sense of the word. However, since the irregularities of turbulent motion are random in nature, implying that the time average of the fluctuations is zero, the definition of steadiness can be extended to include turbulent motions when the time-averaged value of properties within a characteristic time scale is not time dependent. We discussed the nature of these characteristics time scales in Section 2.18. Then a turbulent motion can actually be said to be *quasi steady* or simply steady. This is shown diagrammatically in Fig. 6.2. A detailed discussion on the significance of steadiness in turbulent flows is given in Section 7.12.

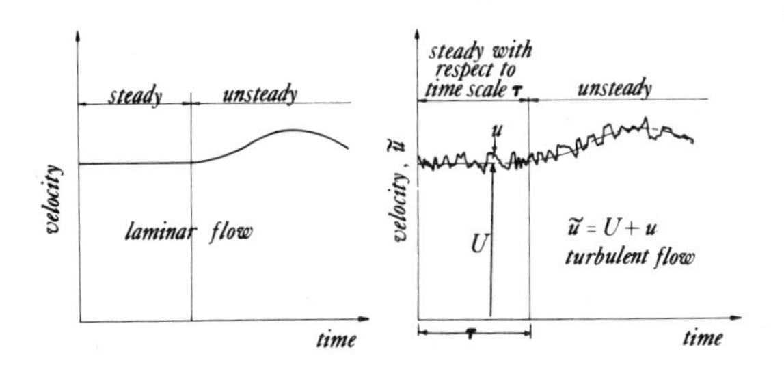

Fig. 6.2 Steady and unsteady laminar and turbulent velocities.

We shall use the following notation to distinguish *instantaneous* values of properties from *mean* and *fluctuating* values: A *tilde* will be placed above the instantaneous value of a property, capital letters will be used for the mean or *temporal average*, and a lower case letter will identify the fluctuating part. For instance, when we speak of velocity in the x direction, we will represent it by

$$\tilde{u}(\boldsymbol{r}, t) = U(\boldsymbol{r}) + u(\boldsymbol{r}, t) \tag{6.1}$$

The averaging process was defined in Eq. (2.67), and is used again in later sections.

By definition, a *one-dimensional motion* is one in which the properties of the motion vary in one direction only. It does not necessarily imply that there is motion in one direction only. The motion and the variation of that motion do not necessarily have to be in the same direction. We shall see later from the conservation of mass point of view that not all variations of velocity can lead to a one-dimensional motion. A one-dimensional motion would, by necessity, imply that on a plane normal to the direction in which the variations are taking place, the motion is *uniform* or constant. One-dimensional flows are not frequently found in the environment.

A *two-dimensional motion* is one in which variations of flow properties occur in two directions. For instance, if we assume that the vertical component and variation of the wind or current are small and negligible compared to the other components and variations, then we have the flow properties varying in two horizontal directions and the motion will be considered two-dimensional. A great number of environmental analyses fall into this category. For example, the flow in a plume out of a stack, being approximately circular in cross section, can be considered *axisymmetric*, implying no variations of properties around the axis of the plume at a given radius from this axis. However, there will be variations of properties with radius and with distance along the axis.

By extension of the preceding, a *three-dimensional motion* permits variations in all three directions. The flow of a shallow and narrow river will give a three-dimensional motion because of the friction layers built on the banks and the bed.

6.3 STREAMLINE, STREAM FILAMENT, STREAM TUBE, AND STREAM SURFACE

A *streamline* is a line drawn in the fluid in such a manner that *all* tangents to this line give the direction of the velocity. Thus by inference this implies that motion normal to a streamline is not possible. A *stream filament* is a bundle of neighboring streamlines forming an infinitesimal cylindrical cross section through which fluid passes. The *stream tube* is a finite-size stream filament with a finite cross section. Updrafts due to local instabilities of the atmosphere, plumes, tornados, and hurricanes can be considered types of stream tubes, provided we can define a *stream surface* across which there is no flow.

Consider Fig. 6.3 in which an environmental flow enters a stream filament at a velocity $\boldsymbol{U}$; after an infinitesimal time dt the volume that has flowed into the filament of cross section dA and unit normal $\boldsymbol{n}$ is

$$(\boldsymbol{n} \cdot \boldsymbol{U})\, dA\, dt = dA_\perp \frac{ds}{dt}\, dt = dA_\perp\, ds$$

where the subscript $\perp$ implies perpendicular to the tube axis and to $\boldsymbol{U}$. The rate of mass flow in the filament is

$$\rho(\boldsymbol{n} \cdot \boldsymbol{U})\, dA = \rho\, |\boldsymbol{U}|\, dA_\perp = \rho U\, dA \cos\theta \tag{6.2}$$

In general, the cross section of the filament may not be perpendicular to the velocity vector and therefore $dA \cos\theta = dA_\perp$, and U is the absolute value or

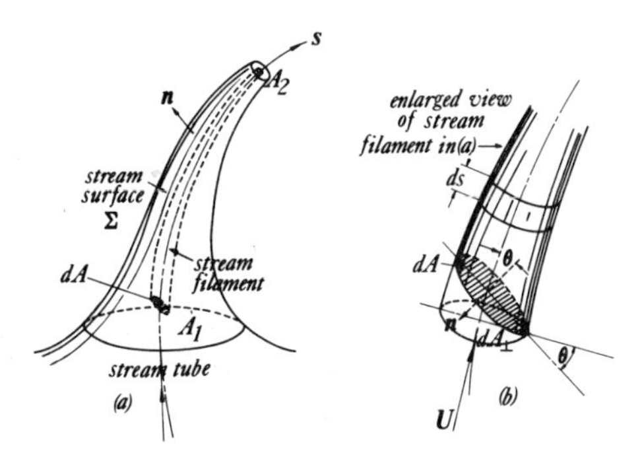

Fig. 6.3 Stream tube, stream filament, and stream surface in an updraft.

magnitude of $\boldsymbol{U}$. For the rate of mass flow entering the entire stream tube we must integrate over the entire cross sectional area

$$\frac{dm}{dt} = \int_{A_1} \rho(\mathbf{n} \cdot \boldsymbol{U})\, dA_1 = \int_{A_1} \rho U \cos\theta\, dA_1 \tag{6.3}$$

The angle θ is measured between $\boldsymbol{U}$ and the perpendicular to the area dA, or the outer normal $\boldsymbol{n}$. One can view this integral as that of $U \cos\theta$ multiplied by dA, or U multiplied by $dA \cos\theta$. In other words, the product is either the projection of $\boldsymbol{U}$ on $\boldsymbol{n}$ multiplied by dA or the projection of $\boldsymbol{n}\, dA$ normal to $\boldsymbol{U}$ and multiplied by U. The density, velocity, and angle can in general be functions of positions in A and therefore must be considered in the integral. Equation (6.3) can also be written for the flow leaving the tube at section 2 (Fig. 6.3). Assuming that the flow is steady, and since there can be no flow across the stream surface Σ, in order to conserve the mass it is necessary that the summation of mass fluxes be zero,

$$\int_{A_1} \rho(\boldsymbol{n} \cdot \boldsymbol{U})\, dA_1 + \int_{A_2} \rho(\boldsymbol{n} \cdot \boldsymbol{U})\, dA_2 = 0 \tag{6.4}$$

or that

$$\int_{A_1} \rho U \cos\theta\, dA_1 = \int_{A_2} \rho U \cos\theta\, dA_2 \tag{6.5}$$

Since $(\boldsymbol{n} \cdot \boldsymbol{U})$ evaluated at A_1 is negative in value (because $\pi/2 < \theta < 3\pi/2$) while it is positive at A_2 the sign difference between Eqs. (6.4) and (6.5) is justified. The stream surface Σ is, by definition, tangent to $\boldsymbol{U}$ everywhere, and thus $(\boldsymbol{n} \cdot \boldsymbol{U})$ on Σ is zero. Then we can add a zero value to Eq. (6.4) and obtain an important integral which represents conservation of mass:

$$\int_{A_1} \rho(\boldsymbol{n} \cdot \boldsymbol{U})\, dA_1 + \int_{A_2} \rho(\boldsymbol{n} \cdot \boldsymbol{U})\, dA_2 + \int_{\Sigma} \rho(\boldsymbol{n} \cdot \boldsymbol{U})\, d\Sigma = 0$$

or by considering the entire closed surface $S = A_1 + A_2 + \Sigma$, we can write the integral form of the conservation of mass as

$$\oint_S \rho(\boldsymbol{n} \cdot \boldsymbol{U})\, dS = 0 \tag{6.6}$$

Here the circled integral sign implies *closed contour integration* on the entire surface S.

When the flow is one-dimensional and $\boldsymbol{U}$ is everywhere normal to A and does not vary in A, and accordingly ρ is uniform in A, then $\boldsymbol{U}$ and ρ can come out of the integral sign so that we obtain

$$\rho_1 A_1 U_1 = \rho_2 A_2 U_2 \tag{6.7}$$

This will be true at each and every cross section of the stream tube. The more general unsteady-state case is treated in the next section.

Illustrative Example 6.1

Because viscous forces reduce the momentum of the gases near the walls of a smokestack and on the outer edges of the plume, the stack gases will have velocity variations along the axis and the radius of the plume. This is shown in Fig. 6.4. In Illustrative Example 4.1 we saw the variation of a characteristic velocity with altitude due to buoyancy, where we had assumed that the velocity was uniform in the stack cross section. In order to generalize the application of the integral form of the conservation of mass, we shall assume that friction near the walls of the stack reduces the momentum there, creating a *boundary layer* and thus a velocity distribution in the radial direction. Axisymmetry in the velocity distribution is assumed.

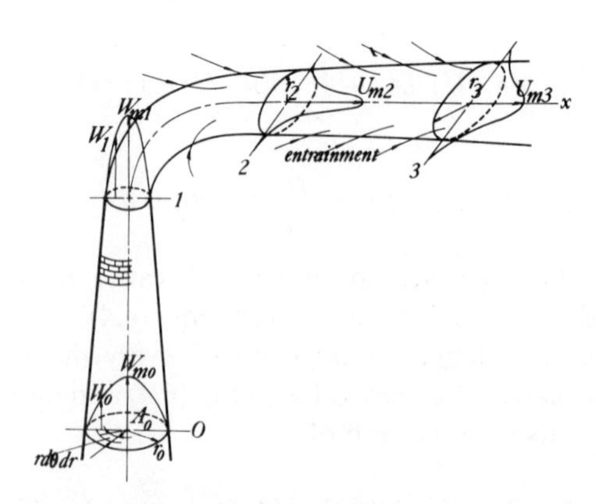

Fig. 6.4 Velocity of effluent in a smokestack and plume.

The velocity distribution in the stack can be expressed as

$$W = W_m[1 - (r/r_*)^n]$$

where W_m is the maximum center velocity and r_* is the radius of the stack. In case of laminar flow in the stack n can be shown to be 2, which for illustrative purposes we shall take to be the case. In the plume[1] the velocity distribution in r can be estimated as

$$U = U_m[1 - (r/r_*)^{3/2}]^2$$

where r_* represents the width of the plume.

A *spatial* mean velocity is calculated at each of the sections 0–3 in Fig. 6.4 by integrating for the mass rate at each cross section and making the result equal to the area times the density times the spatial average of the velocity, W_{av} or U_{av}. The density will vary as the gas cools, so it should be maintained inside the integral, although it is not likely to change much across the flow. Thus at any section

$$\rho \pi r_*^2 W_{av} = \int_0^{2\pi} \int_0^{r_*} \rho(r\, dr\, d\theta) W$$

This is also true for the plume where U and U_{av} can be used instead of W and W_{av}.

For the stack we can substitute the parabolic profile and solve for W_{av} at any of the cross sections. If ρ is constant with r, then

$$W_{av} = \frac{1}{\pi r_*^2} \int_0^{2\pi} \int_0^{r_*} W_m \left[1 - \left(\frac{r}{r_*}\right)^2\right] r\, dr\, d\theta = \frac{W_m}{2}$$

This is valid for every cross section, although W_{av} and W_m are not the same at every cross section. From one cross section to another, since the wall of the stack is a stream surface,

$$\rho_0 r_{*0}^2 W_{av0} = \rho_1 r_{*1}^2 W_{av1} = C$$

where C is the mass rate of flow divided by π, which is constant. Using the relationship of W in terms of ρ and z in Illustrative Example 4.1, we can obtain for W_{av} an expression in terms of z if the slope of the stack wall and the throughflow πC are known:

$$\frac{W_{av}^*}{2gz} - \frac{\rho_a r_*^2}{C} W_{av} + 1 = 0$$

[1] We shall see in Section 10.2.A that the solution $U = U_m \exp[-a(r/r_*)^2]$ is a more appropriate expression, although there is little difference quantitatively with that recommended in this example. The reader could repeat the calculation for this exponential expression.

In the plume the spatial average velocity will be

$$U_{\text{av}} = \frac{1}{\pi r_*^{\ 2}} \int_0^{2\pi} \int_0^{r_*} U_{\text{m}} \left[1 - \left(\frac{r}{r_*} \right)^{3/2} \right]^2 r\, dr\, d\theta$$

and upon integration

$$U_{\text{av}} = 9U_{\text{m}}/35$$

Unlike the case of the stack, the plume extending to r_* is not a stream tube because the plume entrains air from the outside into itself, and consequently the velocity at $r = r_*$ is nearly in the direction along r, as shown in Fig. 6.4. We shall see later that the plume opens up at a linear rate and the axial velocity diminishes as $1/x^3$, where x is the distance downstream. In fact the difference

$$\frac{9\pi}{35} (\rho_3 r_{*3}^2 U_{\text{av}3} - \rho_2 r_{*2}^2 U_{\text{av}2})$$

will represent the entrainment between x_2 and x_3.

6.4 THE INTEGRAL AND DIFFERENTIAL FORMS OF THE CONTINUITY EQUATION

The equation representing the conservation of mass is called the *continuity equation*. We shall consider here the case when the flow is not steady and density is allowed to vary in space and time. Equation (6.6) was developed for a steady flow. Had the flow in and out of the stream tube of Fig. 6.3 been transient, at one instant of time the flow entering the tube would not have been equal to that leaving it. Consequently there would have been accumulation or depletion of mass with time inside the tube. The amount of mass variation inside the tube of fixed volume and surface (control volume) would be

$$-\frac{\partial}{\partial t} \int_{\mathscr{V}} \rho\, d\mathscr{V} = \oint_S \rho(\boldsymbol{n} \cdot \boldsymbol{U})\, dS \tag{6.8}$$

where $\mathscr{V}$ is the volume of the tube section shown in the figure. The term on the right-hand side is the net mass rate crossing the closed boundaries. When it is positive, it implies that more mass is coming out than entering, which must then correspond to a depletion of mass in the volume which will result in a decrease of density. The minus sign in Eq. (6.8) accommodates for this.

Gauss's theorem in Appendix A mathematically relates the surface integral to the volume integral. Applying it to Eq. (6.8) we obtain

$$\int_S \boldsymbol{n} \cdot (\rho \boldsymbol{U})\, dS = \int_{\mathscr{V}} \nabla \cdot (\rho \boldsymbol{U})\, d\mathscr{V} \tag{6.9}$$

and finally

$$\int_{\mathscr{V}} \left[\frac{\partial \rho}{\partial t} + \nabla \cdot (\rho \boldsymbol{U}) \right] d\mathscr{V} = 0 \tag{6.10}$$

The control volume we chose for conservation of mass could have been any arbitrary volume, a stream tube or not. Thus we can say that the result in Eq. (6.10) would be true for any $\mathscr{V}$. Since the integrand is continuously distributed in $\mathscr{V}$, Eq. (6.10) must be zero not because of the special $\mathscr{V}$ we chose but because the integrand is zero. Thus we obtain the differential form of the continuity equation:

$$\frac{\partial \rho}{\partial t} + \nabla \cdot (\rho \boldsymbol{U}) = 0 \tag{6.11}$$

If U, V, and W are the x, y, and z components of the velocity vector in Cartesian coordinates, then (6.11), a scalar equation, can be expanded in component form:

$$\frac{\partial \rho}{\partial t} + \frac{\partial (\rho U)}{\partial x} + \frac{\partial (\rho V)}{\partial y} + \frac{\partial (\rho W)}{\partial z} = 0 \tag{6.12}$$

This partial differential equation is valid at any point and at any time in the flow, with variable density and velocity. This is then the most general form of the continuity equation.

In particular if the flow is *steady*, then Eq. (6.12) becomes

$$\nabla \cdot (\rho \boldsymbol{U}) = \frac{\partial (\rho U)}{\partial x} + \frac{\partial (\rho V)}{\partial y} + \frac{\partial (\rho W)}{\partial z} = 0 \tag{6.13}$$

When the motion is *isochoric* (of constant density), the variations of density in Eq. (6.13) are zero and

$$\nabla \cdot (\rho \boldsymbol{U}) = \rho \nabla \cdot \boldsymbol{U} + \boldsymbol{U} \cdot \nabla \rho$$

and since $\nabla \rho = 0$

$$\nabla \cdot \boldsymbol{U} = \frac{\partial U}{\partial x} + \frac{\partial V}{\partial y} + \frac{\partial W}{\partial z} = 0 \tag{6.14}$$

In *cylindrical coordinates*, the vector equation (6.13) develops into[2]

$$\frac{1}{r} \frac{\partial}{\partial r} (rU) + \frac{1}{r} \frac{\partial V}{\partial \theta} + \frac{\partial W}{\partial z} = 0 \tag{6.15}$$

[2] To obtain this result reference must be made to Appendix A to determine the components of the operator ∇ and the velocity in cylindrical coordinates.

where U, V, and W are the radial, peripheral, and axial components of the velocity vector. For two-dimensional isochoric motion in Cartesian coordinates the continuity equation takes the form[3]

$$\frac{\partial U}{\partial x} + \frac{\partial V}{\partial y} = 0 \tag{6.16}$$

Illustrative Example 6.2

Consider a steady wind whose components parallel to the surface of the earth, U and V, do not vary in magnitude nor direction in a certain region.

(a) If the wind has no vertical motion show that this is a possible motion:

$$\nabla \cdot (\rho \mathbf{U}) = 0$$

$$\rho\left(\frac{\partial U}{\partial x} + \frac{\partial V}{\partial y} + \frac{\partial W}{\partial z}\right) + U\frac{\partial \rho}{\partial x} + V\frac{\partial \rho}{\partial y} + W\frac{\partial \rho}{\partial z} = 0$$

Since U and V are constant and since $W = 0$, the equation of continuity is automatically satisfied and it is a possible motion.

(b) If there is a vertical component of wind velocity and U and V are still constant horizontally, and since the density of the air varies with altitude because of the temperature change and hydrostatics, then

(1) The continuity equation reduces to

$$\rho\frac{\partial W}{\partial z} + W\frac{\partial \rho}{\partial z} = 0$$

Since there are no other variations these partial derivatives can be replaced by total derivatives and

$$\frac{dW}{W} + \frac{d\rho}{\rho} = 0$$

(2) What is the relationship of the vertical component of the velocity to the density? Integrating

$$\ln \rho W = \text{constant} \qquad \text{or} \qquad \rho W = \text{constant}$$

(3) Must this be an updraft or a downdraft and in either case how should it vary with altitude? The preceding equation does not show the sign of the constant. If we consider the differential form and since $d\rho/dz < 0$ in the troposphere, then $d\rho/\rho < 0$ in z. To compensate for this, dW/W must be positive and

[3] See Section 7.12 for turbulent considerations.

equal to the percent density change in absolute value. This implies two possibilities where dW and W must have the same sign. First, for a downdraft $W < 0$ and $dW/dz < 0$, implying that the downdraft velocity increases with z. The other choice is when $W > 0$, an updraft, and $dW/dz > 0$, implying also that the updraft increases with altitude. This corresponds to the geometry of the column shown in Fig. 4.17; the distribution of W and ρ is shown in Fig. 6.5.

(4) In the ocean since $\rho =$ constant, constant horizontal motion must imply constant vertical motion.

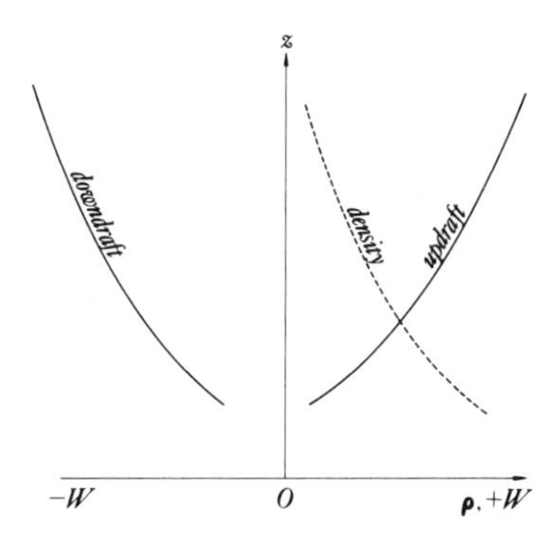

Fig. 6.5 Vertical velocity variation with altitude.

6.5 STREAM FUNCTION IN TWO-DIMENSIONAL FLOWS

By definition, the *stream function* is the mathematical relation representing the geometry of the stream surfaces, at a given time. In a two-dimensional, isochoric, and steady motion, the stream function Ψ can be related to the velocity field $\boldsymbol{U}$ in a simple manner. Since the velocity vector is tangent to the stream function Ψ, then

$$\boldsymbol{U} \cdot \nabla\Psi = 0 \tag{6.17}$$

Then the stream function $\Psi(x, y) =$ constant represents the stream surface, the gradient of Ψ being normal to $\Psi =$ constant; consequently the dot product of $\boldsymbol{U}$ and $\nabla\Psi$ must be zero.

If, for instance, the two-dimensional motion is represented on the x–y plane, then U and V are the velocity components of $\boldsymbol{U}$ as shown in Fig. 6.6. The rate of mass flow crossing the arbitrary line OMP_2 or s_2 can be evaluated according to Eq. (6.2). If we consider a unit dimension normal to the plane of the paper, per unit depth this rate of mass flow is

$$\frac{dm}{dt} = \rho(\boldsymbol{n} \cdot \boldsymbol{U})\, ds$$

where $(\boldsymbol{n} \cdot \boldsymbol{U})$ is $U \cos\theta$ and the infinitesimal area is $dA = 1 \times ds$. The integration of this quantity along the path OMP_2 is in s alone since there are no variations normal to the plane of the paper. The total rate of flow crossing OMP_2 is

$$\int_O^{P_2} \rho(\boldsymbol{n} \cdot \boldsymbol{U})\, ds = \int_O^{P_2} \rho U \cos\theta\, ds = \int_O^{P_2} \rho U_\perp\, ds \tag{6.18}$$

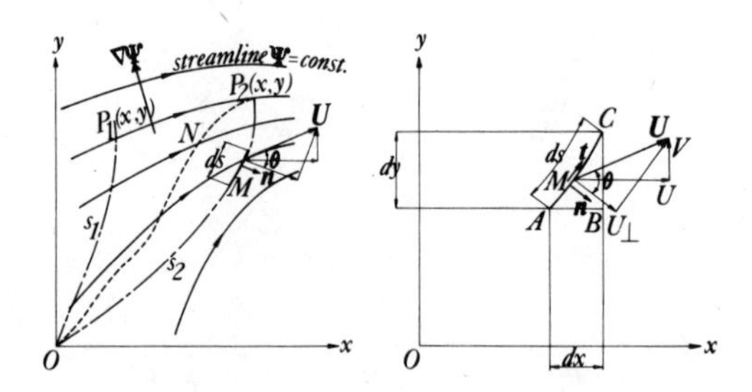

Fig. 6.6 Streamline motion in two-dimensional flow.

Since the density has been postulated to be constant, it can be removed from the integrand and $\Psi(x, y)$ can be defined as the rate of mass flow per unit density and unit depth:

$$d\Psi = (\boldsymbol{n} \cdot \boldsymbol{U})\, ds = U_\perp\, ds = V\, dx - U\, dy$$

$$\Psi_0 - \Psi_{P_2} = \int_O^{P_2} U \cos\theta\, ds = \int_O^{P_2} U_\perp\, ds \tag{6.19}$$

According to Eq. (A.44) (Appendix A) the total change of a scalar function is given by

$$d\Psi = \nabla\Psi \cdot \boldsymbol{ds} = \nabla\Psi \cdot \boldsymbol{t}\, ds$$

where $\boldsymbol{t}$ is the unit vector along OMP_2. Thus

$$\nabla\Psi \cdot \boldsymbol{t} = \boldsymbol{n} \cdot \boldsymbol{U} \tag{6.20}$$

However, the three unit vectors $\boldsymbol{n}$, $\boldsymbol{t}$ and $\boldsymbol{b}$ have the following relationship:

$$-\boldsymbol{t} = \boldsymbol{b} \times \boldsymbol{n}$$

where $\boldsymbol{b}$ is the unit vector normal to the plane of the paper in Fig. 6.6. Thus Eq. (6.20) becomes

$$-\nabla\Psi \cdot (\boldsymbol{b} \times \boldsymbol{n}) = \boldsymbol{n} \cdot \boldsymbol{U} \tag{6.21}$$

The triple scalar product of Section A.10 can be applied to give

$$\nabla\Psi \cdot (\boldsymbol{b} \times \boldsymbol{n}) = \boldsymbol{n} \cdot (\nabla\psi \times \boldsymbol{b}) = -\boldsymbol{n} \cdot (\boldsymbol{b} \times \nabla\psi) \tag{6.22}$$

Then Eq. (6.21) becomes

$$\boldsymbol{b} \times \nabla\Psi = \boldsymbol{U} \tag{6.23}$$

This means that the three vectors $\boldsymbol{b}$, $\nabla\Psi$, and $\boldsymbol{U}$ form an orthogonal set.

In component form Eq. (6.23) is very useful. In Cartesian coordinates ∇ is given by Eq. (A.45), $\boldsymbol{b}$ is $\boldsymbol{k}$, and Eq. (6.23) becomes

$$\mathbf{k} \times \left(\boldsymbol{i}\frac{\partial}{\partial x} + \boldsymbol{j}\frac{\partial}{\partial y} + \boldsymbol{k}\frac{\partial}{\partial z}\right)\Psi = \boldsymbol{i}U + \boldsymbol{j}V + \boldsymbol{k}W$$

or

$$U = -\frac{\partial\Psi}{\partial y}, \qquad V = \frac{\partial\Psi}{\partial x}, \qquad \text{and} \qquad W = 0 \tag{6.24}$$

In axisymmetric *cylindrical coordinates* using the definition of ∇ in Eq. (A.46)

$$U = -\frac{1}{r}\frac{\partial\Psi}{\partial\theta}, \qquad V = \frac{\partial\Psi}{\partial r}, \qquad \text{and} \qquad W = 0 \tag{6.25}$$

where U, V, and W are the r, θ, and z components of $\boldsymbol{U}$. In axisymmetric *spherical coordinates*, where the symmetry is around the radial line, ψ is the latitudinal angle, and θ is the longitudinal angle, W the radial velocity is zero and the two horizontal components U (toward the east) and V (toward the north), taken as standard directions, are

$$U = -\frac{1}{r}\frac{\partial\Psi}{\partial\psi}, \qquad V = \frac{1}{r\cos\psi}\frac{\partial\Psi}{\partial\theta} \tag{6.26}$$

Returning to Fig. 6.6 and Eq. (6.19) it can be verified that for a steady and isochoric flow the integral in the equation is always constant, provided that the points O and P_2 remain unchanged. This is verified by assuming another arbitrary line ONP_2 joining the same two points. Considering then the closed path OMP_2NO if the amount entering it is equal to that leaving, then the statement that the integral is not path dependent is proved. Thus the integrand must be an exact differential i.e., it is $d\Psi$, and the stream function is a property of the motion.

If points P_1 and P_2 are located on a stream surface, the value of the integral still remains the same since there could not be any flow across the segment P_1P_2. Starting from a reference O, as long as P remains on the streamline, Ψ_P remains the same, so the value of Ψ is constant on the streamline.

6.6 LINEAR COMBINATION OF FLOWS

There are a number of applications where superposition of solutions yields the solution of superposed phenomena, i.e., linear phenomena. For example, a *line source* flow is represented by a radial flow emanating from an axis. This is represented in Fig. 6.7. If the velocity field of such a flow is uniform along the axis,

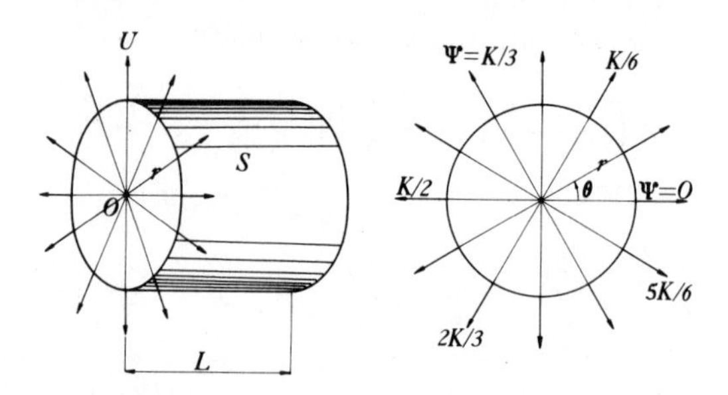

Fig. 6.7 A two-dimensional line source.

then if K is the volume flow rate per unit length of source L and U is the radial velocity, we have

$$KL = 2\pi rLU$$

Solving for U

$$U = \frac{K}{2\pi r}$$

This means that for conservation of mass the radial velocity must decrease with r. Then from the statement of the problem, V in the θ direction and W the axial velocity are zero. In cylindrical coordinates this velocity U is related to the stream function according to Eq. (6.26)

$$U = -\frac{1}{r}\frac{d\Psi}{d\theta} = \frac{K}{2\pi r}$$

Integrating[4] for Ψ we have

$$\Psi = -\frac{K}{2\pi}\theta + \text{constant} \tag{6.27}$$

[4] The partial derivative is replaced by a total derivative since U is the only component of velocity and Ψ can only be $f(\theta)$.

Therefore, the stream function is constant for θ = constant, implying that radial surfaces are the stream surfaces. The constant is absorbed by taking $\Psi = 0$ when $\theta = 0$. Thus

$$\Psi = -\frac{K}{2\pi}\theta \tag{6.28}$$

Another flow whose stream function can be obtained easily is a uniform parallel flow. In Cartesian coordinates, let us orient the flow with velocity $-U_0$ along the $(-x)$ direction. Then $V = 0$ and from Eq. (6.24) we can write

$$-U_0 = -\frac{d\Psi}{dy}$$

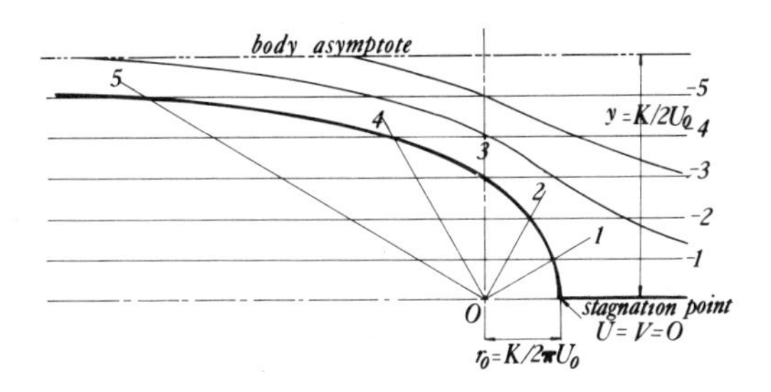

Fig. 6.8 Two-dimensional Rankine half-body.

Since $-U_0$ is the only component of velocity, the total derivative is used, and upon integration

$$yU_0 = \Psi + \text{constant}$$

The constant of integration is absorbed by choosing $\Psi = 0$ when $y = 0$. Thus

$$\Psi = U_0 y \tag{6.29}$$

When we superimposed the line source flow and the parallel flow, since both equations on Ψ for the source and the parallel flow are linear, the sum of the solutions of Eqs. (6.28) and (6.29) will be the solution of the combined flow:

$$\Psi(r, \theta) = U_0 y - \frac{K}{2\pi}\theta = U_0 r \sin\theta - \frac{K}{2\pi}\theta \tag{6.30}$$

For Ψ = constant this equation can be plotted to obtain the streamlines. These are shown in Fig. 6.8. The *stagnation point* is defined as the point in the flow where the flow has zero velocity, or stagnates. This point can be found by evaluating from Eq. (6.30) the velocities U and V in the r and θ directions through Eq. (6.25) and setting them equal to zero. Thus

$$
\begin{aligned}
U &= -\frac{1}{r}\frac{\partial \Psi}{\partial \theta} = \frac{K}{2\pi r} - U_0 \cos\theta = 0 \\
V &= \frac{\partial \Psi}{\partial r} = U_0 \sin\theta = 0
\end{aligned}
\tag{6.31}
$$

The V equation states that the stagnation point must be on $\theta = 0$ or π and the U equation states that the radial distance of this stagnation point from O on the $\theta = 0$ line is $r = r_0 = K/2\pi U_0$. The streamline passing through this point has a zero value and is called the *stagnation streamline*. The equation of this streamline is obtained when Eq. (6.30) is set to zero:

$$r = r_0 \frac{\theta}{\sin\theta}$$

which is also plotted in Fig. 6.8. This flow field can also be solved graphically. Equal amounts of flow can be divided for the line source for equal increments of θ, the entire 360° being unity. The parallel flow can also be divided in equal sets of values (negative coming from the left). The $\Psi = 0$ stagnation streamline, called a Rankine half-body, is obtained by joining the points that have a combined zero value. The resultant -1 streamline will be the line joining the points with a sum of -1, and so on.

Since this flow has no viscous effects, it has no way of differentiating the difference between a streamline in the flow and a solid wall contoured in the shape of the streamline. Thus if we now consider a body shaped as the stagnation streamline, the flow will be unchanged since there is no friction. We use this flow field in the next section to establish pathlines and streaklines.

6.7 PATHLINES AND STREAKLINES

In environmental studies the determination of pathlines and streaklines is important. We define a *pathline* as a line in the flow field tracing the trajectory of a specific fluid parcel. A *streakline* is the locus, at a given time t_0, that connects the instantaneous locations of all the particles that have emanated from a specific contaminating point in the flow field. If the smokestack is considered as a contaminating point, the plume is a streakline.

Only when the flow is *steady* are the streamlines. pathlines, and streaklines represented by the same set of lines. This is the reason why a streak of dye in a steady flow represents a streamline.

streamline, pathline, and streakline become completely different functions although they are related to one another. Consider first a steady motion such as the flow around a ship moving at a constant velocity. A moving ship and two observers, one on the ship and one on a pier, are schematized in Fig. 6.9. To the

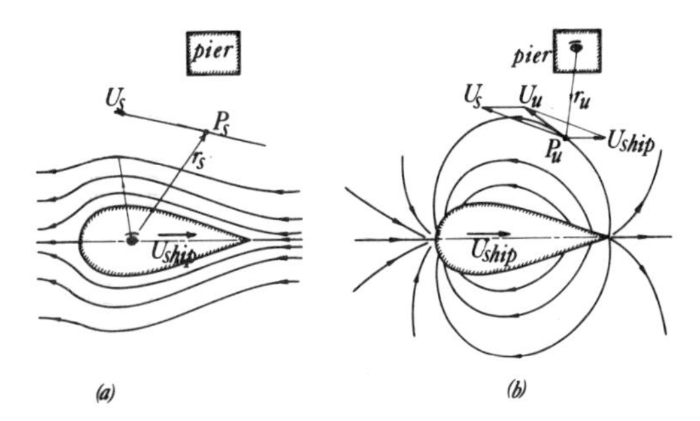

Fig. 6.9 Streamlines for (a) steady and (b) unsteady motion from a uniform displacement of a ship. The symbolic eye represents the position of the observer.

In an unsteady motion, such as environmental motions of large scale, the observer on the ship, the motion of the sea will be toward the ship; the streamlines of this motion as viewed by the observer on the ship are drawn in Fig. 6.9a. Since the speed of the ship is constant, the streamlines the observer sees will be steady, such that at a point r_s from the ship the velocity of the sea relative to the ship will always be U_s. The particles of sea will follow the velocity vector, and since the velocity field is frozen to the ship the pathlines will be the streamlines. If P_s is a contaminating point, all particles passing through P_s will line up along the streamline passing through P_s and the streakline will be the same as the streamline. This is the case of steady motion relative to the observer on the ship.

Consider the observer on the pier. He will not report the same information as the observer on the ship because the velocity at the same point P on the sea will be that reported by the observer on the ship, U_s (steady velocity), plus the velocity of the ship relative to the pier, U_{ship}. The vector addition shown in Fig. 6.9b gives a fluid velocity U_u. This velocity, which is seen by the observer on the pier, is unsteady because as the ship progresses to the right, the velocity U_s at the point P will not be the same. The streamline motion moves with the ship and the streamline pattern reported by the observer on the pier is different than that reported in Fig. 6.9a. Since at different times different streamlines affect the point P_u at a distance r_u from the pier, the direction and magnitude of the velocity at P_u will be varying in time and the flow in this frame will be transient. Thus if particles are emitted at P_u, they will not all move in the same trajectory or pathline, and therefore after a time t_0 will be at different positions connected by a line called the streakline. Let us consider an illustrative example, using the half-body of the preceding section to demonstrate these situations.

Illustrative Example 6.3

Let the ship in Fig. 6.9 be of the form of the Rankine half-body shown in Fig. 6.8. The full body extends identically on the lower side of the abscissa. Let C be the constant $K/2\pi$ of Eq. (6.30).

With reference to O on the ship, at any point on the sea the velocity components are given by Eq. (6.31). Then the total velocity U_s is the vector sum

$$U_s = (U^2 + V^2)^{1/2} = [1 + (C^2/r^2U_0^2) + 2(C/rU_0)\cos\theta]^{1/2}$$

On the surface of the ship we have the special relationship between r and θ given by Eq. (6.31), and when these are substituted, the velocity on the surface is

$$U_s = U_0[1 + \{(\sin\theta)/\theta\}^2 - \{(\sin 2\theta)/\theta\}]^{1/2}$$

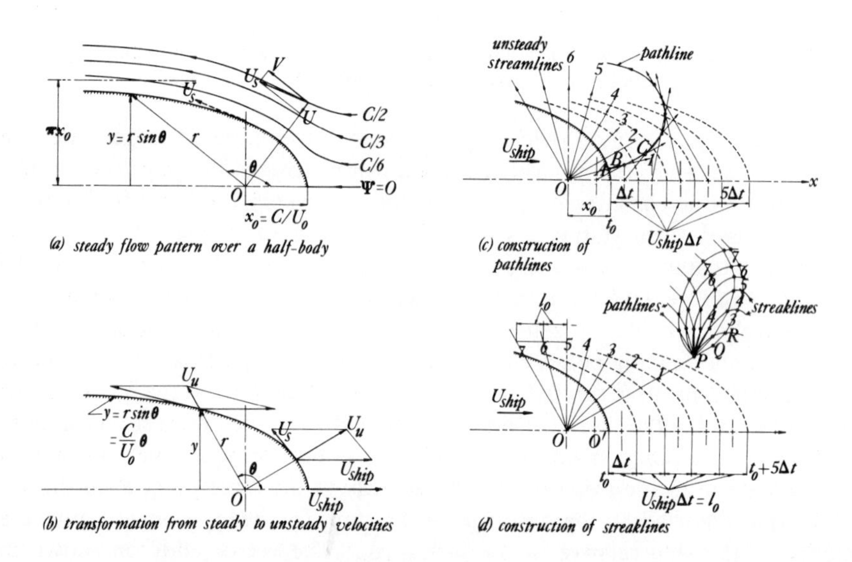

Fig. 6.10 Successive steps in the determination of pathlines and streaklines. (a) Steady flow pattern over a half-body. (b) Transformation from steady to unsteady velocities. (c) Construction of pathlines. (d) Construction of streaklines.

Figure 6.10a shows these velocities plotted for the steady-state case, as viewed from the ship. Since the ship does not move with respect to the observer on it, the fluid will seem to move toward the observer.

To view the unsteady motion as described in Fig. 6.9b the observer must jump on the pier; he will observe a different flow field because the motion he had observed originally was relative to the ship and now the ship has a velocity $U_{\text{ship}} = U_0$ relative to the observer. Therefore the velocity field relative to the pier will be the steady field of Fig. 6.10a plus a vector addition of the velocity of the ship relative to the pier. This new velocity field is shown in Fig. 6.10b. The resultant velocity field is U_u, the unsteady field. The velocity is tangent to the streamline. Thus the new set of unsteady streamlines on the ship's surface is marked U_u; these streamlines are frozen to the ship and move with the ship.

The pathlines are determined in a simple manner. Referring to Fig. 6.10c a particle on the body at a point A is considered at a time t_0. This particle is on streamline 1 and will move along this streamline for an increment of time Δt until it reaches B. During this time the ship together with the streamline pattern has also moved a distance $U_{\text{ship}}\,\Delta t$. Consequently, when the particle has reached B, it will no longer be on streamline 1; streamline 2 will have caught up with it, giving the particle its own direction and speed. For the next increment Δt the particle will move along 2 until streamline 3 catches up with it at C and imparts its own speed and direction, and so on. When the process is continued at infinitesimal times, the final path of the particle will be that of the pathline shown in Fig. 6.10c. Since the velocity of the sea decreases as we move away from the ship new streamlines can always catch up to particles.

The streakline is, as we said, the locus of all the particles that have originated from a given point in space. In Fig. 6.10d the point P has been arbitrarily chosen as the contaminating point. Starting at the instant t_0 streamline 1 will move a dyed particle from P in direction 1 at the corresponding velocity at P. This particle will move along PQ, and the ship with the streamline pattern will move along the x axis. After an interval Δt streamline 2 will be at P and will move another particle from P in the direction of 2 at this new velocity. Shortly thereafter streamline 2 will catch up with the first particle at Q and will impart to it a new velocity in the direction of 2 and move it to QR. At the time $t_0 + 2\,\Delta t$ streamline 3 will cross at P and a third particle will issue in the direction 3. Later, streamline 3 will catch up with particle 2 and give it a now direction and velocity and then it will catch up with 1 and give it a new direction and velocity, and so on.

The process continues indefinitely; the set of moving streamlines will continuously influence all particles originating at P. The locus of all the particles that have started at P at the time t_0 after $5\,\Delta t$ and $6\,\Delta t$ are shown in Fig. 6.10d. They represent the streaklines. The pathlines of each of these particles are also shown. The following section discusses a method of analysis for the determination of pathlines and streaklines in the atmosphere owing to the movement of simple atmospheric disturbances.

6.8 PATHLINES AND STREAKLINES IN THE PRESENCE OF SIMPLE ATMOSPHERIC DISTURBANCES

Consider an arbitrary form of a two-dimensional atmospheric disturbance as shown in Fig. 6.11 and expressed by a disturbance velocity field $\boldsymbol{U}(\boldsymbol{r})$ with respect to its center and moving at a velocity $\boldsymbol{V}$ in an infinite medium which has its own motion defined by $\boldsymbol{U}_\infty$ (see Eskinazi, 1973).

In Fig. 6.11 we consider two frames of reference, namely a frame O fixed to the earth's surface from which the pathlines and streaklines are to be described and with respect to which[5] velocities $\boldsymbol{V}$ and $\boldsymbol{U}_\infty$ are measured, and a frame O' fixed to the moving disturbance from which $\boldsymbol{U}(\boldsymbol{r})$ is measured. In order to keep the energy of the disturbance finite, it must vanish properly in strength at large distances as all finite energy disturbances. Illustrative Example 6.3 was a disturbance of finite category. We shall refer to $\boldsymbol{V}$ as the *propagation* and to $\boldsymbol{U}_\infty$ as the *drift* velocities.

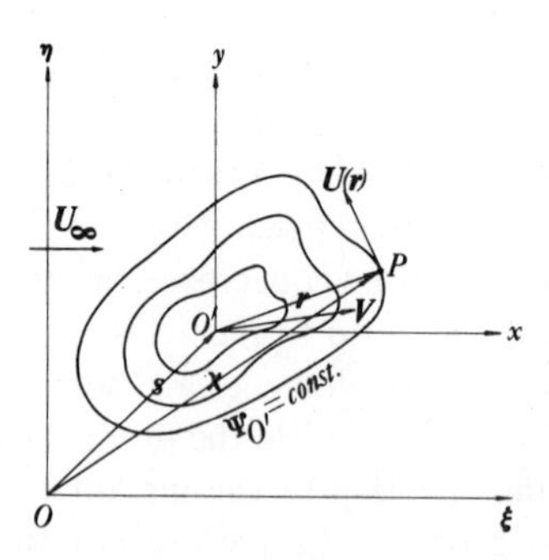

Fig. 6.11 Position and velocity with reference to the earth and an atmospheric disturbance.

In Fig. 6.11 the two frames are separated, at a given instant, by a distance $\boldsymbol{s}$. Attached to O' is the disturbance moving at the velocity $\boldsymbol{V}$ with respect to O and whose streamlines with reference to O' are shown. A particle at a distance $\boldsymbol{r}$ from O' experiences the influences of $\boldsymbol{U}(\boldsymbol{r})$, $\boldsymbol{V}$, and $\boldsymbol{U}_\infty$. Assuming that these velocities are steady the motion viewed from O' is steady, but it is unsteady with reference to O.

The velocity of the particle with reference to O and O' can be related as follows: First we start with O' fixed to O; the velocity of the particle with reference to O, which we call $\boldsymbol{W}_0$, is $\boldsymbol{U}(\boldsymbol{r})$. Second, we make the disturbance move with the velocity $\boldsymbol{V}$. With respect to O' this will look as though the infinite medium comes toward the particle at the velocity $-\boldsymbol{V}$, and thus

$$\boldsymbol{W}_{0'} = (\boldsymbol{U} - \boldsymbol{V}) \tag{6.32}$$

If in addition to the disturbance propagation velocity $\boldsymbol{V}$, the infinite medium has

[5] $\boldsymbol{V}$ is in essence the speed of propagation of the disturbance (storm) owing to the thermodynamic propagation of the pressure field and stream function (as we shall see in Chapter 8) and exclusive of the material drift velocity $\boldsymbol{U}_\infty$ (zonal winds).

its own drift velocity $\boldsymbol{U}_\infty$, then the velocity of the particle with reference to O' is the vector sum

$$\boldsymbol{W}_{O'} = \frac{d\boldsymbol{r}}{dt} = (\boldsymbol{U} - \boldsymbol{V}) + \boldsymbol{U}_\infty \tag{6.33}$$

The streamline pattern of the flow with reference to O' must be derived from this velocity field. However, the velocity of the same particle with reference to O is the velocity of the particle with reference to O', [Eq. (6.33)] plus the velocity of the frame O' with respect to O. Then

$$\boldsymbol{W}_O = \frac{d\boldsymbol{\chi}}{dt} = [(\boldsymbol{U} - \boldsymbol{V}) + \boldsymbol{U}_\infty] + \boldsymbol{V} = \boldsymbol{U}(\boldsymbol{r}) + \boldsymbol{U}_\infty \tag{6.34}$$

In the same manner the position of the particle P with reference to O is

$$\boldsymbol{\chi} = \boldsymbol{s} + \boldsymbol{r} = \boldsymbol{s} + \boldsymbol{V}t + \boldsymbol{r} \tag{6.35}$$

In component form

$$\xi = x + m = x + m_1 + V_x t, \qquad \eta = y + n = y + n_1 + V_y t \tag{6.36}$$

where $\boldsymbol{s}_1$ is the initial location of O' relative to O ($t = 0$) and m_1 and n_1 are the components of $\boldsymbol{s}_1$ in the O system. The quantities m and n are the components of $\boldsymbol{s}$ at any instant of time. The rates of change of coordinates are

$$\frac{d\xi}{dt} = \frac{dx}{dt} + V_x, \qquad \frac{d\eta}{dt} = \frac{dy}{dt} + V_y \tag{6.37}$$

In comparing Eqs. (6.34) and (6.37) we conclude that although $\boldsymbol{W}_0$ is always equal to $\boldsymbol{U}(\boldsymbol{r}) + \boldsymbol{U}_\infty$, it is not in contradiction to Eq. (6.37). In all cases $\boldsymbol{U} \neq d\boldsymbol{r}/dt$ and $\boldsymbol{U}_\infty \neq \boldsymbol{V}$. Furthermore, it may appear from Eq. (6.34) that $\boldsymbol{V}$ is not a factor when the motion is viewed from O. This is not true because $\boldsymbol{V}$ appears also in $\boldsymbol{\chi}$ in Eq. (6.35).

When we consider the stream function, with reference to O' and with $\boldsymbol{V}$ and $\boldsymbol{U}_\infty$ uniform, according to Eq. (6.24),

$$\Psi_{O'} = \Psi_{\text{dist}} - (\boldsymbol{U}_\infty - \boldsymbol{V})_x y + (\boldsymbol{U}_\infty - \boldsymbol{V})_y x \tag{6.38}$$

With reference to O the stream function is

$$\Psi_O = \Psi_{O'} - V_x y + V_y x = \Psi_{\text{dist}} - U_{\infty x} y + U_{\infty y} x \tag{6.39}$$

where V_x, V_y, $U_{\infty x}$, $U_{\infty y}$ are horizontal components of the velocities. Since synoptic weather charts of free air are with reference to O, they are essentially represented by Eq. (6.39).

A Simple Case

Consider the disturbance to be a *cyclonic vortex*, propagating at the velocity $\boldsymbol{V} = \boldsymbol{U}_\infty$. As we shall see in Chapter 7 a cyclonic motion is one in which counterclockwise rotation takes place and which, on the basis of dynamics, corresponds to a low pressure area. As we shall see in Section 6.11 the vortex is a circular flow with a peripheral velocity distribution decreasing inversely as the radius $\boldsymbol{r}$. Thus when $\boldsymbol{V} = \boldsymbol{U}_\infty$ this corresponds to the cyclonic vortex disturbance moving with the prevailing winds of velocity $\boldsymbol{U}_\infty$.

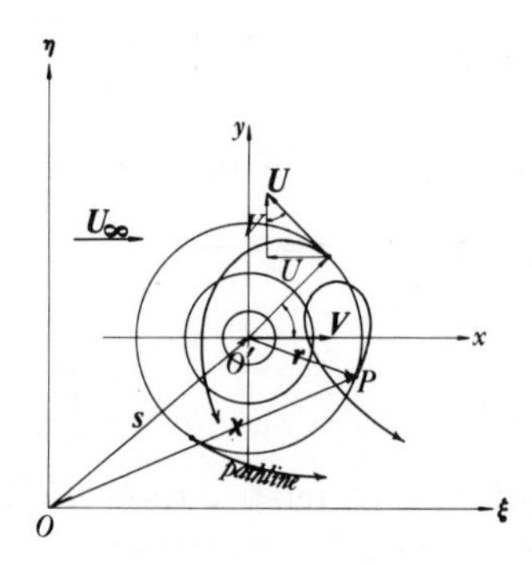

Fig. 6.12 Geometry of velocity components and pathlines in a simple cyclonic vortex.

The vortex model is axisymmetric with a disturbance velocity field $U(r) = (\Gamma/2\pi)(1/r)$ corresponding to a stream function, with reference to O', where $\Gamma/2\pi$ is the strength of the disturbance which is assumed constant.[6] Using Eq. (6.25) the stream functions are

$$\Psi_{O'} = \frac{\Gamma}{2\pi} \ln r \tag{6.40}$$

According to Eq. (6.33) since $\boldsymbol{V} = \boldsymbol{U}_\infty$, the vortex will remain a vortex in the frame O'. It is obvious that for $\boldsymbol{V} \neq \boldsymbol{U}_\infty$ the stream function in the frame O' would be

$$\Psi_{O'} = \frac{\Gamma}{2\pi} \ln r - (\boldsymbol{U}_\infty - \boldsymbol{V})_x y + (\boldsymbol{U}_\infty - \boldsymbol{V})_y x$$

For the case at hand, the x and y components of the disturbance velocity are, as indicated in Fig. 6.12, u and v. From the geometry and the vortex velocity

[6] The quantity Γ is the circulation, which is discussed in Section 6.13.

distribution Eq. (6.37) becomes after using Eq. (6.24) for u and v,

$$\begin{aligned} \frac{d\xi}{dt} &= u + V_x = -\frac{\Gamma y}{2\pi r_0^2} + V_x \\ \frac{d\eta}{dt} &= v + V_y = \frac{\Gamma x}{2\pi r_0^2} + V_y \end{aligned} \tag{6.41}$$

In the O' frame since a vortex remains a vortex and since the streamlines are circular, a particle on a circle with value Ψ_0 and radius r_0 remains, on that circle throughout the motion. Thus although x and y are changing, the sum of the squares remains constant. The two differential equations in (6.41) represent the position of the particle in terms of ξ and η, coordinated in the O frame. Furthermore since $\Gamma/2\pi$ is the strength of the storm, taken as constant in this case, the coefficient $\Gamma/2\pi r_0^2$ is constant for one specific particle (but different constants for different particles initially at different r_0's) and represents the average rotational speed in that circle.

The differential equations in (6.41) have mixed variables. From Eq. (6.35), substituting values of x and y in terms of ξ and η, we have

$$\begin{aligned} \frac{d\xi}{dt} &= -\omega(\eta - n) + V_x \\ \frac{d\eta}{dt} &= \omega(\xi - m) + V_y \end{aligned} \tag{6.42}$$

Differentiating with respect to time and substituting back into Eq. (6.42), recalling that m and n are functions of time according to Eq. (6.36), we have

$$\begin{aligned} \frac{d^2\xi}{dt^2} + \omega^2\xi &= \omega^2 m = \omega^2(m_1 + V_x t) \\ \frac{d^2\eta}{dt^2} + \omega^2\eta &= \omega^2 n = \omega^2(n_1 + V_y t) \end{aligned} \tag{6.43}$$

Subject to the initial condition that the particle originates at $\xi = \xi_0$ and $\eta = \eta_0$ at $t = 0$, the solutions of Eqs. (6.43) are

$$\begin{aligned} \xi &= (\xi_0 - m_1)\cos\omega t - (\eta_0 - n_1)\sin\omega t + m \\ \eta &= (\eta_0 - n_1)\cos\omega t + (\eta_0 - m_1)\sin\omega t + n \end{aligned}$$

or

$$\begin{aligned} \xi &= (\xi_0 - m_1)\cos\omega t - (\eta_0 - n_1)\sin\omega t + m_1 + V_x t \\ \eta &= (\xi_0 - m_1)\sin\omega t + (\eta_0 - n_1)\cos\omega t + n_1 + V_y t \end{aligned} \tag{6.44}$$

where in this case m_1, n_1, V_x, V_y are constants for all particles and ω is constant for each particle. We shall designate ξ_0 and η_0 as the *contaminating point*, fixed in the space of O and analogous to the stack from a power plant, for example.

Equation (6.44) represents the *pathline* originating at the contaminating point based at a given initial location of the storm, and a drift velocity equal to the propagation velocity. The particle will trace the trajectory (ξ, η) in time in the space of the fixed frame O. Looking at Fig. 6.13, we see two successive positions of the storm at O_1' (initially) and at O_2'. The three positions of particle 1 are the contaminating point, p_{12} and p_{13} in the period of two time intervals $2\,\Delta t$. Particle 2 occupies the spaces of the contaminating point and p_{23} in one time interval. Thus the *lifetime* of each particle originating at the same contaminating point is the number of Δt's and is not the same for all particles.

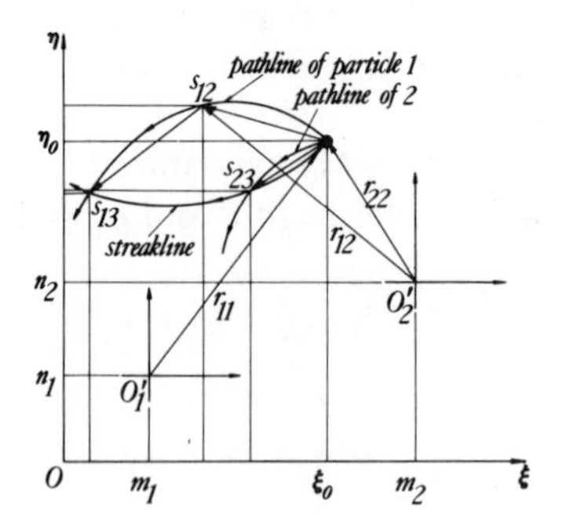

Fig. 6.13 Geometry of pathlines and streaklines.

Since by definition, the *streakline* is the location of all particles that have issued at (ξ_0, η_0) after a time T, the equation for the locus of all such particles must also be included in (6.44). Let us introduce two indices i and j which refer to the particle number and to the number of time interval lapses, respectively. Looking at Fig. 6.13, the coordinates of particle 1 after two time intervals are ξ_{13}, η_{13} and from Eq. (6.44), they are

$$\xi_{13} = (\xi_0 - m_1)\cos\omega_1(2\,\Delta t) - (\eta_0 - n_1)\sin\omega_1(2\,\Delta t) + m_1 + V_x(2\,\Delta t)$$

$$\eta_{13} = (\xi_0 - m_1)\sin\omega_1(2\,\Delta t) + (\eta_0 - n_1)\cos\omega_1(2\,\Delta t) + n_1 + V_y(2\,\Delta t)$$

The same can be written for particles 2, 3, . . . , i. By inspection we can generalize for the ith particle and the jth lapse time:

$$\xi_{ij} = (\xi_0 - m_i)\cos\omega_i(j-i)t - (\eta_0 - n_i)\sin\omega_i(j-i)t + m_i + V_x(j-i)t$$

$$(6.45)$$

$$\eta_{ij} = (\xi_0 - m_i)\sin\omega_i(j-i)t + (\eta_0 - n_i)\cos\omega_i(j-i)t + n_i + V_y(j-i)t$$

where m_i and n_i are locations of O' when the ith particle is ready to issue from (ξ_0, η_0), and

$$\omega_i = \frac{\Gamma}{2\pi} \frac{1}{(\xi_0 - m_{1i})^2 + (\eta_0 - n_{1i})^2} \tag{6.46}$$

$$m_i = m_1 + V_x T, \qquad n_i = n_1 + V_y T$$

m_{1i} and n_{1i} must remain fixed for ω_i once the ith particle originates from (ξ_0, η_0).

Since j is the number of lapse intervals, the quantity $jt = T$ is the maximum lapse time or the life of the first particle that has issued from the contaminating point. The quantity $it = \tau$ is the lifetime of the ith particle from birth, so to speak, at the contaminating point. Thus Eq. (6.45) becomes

$$\begin{aligned} \xi_\tau &= (\xi_0 - m_1 - V_x T) \cos \omega_\tau(T - \tau) - (\eta_0 - n_1 - V_y T) \sin \omega_\tau(T - \tau) \\ &\quad + m_1 + V_x(T - \tau) \\ \eta_\tau &= (\xi_0 - m_1 - V_x T) \sin \omega_\tau(T - \tau) + (\eta_0 - n_1 - V_y T) \cos \omega_\tau(T - \tau) \\ &\quad + n_1 + V_y(T - \tau) \end{aligned} \tag{6.47}$$

where

$$\omega_\tau = \frac{\Gamma}{2\pi} \frac{1}{(\xi_0 - m_1 - V_x T)^2 + (\eta_0 - n_1 - V_y T)^2}$$

Thus after a total lapse time of T the streakline is identified by the position of all ξ_τ and η_τ and thus Eq. (6.47) is the equation of the streakline.

A very special case is when at $t = 0$ the two coordinates O and O' coincide and the drift is only along x, or $V_y = 0$. According to Eq. (6.36) m_1 and n_1 are zero also and

$$\xi = x + V_x t, \qquad \eta = y$$

Then the pathline solution of Eq. (6.44) becomes

$$\begin{aligned} \xi &= \xi_0 \cos \omega t - \eta_0 \sin \omega t + V_x t \\ \eta &= \xi_0 \sin \omega t + \eta_0 \cos \omega t \end{aligned} \tag{6.48}$$

Squaring and rearranging we have the equation of the pathlines

$$\xi^2 + \eta^2 - \xi V_x t = \xi_0{}^2 + \eta_0{}^2 = \text{constant} \tag{6.49}$$

This can be calculated and plotted as in Fig. 6.14. The values assumed for the calculation and the APL program are shown on the same plot. In the same way the streaklines can be plotted for choices of T.

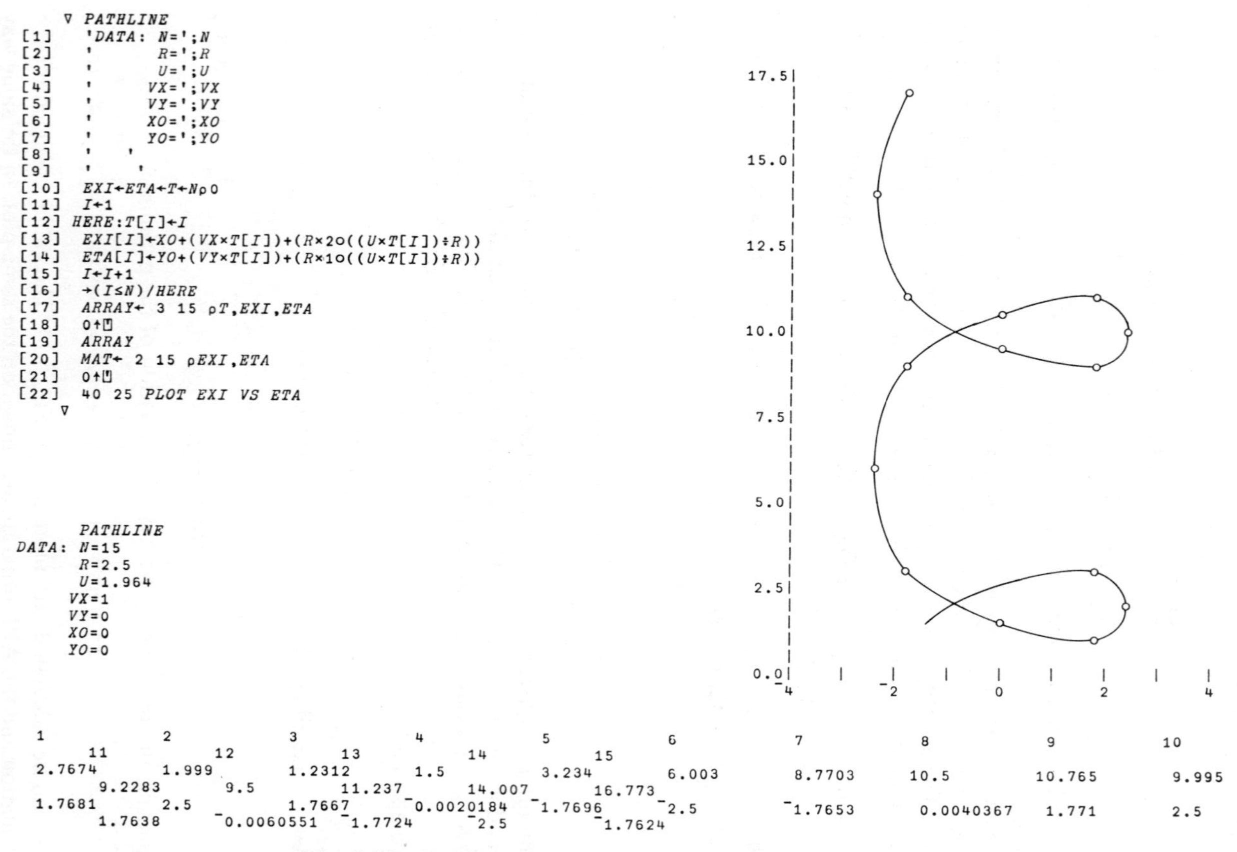

Fig. 6.14 Pathline program and plot for the case of Eq. (6.49).

B The Case of a Row of Alternating Cyclonic and Anticyclonic Disturbances of Equal Strength

Consider a number of N vortices alternating in direction of rotation and each having the same absolute value of the circulation Γ. According to Eq. (6.45) the position of the particle influenced by each of these vortices will be a summation of solutions. However, since the particle originates at ξ_0, η_0 for each disturbance, we must sum the influence $(\xi - \xi_0)$ and $(\eta - \eta_0)$. At the origin of the motion $\xi_0 = m_1 + x_1$. Thus, in Eq. (6.45) m and n before the last term can be replaced by $(\xi_0 - x_1)$ and $(\eta_0 - y_1)$. Thus when the summation takes place we have

$$\begin{aligned} \xi &= \xi_0 + V_x t + \sum_{k=1}^{N} (\xi_0 - m_{1k})(\cos \omega_k t - 1) - (\eta_0 - n_{1k}) \sin \omega_k t \\ \eta &= \xi_0 + V_y t + \sum_{k=1}^{N} (\xi_0 - m_{1k})(\sin \omega_k t) + (\eta_0 - n_{1k})(\cos \omega_k t - 1) \end{aligned} \tag{6.50}$$

where

$$\omega_k = \frac{\Gamma}{2\pi} \frac{1}{(\xi_0 - m_{1k})^2 + (\eta_0 - n_{1k})^2}$$

These are the set of equations for the *pathline* for a row of alternating disturbances.

The same principle applies to the combined effects on the *streakline*. We must sum $(\xi_\tau - \xi_0)$ and $(\eta_\tau - \eta_0)$ by replacing m_1 and n_1 in the last terms as before:

$$\begin{aligned} \xi_\tau &= \xi_0 + V_x(T - \tau) + \sum_{k=1}^{N} (\xi_0 - m_{1k} - V_x \tau)[\cos \omega_{k\tau}(T - \tau) - 1] \\ &\quad - (\eta_0 - n_{1k} - V_y \tau) \sin \omega_{k\tau}(T - \tau) \\ \eta_\tau &= \eta_0 + V_y(T - \tau) + \sum_{k=1}^{N} (\xi_0 - m_{1k} - V_x \tau) \sin \omega_{k\tau}(T - \tau) \\ &\quad + (\eta_0 - n_{1k} - V_y \tau)[\cos \omega_{k\tau}(T - \tau) - 1] \end{aligned} \tag{6.51}$$

and

$$\omega_{k\tau} = \frac{\Gamma}{2\pi} \frac{1}{(\xi_0 - m_{1k} - V_x \tau)^2 + (\eta_0 - n_{1k} - V_y \tau)^2}$$

C Results

Figure 6.15a represents computer solutions of the pathline (dots) of a particle issuing at $t = 0$ from the contaminating point $\eta = 1$, $\xi = 0$ and the streakline at $t = T$ of all particles that have issued from the same contaminating point. The model[7] consists of a row of five vortices alternating in direction of rotation but with equal strength and located on the $\eta = 0$ axis, spaced a distance h apart with the central cyclonic vortex at $\eta = \xi = 0$ at $t = 0$.

In the dimensionless model, at $r/r_0 = 1$ (vortex core) the velocity $U/U_0 = 1$ and the drift velocity $V/U_0 = \frac{1}{2}$. The time when the streakline is computed is $T = tV/r_0 = 7.5$; the time interval between successive positions of the particle is $\Delta\tau = 0.15$, giving a total of 50 points. The vortex spacing $h = 2.5r_0$. The computer program is given in Fig. 6.15b.

In an actual weather system it cannot be hoped that atmospheric disturbances be axisymmetric vortices with equal strength and with $V = U_\infty$. The method is still valid but the analysis becomes more complex.

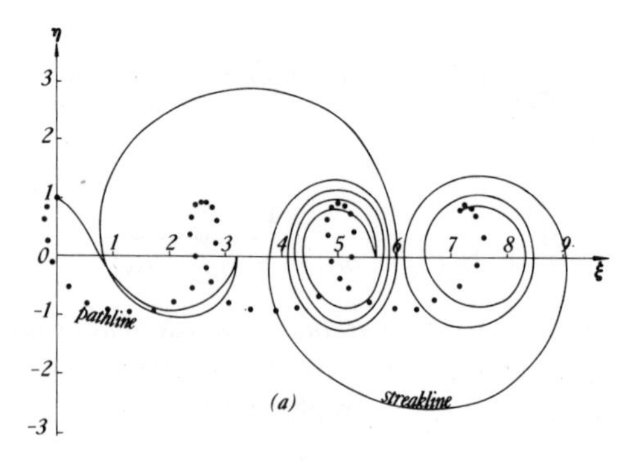

Fig. 6.15a Pathline and streakline in a model of five alternating vortices.

D Curvature Considerations of Streaklines and Pathlines

Let ε be the angle of the wind W_O makes with a given reference axis. Its rate of change with time, instantaneously, is related to the velocity $|W_O|$ and to the radius of curvature of the pathline r_p:

$$\dot{\varepsilon} = \frac{d\varepsilon}{dt} = \frac{W_O}{r_p} \tag{6.52}$$

From the Eulerian point of view and with reference to O, this rate of turn, or veering, to follow the particle is

$$\frac{d\varepsilon}{dt} = \left.\frac{\partial\varepsilon}{\partial t}\right|_O + \boldsymbol{W}_O \cdot \nabla\varepsilon \tag{6.53}$$

[7] The author is indebted to Mr. Robert Silverman and Mr. Z. Moussa for the numerical analysis and computational work.

```
C     STREAKLINE AND PATHLINE PGM--MULTIPLE VORTICES
      DIMENSION AKSI(50,50),ETA(50,50)
      READ(1,4)  NNN
4     FORMAT(I3)
      WRITE(3,2)  NNN
2     FORMAT('1',I3)
      DO 400 II=1,NNN
C     CCCCCCCCCCCCCCCCCCCCCCCCCCCCCCCCCCCCCCCCCCCCCCCCCCCCCCCCC
C     INPUT
      READ(1,5)  VO,RO,N,H,TT,DT,RF,THF,UFS
5     FORMAT(F5.3,F5.3,I3,F5.3,F5.3,F3.2,F5.3,F11.9,F5.3)
      WRITE(3,6)  VO,RO,N,H,TT,DT,RF,THF,UFS
6     FORMAT(' ',F5.3,2X,F5.3,2X,I3,2X,F5.3,2X,F5.3,2X,F3.2,2X,F5.3,
     C2X,F11.9,2X,F5.3)
      PI=3.14159265
C     NUMBER OF ITERATIONS TO GET PTS. ON PATHLINE
      NI=TT/DT
C     K IS PARTICLE INDEX
      DO 300  K=1,NI
C     INITIAL PARTICLE POSITION
      X=RF*COS(THF)
      Y=RF*SIN(THF)
      AKSI0=X
      ETA0=Y
C     ADJUST POSITION OF CONTAMINATING PT. IN MOVING FRAME, K.GT.1
      X=X-((K-1)*UFS*DT)
      Y=Y
      M=K
C     IIIIIIIIIIIIIIIIIIIIIIIIIIIIIIIIIIIIIIIIIIIIIIIIIIIIIIIII
C     CALCULATION OF PATHLINE OF PARTICLE K
C     J IS TIME INDEX
      DO 200  J=M,NI
C     *********************************************************
C     CALCULATE DX AND DY DUE TO N (DDD) VORTICES
      CONST=VO*RO
      DX=0.0
      DY=0.0
C     I IS VORTEX INDEX
      DO 100 I=1,N
C     CENTER LOCATION OF VORTICES
      XX=((-1)*((N-1)/2.)*H)+((I-1)*H)
      YY=0.0
      R=(((XX-X)**2)+((YY-Y)**2))**.5
      TH=ARSIN((ABS(YY-Y))/R)
      IF(XX.GE.X.AND.YY.LE.Y)  TH=PI-TH
      IF(XX.GE.X.AND.YY.GE.Y)  TH=PI+TH
      IF(XX.LE.X.AND.YY.GE.Y)  TH=(2*PI)-TH
      V=CONST/R
      W=(V/R)*((-1)**(I-1))
      TH=TH+(W*DT)
      DX=((R*COS(TH))-X)+DX
100   DY=((R*SIN(TH))-Y)+DY
C     *********************************************************
C     NEW PARTICLE POSITION
      X=X+DX
      Y=Y+DY
      IF(J.EQ.M)  AKSI(K,J)=AKSI0+DX+(UFS*DT)
      IF(J.EQ.M)  ETA(K,J)=ETA0+DY
      IF(J-M) 200,200,25
25    JJ=J-1
      AKSI(K,J)=AKSI(K,JJ)+DX+(UFS*DT)
      ETA(K,J)=ETA(K,JJ)+DY
200   CONTINUE
C     IIIIIIIIIIIIIIIIIIIIIIIIIIIIIIIIIIIIIIIIIIIIIIIIIIIIIIIII
300   CONTINUE
C     OUTPUT
      WRITE(3,10)
10    FORMAT('0','PATHLINE')
      WRITE(3,11)  (AKSI(1,J),ETA(1,J),J=1,NI)
11    FORMAT(' ',F10.4,5X,F10.4)
      WRITE(3,12)
12    FORMAT('0','STREAKLINE')
      WRITE(3,13)  (AKSI(K,NI),ETA(K,NI),K=1,NI)
13    FORMAT(' ',F10.4,5X,F10.4)
C     CCCCCCCCCCCCCCCCCCCCCCCCCCCCCCCCCCCCCCCCCCCCCCCCCCCCCCCCC
400   CONTINUE
      STOP
      END
```

Fig. 6.15b Computer program for Fig. 6.15a.

where the space variation $\boldsymbol{W}_O \cdot \nabla\varepsilon = W_O\, \partial\varepsilon/\partial s = W_O/r_s$. The symbol s refers to the streamline with ds being an infinitesimal length on the streamline and r_s the radius of curvature of the streamline. The quantity $\partial\varepsilon/\partial s$ is the change of angle along the streamline or the radius of its curvature. Thus from Eq. (6.53) we have[8]

$$\frac{\partial\varepsilon}{\partial t} = W_O\left(\frac{1}{r_p} - \frac{1}{r_s}\right) \tag{6.54}$$

The same change of angle, following the particle, could be written with reference to the moving frame O':

$$\frac{d\varepsilon}{dt} = \left.\frac{\partial\varepsilon}{\partial t}\right|_{O'} + \boldsymbol{W}_{O'} \cdot \nabla\varepsilon \tag{6.55}$$

With reference to the moving coordinate we may assume that $\partial\varepsilon/\partial t$ is small compared to the other changes, implying that the disturbance changes little in time with respect to its own frame.

If the angle between[9] $\boldsymbol{W}_O$ and $\boldsymbol{V}$ is γ, then Eq. (6.55) becomes

$$\frac{d\varepsilon}{dt} = [(\boldsymbol{U} + \boldsymbol{U}_\infty) - \boldsymbol{V}] \cdot \nabla\varepsilon$$

or

$$(U + U_\infty)\frac{1}{r_p} = (U + U_\infty)\frac{1}{r_s} - \frac{V\cos\gamma}{r_s}$$

and

$$r_p = \frac{r_s}{1 - (V/W_O)\cos\gamma} \tag{6.56}$$

Petterssen (1956) discusses the implications of Eq. (6.56) when applied to cyclonic and anticylonic disturbances (see Section 8.6 for directions of rotation).

When the speed of propagation $V < W_O$, then both radii of curvature have the same sign in the cyclonic and anticyclonic flows. This is normally the case in most parts of our weather system. For a westerly V, in the first two quadrants $\gamma > \pi/2$ in the cyclone and is negative in the anticyclone. Thus the value of the cosine is negative for the cyclone and positive for the anticyclone. This makes $r_p < r_s$ for the cyclone and the reverse for the anticyclone. Table 6.1 shows this relationship on all four quadrants for $V < W_O$.

[8] The literature cites Blaton (1938) as having been the first to propose this equation.
[9] This angle could be defined as ε, if we wish.

TABLE 6.1

Quadrant	Cyclone			Anticyclone		
	γ	Cosine	Radii of curvature	γ	Cosine	Radii of curvature
1	$\pi/2<\gamma<\pi$	$-$	$r_p<r_s$	$-\pi/2<\gamma<0$	$+$	$r_p>r_s$
2	$\pi<\gamma<3\pi/2$	$-$	$r_p<r_s$	$0<\gamma<\pi/2$	$+$	$r_p>r_s$
3	$\pi/2<\gamma<0$	$+$	$r_p>r_s$	$\pi/2<\gamma<\pi$	$-$	$r_p<r_s$
4	$0<\gamma<\pi/2$	$+$	$r_p>r_s$	$-\pi<\gamma<-\pi/2$	$-$	$r_p<r_s$

Figure 6.12 shows this relationship between the radii of curvature. However, if $V > W_O$, a less frequent occurrence in the immediate vicinity of the center of disturbance or very far from it, the pathlines are curved in the opposite sense to the curvature of the streamlines or isobars.

6.9 ROTATION IN THE ENVIRONMENT

As a consequence of the Helmholtz laws of vortex motion, Sir William Thomson (Lord Kelvin) proved that it was possible to introduce rotation in a given region of the environment through:

(a) Viscosity diffusing rotation already present at other parts of the flow, or viscous deformation at the boundaries such as the surface of the earth.

(b) External body forces that are not conservative, meaning forces that cannot be derived from a potential. Obviously, since the gravitational force is conservative we do not have to be concerned about it creating rotation. However, as we shall see in the next chapter, the Coriolis force, which is a result of a motion on a rotating frame, is a nonconservative body force and will be responsible for the production of rotation in the geofluid

(c) Misalignment of the constant-pressure surfaces with the constant-density surfaces. In other words, the density of the medium does not depend exclusively on the pressure. We have seen all along that this situation is likely to occur in the environment, in the atmosphere because of temperature effects and in the oceans because of temperature and salinity effects. Thus we define a *barotropic* condition in the environment when the density is a function of pressure only. Otherwise we call a situation *baroclinic* when the density depends on other thermodynamic properties besides the pressure.

The rotational circulation system of the atmosphere depends greatly on the density variations from the pole to the equator due to temperature differences, and considerable rotational flows are generated in the oceans, particularly in estuaries as a result of density variations caused by salinity gradients.

Since rotational flows are constantly produced in our geofluid system by all three of the conditions stated above, it is important to define and study all concepts related to rotation of fluids, and to remember that rotation, as all kinematical quantities, is dependent on the frame of reference from which it is measured.

6.10 ROTATION–VORTICITY

In solid mechanics, a *rigid body* is defined as a body in which the respective distances among the particles remain invariant, regardless of the forces or motion to which it is subjected. This is not the case when a fluid is involved. The relative distances among fluid parcels change constantly owing to forces and motion, which is why we call them fluids.

When a rigid body is said to be in *pure rotation*, each particle of the body describes a circle about an axis of rotation. The particle speed is proportional to the radius of rotation. Every particle and every arbitrary line on the rigid body must have the same angular velocity.

Since the relative distances between particles do not necessarily remain constant in a fluid, all parts of the fluid do not have to rotate at the same angular velocity or with respect to the same axis or frame of reference. Consequently, the definition of rotation for solids must be reconsidered so as to make it applicable and consistent for fluids as well.

Let *rotation* at a point be defined as the average angular velocity of two[10] rigid differential linear elements, within a solid or a fluid, originally perpendicular to each other. Rotation, as in rigid body mechanics, will be a vector quantity whose sense is normal to the plane formed by the two elements. By convention, its direction is given by the right-hand rule of coordinate rotation or a right-handed screw.

The discussion that follows is simplified for a two-dimensional motion. The total rotation can be obtained by the vector sum of the rotation on the other planes.

Consider the two rigid linear differential elements 1–2 and 3–4 as shown in Fig. 6.16. The two elements are inside a two-dimensional flow and fixed to be normal at the instant considered. Let $U(x, y)$ and $V(x, y)$ be the two components of the velocity vector $\boldsymbol{U}$. In the configuration shown, the variation in the x component of the velocity will contribute only to rotation of the element 1–2, while variations in y will contribute rotation to the element 3–4 only. At the same time both elements will be translated downstream, but the main concern here is the study of rotation.

[10] Two linear elements are involved so as to give a fluid the freedom it needs to deform.

If the velocity U at the center point 0 is given, the corresponding velocities at points 1, 2, 3, and 4 can be evaluated by a Taylor series expansion [discussed in Eq. (2.1)] provided the lengths of the elements are made very small, so that the second-order terms in the expansion are neglected. For this we need differential lengths. First, consider the element 1–2 and also, by an *a priori* decision, assume that all velocities increase in the increasing coordinate direction. Then the velocities at the endpoints are readily found:

$$U_1 = U_0 + \left(\frac{\partial U}{\partial y}\right)_0 \frac{\Delta l_{1-2}}{2} + \cdots$$

$$U_2 = U_0 - \left(\frac{\partial U}{\partial y}\right)_0 \frac{\Delta l_{1-2}}{2} - \cdots$$

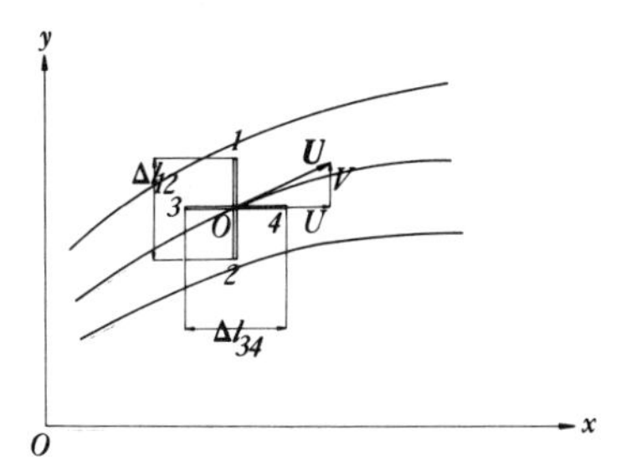

Fig. 6.16 Motion of rigid linear elements in a two-dimensional flow field.

The difference $(U_1 - U_2)$ represents the relative velocity of the point 1 with reference to 2. This relative velocity on the element will contribute a rotation of 1 with respect to 2. The angular velocity will be

$$\omega_{1-2} = \frac{U_1 - U_2}{\Delta l_{1-2}} = 2\left(\frac{\partial U}{\partial y}\right)_0 \frac{\Delta l_{1-2}/2}{\Delta l_{1-2}}$$

$$= \left(\frac{\partial U}{\partial y}\right)_0 \circlearrowright$$

On the basis of Fig. 6.16 element 1–2 rotates in a clockwise direction, and the vector representing the rotation is downward, piercing the plane of the paper.

Element 3–4 in Fig. 6.17 can be analyzed in the same way:

$$V_4 = V_0 + \left(\frac{\partial V}{\partial x}\right)_0 \frac{\Delta l_{3-4}}{2} + \cdots$$

$$V_3 = V_0 - \left(\frac{\partial V}{\partial x}\right)_0 \frac{\Delta l_{3-4}}{2} - \cdots$$

and the angular velocity of that element is

$$\omega_{3-4} = \frac{V_4 - V_3}{\Delta l_{3-4}} = 2\left(\frac{\partial V}{\partial x}\right)_0 \frac{\Delta l_{3-4}/2}{\Delta l_{3-4}}$$
$$= \left(\frac{\partial V}{\partial x}\right)_0^{\circlearrowleft}$$

This is a counterclockwise direction. The rotation of the fluid at 0 is, according to our previous definition, the average rotation of the two elements. Adding, taking into account the directions of rotation and signs, and dividing by 2, the z component of rotation of the motion on the x–y plane is

$$\omega_z = \frac{1}{2}\left(\frac{\partial V}{\partial x} - \frac{\partial U}{\partial y}\right) \tag{6.57}$$

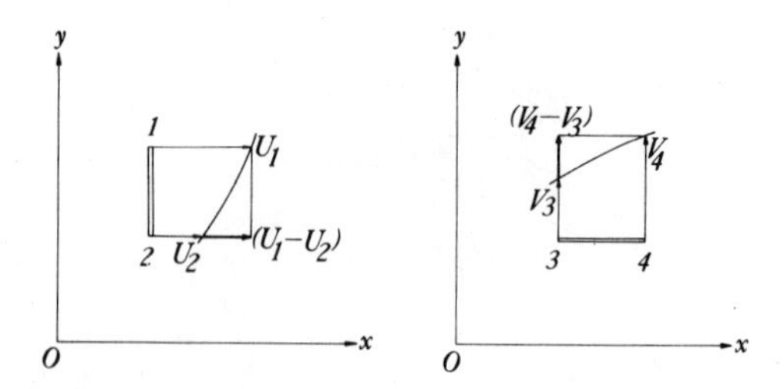

Fig. 6.17 The rigid element in a velocity gradient.

The *vorticity* is defined as twice the value of rotation, and thus the z component of the vorticity in Cartesian coordinates is

$$\zeta_z = \left(\frac{\partial V}{\partial x} - \frac{\partial U}{\partial y}\right) \tag{6.58}$$

By simple deduction, the vorticity components on the other two planes x–z, and y–z are

$$\zeta_y = \left(\frac{\partial U}{\partial z} - \frac{\partial W}{\partial x}\right), \qquad \zeta_x = \left(\frac{\partial W}{\partial y} - \frac{\partial V}{\partial z}\right) \tag{6.59}$$

The total vorticity vector is then the vector sum of its components

$$\boldsymbol{\zeta} = \boldsymbol{i}\zeta_x + \mathbf{j}\zeta_y + \mathbf{k}\zeta_z$$
$$= \boldsymbol{i}\left(\frac{\partial W}{\partial y} - \frac{\partial V}{\partial z}\right) + \boldsymbol{j}\left(\frac{\partial U}{\partial z} - \frac{\partial W}{\partial x}\right) + \boldsymbol{k}\left(\frac{\partial V}{\partial x} - \frac{\partial V}{\partial y}\right)$$
$$= \nabla \times \boldsymbol{U} \tag{6.60}$$

Thus vorticity, which is twice the rotation, is also equal to the *curl* of the velocity vector. This is also developed in Appendix A.

A motion is said to be *irrotational* in a given domain when vorticity is zero at every point in this domain. Thus when vorticity is constant in a region, we can say that the region rotates as a rigid body.

A *vortex line* is a line in the fluid such that the tangent to this line gives the direction of the vorticity at every point on the line. In *cylindrical coordinates* the *del* operator and the velocity vector are

$$\nabla \equiv \boldsymbol{i}\frac{\partial}{\partial r} + \boldsymbol{j}\frac{1}{r}\frac{\partial}{\partial \theta} + \boldsymbol{k}\frac{\partial}{\partial z} \qquad \text{and} \qquad \boldsymbol{U} = \boldsymbol{i}U + \boldsymbol{j}V + \boldsymbol{k}W$$

Thus the cylindrical components of the curl are

$$\nabla \times \boldsymbol{U} = \boldsymbol{i}\left(\frac{1}{r}\frac{\partial W}{\partial \theta} - \frac{\partial V}{\partial z}\right) + \boldsymbol{j}\left(\frac{\partial U}{\partial z} - \frac{\partial W}{\partial r}\right) + \boldsymbol{k}\frac{1}{r}\left[\frac{\partial}{\partial r}(rV) - \frac{\partial U}{\partial \theta}\right]$$

$$= \boldsymbol{i}\zeta_r + \boldsymbol{j}\zeta_\theta + \boldsymbol{k}\zeta_z \tag{6.61}$$

where the unit vectors $\boldsymbol{i}$, $\boldsymbol{j}$, and $\boldsymbol{k}$ and the velocity components U, V, and W are in the directions r, θ, and z, respectively.

In *spherical coordinates*, the conventions for the unit vectors $\boldsymbol{i}$, $\boldsymbol{j}$, and $\boldsymbol{k}$ are the unit vectors in the directions toward the east, the north, and altitude, in other words, along the longitude θ, the latitude ψ, and the elevation z, or R, respectively (see Fig. 6.18). The corresponding velocity components are U, V, and W

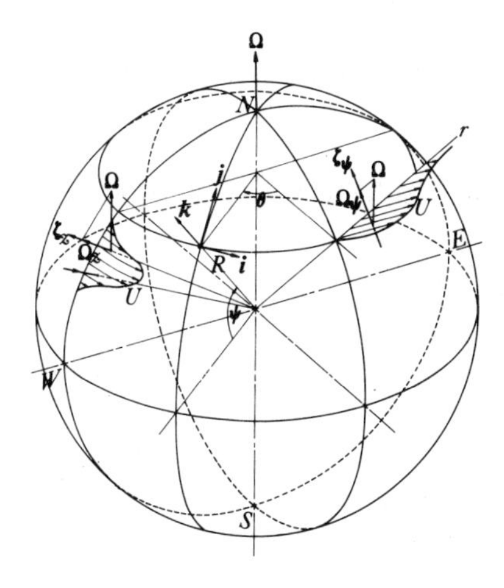

Fig. 6.18 Vorticity components on the earth's surface due to gradients of westerly winds and currents.

and the changes in unit vectors during the differentiation in the curl are

$$\frac{\partial \boldsymbol{i}}{\partial \psi} = 0, \qquad \frac{\partial \boldsymbol{j}}{\partial \Psi} = -\boldsymbol{k}, \qquad \frac{\partial \boldsymbol{k}}{\partial \psi} = \boldsymbol{j}, \qquad \frac{\partial}{\partial R}(\boldsymbol{i}, \boldsymbol{j}, \boldsymbol{k}) = 0,$$

$$\frac{\partial \boldsymbol{i}}{\partial \theta} = -(\boldsymbol{k} \cos \psi + \boldsymbol{j} \sin \psi), \qquad \frac{\partial \boldsymbol{j}}{\partial \theta} = \boldsymbol{i} \sin \psi, \qquad \frac{\partial \boldsymbol{k}}{\partial \theta} = \boldsymbol{i} \cos \psi$$

and the vorticity in its spherical components is

$$\begin{aligned} \nabla \times \boldsymbol{U} &= \boldsymbol{i}\zeta_\theta + \boldsymbol{j}\zeta_\psi + \boldsymbol{k}\zeta_z \\ &= \frac{\boldsymbol{i}}{R}\left[\frac{\partial W}{\partial \psi} - \frac{\partial}{\partial R}(RV)\right] + \frac{\boldsymbol{j}}{R \cos \psi}\left[\frac{\partial}{\partial R}(UR \cos \psi) - \frac{\partial W}{\partial \theta}\right] \\ &\quad + \frac{\boldsymbol{k}}{R \cos \psi}\left[\frac{\partial V}{\partial \theta} - \frac{\partial}{\partial \psi}(U \cos \psi)\right] \end{aligned} \tag{6.62}$$

The *horizontal velocity* is then the vector sum of U and V and is often represented by $\boldsymbol{U}_{\mathrm{h}}$. Depending on the geometry of the problem in question, any of the three representations of the vorticity, Eqs. (6.60), (6.61), or (6.62), may be used.

It can be verified that the rotation of rigid bodies is inclusive in these broader definitions. We have said all along in this chapter that *all* kinematical quantities, including the vorticity, depend very much on the frame from which they are measured. The vorticity equations we have just derived are good for any frame of reference fixed to the surface of the earth. We must realize that the total rotation, or vorticity, of the fluid with reference to the center of the earth is that obtained by these equations to which we add vectorially the rotation of the earth. Illustrative Example 6.4 discusses this question.

6.11 IRROTATIONAL MOTION—THE VELOCITY POTENTIAL

It is a mathematical fact that through Stokes's, theorem in Appendix A [Eq. (A.63)]

$$\oint_C \boldsymbol{U} \cdot d\boldsymbol{c} = \int_S (\nabla \times \boldsymbol{U}) \cdot d\boldsymbol{S}$$

when a motion is said to be irrotational, implying that $\nabla \times \boldsymbol{U} = 0$, in a domain contained in the area $\boldsymbol{S}$ with a limiting contour C, that

$$\oint_C \boldsymbol{U} \cdot d\boldsymbol{c} = 0$$

For a closed line integral of a property $\boldsymbol{U}$ to be zero, the integrand $\boldsymbol{U} \cdot \boldsymbol{dc}$ must be of the form of a total or exact differential $-d\Phi = \boldsymbol{U} \cdot \boldsymbol{dc}$. By definition, the exact differential can always be expressed as $d\Phi = \nabla\Phi \cdot \boldsymbol{dc}$ and one concludes that

$$\boldsymbol{U} = -\nabla\Phi$$

Thus, irrotationality implies that the velocity vector can *always* be expressed in terms of the gradient of a *velocity potential* Φ. This question was also discussed in Section 2.6 with reference to the gravitational acceleration.

In Cartesian components

$$U = -\frac{\partial \Phi}{\partial x}, \qquad V = -\frac{\partial \Phi}{\partial y}, \qquad W = -\frac{\partial \Phi}{\partial z}$$

Illustrative Example 6.4

Consider westerly winds or currents (coming from the west). The westerly (U) is likely to display variations of magnitude with altitude, in R, owing to friction on the earth's surface and also variations with latitude ψ. These variations are shown in Fig. 6.18. Looking at Eq. (6.62) U only appears in the latitudinal ($\boldsymbol{j}$) and vertical ($\boldsymbol{k}$) components of the vorticity. With the variation of U with altitude

$$\zeta_\psi = \frac{1}{R \cos \psi}\left[\frac{\partial}{\partial R}(UR \cos \psi)\right] = \frac{\partial U}{\partial R} + \frac{U}{R} \tag{6.63}$$

are the only terms appearing in the $\boldsymbol{j}$ or latitudinal component of the vorticity ζ_ψ, giving rise to a rotation of the geofluid in that direction because of the motion on the $\boldsymbol{i}$–$\boldsymbol{k}$ surface. At the same point where this vorticity component is calculated, the earth also rotates; it has a component of rotation in the $\boldsymbol{j}$ direction equal to $\Omega_\psi = \Omega \cos \psi$ and a vorticity twice this amount since we have defined it so. Thus the total vorticity component in the $\boldsymbol{j}$ direction with reference to a coordinate fixed to space is

$$\zeta_{t\psi} = \zeta_\psi + 2\Omega \cos \psi$$

where the added subscript t implies total. The senses of ζ_ψ and Ω_ψ must be considered in the addition. When U increases with R, ζ_ψ is in the positive $\boldsymbol{j}$ direction.

If we now consider the variations of U with latitude ψ, we should have ζ_z or the kth component of Eq. (6.62), which gives

$$\zeta_z = -\frac{1}{R \cos \psi}\frac{\partial}{\partial \psi}(U \cos \psi) = \frac{U}{R}\tan \psi - \frac{1}{R}\frac{\partial U}{\partial \psi} \tag{6.64}$$

In the same way, the earth also has a vertical component of rotation, and the total vertical vorticity is

$$\zeta_{tz} = \zeta_z + 2\Omega \sin \psi$$

Vectorially, these total components of vorticity can be added to give the total or absolute vorticity vector:

$$\boldsymbol{\zeta}_t = \boldsymbol{\zeta} + 2\boldsymbol{\Omega} \tag{6.65}$$

where $\boldsymbol{\zeta}$ is the vorticity of the environment with reference to the earth's surface and $2\boldsymbol{\Omega}$ is the vorticity of the earth.

6.12 THE FREE VORTEX

There are a number of interesting flow configurations in the environment that revolve around an axis, and are almost irrotational. For instance, hurricanes, tornados, waterspouts, dust devils, etc., have circular motions without necessarily having vorticity present, except in a small central core. How is it then possible to have circular motion without rotation or vorticity?

Since the examples we gave are cylindrical in form, let us look at the vorticity equation in cylindrical coordinates, Eq. (6.61). For a two-dimensional motion parallel to the surface of the earth, the kth component of the vorticity will be the only component to consider, and even in that component because of axisymmetry there will be no variations with θ. Thus the vorticity in tornado-like flows is simply

$$\zeta_z = \frac{1}{r}\left[\frac{\partial}{\partial r}(rV)\right] \tag{6.66}$$

where r is the radius from the center of the circular flow. For this flow to be irrotational $d(rV)/dr = 0$, where V is the tangential velocity. Thus an integration of this equation gives

$$rV = \text{constant} \tag{6.67}$$

or that the tangential velocity must decrease with increasing radius. We know that the rigid body cannot have such a velocity distribution because it is rotational and it must satisfy $V/r = \text{constant}$, but the fluid can and does perform according to Eq. (6.67) if the conditions are right. It is easy to see that if Eq. (6.67) applies to a region of the environment, then this situation is comparable to conserving the *moment of momentum* or the *angular momentum* of that fluid. This is because $\boldsymbol{r} \times m\boldsymbol{V}$ is the moment of momentum of a fluid parcel. In moving from one $\boldsymbol{r}$ to another if the parcel is to conserve its moment of momentum, for a given mass of the parcel m, the vector product $\boldsymbol{r} \times \boldsymbol{V} = \boldsymbol{k} rV$ must be constant, where $\boldsymbol{k}$ is the unit vector normal to the plane of r and V. Thus moment of momentum is conserved, and the motion is irrotational. This type of flow is called the *free vortex*, and a free vortex is *irrotational*. According to Kelvin's theorem (Section 6.9), if none of the external forces mentioned acts on the vortex, then the vortex can remain irrotational forever.

6.13 TEMPEST IN A TEACUP

Because the motion in a teacup after stirring, has many similarities with severe storms, it is an interesting problem to analyze. Figure 6.19 shows this motion.

During stirring, since the motion of the spoon in the liquid is frictional, according to Kelvin's theorem some of this vorticity would be introduced into

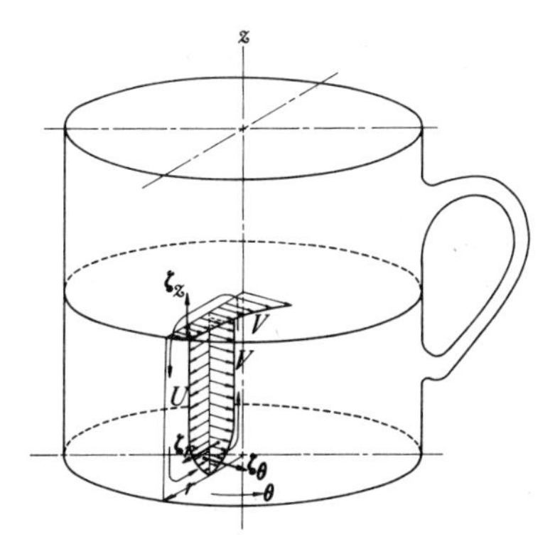

Fig. 6.19 Circumferential motion with friction.

the fluid in the z direction and around OO'. Near the walls of the cup AB the friction at the wall will slow the liquid to zero velocity and thus according to Eq. (6.61) the velocity distribution $V = f(r)$ produced there will show up as a ζ_z vorticity in the $-z$ direction because in the viscous layer V decreases with r.

The equilibrium conditions of a fluid particle moving around a circle of radius r and a velocity V require that the centrifugal force per unit volume $\rho V^2/r$ be balanced by the pressure gradient dp/dr pointed toward the axis OO'. The pressure at $r + dr$ on the particle must be larger than the pressure at r, and this difference in pressure balances the centrifugal force. As we have seen in Chapter 4, the weight of the particle is balanced by the pressure gradient $-dp/dz$ according to the hydrostatic equation.

At the bottom of the cup there is also a viscous layer, as shown on the figure. The viscous layer produces gradients of V in the z direction. Looking again at Eq. (6.61) $\partial V/\partial z$ enters into the ζ_r component of vorticity. The presence of the boundary layer at the bottom causes another effect. Since $\partial p/\partial z$ is not a function of r, $\partial p/\partial r$ cannot be a function of z. But since at the bottom of the cup V decreases in the viscous layer, then $\partial p/\partial r$, which is just sufficient outside the viscous layer to balance the centrifugal force, will be larger than necessary within the viscous layer which has a lower centrifugal force at the same r. This will generate a radial motion toward the axis OO' at the bottom of the cup. This radial flow at the bottom must be replaced by a descending motion along AB and an ascending motion along OO'. This *secondary motion*, generated from the *primary motion* around OO', is counterclockwise and accounts for the generation of vorticity in the θ direction, ζ_θ. Thus according to Kelvin's theorem we have generated rotational motion in θ through nonconservative frictional forces near the solid boundaries.

This radial motion toward the center at the bottom of a cup is observed when tea leaves gather at the center of the cup after stirring because they are too heavy to travel upward along the axis of the cup as the liquid tea does.

Hurricanes and tornados have similar effects near the earth and ocean surface. Although the pumping of air near the center of the hurricane is primarily due to thermodynamic buoyancy, the boundary layer near the surface of the earth contributes to this updraft.

From this experiment we can see how secondary flows in r and z can be generated from motion originally in the θ direction, if the ingredients of Kelvin's theorem are present.

6.14 THE CONCEPT OF CIRCULATION

Given an arbitrary closed region with a surface area S and bound by a closed contour C through which fluid is flowing, the circulation around the contour C is defined as the integral summation of the velocity component tangential to the contour C at every point on the contour times the elemental length of the contour dc. In mathematics, this is the line integral of the velocity vector $\boldsymbol{U}$. Let Γ represent the value of this circulation. Then

$$\Gamma = \oint_C \boldsymbol{U} \cdot \boldsymbol{t}\, dc = \oint_C U \cos\theta\, dc \tag{6.68}$$

To give reality to this definition, consider the tornado-like motion in Fig. 6.20. The circular flow is confined to a finite column with cross-sectional area A. Let the contour C bind the two surfaces A and S, and let $\boldsymbol{U}$ be the velocity everywhere. From what we learned, the vorticity, if any, must be along the axis of the figure, piercing A and S. A vortex line is shown. According to the figure $\boldsymbol{U}$ need not be tangent to the contour C. Thus the projection of $\boldsymbol{U}$ along $\boldsymbol{t}$, the unit vector giving the direction of C, is what needs to be integrated. As we shall see in the course of this book, the circulation Γ is a very important concept in environmental considerations. For instance, if Eq. (6.68) is applied to the free vortex discussed in Section 6.12, the tangential velocity V is everywhere tangent to a contour taken as a circle with radius r, and since V is constant with θ, $\Gamma = 2\pi r V$. Thus the velocity in the vortex is $V = (\Gamma/2\pi)/r$ where $(\Gamma/2\pi)$ is the strength of the vortex.

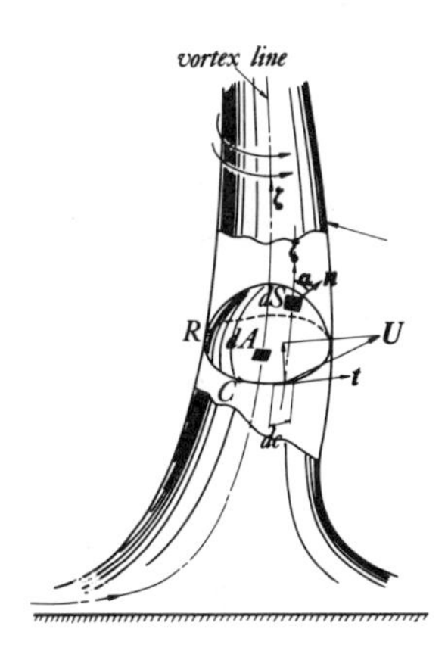

Fig. 6.20 Concept of circulation applied to the flow of a tornado.

According to Stokes's theorem (derived in Appendix A), a closed line contour integral of a vector quantity $\boldsymbol{U}$ is also equal to the integral summation of the vorticity (curl of $\boldsymbol{U}$) on any surface bound by C. In the case of the figure, the contour C binds the surfaces A and S so the theorem should apply to both. Thus,

$$\Gamma = \oint_C \boldsymbol{U} \cdot \boldsymbol{t}\, dc = \int_S (\boldsymbol{\zeta} \cdot \boldsymbol{n})\, dS = \int_S \boldsymbol{n} \cdot (\nabla \times \boldsymbol{U})\, dS \tag{6.69}$$

where $\boldsymbol{n}$ is the unit normal to S or A. Thus we conclude that circulation is also the total summed vorticity in an area. If vorticity is constant in the area S, then $\Gamma = \zeta S$. Conserving Γ in the bounds of an entire circular motion like a tornado implies conserving the angular momentum of the entire flow. From this conservation principle and Eq. (6.69), it is implied that increasing the area of the tornado must decrease the vorticity within it, and vice versa. Actually if we take the height of the vortical flow into account, stretching the flow will decrease the area and increase the rotational speed, and compression will increase the area and decrease the rotational speed [see Section 8.9, Eq. (8.71)]. This is what happens to a skater on an ice rink, when the angular velocity is increased by reducing the area projected on the ground or the moment of inertia with respect to the vertical axis.

Again according to Kelvin's theorem (Section 6.9), the circulation will vary with time if a nonconservative force, such as friction, acts upon the flow. In the atmosphere and the oceans, away from the earth's surface and in regions with small changes in latitude, it is reasonable to think that angular momentum is conserved. The following illustrative example applies these principles to simple circular flows.

Illustrative Example 6.5

A two-dimensional tornado with circular cross section is simulated in Fig. 6.21. The core of the tornado of radius r_0 rotates at a constant angular velocity ω. This means that the core rotates as a rigid body. This approximates what actually happens near the center of a tornado. Thus in the core

$$V = \omega r$$

Outside the core, $r > r_0$, the motion is that of a free vortex; thus

$$V = C/r$$

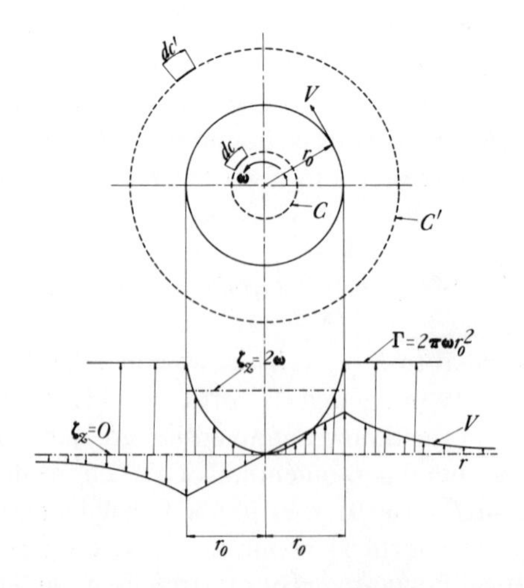

Fig. 6.21 Velocity, vorticity, and circulation in tornado-like motion.

Since at $r = r_0$ the velocity should have a single value (no slip), the constant can be evaluated through

$$\omega r_0 = C/r_0, \qquad C = \omega r_0^2$$

The axial (updraft) velocity component is of no concern in this problem since the relations expressed so far involve the tangential component of the velocity. The vorticity in the core is constant everywhere in that core and equal to 2ω. The vorticity in the outer part is zero, according to Section 6.12.

Figure 6.21 shows the distribution of V and ζ_z with r.

The circulation Γ can be computed either from the line integral of the velocity or the area integral of the vorticity in Eq. (6.69). *Thus in the core, $r < r_0$*, giving a *line integral*

$$\Gamma_C = \oint V\,dc = \int_0^{2\pi} Vr\,d\theta$$

$$= \int_0^{2\pi} (\omega r) r\,d\theta = 2\pi\omega r^2$$

and an *area integral*

$$\Gamma_C = (2\omega)\pi r^2 = 2\pi\omega r^2$$

Outside the core, $r > r_0$, the *line integral*

$$\Gamma_{C'} = \oint V\,dc' = \int_0^{2\pi} Vr\,d\theta$$

$$= \int_0^{2\pi} \frac{\Gamma}{2\pi r}\,d\theta = 2\pi\omega r_0^2 = \text{constant}$$

For the *area integral*, since the vorticity is zero everywhere outside the core, the area bounded by the contour C' has vorticity only in the portion bounded by C and thus circulation will cease to increase beyond r_0. Figure 6.21 shows this in the plot of Γ.

6.15 INFLUENCE OF THE EARTH'S ROTATION ON THE ROTATION OF THE GEOFLUID

Since the geofluid is free to move on the earth's surface, subject of course to existing forces, let us examine the influence of the earth's rotation and the variations of the components of the earth's rotation on the fluid.

In considering the diagram of Fig. 6.22, the angular velocity of the earth $\boldsymbol{\Omega}$ is a free vector which applies everywhere on the surface of the earth, maintaining its magnitude and direction. As we shall see in Chapter 7 there are two forms of acceleration that are due to the earth's rotation, the absolute magnitudes of which are the centrifugal acceleration $|\Omega^2 \boldsymbol{r}| = \Omega^2 R \cos \psi$ and the Coriolis acceleration $|2\boldsymbol{\Omega} \times \boldsymbol{U}| = 2U\Omega \sin \alpha$, where α is the angle between the resultant velocity of the wind or current $\boldsymbol{U}$ and the angular velocity vector $\boldsymbol{\Omega}$. For a horizontal motion this angle can vary as $\psi < \alpha < 180 - \psi$ and the Coriolis acceleration is zero at the equator where $\psi = 0$ and maximum at the poles for the same fluid velocity. The radius $\boldsymbol{r}$ is that of the latitude circle, and is equal to $R \cos \psi$.

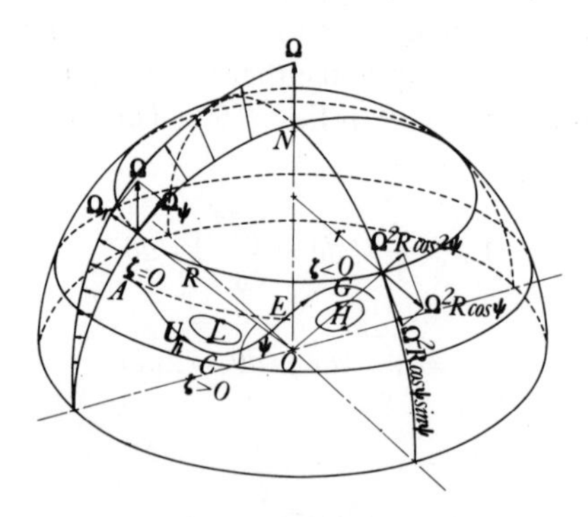

Fig. 6.22 Effects of change of latitude in nearly geostrophic flows in the Northern Hemisphere.

These two accelerations have components in the vertical and horizontal directions. Assuming that $\boldsymbol{U} = \boldsymbol{U}_h$ is everywhere horizontal, the following relationships become evident:

Vertical direction

(a) earth's rotation, $\Omega_z = \Omega \sin \psi$;
(b) centrifugal acceleration, $\Omega^2 R \cos^2 \psi$;
(c) Coriolis acceleration, $2\Omega U_h \sin \alpha \cos \psi = 2\Omega_\psi U_h \sin \alpha$.

Horizontal direction

(a) earth's rotation, $\Omega_\psi = \Omega \cos \psi$;
(b) centrifugal acceleration, $\Omega^2 R \cos \psi \sin \psi$ (toward south);
(c) Coriolis acceleration, $2\Omega_z U_h = 2\Omega U_h \sin \psi$ (normal to U_h).

When these accelerations are dominant in the motion of the geofluid, then the rotation of the earth is the motor force for the environment, balancing, of course, established pressure gradients. The motion is said to be *geostrophic*, i.e., under the exclusive influence of the earth's rotation.

Looking at Fig. 6.22 consider a horizontal velocity $\boldsymbol{U}_h$ to be westerly at the point A and let this motion originally be irrotational relative to the earth's surface. According to the acceleration components listed, this motion will soon be under the influence of the *Coriolis force*[11] applicable to the latitude of A and the magnitude of U_h. This influence will be felt: (a) in the horizontal direction with a magnitude $2\Omega U_h \sin \psi$ and normal to $\boldsymbol{U}_h$. When facing $\boldsymbol{U}_h$ the Coriolis force veers the motion to the left. As discussed in Section 7.9, the Coriolis force is defined as $2\boldsymbol{U}_h \times \boldsymbol{\Omega}$, opposite in direction to the Coriolis acceleration. This means that the original motion will veer toward the south. (b) In the vertical upward direction the influence of the Coriolis will be $2\Omega U_h \cos \psi$ since at A the angle $\alpha = \pi/2$. As the wind or current veers south, the following effects take place:

(1) The latitude angle ψ decreases, and as a result all terms in $\sin \psi$ decrease while the terms in $\cos \psi$ increase.

(2) The vertical component of the earth's rotation $\Omega_z = \Omega \sin \psi$ decreases as a consequence of (1). The wind or current originally at A with the same angular momentum as that of the earth, will move at lower latitudes where the vertical angular momentum of the earth or rotation is smaller. The motion of the environment as it tries to conserve its angular momentum will find itself developing a vertical vorticity relative to the surface of the earth as it veers south. The farther south it moves, the larger will be the relative vorticity of the stream. As indicated on Fig. 6.22 the geofluid will have $\zeta > 0$ at the point C. This means that through Coriolis, a relative rotation has been generated. Since the environment has a larger counterclockwise vorticity than the earth, its vorticity is positive.[12] This counterclockwise motion is called a *cyclonic flow*. We shall see in dynamics that this corresponds to a low pressure area.

(3) The vertical components of the centrifugal and Coriolis accelerations also increase when the stream veers to the south. This means that an extension or acceleration upward of the vortex filaments will take place,[13] reducing their cross-sectional areas. As we discussed after Eq. (6.69), in order to conserve the moment of momentum while decreasing cross-sectional area, the vorticity in the column must increase. As we shall see later the iteraction of dW/dz and ζ may increase ζ beyond conservation. This will intensify the results of (2). However, although the vertical component of the Coriolis will increase with decreasing ψ, the angle α, originally $\pi/2$, will become acute from A to C, tending to reduce the acceleration as the column veers south. Naturally, the final balance will depend on dynamics and not on this kinematical qualitative explanation.

(4) We have established that the horizontal component of the Coriolis force is always to the left of the motion when facing the velocity vector. This horizontal component of the centrifugal acceleration was given by $\Omega^2 R \cos \psi$

[11] The distinction between a Coriolis acceleration and a Coriolis force is made in Sections 7.3 and 7.9. The difference is simply that one is the negative of the other.

[12] The opposite takes place in the Southern Hemisphere.

[13] This phenomenon is considered quantitatively in Section 8.9, Eq. (8.71).

sin ψ. At $\psi = 45°$ the rates of change of the sine and the cosine are the same. However, below 45° the sine varies faster than the cosine and therefore a movement towards the south below $\psi = 45°$ reduces the magnitude of the horizontal component.

(5) Because of (4) and because the angular velocity of the environment increases relative to the earth's surface there will be a tendency for the stream to follow the path *ACE*. Return to *E* indicates return to the same kinematical conditions as those at *A*.

(6) At *E* the stream has a NE direction, after having generated a low pressure area at *C*. The horizontal component of the Coriolis force is again to the left facing the velocity vector. This will tend to veer the flow toward the east as it proceeds from *E* to *G*. Trying to maintain its angular momentum, as the stream proceeds to higher latitudes, its vorticity relative to the earth's surface will decrease and thus become negative, or clockwise with reference to the earth's surface. This will be an *anticyclonic* region rotating clockwise and with higher barometric pressures as we shall see in Section 8.6. In turn the vertical components of the centrifugal and Coriolis will decrease, since both contain cos ψ. The result will be the opposite of what occurred at (3), a compression of the column, increase of cross-sectional areas of vortex filaments, and reduction of vorticity.

This model of geostrophic motion of *zonal patterns* that are veered to the left by the horizontal component of the Coriolis force and the expansion and contraction of the column as the stream changes latitude, which results in the increase and decrease of vorticity relative to the earth's surface, is what actually takes place in our atmospheric environment at mid-latitudes and is responsible for the control of our weather system. Although ocean currents are also susceptible to the same influences, the surface wind stresses play a dominant role on them.

These waves of secondary motion called zonal patterns, which are successions of high and low pressure areas due to the alignment of the centrifugal and Coriolis forces acting on the existing winds and currents, are noticeable in any global-size weather map such as the sample shown in Fig. 6.23. For pure geostrophic motions the wavelength of these meandering streams[14] is given by

$$\lambda = 2\pi(U_h/\beta)^{1/2}$$

where U_h is the horizontal wind speed, and β is twice the rate of change with latitude of the vertical component of the earth's rotation. The units of β are radians per distance along the longitude, or $(2/R)(d\Omega_z/d\psi)$. Thus

$$\lambda = 2\pi\left(\frac{U_h R}{2\Omega \cos \psi}\right)^{1/2} \tag{6.70}$$

[14] This wave phenomenon is treated formally in Section 8.8.

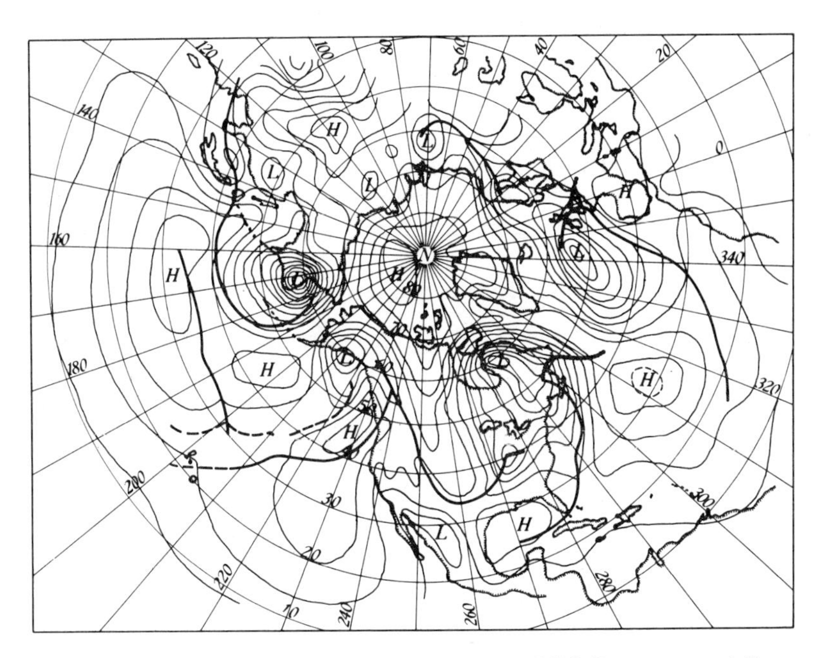

Fig. 6.23 Hemispherical weather map. Courtesy of US Department of Commerce, Environmental Science Services Administration.

Since the cosine is maximum at the equator, the wavelength will be minimum there. Actually these patterns are confined to mid-latitudes.

Illustrative Example 6.6

Calculate the wavelength or the distance from one low pressure area to the next, at a latitude of 45° and with a horizontal wind speed of 50 km/hr:

$$\lambda = 2\pi\left(\frac{50 \times 6370}{0.52 \times 0.7}\right)^{1/2}$$

$$\simeq 6000 \quad \text{km}$$

where $\Omega = 0.26$ rad/h and $\cos\psi \simeq 0.7$. This wavelength is of the order of the radius of the earth. See also Table 8.1.

Illustrative Example 6.7

Determine the criteria for a southeasterly to develop into a cyclonic motion. With a southeasterly latitude increases in the direction of the motion; the stream moves into regions of increasing vertical rotation of the earth. In order for a cyclonic motion to develop, the stretching of the vortex lines with increasing ψ must cause vorticity to increase faster than the earth's vertical rotation. Let us examine these conditions.

For the southeasterly, the total vertical acceleration due to the earth's rotation is the sum of the vertical components of the centrifugal and Coriolis accelerations, excluding g, which is considered constant in the displacement of this problem:

$$a_z = \Omega^2 R \cos^2 \psi - 2\Omega U_h \cos \psi \sin \alpha$$

Except near the polar regions, the centrifugal part in this expression is considerably larger than the Coriolis component. Even for 100-km/h wind the centrifugal part is about five times greater than the vertical component of the Coriolis. We shall neglect the second term and start with

$$a_z = \Omega^2 R \cos^2 \psi$$

In moving toward higher latitudes, if the increase in a_z can generate an extension of an atmospheric column with area A and a vertical velocity W, considering that the average density of the column remains unchanged, the rate of volume extension of this column will be

$$Q = WA$$

If angular momentum is conserved during this extension, then according to Eq. (6.69) the circulation in the column with uniform vorticity will be

$$\Gamma = \zeta A$$

Thus the rate of stretching, or updraft, is

$$W = \frac{Q}{\Gamma} \zeta$$

and the acceleration $a_z = dW/dt$ is

$$a_z = \frac{Q}{\Gamma} \frac{d\zeta}{dt}$$

Equating the two values of a_z we arrive at

$$\frac{d\zeta}{dt} = \frac{\Gamma}{Q} \Omega^2 R \cos^2 \psi$$

But in the motion of the column dt can be replaced by $R\, d\psi / U_h$. Although $R\, d\psi$ is not the total distance covered, it is the northern component of the distance that will contribute to changes in ζ. Thus

$$\frac{d\zeta}{d\psi} = \frac{\Gamma}{U_h Q} \Omega^2 R^2 \cos^2 \psi$$

This then is the rate of change of vorticity of the stream relative to the surface of the earth and with respect to change of latitude. The fact that this is a scalar equation implies that the change of magnitude of ζ is positive if Γ is positive. However, if the circulation is negative at the start, it will become more negative. Both cases imply that the absolute value of the vorticity will increase.

In the meantime, the earth's rotation (vertical component) will also change with ψ. If we let ζ_e represent the earth's vertical component of vorticity

$$\zeta_e = 2\Omega \sin \psi$$

then

$$\frac{d\zeta_e}{d\psi} = 2\Omega \cos \psi$$

which also increases with increasing ψ because Ω and the cosine are positive quantities. Thus in dividing the two rates of changes of vorticity we have

$$\frac{d\zeta}{d\zeta_e} = \frac{\Gamma}{2U_h Q} \Omega R^2 \cos \psi$$

This is the result sought. The following conclusions can be derived from this expression. Taking the original statement of the problem, for a cyclonic pattern to develop from a southeasterly, the stream vorticity must increase faster than the earth's vertical vorticity. Thus the ratio of changes above must be positive and if we make it larger than unity, then

$$\frac{\Gamma}{2U_h Q} \Omega R^2 \cos \psi > 1$$

The quantity $\Omega R \gg U_h$ because for $\Omega = 7.3 \times 10^{-5}$ rad/s and $R = 6370$ km we obtain a peripheral velocity of the earth of $\Omega R = 1670$ km/h. This is at least 10 times larger than any wind we might consider. Only near the poles will $\cos \psi$ tend to zero. Even if we let $\Omega R \cos \psi \simeq U_h$, $\Gamma R/2Q$ must still be larger than unity. Replacing Γ/Q by ζ/W we conclude that $\zeta R/2W > 1$ for a cyclonic pattern to develop. With R being so large it is difficult to imagine on the basis of this simple analysis that conditions for a cyclonic pattern will not be met. Since most hurricanes in the eastern part of the US develop through southeasterlies, this problem is discussed in more detail in Section 10.3.

7

Dynamics of the Environment

7.1 INTRODUCTION

According to Kelvin's theorem, discussed in Section 6.9, only nonconservative forces could set and keep an environment in rotational motion, if it ever started from rest. In the case of the atmosphere and oceans, the earth's rotation is a major motor force that keeps the environment in motion either through friction at the earth's surface, or through Coriolis, or through its centrifugal acceleration.

From the start we shall eliminate considerations of the centrifugal acceleration, not because it is small compared to the Coriolis but because *in the definition of the vertical direction we considered the plumb line as the geometric altitude*, as discussed in Section 2.5. The vertical direction is the direction opposite the local gravitational acceleration, which is the vector sum of $\boldsymbol{g}^* + \Omega^2 \boldsymbol{r}$, where $\boldsymbol{r}$ is the radius of the latitude circle. Thus the Coriolis force will be a motor force of the earth. The earth's rotation is not the only motor force. Through Kelvin's theorem we know that the environment can be set in motion when it is *baroclinic* as a result of the uneven heat from the sun, on the curved surface of the earth, which produces misalignment of density and pressure gradients, or misalignment of salinity and pressure gradients in the sea. In general, we conclude that the forces in the environment are due to the earth's rotation, the uneven heating from the sun, and friction.

We have termed motions that are due exclusively to the earth's rotation as *geostrophic motions.*

7.2 THE ACCELERATION ON A ROTATING EARTH

We said in the preceding chapter that *all* kinematic relations depend on the frame of reference through which we see and measure them. In most dynamical applications on the earth's surface, velocities and accelerations are expressed in terms of the surface of the earth, but Newton's law of inertia applies to a fixed, inertial, or absolute frame of reference. It is true that we do not know of a frame of reference in our planetary system that is not moving. Even the earth's center rotates with respect to the sun, and the sun belongs to a galaxy that has its own movement. Therefore, the practical question is not one of finding *the* absolute frame of reference but the amount of error introduced in choosing one moving frame in lieu of another.

In Illustrative Example 6.6 we saw that the large scale of the motion of the environment resulting from the earth's rotation is approximately that of the earth's radius. In half that distance, the velocity of the environment reverses direction; that is, it goes from clockwise to counterclockwise. In a quarter of this radial distance it may vary from a value of U_h to zero before reversing direction. The time it takes to travel $\lambda/4$ distance at a velocity U_h is $t = \lambda/4U_h$ and the acceleration of the environment in that lapse of time is approximately $4U_h^2/\lambda$. In comparing this acceleration with the Coriolis $2\Omega U_h$, the ratio of the fluid acceleration to the Coriolis, which is called the *Rossby number* $\mathcal{R}o$, is $U_h/\Omega L$, where L appears as the half-wavelength. In comparisons, the value of the Rossby number ought to tell us whether or not the rotation of the surface of the earth is important. When the number is large we can neglect the effect of the earth's rotation on the fluid acceleration, and when it is of the order of unity or less, we cannot ignore the rotation of the earth.

We know $\Omega = 0.26$ rad/h and for the values given in Illustrative Example 6.6 of $U_h = 50$ km/h and with the half-wavelength $L = 3000$ km the Rossby number is about 1/17, meaning that the Coriolis is 17 times larger than any fluid acceleration that may develop.[1] Therefore, we conclude that in examining the acceleration of a fluid parcel we must not ignore the contribution from the rotation of the earth, by choosing the axis of reference on the surface of the earth. Thus we must choose our coordinates at the center of the earth.

Consider a parcel of the environment P (Fig. 7.1) moving and located instantly at a distance $\boldsymbol{s}$ relative to a frame O on the surface of the earth at a latitude ψ and longitude θ. The coordinate axes x, y, and z in the direction of unit vectors $\boldsymbol{i}, \boldsymbol{j}$, and $\boldsymbol{k}$, respectively, are located on the earth's surface at O and constitute the reference axes through which an observer on the earth sees the

[1] This would not be true in most small-scale industrial applications to fluid motion because the same reduction of velocity in the motion can occur in much smaller distances. Since the rotation of the center of the earth relative to the sun is 1/365 of the earth's rotation, the effect of the earth's rotation around the sun can be ultimately neglected.

motion of P. Thus, according to convention the x and y coordinates are horizontal, directed toward the east and north, and z is the vertical axis (perpendicular to the geopotential surfaces or along the plumb line $\boldsymbol{g}$ which is the vector sum of the gravitational acceleration of the earth and the centrifugal acceleration).

Since O moves relative to O' at the center of the earth, it is clear that observers at O and O' will not describe the motion of P in the same manner. From vector algebra, the position ξ of P relative to O' is

$$\xi = \boldsymbol{R} + \boldsymbol{s} \tag{7.1}$$

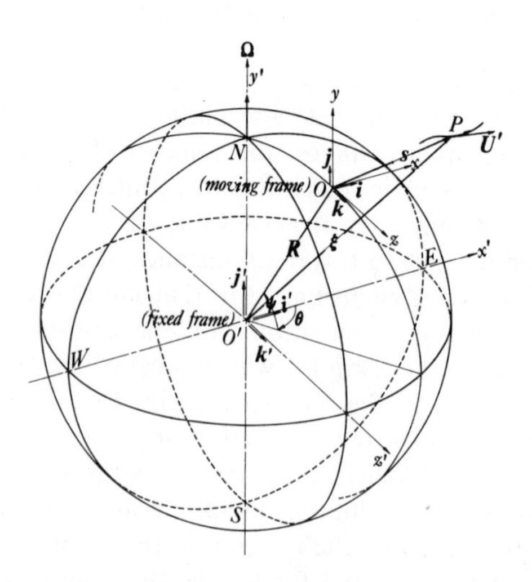

Fig. 7.1 Motion of P with reference to the surface of the earth and the center of the earth.

where $\boldsymbol{R}$ is the radius of the earth at O. This equation *transforms positions* of parcels in the environment from one coordinate system to the other. The transformation of velocity from one coordinate system to the other can also be found when we begin by expressing $\boldsymbol{s}$ in component form:

$$\boldsymbol{s} = \boldsymbol{i}x + \boldsymbol{j}y + \boldsymbol{k}z$$

The rate of change of this position vector[2] with reference to O', the center of the Earth, is

$$\frac{d\boldsymbol{s}}{dt} = \frac{d}{dt}(\boldsymbol{i}x + \boldsymbol{j}y + \boldsymbol{k}z)$$

[2] It should be clear that this is not the velocity of P relative to O' which is given by $d\xi/dt$. It is simply the rate of change of any vector observed from O and relative to O'.

Since the unit vectors $\boldsymbol{i}, \boldsymbol{j}$, and $\boldsymbol{k}$ vary in orientation with time as the earth moves, and since the coordinates of P move in time with the motion of the environment, then in the rate of change we must consider both the orientation of the unit vectors and the change of the coordinates of P. Thus we must differentiate by parts the previous expression:

$$\left(\frac{d\boldsymbol{s}}{dt}\right)_{O'} = \boldsymbol{i}\frac{dx}{dt} + \boldsymbol{j}\frac{dy}{dt} + \boldsymbol{k}\frac{dz}{dt} + x\frac{d\boldsymbol{i}}{dt} + y\frac{d\boldsymbol{j}}{dt} + z\frac{d\boldsymbol{k}}{dt} \tag{7.2}$$

where subscript O' implies with reference to the center of the earth. We have said that $\boldsymbol{i}, \boldsymbol{j}$, and $\boldsymbol{k}$ can only vary in orientation because their magnitudes will always remain unity in *all* positions of O relative to the earth's center as the earth rotates. Then since the unit vectors only rotate, their rates of change with time are the velocities expressed by

$$\frac{d\boldsymbol{i}}{dt} = \boldsymbol{\Omega} \times \boldsymbol{i}, \qquad \frac{d\boldsymbol{j}}{dt} = \boldsymbol{\Omega} \times \boldsymbol{j}, \qquad \frac{d\boldsymbol{k}}{dt} = \boldsymbol{\Omega} \times \boldsymbol{k}$$

where $\boldsymbol{\Omega}$ is the angular velocity of the frame O with reference to O', or simply the angular velocity of the earth since O is fixed on it. Then after summing components Eq. (7.2) can be written

$$\left(\frac{d\boldsymbol{s}}{dt}\right)_{O'} = \left(\frac{d\boldsymbol{s}}{dt}\right)_{O} + \boldsymbol{\Omega} \times \boldsymbol{s} = \frac{D\boldsymbol{s}}{Dt} + \boldsymbol{\Omega} \times \boldsymbol{s} \tag{7.3}$$

In order to avoid carrying the subscripts O and O' to imply with reference to O and O', we use d/dt to imply with reference to O' (total) and D/Dt to imply with reference to O relative to the observer on the surface of the earth. Thus as Eq. (7.1) transforms positions in space from O to O', Eq. (7.3) transforms the rates of change of *any* vector $\boldsymbol{A}$ (position, velocity, acceleration, vorticity, etc.), from one reference to the other

$$\frac{d\boldsymbol{A}}{dt} = \frac{D\boldsymbol{A}}{Dt} + \boldsymbol{\Omega} \times \boldsymbol{A} \tag{7.4}$$

If we take the derivative of Eq. (7.1) with time which naturally would imply with reference to O' since ξ is measured from that reference, we have

$$\frac{d\boldsymbol{\xi}}{dt} = \frac{d\boldsymbol{R}}{dt} + \frac{d\boldsymbol{s}}{dt}$$

Substituting Eq. (7.3) for $d\boldsymbol{s}/dt$ we obtain the velocity of the parcel P relative to O', the center of the earth,

$$\frac{d\boldsymbol{\xi}}{dt} = \frac{d\boldsymbol{R}}{dt} + \frac{D\boldsymbol{s}}{Dt} + \boldsymbol{\Omega} \times \boldsymbol{s} \tag{7.5}$$

or

$$\boldsymbol{U}' = \boldsymbol{V}_O + \boldsymbol{U} + \boldsymbol{\Omega} \times \boldsymbol{s} \tag{7.6}$$

This is the equation that *transforms velocities* from one frame to another, simply because $\boldsymbol{U}' = d\xi/dt$ is the velocity of P relative to O', which we cannot measure directly, $\boldsymbol{U} = Ds/Dt$ is the velocity of P relative to O, which we can measure, while $V_O = d\boldsymbol{R}/dt$ and $\boldsymbol{\Omega} \times \boldsymbol{s}$ combine to give the total velocity of the frame O relative to O'. For the case of the earth with approximately constant radius and for the frame O on the surface of the earth, there will be no need to carry the quantity $d\boldsymbol{R}/dt$ in the following equations[3] as such; we can replace it by $(\boldsymbol{\Omega} \times \boldsymbol{R})$.

Now we are ready to determine the relationship between the accelerations in the two frames. Designating by $\boldsymbol{a}'$ the inertial acceleration with reference to the assumed fixed frame O', after differentiating Eq. (7.6) with time we have

$$\boldsymbol{a}' = \frac{d\boldsymbol{U}'}{dt} = \frac{d}{dt}[(\boldsymbol{\Omega} \times \boldsymbol{R}) + \boldsymbol{U} + (\boldsymbol{\Omega} \times \boldsymbol{s})]$$

$$= \frac{d\boldsymbol{U}}{dt} + \frac{d}{dt}(\boldsymbol{\Omega} \times \boldsymbol{R}) + \frac{d}{dt}(\boldsymbol{\Omega} \times \boldsymbol{s}) \tag{7.7}$$

In order to introduce the acceleration of P with reference to O, which is $D\boldsymbol{U}/Dt$, we must apply the transformation rule (7.4) to all the terms in Eq. (7.7). Listing them in order[4]

$$\frac{d\boldsymbol{U}}{dt} = \frac{D\boldsymbol{U}}{Dt} + \boldsymbol{\Omega} \times \boldsymbol{U} = \boldsymbol{a} + \boldsymbol{\Omega} \times \boldsymbol{U} \tag{7.8}$$

$$\frac{d}{dt}(\boldsymbol{\Omega} \times \boldsymbol{R}) = \frac{D}{Dt}(\boldsymbol{\Omega} \times \boldsymbol{R}) + \boldsymbol{\Omega} \times (\boldsymbol{\Omega} \times \boldsymbol{R})$$

$$= \boldsymbol{\Omega} \times \frac{D\boldsymbol{R}}{Dt} + \frac{D\boldsymbol{\Omega}}{Dt} \times \boldsymbol{R} + \boldsymbol{\Omega} \times (\boldsymbol{\Omega} \times \boldsymbol{R})$$

$$= \boldsymbol{\Omega} \times (\boldsymbol{\Omega} \times \boldsymbol{R}) \tag{7.9}$$

$$\frac{d}{dt}(\boldsymbol{\Omega} \times \boldsymbol{s}) = \frac{D}{Dt}(\boldsymbol{\Omega} \times \boldsymbol{s}) + \boldsymbol{\Omega} \times (\boldsymbol{\Omega} \times \boldsymbol{s})$$

$$= \boldsymbol{\Omega} \times \frac{D\boldsymbol{s}}{Dt} + \boldsymbol{\Omega} \times (\boldsymbol{\Omega} \times \boldsymbol{s})$$

$$= \boldsymbol{\Omega} \times \boldsymbol{U} + \boldsymbol{\Omega} \times (\boldsymbol{\Omega} \times \boldsymbol{s}) \tag{7.10}$$

[3] In case O is not fixed to the earth and has its own movement or change of distance $\boldsymbol{R}$, then $d\boldsymbol{R}/dt$ should be retained until $\boldsymbol{R}$ is prescribed as a function of time.

[4] Since O is fixed to the earth $D\boldsymbol{R}/Dt = 0$ and we can show through Eq. (7.4) that $d\boldsymbol{\Omega}/dt = D\boldsymbol{\Omega}/Dt$ because $\boldsymbol{\Omega} \times \boldsymbol{\Omega} = 0$. For the case of the earth $D\boldsymbol{\Omega}/Dt = 0$, otherwise $(D\boldsymbol{\Omega}/Dt) \times \boldsymbol{R}$ and $(D\boldsymbol{\Omega}/Dt) \times \boldsymbol{s}$ give what is known as *Euler's acceleration*.

Thus reconstructing Eq. (7.7) with the details of Eqs. (7.8)–(7.10) we have[5]

$$\boldsymbol{a}' = \boldsymbol{a} + 2\boldsymbol{\Omega} \times \boldsymbol{U} + \boldsymbol{\Omega} \times (\boldsymbol{\Omega} \times \boldsymbol{R}) + \boldsymbol{\Omega} \times (\boldsymbol{\Omega} \times \boldsymbol{s}) \tag{7.11}$$

As Eqs. (7.1) and (7.6) transformed positions and velocities from one coordinate system to the other, Eq. (7.11) *transforms accelerations.* The symbol $\boldsymbol{a}'$ represents the absolute or *total acceleration* of P with reference to O', $\boldsymbol{a}$ represents the *acceleration of P as we measure it* from the surface of the earth, $\boldsymbol{\Omega} \times (\boldsymbol{\Omega} \times \boldsymbol{R})$ represents the *centripetal acceleration of the earth*, and $\boldsymbol{\Omega} \times (\boldsymbol{\Omega} \times \boldsymbol{s})$ represents the *added centripetal acceleration due to the position of the parcel above the surface of the earth*, and finally $2\boldsymbol{\Omega} \times \boldsymbol{U}$, which comes partly from the contribution of Eq. (7.8) and partly from Eq. (7.10) is an *apparent acceleration*[6] known as the *Coriolis acceleration* which exists only if there is motion with reference to a moving frame such as the earth.

In the first chapter we established that $\boldsymbol{s}$ is very small compared to $\boldsymbol{R}$ and that it will contribute little to Eq. (7.11). Here we see very well that when the vertical direction $\boldsymbol{k}$ is defined along the plumb line, then the centripetal acceleration[7] of the earth $\boldsymbol{\Omega} \times (\boldsymbol{\Omega} \times \boldsymbol{R})$ combined with $\boldsymbol{g}^*$, the gravitational pull of the earth, can be taken as the combined body force when we consider the dynamical equilibrium of the parcel P. This is what is normally done and the centripetal acceleration, which is included in $\boldsymbol{g}$, never appears by itself. Finally we conclude that Eq. (7.11) can be written as

$$\boldsymbol{a}' = \boldsymbol{a} + 2\boldsymbol{\Omega} \times \boldsymbol{U} \tag{7.12}$$

7.3 THE GEOMETRY OF THE CORIOLIS ACCELERATION

To become more familiar with the Coriolis acceleration let us consider the earth's geometry again in Figs. 7.2a and 7.2b. Picking points O_1 and O at a latitude ψ on the surface of the earth, where there is an atmospheric or oceanic motion with velocity $\boldsymbol{U}$, the motion is shown to be easterly at O and westerly at O_1. Since $\boldsymbol{\Omega}$, the vector representing the rotation of the earth, is a free vector, it has the same direction on the entire earth and is always perpendicular to the planes of the latitude circles.

[5] Euler's acceleration $(d\boldsymbol{\Omega}/dt) \times \boldsymbol{\xi}$ has been omitted for the case of the earth as the moving frame. It may be important in transient turbomachinery applications, for instance. Also $D\boldsymbol{R}/Dt$ has been ignored for the O frame of the earth.

[6] According to Neumann (1968), this apparent acceleration was known to Hadley in 1735 when he used it qualitatively in his meteorological work and by Maclaurin in 1740 in his work on ocean currents. The mathematical expression for the horizontal component of this apparent acceleration was obtained by P. S. Laplace (1812) in 1775. For some reason, this apparent acceleration became identified with G. Coriolis (1835). The thorough explanation of this apparent acceleration is attributed to S. D. Poisson (1837).

[7] This is also $\Omega^2 \boldsymbol{r}$, where $\boldsymbol{r}$ is the radius of the latitude circle.

The Coriolis acceleration[8] $2\boldsymbol{\Omega} \times \boldsymbol{U}$ must be perpendicular to the plane formed by the vectors $\boldsymbol{U}$ and $\boldsymbol{\Omega}$. In other words, it must be normal to $\boldsymbol{U}$ and to $\boldsymbol{\Omega}$. Since $\boldsymbol{\Omega}$ is normal to the latitude circles, we conclude that the Coriolis acceleration must lie on the plane of the latitude circle, regardless of the direction of $\boldsymbol{U}$. For the special cases when $\boldsymbol{U}$ is easterly or westerly, meaning tangent to the latitude circle, the Coriolis acceleration is along the radial line of the latitude circle and intersects the axis of rotation of the earth. At any rate, the Coriolis acceleration $\boldsymbol{C} = 2\boldsymbol{\Omega} \times \boldsymbol{U}$ can have horizontal components C_x and C_y and a vertical component C_z. Again, in the case of motion along the latitude circle at O, the horizontal

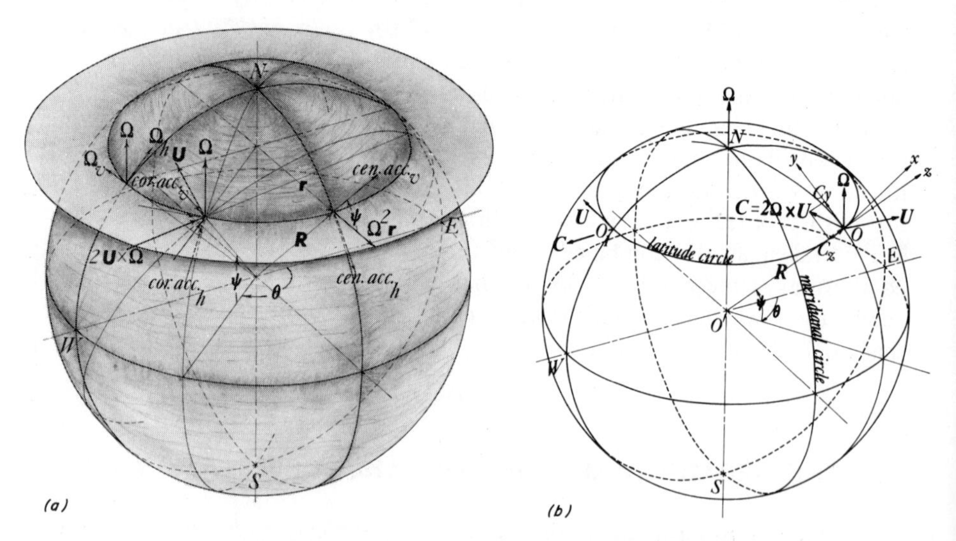

Fig. 7.2 Components of the Coriolis acceleration.

component C_y is toward the north and C_x is zero. The vertical component is downward or toward the center of the earth. In the easterly motion at O_1 the component directions are reversed. Naturally, for a given $\boldsymbol{U}$ the magnitude of the components of the Coriolis will depend on the latitude angle ψ.

From vector algebra the cross product can be put in terms of its coordinates x, y, and z:

$$2\boldsymbol{\Omega} \times \boldsymbol{U} = 2\boldsymbol{i}(\Omega_y W - \Omega_z V) + 2\boldsymbol{j}(\Omega_z U - \Omega_x W) + 2\boldsymbol{k}(\Omega_x V - \Omega_y U) \quad (7.13)$$

[8] We consider in Section 7.9 a *Coriolis deflection force* which is defined as the negative value of the Coriolis acceleration, or $2\boldsymbol{U} \times \boldsymbol{\Omega}$. Looking at the vector geometry, when facing the motion $\boldsymbol{U}$, the Coriolis force is to the left of $\boldsymbol{U}$ while the Coriolis acceleration is to the right. We have already referred to the Coriolis force in Section 6.14.

where Ω_x, Ω_y, Ω_z, U, V, and W are the x, y, and z components of the earth's rotation and the velocity of the environment. From the geometry of Fig. 7.2b we see that

$$\Omega_x = 0, \qquad \Omega_y = \Omega \cos \psi, \qquad \Omega_z = \Omega \sin \psi$$

Then Eq. (7.13) becomes

$$2\boldsymbol{\Omega} \times \boldsymbol{U} = 2\boldsymbol{i}(W\Omega \cos \psi - V\Omega \sin \psi) + 2\boldsymbol{j}U\Omega \sin \psi - 2\boldsymbol{k}U\Omega \cos \psi \tag{7.14}$$

and the components of $\boldsymbol{C}$ are

$$C_x = 2(W\Omega \cos \psi - V\Omega \sin \psi)$$
$$C_y = 2U\Omega \sin \psi$$
$$C_z = -2U\Omega \cos \psi$$

For ordinary wind speeds the vertical component of the Coriolis acceleration C_z is less than one-thousandth of the gravitational acceleration; in other words, the pressure increase or decrease on the ground resulting from the vertical component of the Coriolis is of the order of a few tenths of a millibar, and this effect is overwhelmed by the magnitude of the gravitational acceleration and the pressure of about 1000 mb it imposes on the ground. Since C_z is due to the product of Ω_y and U, we can conclude that Ω_y has negligible influence. The term C_x along the east also depends on the product of Ω_y and W, and since the vertical component of the velocity is at least one order smaller than the other components, $2W\Omega_y$ can also be ignored. Thus for most applications in meteorology and oceanology Ω_y has little influence on the environment and Eq. (7.14) takes the form

$$2\boldsymbol{\Omega} \times \boldsymbol{U} = 2\boldsymbol{j}U\Omega \sin \psi - 2\boldsymbol{i}V\Omega \sin \psi \tag{7.15}$$

where both terms on the right-hand side contain $\Omega_z = \Omega \sin \psi$. If we let the symbol $f = 2\Omega \sin \psi$ be called the *Coriolis parameter*, the final approximate expression for the Coriolis acceleration in terms of its components is

$$\boldsymbol{C} = 2\boldsymbol{\Omega} \times \boldsymbol{U} = \boldsymbol{j}fU - \boldsymbol{i}fV \tag{7.16}$$

Returning to Eq. (7.12) the absolute acceleration of an environmental parcel with reference to the center of the earth is

$$\boldsymbol{a}' = \boldsymbol{a} - \boldsymbol{i}fV + \boldsymbol{j}fU \tag{7.17}$$

This finally confirms that the influence of the Coriolis acceleration is confined to horizontal directions.

We are now ready to develop the dynamics of the environment by applying Newton's second law to an inertial or absolute frame of reference. It is still important, however, to examine the ways in which we can express the rate of change with respect to the time implied in $\boldsymbol{a}$, the parcel's acceleration, viewed from the surface of the earth. To distinguish the rates of change with reference to O and O' we used the symbols D/Dt and d/dt, respectively. However, since from now on we shall be using the right-hand side of Eq. (7.17) in which $\boldsymbol{a}$ will have to be expressed as a rate of change of velocity in the frame O, we shall adopt the symbol d/dt instead of D/Dt, for $\boldsymbol{a} = dU/dt$ to imply the O frame. This shift of symbols is done because scientific convention in almost all other works recognizes the symbol d/dt as the rate of change with time, and since the desired result in Eq. (7.17) has been obtained it is hoped that this shift of symbols will not be confusing from here on.

7.4 THE TIME RATE OF CHANGE

To describe the value a property assumes in the environment, we often need space as well as time coordinates. This representation conforms to observational methods used in practice. We, as observers, see ourselves located in a geometrical space and from that space we judge the progress in time. In other words, time and position space is the most natural to us.

Let us say that a vector ξ representing a physical property of the motion is a function of position vector $\boldsymbol{s}_i$ and time t as shown in Fig. 7.3, such that

$$\xi = \xi(\boldsymbol{s}_i, t) \tag{7.18}$$

Before evaluating the rate of change of ξ let us look at the meaning we wish to attach to $\boldsymbol{s}_i$ and t

First, let the vector $\boldsymbol{s}_i$ represent the position that an infinitesimal parcel of the environment occupies at the time $t = t_0$. If the property ξ is to represent the position of the parcel $\boldsymbol{r}$ at all times t, then at the origin of time $t = t_0$ we

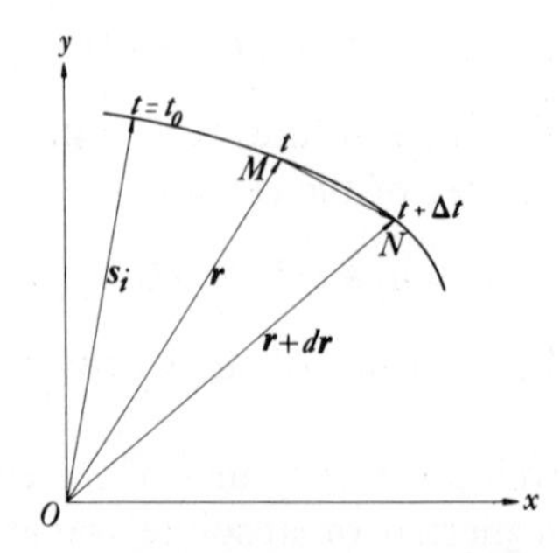

Fig. 7.3 Time history and location of a particle.

must say that $\boldsymbol{r} = \boldsymbol{s}_i$, identifying the parcel in question. Thus the subscript i identifies the parcel in question occupying a position $\boldsymbol{s}_i$ at time t_0. In this description, once the original position of the said parcel has been specified, then according to Eq. (7.18) $\boldsymbol{s}_i$ becomes a parameter and time is the only independent variable. This implies that for $\xi = \boldsymbol{r}$, we follow the progress in time of the parcel $\boldsymbol{s}_i$, *following the parcel*. This is the *Lagrangian description* we discussed in Section 6.1. Then the rate of change of the position vector $\boldsymbol{r}$ for parcel $\boldsymbol{s}_i$ is

$$\boldsymbol{U} = \frac{d\boldsymbol{r}}{dt} = \lim_{\Delta t \to 0} \left(\frac{\Delta \boldsymbol{r}}{\Delta t} \right)_{\boldsymbol{s}_i} \tag{7.19}$$

The basis of *following the parcel* is very important because Newton's second law was enunciated on the basis of a parcel representing a *system* and if we are to use this law, we must comply with its conditions. In the Lagrangian representation, differentiation with $\boldsymbol{s}_i$ has no useful significance and consequently the derivative with time keeping $\boldsymbol{s}_i$ constant is the only rate of change possible; thus Eq. (7.19) is the total rate of change with time. The same can be said for the acceleration and rate of change of any other property.

In particle and rigid body mechanics either because the relative positions of the parcels do not vary or because with a finite number of particles the particles can be identified easily, the Lagrangian description is a useful method of representation. However, in a deformable continuum such as the environment, this method of description becomes unmanageable because of the infinity of parcels that do not keep their relative positions fixed, which makes their identification and follow-up impossible.

Instead, it is best to concentrate on what happens to the *material* at various positions $\boldsymbol{r}$ in time t. Looking at Fig. 7.3, if at M we consider a scalar property p of a parcel that happens to be there at the time t, then

$$p = p(\boldsymbol{r}, t) \tag{7.20}$$

is the *Eulerian description*. Now to evaluate the rate of change of p of the parcel, *following the parcel* as in Eq. (7.19), in the Eulerian description as the motion proceeds from M to N and with the position $\boldsymbol{r}$ varying in time,

$$p + \Delta p = p\left(\boldsymbol{r} + \frac{d\boldsymbol{r}}{dt} \Delta t, t + \Delta t\right)$$

Expanding in a Taylor series to the first order changes since a limit will be taken:

$$p + \Delta p = p(\boldsymbol{r}, t) + \frac{d\boldsymbol{r}}{dt} \Delta t \frac{\partial p}{\partial \boldsymbol{r}} + \Delta t \frac{\partial p}{\partial t} + \cdots$$

Subtracting Eq. (7.20) and taking the limit we obtain

$$\lim_{\Delta t \to 0} \frac{\Delta p}{\Delta t} = \frac{dp}{dt} = \frac{\partial p}{\partial t} + \frac{d\boldsymbol{r}}{dt} \cdot \frac{\partial p}{\partial \boldsymbol{r}}$$

According to Appendix A, $\partial p/\partial \boldsymbol{r}$ is the gradient of p and $d\boldsymbol{r}/dt = \boldsymbol{U}$. Thus

$$\frac{dp}{dt} = \frac{\partial p}{\partial t} + (\boldsymbol{U} \cdot \nabla)p \tag{7.21}$$

The total rate of change of p (following the parcel) in the Eulerian description is composed of a change in time at M and the change of the property of the parcel in its motion from M to N. The first change $\partial/\partial t$ will determine if the motion of the material and not the parcel varies in time at one coordinate position and the second term $(\boldsymbol{U} \cdot \nabla)p$ is the *convective part* of the rate of change. The two constitute the total rate of change of p with time of the parcel, following the parcel, but in Eulerian representation. This total rate of change following the parcel is symbolically represented as

$$\frac{d}{dt} = \frac{\partial}{\partial t} + (\boldsymbol{U} \cdot \nabla) \tag{7.22}$$

7.5 THE ACCELERATION

Now it is easy to apply the Eulerian representation to obtain the acceleration vector from the velocity vector. From Eq. (7.22)

$$\boldsymbol{a} = \frac{d\boldsymbol{U}}{dt} = \frac{\partial \boldsymbol{U}}{\partial t} + (\boldsymbol{U} \cdot \nabla)\boldsymbol{U} \tag{7.23}$$

Since $\partial \boldsymbol{U}/\partial t$ represents the change of velocity with time at M for steady motion, it will be zero, while $(\boldsymbol{U} \cdot \nabla)\boldsymbol{U}$ represents the change of velocity of the parcel with time as it moves from M to N.

Having established that U, V, and W are the velocity components in the x, y, and z directions fixed to the surface of the earth, the Cartesian components of the acceleration due to the rate of change of the velocity $\boldsymbol{U}$ are

$$\begin{aligned}
\boldsymbol{a} &= \frac{\partial}{\partial t}(\boldsymbol{i}U + \boldsymbol{j}V + \boldsymbol{k}W) \\
&\quad + \left[(\boldsymbol{i}U + \boldsymbol{j}V + \boldsymbol{k}W) \cdot \left(\boldsymbol{i}\frac{\partial}{\partial x} + \boldsymbol{j}\frac{\partial}{\partial y} + \boldsymbol{k}\frac{\partial}{\partial z}\right)\right](\boldsymbol{i}U + \boldsymbol{j}V + \boldsymbol{k}W) \\
&= \boldsymbol{i}a_x + \boldsymbol{j}a_y + \boldsymbol{k}a_z
\end{aligned}$$

Performing the operation and separating the components we obtain

$$a_x = \frac{dU}{dt} = \frac{\partial U}{\partial t} + U\frac{\partial U}{\partial x} + V\frac{\partial U}{\partial y} + W\frac{\partial U}{\partial z}$$

$$a_y = \frac{dV}{dt} = \frac{\partial V}{\partial t} + U\frac{\partial V}{\partial x} + V\frac{\partial V}{\partial y} + W\frac{\partial V}{\partial z} \tag{7.24}$$

$$a_z = \frac{dW}{dt} = \frac{\partial W}{\partial t} + U\frac{\partial W}{\partial x} + V\frac{\partial W}{\partial y} + W\frac{\partial W}{\partial z}$$

Entering this information into the absolute acceleration of Eqs. (7.12) and (7.17) we have

$$\boldsymbol{a}' = \frac{\partial \boldsymbol{U}}{\partial t} + (\boldsymbol{U}\cdot\nabla)\boldsymbol{U} + 2\boldsymbol{\Omega}\times\boldsymbol{U} \tag{7.25}$$

The x *component toward the east* is

$$a_x' = \frac{\partial U}{\partial t} + U\frac{\partial U}{\partial x} + V\frac{\partial U}{\partial y} + W\frac{\partial U}{\partial z} - fV \tag{7.26}$$

The y *component toward the north* is

$$a_y' = \frac{\partial V}{\partial t} + U\frac{\partial V}{\partial x} + V\frac{\partial V}{\partial y} + W\frac{\partial V}{\partial z} + fU \tag{7.27}$$

The z *component vertical* is

$$a_z' = \frac{\partial W}{\partial t} + U\frac{\partial W}{\partial x} + V\frac{\partial W}{\partial y} + W\frac{\partial W}{\partial z} \tag{7.28}$$

Again, at the risk of repetition, the centrifugal acceleration of the earth which normally should appear in the z component of the environment will be included with the gravitational acceleration $\boldsymbol{g}^*$ to give a total force of the earth per unit mass $\boldsymbol{g}$ in the direction of the vertical.

Another form of Eq. (7.25) will also be used frequently in this text. It can be shown (Eskinazi, 1967) from the triple cross-product expansion that

$$(\boldsymbol{U}\cdot\nabla)\boldsymbol{U} = \tfrac{1}{2}\nabla\boldsymbol{U}^2 - \boldsymbol{U}\times(\nabla\times\boldsymbol{U}) = \tfrac{1}{2}\nabla\boldsymbol{U}^2 - \boldsymbol{U}\times\boldsymbol{\zeta} \tag{7.29}$$

The quantity $\boldsymbol{\zeta} = \nabla\times\boldsymbol{U}$ is the vorticity of the environment relative to the earth. Thus the absolute acceleration of Eq. (7.25) becomes

$$\boldsymbol{a}' = \frac{\partial \boldsymbol{U}}{\partial t} + \frac{1}{2}\nabla\boldsymbol{U}^2 - \boldsymbol{U}\times\boldsymbol{\zeta} + 2\boldsymbol{\Omega}\times\boldsymbol{U}$$

$$= \frac{\partial \boldsymbol{U}}{\partial t} + \frac{1}{2}\nabla\boldsymbol{U}^2 + (2\boldsymbol{\Omega} + \boldsymbol{\zeta})\times\boldsymbol{U} \tag{7.30}$$

This expression combines ζ, the vorticity of the environment relative to the earth's surface, and the vorticity of the earth $2\mathbf{\Omega}$ relative to the absolute or inertial frame. Thus $(2\mathbf{\Omega} + \zeta)$ is the absolute vorticity of the environment discussed in Illustrative Example 6.4.

7.6 DYNAMICAL EQUATIONS FOR THE ENVIRONMENT

Having found the expression for the absolute acceleration of the environment in terms of velocities measured with reference to the surface of the earth, we are ready to apply Newton's inertial law, also called Newton's second law. This law states that the sum of all forces acting on an environmental parcel must be equal to the product of the mass of the parcel and its absolute or inertial acceleration. We have established the absolute acceleration in the preceding section. The types of forces were discussed in Chapter 4 and were classified as body and surface forces. The surface tension forces discussed in Chapter 5, although important in the formation and collapse of gas, liquid, and ice drops, contribute negligibly to the overall dynamics of the atmosphere and oceans. For this reason surface tension forces will not take part in the dynamics of large masses.

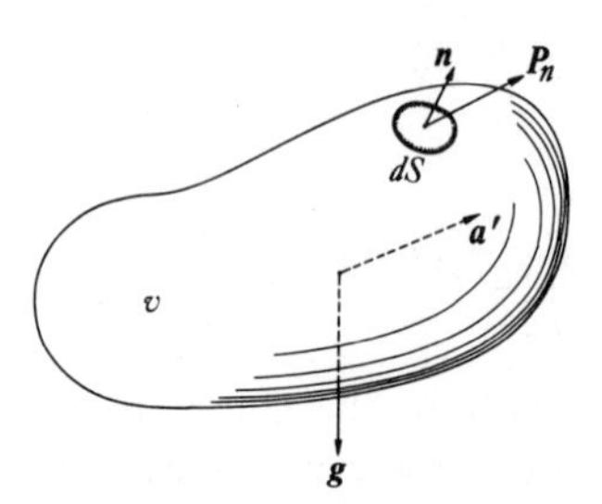

Fig. 7.4 Forces in the motion of an arbitrary volume of an environmental substance.

Through Newton's law of inertia, d'Alembert in 1758 extended the concept of static equilibrium discussed in Chapter 4 to *dynamical equilibrium* by stating that if the absolute acceleration following the particle was subtracted from the body force per unit mass $\boldsymbol{g}$, then the resulting form of Eq. (4.12) would constitute dynamical equilibrium. Thus

$$\int_{\mathscr{V}} (\boldsymbol{g} - \boldsymbol{a}')\rho \, d\mathscr{V} + \oint_S \boldsymbol{n} \cdot \mathscr{S} \, dS = 0 \tag{7.31}$$

Figure 7.4 is a modified version of Fig. 4.1, i.e., the acceleration of the environment has been included. The remaining terms in Eq. (7.31) were defined in Section 4.2 where ρ is the density of the environment, $\mathscr{S}$ is the stress tensor, $\boldsymbol{n}$ is the unit external normal, and S and $\mathscr{V}$ are the surface area and the volume of the arbitrary volume in the figure. With the use of Gauss's theorem we have again

$$\int_{\mathscr{V}} [(\boldsymbol{g} - \boldsymbol{a}')\rho + \nabla \cdot \mathscr{S}]\, d\mathscr{V} = 0 \tag{7.32}$$

Because $\mathscr{V}$ in Eq. (7.32) is completely arbitrary, we conclude that the integrand must be zero. Thus

$$\rho \boldsymbol{a}' = \rho \boldsymbol{g} + \nabla \cdot \mathscr{S} \tag{7.33}$$

This is *Cauchy's equation of motion* for any deformable substance such as the environment in motion. This dynamical equation constitutes the basis for d'Alembert's type of equilibrium when there is motion.

The vector $\boldsymbol{a}'$ and its components were developed in Eqs. (7.25)–(7.28) and (7.30). The gravitational acceleration $\boldsymbol{g}$, which includes the centrifugal acceleration of the earth, was discussed thoroughly in Chapter 2 and is considered throughout this one. It constitutes the only body force in the environment.[9] The components of the stress tensor $\mathscr{S}$ were given in Eq. (4.3) without being related to the rates of deformation of the environmental fluid.

Using Eq. (7.25) for $\boldsymbol{a}'$ and Eqs. (4.3) and (4.7) for the components of the stress tensor, Cauchy's equation of motion becomes

$$\frac{\partial \boldsymbol{U}}{\partial t} + (\boldsymbol{U} \cdot \nabla)\boldsymbol{U} + 2\boldsymbol{\Omega} \times \boldsymbol{U} = -\boldsymbol{g} + \frac{1}{\rho}\left(\frac{\partial \boldsymbol{P}_x}{\partial x} + \frac{\partial \boldsymbol{P}_y}{\partial y} + \frac{\partial \boldsymbol{P}_z}{\partial z}\right) \tag{7.34}$$

where the terms in parentheses on the right-hand side equal $\nabla \cdot \mathscr{S}$. In Cartesian components this vector equation takes the following forms:

The x *component toward the east* is

$$\frac{\partial U}{\partial t} + U\frac{\partial U}{\partial x} + V\frac{\partial U}{\partial y} + W\frac{\partial U}{\partial z} - fV = \frac{1}{\rho}\left(\frac{\partial \sigma_{xx}}{\partial x} + \frac{\partial \tau_{xy}}{\partial y} + \frac{\partial \tau_{xz}}{\partial z}\right) \tag{7.35}$$

The y *component toward the north* is

$$\frac{\partial V}{\partial t} + U\frac{\partial V}{\partial x} + V\frac{\partial V}{\partial y} + W\frac{\partial V}{\partial z} + fU = \frac{1}{\rho}\left(\frac{\partial \tau_{xy}}{\partial x} + \frac{\partial \sigma_{yy}}{\partial y} + \frac{\partial \tau_{yz}}{\partial z}\right) \tag{7.36}$$

The z *component vertical* is

$$\frac{\partial W}{\partial t} + U\frac{\partial W}{\partial x} + V\frac{\partial W}{\partial y} + W\frac{\partial W}{\partial z} = -g + \frac{1}{\rho}\left(\frac{\partial \tau_{xz}}{\partial x} + \frac{\partial \tau_{yz}}{\partial y} + \frac{\partial \sigma_{zz}}{\partial z}\right) \tag{7.37}$$

[9] In these general considerations, the buoyant force due to temperature stratification and in the direction of $-\boldsymbol{g}$ is not identified. See Section 7.8 for the influence of buoyant forces.

The Coriolis parameter f and the reasons for including only fV and fU in the horizontal components were explained in detail in Section 7.3.

These component equations are still much too general for the majority of applications in meteorology and oceanology and many local environmental motions. Depending on the applications, we shall see that all the terms in these equations are not of the same order and do not need to be conserved. We shall simplify these equations as we proceed.

From the point of view of closure alone when we count the number of dependent variables in the three component equations, they are U, V, W, ψ, ρ, σ_{xx}, σ_{yy}, σ_{zz}, τ_{xy}, τ_{xz}, τ_{yz}, a total of 11. This number has already been reduced by three because we considered the stress tensor to be symmetric,[10] that is, $\tau_{xy} = \tau_{yx}$, $\tau_{xz} = \tau_{zx}$, $\tau_{yz} = \tau_{zy}$, because of the fact that the geofluid substance cannot sustain internal moments. Still, 11 dependent variables will have to be resolved in terms of four independent variables x, y, z, and t. The angular velocity of the earth and the gravitational acceleration have been considered constant in this count. It is obvious that 11 independent equations would be necessary to solve for each and every dependent variable.

The number of independent equations we have so far are the three component equations of dynamics, Eqs. (7.35)–(7.37), and one scalar equation for the conservation of mass, Eq. (6.12). This makes a total of four independent equations, so far. In 1686 Newton postulated a law of viscosity for fluids in which he related the stresses to the velocities of deformation expressed in terms of space derivatives of the velocity components of the motion. Since the environmental fluid is a Newtonian fluid, six additional relations occur for the σ's and the τ's in terms of velocity derivatives, but a new dependent variable,[11] the pressure, is introduced, which did not appear in any of the dynamical equations. Thus, Newton's viscosity postulate increases the number of independent equations to 10 but also increases the number of dependent variables to 12. The difference has narrowed down to two, however. One of these two can be found simply by establishing a relationship between the latitude angle ψ and the coordinates x, y, and z on the earth. The final independent relation involving the pressure must come through thermodynamics. Here we have two choices. In *barotropic* cases (defined in Section 6.9), since the density is a unique function of pressure, this relationship, if known, is sufficient to close the problem mathematically. Otherwise, in *baroclinic* processes the use of the equation of state will introduce a new independent relation but also a new dependent variable, the temperature T in the atmosphere and salinity in the sea. Then to close the problem we must appeal to the first law of thermodynamics in such a way as not to introduce new dependent variables.

[10] Consult Eskinazi (1967, p. 196).
[11] See, for instance, Eskinazi (1968).

The vertical gradients of velocities are the most important in the environment because of the friction introduced either by the presence of rigid surfaces at the earth's surface or by density stratification. This can be seen in the comparison of the terms in Table 7.1. Thus Newton's viscosity postulate simplifies to

$$\sigma_{xx} = \sigma_{yy} = \sigma_{zz} = -p \tag{7.38}$$

and

$$\begin{aligned} \tau_{xy} &= \tau_{yx} = \mu\left(\frac{\partial U}{\partial y} + \frac{\partial V}{\partial x}\right) \\ \tau_{xz} &= \tau_{zx} = \mu\left(\frac{\partial U}{\partial z} + \frac{\partial W}{\partial x}\right) \\ \tau_{yz} &= \tau_{zy} = \mu\left(\frac{\partial V}{\partial z} + \frac{\partial W}{\partial y}\right) \end{aligned} \tag{7.39}$$

where μ is the *viscosity coefficient* or *dynamic viscosity* of the environment. If only the vertical derivatives are important, as is the case for large-scale environmental motions, then the first component of the *shear stress* τ_{xy} can be neglected. The second component of the shear stress $\tau_{xz} = \mu\, dU/dz$, and the third component $\tau_{yz} = \mu\, dV/dz$.

Since we introduced Newton's viscosity postulate and defined the pressure as the average normal stress, it can be shown[12] that the divergence of the stress tensor appearing in Eq. (7.33) is equivalent to $\nabla \cdot \mathscr{S} = -\nabla p + \mu \nabla^2 \boldsymbol{U}$, and Eq. (7.34) can be written as

$$\frac{\partial \boldsymbol{U}}{\partial t} + (\boldsymbol{U} \cdot \nabla)\boldsymbol{U} + 2\boldsymbol{\Omega} \times \boldsymbol{U} = -\frac{1}{\rho}\nabla p + \boldsymbol{g} + \frac{\mu}{\rho}\nabla^2 \boldsymbol{U} \tag{7.40}$$

In Cartesian component form this vector equation becomes

$$\begin{aligned} \frac{\partial U}{\partial t} + U\frac{\partial U}{\partial x} + V\frac{\partial U}{\partial y} + W\frac{\partial U}{\partial z} - fV &= -\frac{1}{\rho}\frac{\partial p}{\partial x} + \nu\left(\frac{\partial^2 U}{\partial x^2} + \frac{\partial^2 U}{\partial y^2} + \frac{\partial^2 U}{\partial z^2}\right) \\ \frac{\partial V}{\partial t} + U\frac{\partial V}{\partial x} + V\frac{\partial V}{\partial y} + W\frac{\partial V}{\partial z} + fU &= -\frac{1}{\rho}\frac{\partial p}{\partial y} + \nu\left(\frac{\partial^2 V}{\partial x^2} + \frac{\partial^2 V}{\partial y^2} + \frac{\partial^2 V}{\partial z^2}\right) \\ \frac{\partial W}{\partial t} + U\frac{\partial W}{\partial x} + V\frac{\partial W}{\partial y} + W\frac{\partial W}{\partial z} &= -g - \frac{1}{\rho}\frac{\partial p}{\partial z} + \nu\left(\frac{\partial^2 W}{\partial x^2} + \frac{\partial^2 W}{\partial y^2} + \frac{\partial^2 W}{\partial z^2}\right) \end{aligned} \tag{7.41}$$

where $\nu = \mu/\rho$ is the kinematic viscosity. These are known as the *Navier–Stokes equations.*

In the following section we evaluate the relative importance of these terms.

[12] See, for instance, Eskinazi (1967). In Newton's postulate, the motion is considered laminar. Additional turbulent Reynolds stresses appear when the motion is turbulent. These additional stresses are developed later.

7.7 CRITERIA FOR ORDERS OF MAGNITUDE CONSIDERATION

Just by looking at Eqs. (7.41) we can conclude that it would be very difficult to obtain a general solution if we had to keep all the terms in them. Naturally, it would be logical to proceed to an order of magnitude analysis to see if all the terms in the equations are of the same importance. In other words, we wish to find the terms that could be neglected in certain specific applications. Each application will have to be analyzed separately for order of magnitude because it could have its own order of relative magnitudes.

Table 7.1 gives an order of magnitude comparison of the quantities in Eq. (7.41) for large-scale atmospheric movements (Belinskii, 1967). In this assessment the derivatives are approximated by finite differences, where dx and dy are of the order of the zonal patterns, or approximately 10^4 m, dz is of the order of 10^2 m, the size of the viscous layer near the ground, and dt is of the order of 1 h or 3600 s. The numbers in the table are in the MTS (meter, ton, second) system of units. If we consider the pressure on the surface of the earth to be 10^3 mb, it will correspond to 10^2 tons$\cdot$m/m$^2\cdot$s^2. This is the number given to p in the table. The specific volume, which is the reciprocal of the density, is taken at the earth's surface as 10^3 m^3/ton, and the absolute temperature, since it ranges from 200 to 300°K in the atmosphere, is entered in the Table as 10^2. For the ocean, the specific volume will be of the order of unity, but d/dz will be about 10^3 times larger than that of the atmosphere.

In approaching certain problems of the environment it is useful to prepare such tables and to determine the order of importance of terms.

As we can see from Table 7.1, it is not always necessary that all the terms in the dynamical equations (7.41) be of the same order of importance. Thus, it is possible to classify problems on the basis of the order of importance of the terms in the equations. We have defined the *Rossby number* $\mathcal{R}o$ (Section 7.2) to be representative of the ratio of the inertial forces to the Coriolis force. When this number becomes smaller than unity, the inertia tends to become insignificant. Geostrophic motions are in this category. These motions occur outside the atmospheric and oceanic boundary layers.

Since viscosity plays an important role in the boundary layers, we can consider two other parameters involving the viscous forces, namely the *Ekman number*[13] $\mathcal{E}$ and the *Reynolds number* $\mathcal{R}$: these show the relative importance of the viscous to the Coriolis and inertia to viscous, respectively. Thus

$$\mathcal{R}o = U/L\Omega, \qquad \mathcal{E} = \nu/L^2\Omega, \qquad \mathcal{R} = UL/\nu \tag{7.42}$$

It follows then that the Rossby number is the product of the Ekman and Reynolds numbers. The characteristic velocity and length must be chosen so

[13] The reciprocal of the Ekman number is also referred to as the *Taylor number*.

TABLE 7.1

Order of Magnitudes in the Dynamics of the Atmosphere

	U, V	W	p	$1/\rho$	T
	$1–10$	$10^{-2}–10^{-1}$[a]	10^{2}	10^{3}	10^{2}
$\partial/\partial x$, $\partial/\partial y$	$10^{-5}–10^{-4}$	$10^{-7}–10^{-6}$[a]	$10^{-7}–10^{-6}$	$10^{-5}–10^{-4}$	$10^{-5}–10^{-4}$
$\partial/\partial z$	$10^{-3}–10^{-2}$	$10^{-5}–10^{-4}$	10^{-2}	10^{-1}	$10^{-3}–10^{-2}$
$\partial/\partial t$	$10^{-4}–10^{-3}$	$10^{-6}–10^{-5}$[a]	$10^{-5}–10^{-4}$	$10^{-4}–10^{-3}$	$10^{-4}–10^{-3}$
$\partial^2/\partial x^2$, $\partial^2/\partial y^2$	$10^{-10}–10^{-9}$	$10^{-10}–10^{-9}$	$10^{-12}–10^{-11}$	$10^{-10}–10^{-9}$	$10^{-10}–10^{-9}$
$\partial^2/\partial z^2$	$10^{-5}–10^{-4}$	$10^{-8}–10^{-7}$	10^{-6}	10^{-5}	$10^{-5}–10^{-4}$
$\partial^2/\partial x\,\partial z$, $\partial^2/\partial y\,\partial z$	$10^{-8}–10^{-7}$	$10^{-10}–10^{-9}$	$10^{-10}–10^{-9}$	$10^{-8}–10^{-7}$	$10^{-8}–10^{-7}$
$\partial^2/\partial t^2$	$10^{-8}–10^{-7}$	$10^{-10}–10^{-9}$[a]	$10^{-9}–10^{-8}$	$10^{-8}–10^{-7}$	$10^{-8}–10^{-7}$
$\partial^2/\partial x\,\partial t$, $\partial^2/\partial y\,\partial t$	$10^{-8}–10^{-7}$	$10^{-11}–10^{-10}$[a]	$10^{-10}–10^{-9}$	$10^{-9}–10^{-8}$	$10^{-9}–10^{-8}$
$\partial^2/\partial z\,\partial t$	$10^{-7}–10^{-6}$	$10^{-9}–10^{-8}$	$10^{-9}–10^{-8}$	$10^{-7}–10^{-6}$	$10^{-7}–10^{-6}$
$U\,\partial/\partial x$, $V\,\partial/\partial y$	$10^{-4}–10^{-3}$	$10^{-6}–10^{-5}$[a]	$10^{-6}–10^{-5}$	$10^{-4}–10^{-3}$	$10^{-4}–10^{-3}$
$W\,\partial/\partial z$	$10^{-4}–10^{-3}$[a]	$10^{-6}–10^{-5}$[a]	$10^{-4}–10^{-3}$[a]	$10^{-3}–10^{-2}$[a]	$10^{-4}–10^{-3}$[a]
d/dt	$10^{-4}–10^{-3}$[a]	$10^{-6}–10^{-5}$[a]	$10^{-4}–10^{-3}$[b]	$10^{-3}–10^{-2}$	$10^{-4}–10^{-3}$[a]

	Divergence	Vorticity	
	$\nabla \cdot U$	$(\nabla \times U)_{x \text{ or } y}$	$(\nabla \times U)_z$
	$10^{-6}–10^{-5}$	$10^{-3}–10^{-2}$	$10^{-5}–10^{-4}$
$\partial/\partial x$, $\partial/\partial y$	$10^{-11}–10^{-10}$	$10^{-8}–10^{-7}$	$10^{-10}–10^{-9}$
$\partial/\partial z$	$10^{-9}–10^{-8}$	$10^{-5}–10^{-4}$	$10^{-8}–10^{-7}$
$\partial/\partial t$	$10^{-10}–10^{-9}$	10^{-7} 10^{-6}	$10^{-9}–10^{-8}$
d/dt	$10^{-10}–10^{-9}$	$10^{-7}–10^{-6}$	$10^{-9}–10^{-8}$

Ω_x, Ω_y	Ω_z	μ	g	$\partial g/\partial x$, $\partial g/\partial y$	$\partial g/\partial z$
10^{-4}	10^{-4}	$10^{-3}–10^{-2}$	10	10^{-8}	10^{-6}

[a] These values should be decreased by a factor of 10 to 100 times for altitudes under 500 m.
[b] This value should be decreased by a factor of 10 for altitudes under 500 m.

that they are representative of the motion involved. We conclude that when the Rossby number is significant and the Ekman number is very small, we deal with geostrophic motion. Conversely, when the Rossby number is very small and the Ekman number significant, we have an Ekman-type boundary layer to consider, which we develop in the following sections.

In the dynamical equations (7.41) buoyant forces were not considered. Buoyancy forces appear in the environment, in the direction along the resultant gravitational acceleration, as a result of density variations other than those specified by an adiabatic environment. The buoyancy acceleration was discussed in Chapter 4. In the baroclinic environment the density variations can be independently influenced by thermal and humidity concentrations, in addition to pressure variations. When the buoyancy force is significant, as we shall encounter in certain applications, its relative order of importance is weighed either with reference to the inertial force or the viscous force as shown in the following section. When the buoyant force is referred to the viscous force, the dimensionless number obtained is the *Grashof number* $\mathcal{G}$, which is discussed in detail in Section 7.8. In environmental studies, the *Rayleigh number* $\mathcal{Ra}$ is often preferred to the Grashof number and the relationship between them is as follows:

$$\mathcal{G} = \frac{g\beta_1 L^3 \, \Delta T}{\nu^2}$$

$$\mathcal{Ra} = \mathcal{G} \cdot \mathcal{Pr} = \frac{g\beta_1 L^3 \, \Delta T}{\nu D} \tag{7.43}$$

where $\mathcal{Pr}$ is the *Prandtl number*, defined as ν/D, the ratio of the viscosity ν to the heat diffusivity D. In these relations involving the buoyancy force, g is the gravitational constant, β_1 is the coefficient of thermal expansion as defined in Section 2.11, ΔT is the characteristic temperature difference responsible for the buoyancy effects, and L is a characteristic buoyancy length, which is not always easy to define, so that it is truly characteristic of buoyant convection.

7.8 BUOYANCY EFFECTS

In Section 4.12 and Eq. (4.44) we established that the buoyant acceleration was $g(\rho - \rho_0)/\rho_0$ where ρ_0 is a reference density, often taken as the adiabatic value at the same elevation. Since physically for a positive updraft $\rho < \rho_0$ we define the buoyancy force per unit mass

$$b = -g\frac{\rho - \rho_0}{\rho_0} = -g\beta_1 \, \Delta T \tag{7.44}$$

if the density variation is primarily due to thermal effects. The coefficient of thermal expansion β, was defined in Eq. (2.32).

Experimental evidence shows (McBean *et al.*, 1971) that the contribution of humidity stratification to the buoyant energy is less than 10% of that contributed by thermal stratification in stable and neutral environments. Thus Eq. (7.44) is valid for most stable conditions.

In the case of the ocean, even though salinity gradients will affect the buoyancy b, it is always of interest to compare the actual results with those of constant entropy and salinity.

For an isentropic gaseous environment, according to Eq. (4.48),

$$p\rho^{-\gamma} = \text{constant} \tag{7.45}$$

and thus after differentiations and using Eq. (7.45) again

$$c^2 \, d\rho = dp \tag{7.46}$$

where $c^2 = \gamma RT$ is the square of the velocity of sound in the particular environment. Equation (7.46) is also valid for liquids although not for the reasons of Eq. (7.45).

From the conservation of mass [Eq. (6.11)]

$$\frac{\partial \rho}{\partial t} + \nabla \cdot (\rho \boldsymbol{U}) = 0$$

After expanding the divergence term and using the definition of Eq. (7.22) for the total rate of change

$$\frac{d\rho}{dt} + \rho(\nabla \cdot \boldsymbol{U}) = 0 \tag{7.47}$$

But from Eq. (7.46), $d\rho/dt = (1/c^2) \, dp/dt$, and since the pressure is locked strongly to the hydrostatic equation,

$$\frac{d\rho}{dt} = -\frac{\rho g}{c^2} \frac{dz}{dt} = -\frac{\rho g}{c^2} W \tag{7.48}$$

and

$$\nabla \cdot \boldsymbol{U} = \frac{g}{c^2} W \tag{7.49}$$

Then,

$$\frac{db}{dt} = -\frac{g}{\rho_0} \frac{d\rho}{dt} + \frac{g}{\rho_0} \frac{d\rho_0}{dt}$$

where both the local instantaneous density and the reference or mean density may vary with time. Substituting Eq. (7.48) for $d\rho/dt$ and letting $d\rho_0/dt = (d\rho_0/dz)(dz/dt)$, we have approximately the total change of the buoyant force

$$\frac{db}{dt} = \left(\frac{g}{\rho_0} \frac{d\rho_0}{dz} + \frac{g^2}{c^2} \right) W = -N^2 W \tag{7.50}$$

where

$$N = -\left(\frac{g}{\rho_0} \frac{d\rho_0}{dz} + \frac{g^2}{c^2} \right)^{1/2} \tag{7.51}$$

is known as the Brunt–Väisälä frequency or the natural frequency of oscillation of a vertical column of fluid given a small displacement from its equilibrium position.[14] The fluid is statically stable when N is real.

[14] See also Phillips (1969, p. 17). This is developed in Eq. (7.54) in terms of the temperature gradient.

If this buoyant acceleration is introduced in the z component of the equation of motion and given a characteristic buoyant length scale L_M, which we shall define as the *Monin–Obukhov length scale* derived from energy considerations in the atmospheric boundary layer (Monin and Obukhov, 1954), the ratio of the buoyant force per unit mass of Eq. (7.44) to that of inertia per unit mass U^2/L_M is $g\beta_1 L_M \, \Delta T/U^2$. When this ratio is multiplied by the square of the Reynolds number based on the same characteristic dimensions we obtain the Grashof number

$$\mathscr{G} = \frac{g\beta_1 L_M \, \Delta T}{U^2} \cdot \left(\frac{UL_M}{\nu}\right)^2 = \frac{g\beta_1 {L_M}^3 \, \Delta T}{\nu^2}$$

The inertia effect has canceled out in this product and thus $\mathscr{G}$ is the ratio of the buoyancy to the viscosity. It is known that in such thermal problems the heat diffusion also has a control on the density stratification which engenders the buoyancy forces. Thus the Rayleigh number, as defined in Eq. (7.43), is a more representative criterion for the determination of the order of magnitude of the buoyancy effects.

The question arises as to whether compressibility effects generating elastic forces are important and should have been taken into consideration in the development of Eq. (7.41). When the normal stresses were developed in Eq. (7.38) the elasticity of the environment was not considered. Had it been taken into account the normal stresses would have included[15] in addition to pressure, a term involving the product of the viscosity $\frac{2}{3}\mu$ and the divergence $\nabla \cdot \boldsymbol{U}$. The fact that this elastic force was not included in the development of the dynamical equations implies that it must be negligible in all dynamical aspects of the environment, even when the density variations are caused by thermal, humidity, or salinity variations, as the case may be.

From Eq. (7.46) it can be deduced that the quantity $d\rho/\rho$ in the buoyancy is equal to $(1/c^2)\, dp/\rho$, where dp is the pressure change brought about by the density change. This pressure change is nearly equal to the momentum change ρu^2 due to buoyant effects. Thus $d\rho/\rho$ is of the order of u^2/c^2, which is the square of the *Mach number*, $\mathscr{M} = u/c$.

In the units of Table 7.1, c^2 can be evaluated to be of the order of 10^4 to 10^6 for most of the environment, and the Mach number is of the order of 10^{-1} to 10^{-3} which is very small indeed. In air it is found that compressibility begins to play a role when the Mach number is greater than 0.4. Thus although the buoyancy forces may be appreciable, the secondary effect of inducing elastic forces is very small.

Looking at this another way, the divergence of the velocity in Eq. (7.49) is of the order of 10^{-4} (see Table 7.1) and $\frac{2}{3}\mu\nabla \cdot \boldsymbol{U}$, which would have been added

[15] See, for instance, Eskinazi (1967).

to p in Eq. (7.38), is of the order of 10^{-6} to 10^{-7}, while p is of the order of 10^2. Again from Eq. (7.49) this implies that the length scale characterizing the density field c^2/g is much larger than the depth scale, which is the most important scale in the divergence. In the case of the ocean c^2/g is of the order of 200 km while the depth of the ocean is of the order of 5 km aι most; and in the atmosphere c^2/g is of the order of 10 km, or roughly the height of the troposphere, while the height of the boundary layer in which the divergence is significant is only 1 km. While this shows that the influence of compressibility in the atmosphere is more important than that in the ocean, we conclude that it is still too small to involve elastic effects arising from density variations in the dynamical equations of the environment.

Even though Eqs. (4.37), (4.38), and (4.40) have been established to determine the criteria for vertical instability, ultimately, in order for buoyant energy to be transformed into an eddy-like turbulent motion, the ratio of the buoyant energy production to the eddy kinetic energy production, defined as the *Richardson number* $\mathscr{Ri}$ (Richardson, 1920), must be the significant criterion for being able to maintain this secondary turbulent motion. This ratio is approximated as

$$\mathscr{Ri} = \frac{g}{T_{\mathrm{ad}}} \frac{\partial T}{\partial z} \Big/ \left(\frac{\partial U}{\partial z}\right)^2 \tag{7.52}$$

The numerator in the Richardson expression is proportional to the energy necessary, against gravity, in lifting the mass of an air parcel, while the denominator represents the energy produced in turbulence or that extracted from the mean motion to sustain turbulence. It can be said then that when $\mathscr{Ri} = 0$ the dynamic atmosphere is vertically neutral in the absence of buoyant forces; when $\mathscr{Ri} < 0$ turbulence increases because the temperature difference with the adiabatic is negative and thus unstable; and when $\mathscr{Ri} > 1$ turbulence is suppressed because $T > T_{\mathrm{ad}}$. If one takes into account the turbulent heat diffusion coefficient γ_{T} and the momentum diffusion coefficient (eddy viscosity) ν_{T}, a new *flux Richardson number* can be defined such that

$$\mathscr{Ri}' = \frac{\gamma_{\mathrm{T}}}{\nu_{\mathrm{T}}} \mathscr{Ri} \tag{7.53}$$

Observations show that for $\mathscr{Ri}' > 0.2$ turbulence cannot be maintained. This result agrees well with Taylor's analysis using the method of oscillations (Taylor, 1931).

The numerator in Eq. (7.52) has the dimensions of one over the square of the time scale. It can easily be shown that this numerator is N^2, where N was defined as the Brunt–Väisälä frequency. This can be shown as follows: From the definition of Eq. (7.44)

$$b = -g \frac{dT}{T_{\mathrm{ad}}}$$

Thus db/dt, which was found to equal $-N^2\, dz/dt$ in Eq. (7.50), is also

$$\frac{db}{dt} = -\frac{g}{T_{\mathrm{ad}}}\frac{dT}{dt} = -\frac{g}{T_{\mathrm{ad}}}\frac{\partial T}{\partial z}\frac{dz}{dt}$$

It follows then that

$$N^2 = \frac{g}{T_{\mathrm{ad}}}\frac{\partial T}{\partial z} \tag{7.54}$$

If $(\partial T/\partial z) > 0$, we know that the column of the environment is statically stable and Eq. (7.54), which defines the frequency of the gravity wave, has significance. However, when for the values of $(\partial T/\partial z) < 0$ the environment is unstable, we can define a *buoyancy time scale* t_{b} such that

$$\frac{1}{t_{\mathrm{b}}{}^2} = -\frac{g}{T_{\mathrm{ad}}}\frac{\partial T}{\partial z} \tag{7.55}$$

The mean wind gradient $\partial U/dz$ in the Richardson number has the dimensions of one over time. If we define the *shearing time* t_{s} as

$$\frac{1}{t_{\mathrm{s}}} = \frac{\partial U}{\partial z} \tag{7.56}$$

then it becomes clear that the Richardson number is the square of the time ratios

$$\mathscr{R}i = \begin{cases} (Nt_{\mathrm{s}})^2 & \text{for } dT/dz > 0 \\ -(t_{\mathrm{s}}/t_{\mathrm{b}})^2 & \text{for } dT/dz < 0 \end{cases} \tag{7.57}$$

Following a similar treatment Tennekes and Lumley (1972, p. 99) conclude that in a sunny unstable atmosphere t_{b} is typically of the order of a few minutes. In a neutral atmosphere, however, t_{b} tends to infinite values and the frequency of oscillation tends to zero.

Finally, the buoyancy length scale which we defined earlier in this section as the Monin–Obukhov scale can be obtained from Eq. (7.52) by introducing the heat flux from the surface $H = -\rho c_p\, \partial T/\partial z$ and an approximation for the mean shear rate $(\partial U/\partial z)^2 = u_*{}^3/\kappa z$ where $u_*{}^2$ is defined as the square of the *shearing velocity* near the surface and thus equal to the ratio of the shear stress at the surface divided by the density. The quantity κ is the so-called *von Karman constant* which is developed in the boundary layer treatment (Section 9.4). Thus

$$\mathscr{R}i = -\frac{\kappa g H}{\rho u_*{}^3 T_{\mathrm{ad}}\, c_p}\, z$$

The coefficient before z in the Richardson number must have the units of reciprocal length; this length is defined as the Monin–Obukhov length, which is characteristic of convective motions. The absolute value of L_M is seldom less than 10 m:

$$L_M = -\frac{\rho u_*^3 T_{ad} c_p}{\kappa g H} \tag{7.58}$$

and

$$\mathcal{Ri} = \frac{z}{L_M} \tag{7.59}$$

7.9 THE CORIOLIS ACCELERATION AND THE DEFLECTING FORCE

The centrifugal acceleration on the earth's surface is a function of latitude. The Coriolis acceleration for a given velocity on the earth's surface lies on a plane perpendicular to the axis of the earth; its magnitude and direction on that plane depend on the angle between $\boldsymbol{\Omega}$ and $\boldsymbol{U}$. An east–west motion makes this angle 90° at all latitudes, and consequently maximal.

As given in Eq. (7.40) the term $2\boldsymbol{\Omega} \times \boldsymbol{U}$ on the left-hand side is the Coriolis acceleration shown together with the Eulerian acceleration of the medium. The right-hand side of the equation contains the summation of forces per unit mass responsible for this total acceleration. Often, the Coriolis acceleration in this equation is transposed to the right-hand side by changing the order of the product $2\boldsymbol{U} \times \boldsymbol{\Omega}$ and calling it the *deflecting force* because, as we know, this force is normal to $\boldsymbol{U}$.

In the Northern Hemisphere, the geometry of the vectors $\boldsymbol{U}$ and $\boldsymbol{\Omega}$ is such that *when we face the motion* $\boldsymbol{U}$ *(standing upright) the horizontal component of the deflecting force acts to the left of the motion.* In the Southern Hemisphere the result is opposite.

7.10 ZONAL FLOW PARALLEL TO LATITUDE CIRCLE

The arguments of Section 6.15 can be renewed in order to have a better understanding of *zonal flows*. Viscous forces need not enter this discussion because they do not add to the qualitative understanding of these motions.

Assume that the motion of the environment relative to the earth starts with a velocity $\boldsymbol{U} = d\boldsymbol{r}/dt = (\boldsymbol{\zeta} \times \boldsymbol{r})/2$, as in a rigid body where $\boldsymbol{\zeta}$ is the vorticity of the environment relative to the earth and $\zeta/2$ is its angular velocity $r = R \cos \psi$, as shown in Fig. 7.5. It is of interest to see what happens to this motion through this special equilibrium. The acceleration of the environment is given by $dU/dt = -\zeta^2 r/4$ and the Coriolis force is $2(U \times \Omega) = \Omega\zeta r$. Substituting into Eq. (7.40), omitting the viscous force, and taking into account this special form of acceleration for the special zonal flow, we have for equilibrium

$$(\Omega\zeta + \tfrac{1}{4}\zeta^2)\boldsymbol{r} + \boldsymbol{g} - \frac{1}{\rho}\nabla p = 0$$

First consider the case of the environment stationary with respect to the earth, i.e., $\zeta = 0$. From the equilibrium equation we are left with the pressure gradient balancing the gravitational force. Thus we have hydrostatic equilibrium and the isobaric and equipotential surfaces coincide, by being parallel to each other. The result of this equilibrium is shown in Fig. 7.5a.

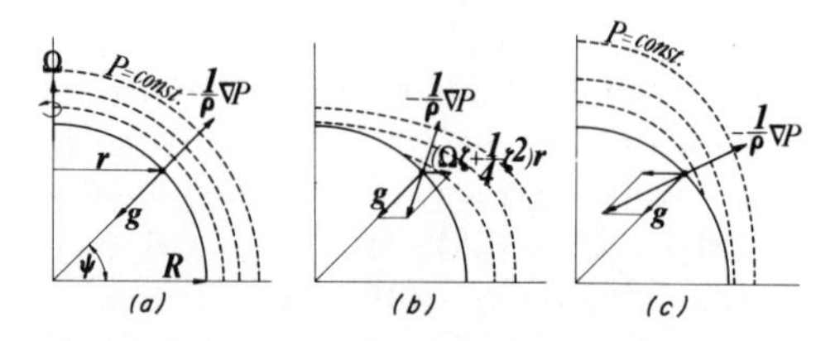

Fig. 7.5 Equilibrium of simple zonal motions: (a) $\zeta = 0$; (b) $\zeta > 0$; (c) $\zeta < 0$.

Second consider $\zeta > 0$, a westerly motion relative to the earth. In addition to pressure and gravity we have a force $(\Omega\zeta + \frac{1}{4}\zeta^2)\boldsymbol{r}$ in the positive $\boldsymbol{r}$ direction which when combined first with $\boldsymbol{g}$ will give a new orientation to the pressure gradient which must balance that addition. This new orientation of $-(1/\rho)\nabla p$ is inclined away from the vertical toward the north. This is shown in Fig. 7.5b. We conclude that for $\zeta > 0$ the isobaric surfaces are no longer parallel to equipotential surfaces, making the sea-level pressure toward the equator higher than that toward the north pole. So a low pressure region will prevail north of the westerly.

Third, consider $\zeta < 0$, an easterly. The vector $(-\Omega\zeta + \frac{1}{4}\zeta^2)\boldsymbol{r}$ will be in the direction of $-\boldsymbol{r}$ since $|\Omega|$ is usually larger than $|\zeta|$. The opposite argument will prevail, making the pressure at a fixed altitude toward the pole higher than that toward the equator at the same altitude. So a high pressure region will prevail north of the easterly.

This account is a kind of quasi-static consideration of the environment. In reality a westerly will not remain a westerly but will meander toward the south and then toward the north, producing alternating lows and highs as described in Section 6.15.

The stability of these zonal motions is discussed in Section 8.8 and an ensuing wave equation will predict this meandering flow.

Illustrative Example 7.1

Let us consider a simple example showing the influence of the Coriolis deflecting force on a mass descending in free fall.

As shown in Fig. 7.6 let the mass be initially located at the point P represented by the coordinates x_0, y_0, and z_0, recalling that by convention x is toward the east, y is toward the north, and z is along the plumb line.

From the dynamical equilibrium of the falling mass we can write

$$\frac{d\boldsymbol{W}}{dt} = \frac{d^2\boldsymbol{s}}{dt^2} = -2\boldsymbol{\Omega} \times \frac{d\boldsymbol{s}}{dt} + \frac{\boldsymbol{F}}{m}$$

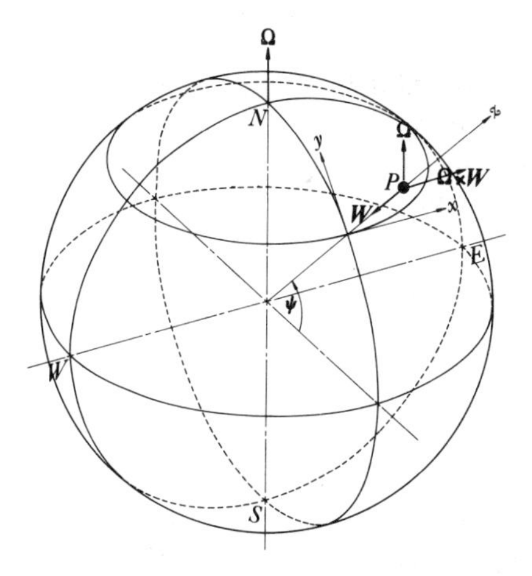

Fig. 7.6 A mass falling from a point P.

where $\boldsymbol{s}$ is the position vector of the mass measured from the initial position, $\boldsymbol{W}$ is the velocity of the falling mass along the plumb line, and $\boldsymbol{F}$ is the sum of all external forces, namely $\boldsymbol{g}$ and the drag, and m is the mass.

Since the deflecting force must be normal to $\boldsymbol{W}$ and $\boldsymbol{\Omega}$ and thus nearly along x, the x component of the equation of motion is

$$\frac{d^2x}{dt^2} = -2\left(\boldsymbol{\Omega} \times \frac{d\boldsymbol{s}}{dt}\right)_x + \left(\frac{\boldsymbol{F}}{m}\right)_x$$

At the onset of the motion, if the gravitational force is the only force on the mass, then initially the drag component $(\boldsymbol{F}/m)_x$ can be neglected. So initially $ds/dt = -\boldsymbol{k}gt + \boldsymbol{i}\, dx/dt$ and $\boldsymbol{\Omega} = \boldsymbol{k}\Omega \sin \psi + \boldsymbol{j}\Omega \cos \psi$.

Thus,

$$2\left(\boldsymbol{\Omega} \times \frac{ds}{dt}\right)_x = -2\Omega gt \cos \psi$$

and

$$\frac{d^2x}{dt^2} = 2\Omega gt \cos \psi$$

It is a good approximation to assume the latitude constant during the fall, particularly since the deflection is toward the east. Integration yields

$$x = \frac{\Omega g t^3}{3} \cos \psi$$

This is the amount of eastward deflection as a function of time. However, after a period of free fall, the terminal velocity W_0 will be such that the frictional forces on the mass will just balance the force of gravity and the mass will fall at constant velocity. Beyond that time $t > t_0$, $ds/dt = -\boldsymbol{k}W_0 + \boldsymbol{i}\, dx/dt$ and the x component is

$$\frac{d^2x}{dt^2} = 2\Omega W_0 \cos \psi$$

and

$$x = \Omega W_0 t^2 \cos \psi$$

Given sufficient time and altitude, the deflection can be as large as we wish. Thus we see that initially the mass will deflect as t^3 and when free fall conditions are reached the deflection will be proportional to t^2.

7.11 THE EQUATION OF MOTION IN TERMS OF VORTICITY

Equation (7.40) is essentially a pressure and velocity equation representing the equilibrium at a point in the environment. In Section 7.8 we discovered that this equation must be modified by a term $\boldsymbol{g}(T - T_{ad})/T_{ad}$ when acted upon by buoyant forces. As we shall see the inclusion of this buoyant term in the equation of motion is not important when Eq. (7.40) is transformed into a vorticity equation.

To perform this transformation it is first necessary to take the curl of the entire equation so that vorticity appears instead of velocity. First we have seen in Eq. (7.30) that the total acceleration of the fluid can be written in terms of the total or absolute vorticity $\boldsymbol{\zeta}_t = (2\boldsymbol{\Omega} + \boldsymbol{\zeta})$. Thus we start from the equation

$$\frac{\partial \boldsymbol{U}}{\partial t} + \frac{1}{2}\nabla U^2 + \boldsymbol{\zeta}_t \times \boldsymbol{U} = \boldsymbol{g} - \frac{1}{\rho}\nabla p + \nu \nabla^2 \boldsymbol{U} + \frac{\boldsymbol{g}}{T_{ad}}(T - T_{ad}) \tag{7.60}$$

and take the curl, term by term, keeping in mind that all the gradient terms in this process will go to zero, because the curl of the gradient is always zero. We discussed in Chapter 1 that the gravitational acceleration can be put in terms of the gradient of a geopotential function. Thus the term in $\boldsymbol{g}$ (including the buoyant term) will disappear in the curl and we will have

$$\nabla \times \frac{\partial \boldsymbol{U}}{\partial t} + \nabla \times (\boldsymbol{\zeta}_t \times \boldsymbol{U}) = -\nabla \times \left(\frac{1}{\rho}\nabla p\right) + \nabla \times \left(\frac{\mu}{\rho}\nabla^2 \boldsymbol{U}\right) \tag{7.61}$$

Each one of these terms can be developed further. Since time and space are the independent variables and are also independent of each other, then the space differentiation involved in the curl of the first term can be interchanged with the time differentiation, thus making

$$\nabla \times \frac{\partial \boldsymbol{U}}{\partial t} = \frac{\partial}{\partial t}(\nabla \times \boldsymbol{U}) = \frac{\partial \boldsymbol{\zeta}}{\partial t}$$

and since the earth's rotation does not vary in time, this rate of change of the vorticity of the fluid relative to the surface of the earth can be replaced by $\partial \boldsymbol{\zeta}_t/\partial t$ which is the rate of change of the absolute vorticity.

The second term on the left, through the triple vector product given in Appendix A, can be developed as follows:

$$\begin{aligned}\nabla \times (\boldsymbol{\zeta}_t \times \boldsymbol{U}) &= (\boldsymbol{U} \cdot \nabla)\boldsymbol{\zeta}_t - \boldsymbol{U}(\nabla \cdot \boldsymbol{\zeta}_t) + \boldsymbol{\zeta}_t(\nabla \cdot \boldsymbol{U}) - (\boldsymbol{\zeta}_t \cdot \nabla)\boldsymbol{U} \\ &= (\boldsymbol{U} \cdot \nabla)\boldsymbol{\zeta}_t - (\boldsymbol{\zeta}_t \cdot \nabla)\boldsymbol{U}\end{aligned}$$

Two terms are usually dropped in this development. The divergence of the curl is always zero (see Appendix A). The divergence of the velocity is often small even with temperature stratification, for the reasons discussed in Section 7.8. Otherwise, the term $\boldsymbol{\zeta}_t(\nabla \cdot \boldsymbol{U})$ represents addition of vorticity in the same direction as existing vorticity. Thus nonisochoric flows can generate vorticity through this term.

The first term on the right-hand side of Eq. (7.61) has to be developed according to the rule for the curl of a scalar times a vector. This gives

$$\begin{aligned}\nabla \times \left(\frac{1}{\rho}\nabla p\right) &= \frac{1}{\rho}\nabla \times \nabla p + \nabla\left(\frac{1}{\rho}\right) \times \nabla p \\ &= \nabla\left(\frac{1}{\rho}\right) \times \nabla p\end{aligned}$$

recalling that the curl of the gradient of the pressure must always be zero.

The last term in Eq. (7.61) must be developed in the same manner because $(1/\rho)$ can vary in space while the viscosity can be maintained constant:

$$\mu\left[\nabla \times \left(\frac{1}{\rho} \nabla \times \zeta\right)\right] = \mu\left[\frac{1}{\rho} \nabla \times (\nabla \times \zeta) + \nabla\left(\frac{1}{\rho}\right) \times (\nabla \times \zeta)\right]$$

and

$$\nabla \times (\nabla \times \zeta) = \nabla(\nabla \cdot \zeta) - \nabla^2 \zeta = -\nabla^2 \zeta$$

Thus

$$\mu\left[\nabla \times \left(\frac{1}{\rho} \nabla \times \zeta\right)\right] = -\nu \nabla^2 \boldsymbol{U} + \mu \nabla\left(\frac{1}{\rho}\right) \times (\nabla \times \zeta)$$

Finally Eq. (7.61) becomes

$$\frac{\partial \zeta_t}{\partial t} + (\boldsymbol{U} \cdot \nabla)\zeta_t = (\zeta_t \cdot \nabla)\boldsymbol{U} - \nabla\left(\frac{1}{\rho}\right) \times \nabla p + \nu \nabla^2 \zeta_t - \mu \nabla\left(\frac{1}{\rho}\right) \times (\nabla \times \zeta_t)$$

$$= \frac{d\zeta_t}{dt} \tag{7.62}$$

This is the expression sought. It is apparent that ζ has been replaced by ζ_t since the addition of the rotation of the earth in the space derivatives contributes nothing.[15a]

This total rate of change of vorticity equation, developed from the dynamical equilibrium, implies that absolute vorticity of an environmental parcel is augmented or reduced in time if any or all of the terms to the right of the equality sign are nonzero. The first term $(\zeta_1 \cdot \nabla)\boldsymbol{U}$ is called the *advection of vorticity* with the fluid. In other words, it can be shown to represent the fact that a ζ_t line moves with the fluid and that the fluid material on that line is always the same fluid material. In the process of this advection, however, the components of $\nabla\boldsymbol{U}$ parallel to ζ_t extend and contract the vortex lines, changing the vorticity (see Illustrative Example 7.2). Part of the change of ζ_t comes from the rigid rotation of the vortex line due to the components of $\nabla\boldsymbol{U}$ perpendicular to ζ_t. Sometimes this term is also called the *vortex stretching*, and as we can see it is an interaction between the available local vorticity ζ_t and the gradient of the velocity vector $\nabla\boldsymbol{U}$ which is a tensor giving the total deformation of the fluid.[16] This

[15a] If we include the divergence term $\zeta_t(\nabla \cdot \boldsymbol{U})$ dropped on the previous page and consider two-dimensional isentropic surfaces where $p = f(\rho)$ and no friction, Eq. (7.62) becomes

$$d\zeta_t/dt = -\zeta_t (\nabla \cdot \boldsymbol{U})$$

[16] See, for instance, Eskinazi (1967, pp. 127–139) and Tennekes and Lumley (1972, p. 83).

interaction gives a vector that represents the stretching of vortex lines followed by an increase of vorticity, or the convergence of vortex lines followed by a decrease in absolute vorticity. In this process, angular momentum may or may not be conserved, depending on the losses due to viscosity.

The second term to the right of the equality sign in Eq. (7.62) represents vorticity change due to *baroclinic conditions in the environment.* Baroclinicity was discussed in Section 6.9. We can see immediately from the geometry of this vector that if the $p =$ constant (isobars) lines are not parallel to the $\rho =$ constant (isochors) lines, the gradients of these surfaces will not be parallel, and their cross product will be nonzero. We know that density varies with pressure, but it must vary with temperature, humidity, or salinity as in the case of the ocean, in order for this baroclinic effect to occur. This constitutes the basis of atmospheric circulation due to temperature gradients from the north pole to the equator, and in the case of estuaries salinity gradients from river to ocean cause added vorticity in the water.

The third term to the right of the equality sign in Eq. (7.62) gives the *diffusion of absolute vorticity* in the environment through viscous action. Thus diffusion is primarily important in boundary layers near the ground for the atmosphere and near the ocean surface.

The last term in the equation is a viscous interaction between the density gradients and the variation of vorticity. This term may not be significant unless the vorticity is distributed as in the boundary layers and with large curls.

Various applications are selected in the course of this book, to involve the effects contained in Eq. (7.62).

Illustrative Example 7.2

As shown in Fig. 7.7 let us consider a vertical convergence in which initially the environment has two components of vorticity, namely ζ_x and ζ_z initially acted upon by two components of the strain-rate field represented by the tensor $\nabla \boldsymbol{U}$. The strain-rate field has a total of nine components, namely three that are normal and six that are a combination of shear and rotation. For simplicity, let the two components of the normal strain rate $\partial W/\partial z$ and $\partial U/\partial x$ be equal in absolute value, such that

$$\frac{\partial W}{\partial z} = m_{zz} = m \qquad \text{and} \qquad \frac{\partial U}{\partial x} = -m_{xx} = -m$$

As the motion progresses from (a) to (b) in the updraft, on the basis of conservation of angular momentum the vorticity component ζ_z in the direction of the positive strain rate is amplified while the vorticity ζ_x is attenuated according to the first term to the right of the equality sign in Eq. (7.62).

If we assume that this *vortex stretching* is the only important term in the rate of change of the vorticity, then

$$\frac{d\zeta_t}{dt} = \zeta_t \cdot \nabla \boldsymbol{U}$$

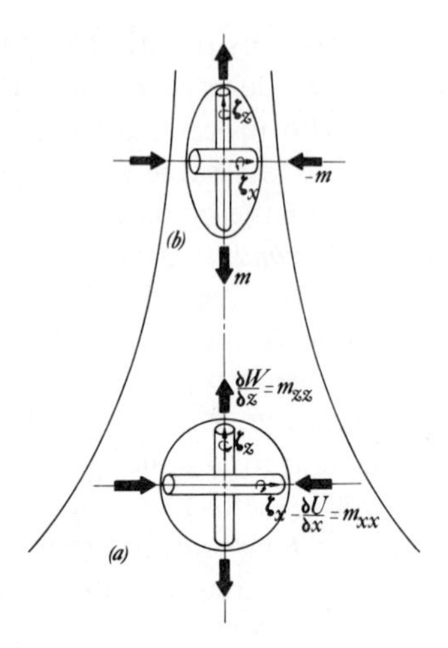

Fig. 7.7 Vortex stretching in a strain field: (a) initial strain and vorticity; (b) change of vorticity due to stretch.

and each component will obey the following differential equations:

$$\frac{d\zeta_z}{dt} = m\zeta_z \qquad \text{and} \qquad \frac{d\zeta_x}{dt} = -m\zeta_x$$

If at time $t = t_0$ all vorticity components $\zeta_x = \zeta_z = \zeta_0$, then the solution of these two equations yields

$$\zeta_z = \zeta_0 e^{mt} \qquad \text{and} \qquad \zeta_x = \zeta_0 e^{-mt}$$

and after combining, the total vorticity will be

$$\zeta = (\zeta_z^2 + \zeta_x^2)^{1/2} = (2\zeta_0^2 \cosh 2mt)^{1/2}$$

which indicates that the total amount of vorticity increases with increasing values of mt. This is shown qualitatively in Fig. 7.8. The amount of increase of ζ_z under a positive normal strain rate becomes larger in time than the rate of decrease of ζ_x owing to a negative influence of the normal strain rate.

We must not forget that in this simple illustrative example all other terms in Eq. (7.62) have been omitted in the considerations and that the other components of the strain rate $m_{yy} = m_{xy} = m_{xz} = m_{yz} = m_{yx} = m_{zx} = m_{zy} = 0$.

It is of interest to find out what happened to the total kinetic energy of the vortex stretching process (Tennekes and Lumley, 1972, p. 257). Since work is necessary to stretch a vortex in motion, the rate of energy change in the process can be computed. It can be shown that the energy per unit mass and unit time involved in each stretching is the product of the strain rate m and the instantaneous kinetic energy. Thus the net rate of energy exchanged is

$$\dot{E} = m(u^2 - w^2)$$

where u and w are the x and z components of the circular velocities of the two vortex elements. We know that ζ_z has increased in time, which corresponds, according to Eq. (6.58), to an increase in u and v. However, since ζ_x has decreased, according to Eq. (6.59), it is likely that both v and w have decreased. Thus we cannot be too sure about the outcome of v since it increased in one case and decreased in the other. But since we are fairly certain that u increased and w decreased, we can conclude from the rate of energy exchanged $\dot{E}$ that it is positive and that the strain rate has performed work.

This simple example of vortex stretching is fundamental in the transfer of turbulent energy from one eddy to another.

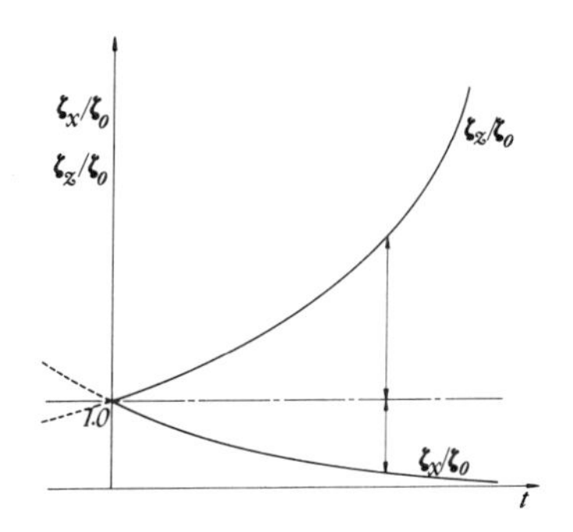

Fig. 7.8 Time variation of vorticity.

7.12 BASIC PRINCIPLES OF TURBULENCE

It is not the purpose of this review to go into fundamental details on the vast treatment of the theories of turbulence[17] and to dwell on the physical aspects of its generation, maintenance, and decay in environmental flows. It suffices to know, for the relevance of the treatment that follows, that turbulence is an irregular state of the motion, with respect to time and space, which occurs when the ratio of inertia to viscous forces (Reynolds number) reaches a certain level. It

[17] See, for instance, Hinze (1959) or Tennekes and Lumley (1972), and Panchev (1971).

is important to note that turbulence is not a state of the fluid but a state of the motion. Thus turbulent properties depend on the state of the motion and not directly on the fluid. (The Reynolds number was discussed in Section 7.7.) In the motion of fluids, at large Reynolds numbers, internal dynamical instabilities occur which generate property fluctuations with respect to time and space even though the motion was kept steady and uniform prior to the start of turbulence. These fluctuations of flow properties are irregular and random in nature with respect to time and space.

Generally speaking, turbulence has the following basic physical characteristics: It is irregular, diffusive, three-dimensional (even when the mean motion is one- or two-dimensional), dissipative, continuous in time and space, and finally a property of the motion and not of the fluid. It can be shown that because of the dissipative character of turbulence, which arises from viscous friction, it can only maintain itself in a quasi steady state in flows that display mean velocity gradients or flows in which there exists a mean strain rate. The boundary layer of the atmosphere near the surface of the earth and that on the surface of the ocean display this irregular turbulence characteristic. This does not preclude that local instabilities resembling turbulence cannot be generated owing to baroclinicity, for instance, in layers outside the boundary layer where the mean strain rate is negligible. These baroclinic instabilities will dissipate without the turbulent energy production generated by the action of the strain rate. In a very simple way, Illustrative Example 7.2 has alluded to this phenomenon. Thus turbulence will be primarily associated with boundary layer regions.

Speaking of turbulence implies irregularities in *all* the vector and scalar properties of the motion: Pressure, temperature, density, velocity, humidity, and so on, will fluctuate with respect to time and space at large enough Reynolds numbers. We can define a *critical Reynolds number* in terms of a characteristic velocity, a length scale, and a viscosity at which point turbulence will set in. Insofar as atmospheric boundary layer flows are concerned, the Reynolds number is almost always larger than this critical value. Based on the thickness of the boundary layer the critical Reynolds number is of the order of 10^3 to 10^4.

In a rotating frame of reference such as the earth, the horizontal motion of the environment is significantly caused by the Coriolis force and since this force depends on the vertical component of the earth's angular velocity, which is equal to approximately $10^{-4}\ \mathrm{s}^{-1}$, its time scale t is of the order of 10^4 s.

If the atmospheric boundary layer were laminar, its thickness would be governed by molecular viscous diffusion so that its thickness would be proportional according to $\delta_l^2 \sim \nu t$. With $\nu = 0.15\ \mathrm{cm}^2/\mathrm{s}$ and $t = 10^4$ s, the thickness of the boundary layer would be approximately 40 cm. Measurements of the turbulent atmospheric boundary layer predict thicknesses of the order of 1 km. Then for a turbulent state we cannot use this laminar relation but must relate the thickness to a characteristic velocity U multiplied by the same time scale t. Thus $\delta_t \sim Ut$. The ratio between these two different length scales

$$\frac{\delta_t}{\delta_l} \sim \frac{Ut}{(vt)^{1/2}} = \left(\frac{U^2}{\Omega v}\right)^{1/2}$$

According to the definitions of Eq. (7.42) this ratio of thicknesses is the square root of the product of the Reynolds number and the Rossby number. For Rossby number nearly unity, δ_t/δ_l can be interpreted as the square root of the Reynolds number. For normal wind velocities in the atmospheric boundary layer the Reynolds number is of the order of 10^7, which puts it in the turbulent regime.

The irregular turbulent fluctuation of properties is random in nature and averages to zero in time at all points of the flow. Although the instantaneous magnitudes of these fluctuations may be small compared to their mean values, the dynamic and thermodynamic influence of these fluctuations when they are interdependent (correlated) can be of considerable influence on the motion and in diffusion when compared to the laminar behavior. For this reason turbulence cannot be neglected in the boundary layer or in regions with high rates of shear.

The mathematical treatment of random quantities is the subject of statistics. Thus the presence of turbulence in the motion brings the following complications to the type of analysis we have followed so far:

(1) *Strictly speaking* it is impossible to treat turbulence as a steady-state phenomenon.

(2) Since the time dependence of turbulence is random, it is no longer possible to obtain deterministic solutions for the instantaneous properties.

We shall speak instead of average behavior. Mathematically, an averaging process of an instantaneous value of a property $\tilde{p}$ involves an integral with respect to the independent variable through which it fluctuates. At a given point in space since the property fluctuations are with respect to time, the *time average* is defined as

$$P = \bar{\tilde{p}} = \lim_{T \to \tau} \frac{1}{T} \int_0^T \tilde{p}(t + t_0)\, dt \qquad (7.63)$$

where t_0 is the initial time when the averaging process begins, T is the period of averaging, and τ is the largest time scale in the phenomenon that is responsible for the fluctuation $\tilde{p}$. This was discussed in Section 2.18.

The time scale τ in the average process must be identified in order to attach the proper significance to P. Fluctuations in the environment are caused first by the randomness of molecular effects (particularly at very high altitudes). Molecular effects have their own characteristic time scale τ_m and averages pertaining to molecular effects must be taken in periods comparable to τ_m. If turbulence effects on the environment are of interest, their characteristic time scale $\tau_t \gg \tau_m$ should be considered in the time average. When times larger than τ_t are considered, i.e., contributions from zonal and seasonal fluctuations are included in the averages.

Figure 7.9 shows the importance of the characteristic averaging time in the average values. In this figure we qualitatively consider the time trace of $\tilde{p}(t)$ in a seasonal period τ_s of the order of many days. In this trace we show three different average values for the three different periods: τ_t in seconds for turbulence, τ_z in a day, and τ_s in a month. Whatever the integration time τ, the instantaneous value of p, written as $\tilde{p}$, can be decomposed into an average value P, which by definition, within τ, should not be a function of time, and a fluctuating part p. Therefore, as in Eq. (6.1)

$$\tilde{p}(\mathbf{r}, t) = P(\mathbf{r}) + p(\mathbf{r}, t) \tag{7.64}$$

It is clear that applying Eq. (7.63) to this decomposition yields $\bar{\tilde{p}} = P$ and $\bar{p} = 0$.

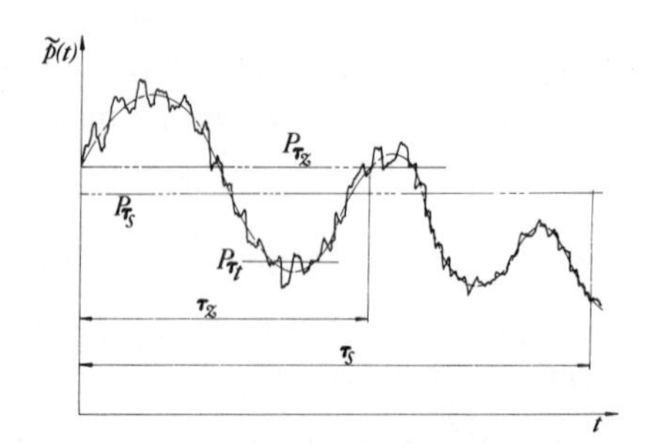

Fig. 7.9 Decomposition of fluctuating and mean parts of an instantaneous value of a property.

This averaging process, when applied to functional relationships of properties in the physical laws we have been discussing, gives additional statistical contributions when the laws involve products and powers. Let us consider first the averaging of a *sum* of fluctuating properties. Consider the total pressure of a mixture as discussed in Section 2.13. If the total pressure fluctuates in time, then its instantaneous value is the sum of the instantaneous values of the partial pressures of the constituents:

$$\tilde{p}(t) = \tilde{p}_1(t) + \tilde{p}_2(t) + \tilde{p}_3(t) \tag{7.65}$$

and since each of these instantaneous values can be split according to Eq. (7.64)

$$P + p = (P_1 + p_1) + (P_2 + p_2) + (P_3 + p_3)$$

The time average of this equation yields simply

$$P = P_1 + P_2 + P_3$$

because $\bar{p}_1 = \bar{p}_2 = \bar{p}_3 = 0$, by definition.

However, if instead of *Dalton's law* we apply the averaging to the instantaneous perfect gas law $\tilde{p} = R\tilde{\rho}\tilde{T}$, and after substituting the decomposition of Eq. (7.64) such that $\tilde{\rho} = \bar{\rho} + \varepsilon$ and $\tilde{T} = \Theta + \vartheta$, we have

$$\begin{aligned} P + p &= R(\bar{\rho} + \varepsilon)(\Theta + \vartheta) \\ &= R(\bar{\rho}\Theta + \bar{\rho}\vartheta + \varepsilon\Theta + \varepsilon\vartheta) \end{aligned}$$

If we take the time average of this expression, we will have

$$P = R(\bar{\rho}\Theta + \overline{\varepsilon\vartheta}) \tag{7.66}$$

because $\bar{p} = \bar{\varepsilon} = \bar{\vartheta} = 0$. However, the average of the product $\overline{\varepsilon\vartheta}$ will not be zero, particularly if ε and ϑ are correlated, that is, if part or all of the density fluctuations ε are due to the temperature fluctuation ϑ. So we see that in the average of products a statistical correlation term is added to the product of the means. In other words, the average of the product *is not* always the product of the averages. Since fluid dynamical and thermodynamical principles have product and power relations, these statistical correlation terms become important. In particular since the inertia terms in Eqs. (7.35)–(7.37) are not linear, we expect to obtain correlation terms when we expand these equations, and their significance will have to be explained.

If we decompose the velocity components of the motion into mean and fluctuating parts, as in Eq. (7.64), and introduce them into the three components of Eqs. (7.35)–(7.37) and then take the time average of the three components, we will have nine additional terms as derivatives of correlations. These additional terms can be represented in the following matrix:

$$\left\{\begin{matrix} \dfrac{\partial \overline{u^2}}{\partial x}, & \dfrac{\partial \overline{vu}}{\partial y}, & \dfrac{\partial \overline{wu}}{\partial z} \\[2ex] \dfrac{\partial \overline{uv}}{\partial x}, & \dfrac{\partial \overline{v^2}}{\partial y}, & \dfrac{\partial \overline{wv}}{\partial z} \\[2ex] \dfrac{\partial \overline{uw}}{\partial x}, & \dfrac{\partial \overline{vw}}{\partial y}, & \dfrac{\partial \overline{w^2}}{\partial z} \end{matrix}\right\} \tag{7.67}$$

where the instantaneous values of the velocities should be replaced in Eqs. (7.35)–(7.37) and the decompositions

$$\tilde{u} = U + u, \qquad \tilde{v} = V + v, \qquad \tilde{w} = W + w \tag{7.68}$$

applied before the average is performed. When each of the terms in the matrix (7.67) is transposed to the right-hand side of the equations of motion (7.35)–(7.37) and combined with the stresses already there, these equations will take the form

$$\begin{aligned}
&\frac{\partial U}{\partial t} + U\frac{\partial U}{\partial x} + V\frac{\partial U}{\partial y} + W\frac{\partial U}{\partial z} - fV \\
&\qquad = \frac{1}{\rho}\left[\frac{\partial}{\partial x}(\sigma_{xx} - \rho\overline{u^2}) + \frac{\partial}{\partial y}(\tau_{xy} - \rho\overline{vu}) + \frac{\partial}{\partial z}(\tau_{xz} - \rho\overline{wu})\right] \\
&\frac{\partial V}{\partial t} + U\frac{\partial V}{\partial x} + V\frac{\partial V}{\partial y} + W\frac{\partial V}{\partial z} + fU \\
&\qquad = \frac{1}{\rho}\left[\frac{\partial}{\partial x}(\tau_{xy} - \rho\overline{uv}) + \frac{\partial}{\partial y}(\sigma_{yy} - \rho\overline{v^2}) + \frac{\partial}{\partial z}(\tau_{yz} - \rho\overline{wv})\right] \\
&\frac{\partial W}{\partial t} + U\frac{\partial W}{\partial x} + V\frac{\partial W}{\partial y} + W\frac{\partial W}{\partial z} \\
&\qquad = -g + \frac{1}{\rho}\left[\frac{\partial}{\partial x}(\tau_{xz} - \rho\overline{uw}) + \frac{\partial}{\partial y}(\tau_{yz} - \rho\overline{vw}) + \frac{\partial}{\partial z}(\sigma_{zz} - \rho\overline{w^2})\right]
\end{aligned} \tag{7.69}$$

This is known as the *Reynolds equation*, since Reynolds first introduced the decomposition equations (7.68) into the equations of motion. Notice that the density has been kept constant in this decomposition. The quantities $-\rho\overline{u^2}$, $-\rho\overline{uv}$, etc., are called the *Reynolds stresses* because as shown in Eq. (7.69) they are additive to the viscous stresses. As a matter of fact experiments show that these Reynolds stresses could be many orders of magnitude larger than the viscous stresses in the boundary layer.

The closure problem of the laminar form of the equations of motion (7.41) is discussed thoroughly in many books.[18] The conclusion we arrive at is that for laminar flows there are as many independent equations as there are unknowns. However, since we have added *six* new variables with the Reynolds stresses (the stresses are symmetrical, $\overline{uv} = \overline{vu}$, etc., reducing the number from nine to six), the turbulence problem remains open to this day. Various hypotheses and approximations about the Reynolds stresses have been used to complete the mathematical closure of the problem.

Fortunately, after looking at Table 7.1 we realize that the second derivatives with respect to z (altitude) of the velocities are many orders larger than those in the x and y directions (horizontal). Thus for the stresses in Eq. (7.69) we need

[18] See, for instance, Eskinazi (1967, p. 197).

to keep only the terms in $\partial/\partial z$ and of course the gradients of mean pressure in the normal stresses. Furthermore, for a thermally stable atmosphere since W is zero, we can assume for most cases horizontal motion. Thus Eq. (7.69) simplifies for most applications to

$$U\frac{\partial U}{\partial x} + V\frac{\partial U}{\partial y} - fV = -\frac{1}{\rho}\frac{\partial P}{\partial x} + \frac{\partial}{\partial z}\left(\nu\frac{\partial U}{\partial z} - \overline{uw}\right)$$

$$U\frac{\partial V}{\partial x} + V\frac{\partial V}{\partial y} + fU = -\frac{1}{\rho}\frac{\partial P}{\partial y} + \frac{\partial}{\partial z}\left(\nu\frac{\partial V}{\partial z} - \overline{vw}\right) \tag{7.70}$$

$$g = -\frac{1}{\rho}\frac{\partial P}{\partial z}$$

and in the case of buoyancy g must be relieved by the buoyancy acceleration discussed in Section 7.8. These simplified equations will become the fundamental equations for most environmental applications.

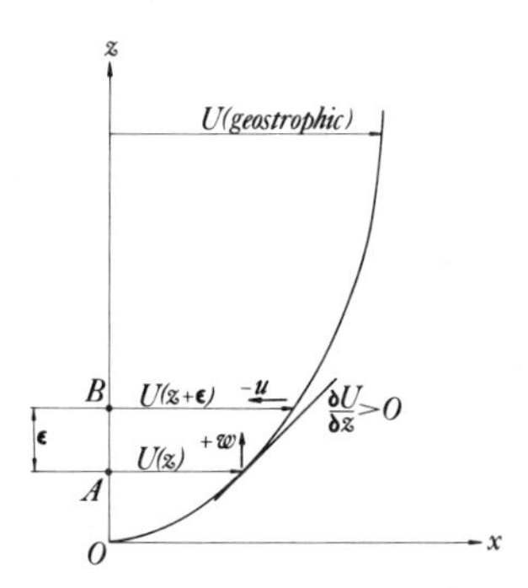

Fig. 7.10 Significance of the Reynolds stress.

The diagram in Fig. 7.10 illustrates the physical significance of the Reynolds stresses. Consider an atmospheric boundary layer where the x component of the mean velocity increases with altitude, $\partial U/\partial z > 0$. If a small parcel of air A at altitude z has a horizontal mean velocity U and if at the time of consideration the z component of the velocity fluctuation[19] of A is $+w$ in the positive z direction, this instantaneous movement will instantaneously carry a mass flow per unit area equal to ρw at a level B positioned at $(z + \varepsilon)$. This mass had an x momentum equal to $\rho w U$ before leaving A, and at the new position B it will find itself deficient of momentum per unit mass since the fluid already there

[19] Turbulence is always three-dimensional in character even if the mean motion is one-dimensional. This was discussed early in this section.

has higher x momentum by virtue of $U_B > U_A$. When the mass originally at A arrives at B and mixes with it, by virtue of its lower momentum it will retard the flow at B by an amount $-u$ and the retardation will be caused by the Reynolds stress equal to the mean momentum deficiency $\rho\overline{w(-u)} = -\rho\overline{wu}$. This retardation is in the same direction as the viscous stress $\tau = \mu\, \partial U/\partial z$. The same argument holds for the other components and even if the gradient of the mean velocity or w is reversed. Thus

$$\frac{\partial U}{\partial z} > 0; \qquad \left\{\begin{matrix} +w \Rightarrow -u \\ -w \Rightarrow +u \end{matrix}\right\}; \qquad -\overline{uw} > 0$$

$$\frac{\partial U}{\partial z} < 0; \qquad \left\{\begin{matrix} +w \Rightarrow +u \\ -w \Rightarrow -u \end{matrix}\right\}; \qquad \overline{uw} < 0$$

The other correlations follow the same logic.

The conservation of mass or continuity equation in turbulent flows remains unchanged because starting from Eq. (6.14), written for the instantaneous velocity, and applying the decomposition equation (7.68), we have

$$\frac{\partial \tilde{u}}{\partial x} + \frac{\partial \tilde{v}}{\partial y} + \frac{\partial \tilde{w}}{\partial z} = \left(\frac{\partial U}{\partial x} + \frac{\partial V}{\partial y} + \frac{\partial W}{\partial z}\right) + \left(\frac{\partial u}{\partial x} + \frac{\partial v}{\partial y} + \frac{\partial w}{\partial z}\right) = 0$$

After taking the average of this relation we conclude that the terms in both parentheses must be zero independently:

$$\frac{\partial U}{\partial x} + \frac{\partial V}{\partial y} + \frac{\partial W}{\partial z} = \frac{\partial u}{\partial x} + \frac{\partial v}{\partial y} + \frac{\partial w}{\partial z} = 0 \qquad (7.71)$$

7.13 DYNAMICAL MODELS OF THE ATMOSPHERE AND OCEANS

The partial differential equations developed so far for the environment seem very complicated to solve in their complete form. It is fortunate, however, for those interested in solutions, that each and every term in these equations is not equally important in all layers of the environment. Some terms may be important in certain regions while at the same time be unimportant in others.

Figure 7.11 can be used to demonstrate the relative order of importance of the various terms in the dynamical equations in different regions of the atmosphere and ocean. We know, so far, that the horizontal component of the motion is the most significant in most cases. This horizontal motion when flowing over the earth's surface and over the ocean surface, is subjected to maximum frictional forces at atmosphere–ocean interface. This frictional force will be the cause for the formation of a *boundary layer* in which the velocity of air at the surface of the earth will be zero relative to the earth, and have a minimum velocity at the surface of the ocean equal to the surface velocity of the water. In turn, if the ocean current is primarily generated by the wind, the water velocity at the surface will be maximum and decrease in depth again as a result of frictional forces. In both cases, the boundary layer is defined as the layer in which frictional forces cause strong velocity gradients and vorticity.

The size of the atmospheric boundary layer depends of course on the wind speed and on the type of terrain, but it is usually of the order of 1000 m while the thickness of the oceanic boundary layer is roughly 1/30 of this thickness or 30 m. Beyond this layer is an environment free of viscous or turbulent shear; it is called the *free layer*. In the absence of shear stresses, the motion of the

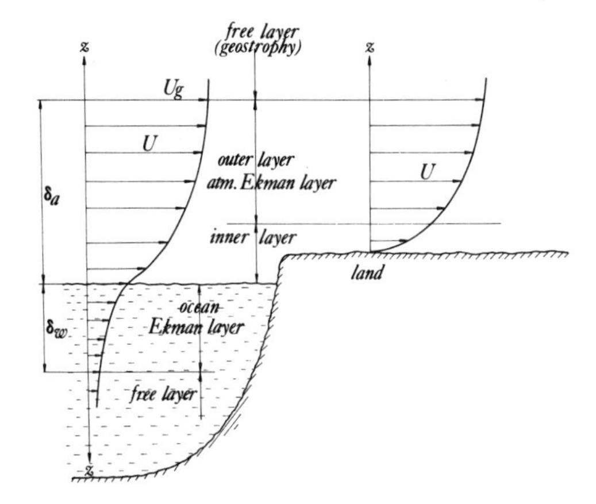

Fig. 7.11 Velocity distribution in the atmosphere and ocean.

free layer is strongly governed by the rotation of the earth and consequently is given the name of *geostrophic layer* where the motion is called geostrophic. From the dimensions of the boundary layers given, it is apparent that the largest portion of the troposphere and of the ocean is under geostrophic conditions.

The whole of the boundary layers is not influenced the same way by the same forces either. Near the surface in a layer a few meters high the mean velocities are small and consequently the inertia and the Coriolis force, which depends on the velocity, are negligible compared to the shear forces (viscous and Reynolds) which appear to be constant in this *inner layer*. In the bulk of the boundary layer, called the *outer layer* or the *Ekman layer*, inertia is still negligible and the viscous forces become negligibly small. This part of the boundary layer is in equilibrium through a balance of the Coriolis, pressure, and Reynolds stresses.

On the basis of this physical model these various regions are developed in the chapters that follow.

8

Geostrophic Motion and Applications

8.1 INTRODUCTION

In this chapter we consider the inviscid motion of the atmospheric and oceanic *free layers*. The geometric and physical aspects of this layer were discussed in Section 7.12 in context with the rest of the environment.

Since approximately nine-tenths of the troposphere satisfies the conditions of the inviscid free layer, because most cloud formations are in the free layer, and since our weather physics is substantially shaped in this layer, the field of meteorology makes considerable use of the study of free layers. In oceanography, however, the free layer of the ocean is of minor influence and interest to the practical needs of man. The surface or Ekman layer of the ocean is where the action is, so to speak, since it is the part that provides the atmosphere with considerable amounts of energy in the form of latent heat in vapor and thus contributes in a direct way to our weather physics. Furthermore, it is the layer in which larger horizontal and vertical currents occur, thus creating its importance in navigation, food supply for fish, and even pollution.

8.2 FRICTIONLESS FLOW—EULER'S EQUATION AND ITS INTEGRATION

We have seen in Eq. (7.39) that frictional stresses are a product of viscosity and velocity derivatives. Since air and water in our environment have finite viscosities, the assumption of frictionless motion can only apply to regions with negligible velocity derivatives. We have seen in Section 7.11 that turbulence cannot sustain itself in regions without velocity gradients in space (shear rates) as it is through shearing motion that it can receive energy from the mean flow to compensate for losses due to frictional dissipation. Thus, in layers without friction even with the presence of turbulence spots (clear air turbulence) resulting from baroclinic conditions due to thermal or humidity stratification, the Reynolds stresses in Eq. (7.67) play a minor role in dynamics.

In the absence of shearing stresses (viscous and Reynolds stresses), Eqs. (7.40) and (7.70) become

$$\frac{\partial \boldsymbol{U}}{\partial t} + (\boldsymbol{U} \cdot \nabla)\boldsymbol{U} + 2\boldsymbol{\Omega} \times \boldsymbol{U} = -\frac{1}{\rho}\nabla P + \boldsymbol{g} \tag{8.1}$$

and in component form

$$\begin{aligned} \frac{\partial U}{\partial t} + U\frac{\partial U}{\partial x} + V\frac{\partial U}{\partial y} - fV &= -\frac{1}{\rho}\frac{\partial P}{\partial x} \\ \frac{\partial V}{\partial t} + U\frac{\partial V}{\partial x} + V\frac{\partial V}{\partial y} + fU &= -\frac{1}{\rho}\frac{\partial P}{\partial y} \\ g &= -\frac{1}{\rho}\frac{\partial P}{\partial z} \end{aligned} \tag{8.2}$$

Equation (7.41) was called the Navier–Stokes equation; Eq. (7.69), in which the Reynolds stress appears, was called the Reynolds equation; and the inviscid form represented in Eqs. (8.1) and (8.2) is called the *Euler equation.*

Four cases are discussed for which the Euler equations can be directly integrated. These are A, steady motion along a streamline; B, steady motion along a vortex line; C, Beltrami flow; and D, two-dimensional steady flow.

A Steady Motion along a Streamline

The horizontal components of the velocity make up the inertia terms of Eqs. (8.1) and (8.2) and therefore the streamlines and stream surfaces $\Psi(x, y)$ are parallel to the surface of the earth. Figure 8.1 shows this geometry. Thus along a streamline we have a special relationship between the coordinates of the streamline and the components of the horizontal velocity $\boldsymbol{U}_{\mathrm{h}}$,

$$\frac{U}{dx} = \frac{V}{dy} \tag{8.3}$$

or $U\,dy = V\,dx$. To make use of this relationship we multiply the x component[1]

[1] For steady motion the transient term has been omitted.

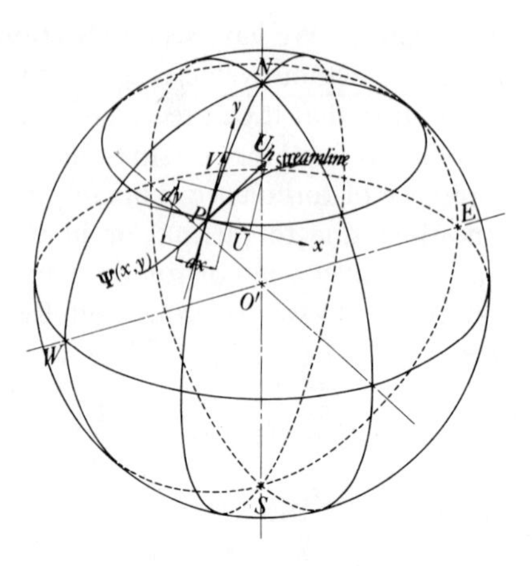

Fig. 8.1 Streamline geometry of wind and current.

of Eq. (8.2) by dx, the y component by dy, and the z component by dz:

$$\begin{aligned} U\frac{\partial U}{\partial x}\,dx + V\frac{\partial U}{\partial y}\,dx - fV\,dx &= -\frac{1}{\rho}\frac{\partial P}{\partial x}\,dx \\ U\frac{\partial V}{\partial x}\,dy + V\frac{\partial V}{\partial y}\,dy + fU\,dy &= -\frac{1}{\rho}\frac{\partial P}{\partial y}\,dy \\ g\,dz &= -\frac{1}{\rho}\frac{\partial P}{\partial z}\,dz \end{aligned} \tag{8.4}$$

If we use the special relationship of Eq. (8.3) valid only along the streamline, Eqs. (8.4) take the form

$$\begin{aligned} U\frac{\partial U}{\partial x}\,dx + U\frac{\partial U}{\partial y}\,dy - fV\,dx &= -\frac{1}{\rho}\frac{\partial P}{\partial x}\,dx \\ V\frac{\partial V}{\partial x}\,dx + V\frac{\partial V}{\partial y}\,dy + fU\,dy &= -\frac{1}{\rho}\frac{\partial P}{\partial y}\,dy \\ g\,dz &= -\frac{1}{\rho}\frac{\partial P}{\partial z}\,dz \end{aligned} \tag{8.5}$$

Since U and V are functions of x and y only, Eqs. (8.5) when added first horizontally and then vertically yield

$$\tfrac{1}{2}\,d(U^2 + V^2) + g\,dz = -\frac{1}{\rho}\,dP \tag{8.6}$$

This is an ordinary differential equation which when integrated between any two points *along a streamline*,[2] after replacing for the horizontal velocity ${U_h}^2 = U^2 + V^2$, yields

$$\tfrac{1}{2}U_{h2}^2 + gz_2 + \int_1^2 (dP/\rho) = \tfrac{1}{2}U_{h1}^2 + gz_1 \tag{8.7}$$

This is the *Bernoulli* equation and it can be evaluated between any two points along the streamline provided ρ is known in terms of P. Thus barotropic flows are easily integrable. Given a process equation $P = f(\rho)$ special cases of Eq. (8.7) can be obtained. For an isothermal process along a streamline the Bernoulli equation becomes

$$\tfrac{1}{2}U_{h2}^2 + gz_2 + RT\ln P_2 = \tfrac{1}{2}U_{h1}^2 + gz_1 + RT\ln P_1 = \text{constant} \tag{8.8}$$

If the density does not vary horizontally along a streamline, we have the special relation

$$\tfrac{1}{2}U_{h2}^2 + gz_2 + (P_2/\rho) = \tfrac{1}{2}U_{h1}^2 + gz_1 + (P_1/\rho) = \text{constant} \tag{8.9}$$

B Steady Motion along a Vortex Line

This can be best demonstrated by starting from the vector equation (8.1), replacing the acceleration and the Coriolis from Eq. (7.30), and dropping the transient term:

$$\tfrac{1}{2}\nabla U^2 + (2\boldsymbol{\Omega} + \boldsymbol{\zeta}) \times \boldsymbol{U} = -\frac{1}{\rho}\nabla P + \boldsymbol{g} \tag{8.10}$$

It is already known that the gravitational acceleration has a potential defined in Section 2.6 as Φ_E. Thus if all the gradient terms in Eq. (8.10) are combined

$$(2\boldsymbol{\Omega} + \boldsymbol{\zeta}) \times \boldsymbol{U} = -\nabla\left[\tfrac{1}{2}U^2 + \int (dP/\rho) + \Phi_E\right] \tag{8.11}$$

For barotropic conditions the term $(1/\rho)\,\nabla P = \nabla \int (1/\rho)\,dP$ because $d\int (1/\rho)\,dP = (1/\rho)\,dP$. Therefore, from the definition of the gradient

$$\frac{1}{\rho}\nabla P \cdot d\boldsymbol{r} = \nabla\left(\int \frac{\partial P}{\rho}\right) \cdot d\boldsymbol{r}$$

Equation (8.11) is often known as *Crocco's or Lamb's equation.* Its vectorial geometry gives an important relationship between the velocity, the absolute

[2] Integration should be along a streamline because of the use of the special relationship (8.3).

vorticity vector, and the surface of constant mechanical energy per unit mass ($\int (dP/\rho) + \frac{1}{2}U^2 + \Phi_E$), otherwise known as the *total head.*

From the property of the cross product, Eq. (8.11) states that the vector $(2\boldsymbol{\Omega} + \boldsymbol{\zeta}) \times \boldsymbol{U}$ is normal to the surface $\int (dP/\rho) + \frac{1}{2}U^2 + \Phi_E = \text{constant}$. Since streamlines are tangent to $\boldsymbol{U}$ and vortex lines are tangent to $(2\boldsymbol{\Omega} + \boldsymbol{\zeta})$ the network of streamlines and that of vortex lines must lie on this total head surface. Thus we conclude that along a streamline (as we discovered in Section 8.2.A) and along vortex lines the Bernoulli equation must be valid:

$$\int (dP/\rho) + \tfrac{1}{2}U^2 + gz = \text{constant}$$

Figure 8.2 illustrates the geometrical relationship of Eq. (8.11).

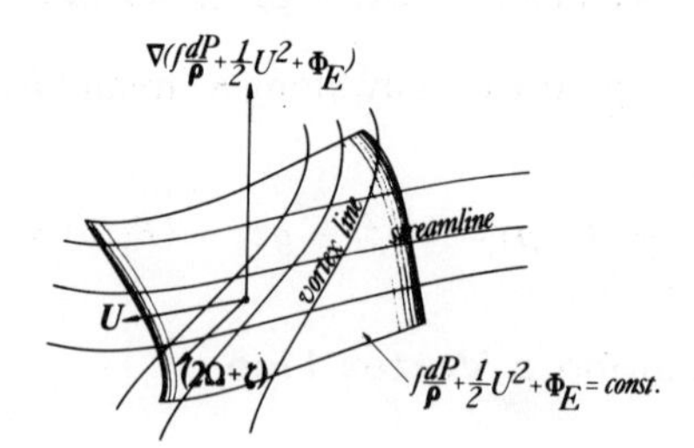

Fig. 8.2 Geometry of streamline and vortex line.

When we cannot assume a barotropic environment, which is often the case, from the combined first and second laws of thermodynamics [Eq. (2.57)] we have

$$dh = T\,ds + \frac{dP}{\rho}$$

or

$$\nabla h = T\,\nabla s + \frac{\nabla P}{\rho} \tag{8.12}$$

where h is the enthalpy, T is the absolute temperature, and s is the entropy. Replacing $\nabla P/\rho$ in Eq. (8.10) and defining the *stagnation enthalpy* $h^0 = h + \frac{1}{2}U^2 + gz$, we obtain

$$T\,\nabla s = \nabla h^0 + (2\boldsymbol{\Omega} + \boldsymbol{\zeta}) \times \boldsymbol{U} \tag{8.13}$$

This final relation implies that if the stagnation enthalpy is uniform, an irrotational motion implies an isentropic process. Even if $\boldsymbol{\zeta}$, the vorticity of the environment relative to the earth's surface, is zero, the Coriolis contribution will always be there, and even if the enthalpy remains uniform, it is not possible to maintain isentropic conditions in the environment. We have seen in Chapter 4 and in Eq. (4.35) that the entropy must increase with altitude for a vertical mechanical equilibrium. Here we see that rotationality also increases the entropy in a direction normal to the plane of the streamlines and vortex lines. *For horizontal motion the plane normal to the earth's axis, in other words the plane of the latitude circle, is the plane on which maximum entropy change takes place.*

C Beltrami Flow

A Beltrami flow is defined as one in which vorticity is in the same direction as the velocity. In planetary motions this kind of special flow will be difficult to encounter since the dominant motion is nearly horizontal and the vorticity is partly vertical due to the earth's rotation and partly horizontal but normal to $\boldsymbol{U}_h$ in the boundary layer. However, in local environmental conditions such as found in plumes after they have bent in the direction of the wind, in river bends, and so on, this condition may be encountered. Then from Eq. (8.11) we could see that the vorticity–velocity term will drop out and

$$\tfrac{1}{2}U^2 + \int (dP/\rho) + gz = \text{constant}$$

which is again Bernoulli's equation.

D Two-Dimensional Steady Flow

By introducing the stream function defined in Section 6.5, it is possible to integrate between *any two points* the steady Euler's equation. We begin again with Eq. (8.11), multiply it by a differential distance $d\boldsymbol{r}$, and integrate:

$$\int [(2\boldsymbol{\Omega} + \boldsymbol{\zeta}) \times \boldsymbol{U}] \cdot d\boldsymbol{r} = \int (dP/\rho) + \tfrac{1}{2}U^2 + \Phi_E \tag{8.14}$$

We need first to evaluate the triple scalar product in the integrand of the left-hand integral:

$$[(2\boldsymbol{\Omega} + \boldsymbol{\zeta}) \times \boldsymbol{U}] \cdot d\boldsymbol{r} = \boldsymbol{U} \cdot [(2\boldsymbol{\Omega} + \boldsymbol{\zeta})] \times d\boldsymbol{r}$$

Now applying the condition for two-dimensionality (horizontal motion), and calling the vertical component of the total vorticity $\zeta_{tz} = \boldsymbol{k} \cdot (2\boldsymbol{\Omega} + \boldsymbol{\zeta})$, the right-hand side of the preceding equation becomes

$$\boldsymbol{U} \cdot (\boldsymbol{k}\zeta_{tz} \times d\boldsymbol{r}) = \zeta_{tz}(V\,dx - U\,dy)$$

and using the stream function relationship for the velocities

$$\int (\zeta_t \times U) \cdot dr = \int \zeta_{tz}\left(\frac{\partial \Psi}{\partial x}\, dx + \frac{\partial \Psi}{\partial y}\, dy\right) = \int \zeta_{tz}\, d\Psi$$

where $\zeta_{tz} = (\Omega_z + \zeta_z)$. But ζ_z, the relative vorticity, can also be expressed in terms of the stream function if we start by taking the curl of Eq. (6.23) and use the triple vector product found in Appendix A, Eq. (A.30):

$$\nabla \times (\boldsymbol{k} \times \nabla\Psi) = \nabla \times \boldsymbol{U}$$

and

$$\nabla \times (\boldsymbol{k} \times \nabla\Psi) = [\nabla \cdot (\nabla\Psi)]\boldsymbol{k} - (\nabla \cdot \boldsymbol{k})\, \nabla\Psi$$

Since $\nabla \cdot \boldsymbol{k} = 0$

$$\zeta_z = \nabla \times \boldsymbol{U} = \boldsymbol{k}\, \nabla^2\Psi \tag{8.15}$$

Substituting back into Eq. (8.14) we have

$$\int (\zeta_t \times U) \cdot dr = \int (2\Omega_z + \nabla^2\Psi)\, d\Psi$$

Thus

$$\int (dP/\rho) + \tfrac{1}{2}U^2 + gz = 2\Omega_z \Psi + \int \nabla^2\Psi\, d\Psi \tag{8.16}$$

With prior knowledge of the stream function the left-hand side can be evaluated from point to point in a two-dimensional steady flow.

Illustrative Example 8.1

Since concentrated vorticity flows such as a free vortex can be found in atmospheric motion in the form of tornados, hurricanes, and so on, it would be interesting to consider the flow discussed in Illustrative Example 6.5 and apply the principles for the integration of Bernoulli's equation to this vortex motion. Excluding the center part (the eye) of a tornado where considerable updrafts take place, in the remaining parts of a tornado we can assume that the peripheral velocity V varies inversely as the radial distance and that axial symmetry prevails.

According to Eq. (6.25) the peripheral velocity in terms of the stream function is given as

$$V = \frac{d\Psi}{dr}$$

and we also know according to Illustrative Example 6.5 that this velocity is

$$V = \frac{\Gamma}{2\pi}\frac{1}{r}$$

where $\Gamma/2\pi$ is the strength of the tornado and Γ is its circulation, as defined in Section 6.14. The total derivative has been used in the stream function relation because we assumed the other velocity components to be negligible. Thus from these two relations, after integration,

$$\Psi = \frac{\Gamma}{2\pi}\ln r$$

Since the logarithm is involved here the lower limit of the integration was set at an arbitrary unit circle with radius 1.0, the edge of the eye. It can be shown readily that $\nabla^2\Psi = 0$ for this flow. Thus using Eq. (8.16) we conclude that the Bernoulli equation applied to the tornado, excluding the eye and the region very near the ground, becomes

$$\frac{1}{2}U^2 + gz + \int\frac{dP}{\rho} = \frac{\Gamma\Omega_z}{\pi}\ln r$$

This result, which is obtained from the application to the two-dimensional flow considerations of Section 8.2.D, also satisfies the conditions of Section 8.2.A, because along a streamline (r = constant in this case), the right-hand side of the preceding equation is constant. Furthermore, this last expression implies that Bernoulli's equation will be valid for *all* streamlines provided a new constant proportional to $\ln r$ is evaluated for every streamline. Finally, according to Crocco's or Lamb's equation (8.13) the stagnation enthalpy is not constant with r, and consequently has a definite gradient in the radial direction.

8.3 PRESSURE IN THE CONTINUITY EQUATION OF THE ENVIRONMENT

Because of the nearly exclusive influence of the gravitational force in the vertical equilibrium of the environment the instantaneous value of the pressure p can be computed from the integration of the hydrostatic equation

$$\tilde{p} = \int_z^\infty \tilde{\rho}g\,d\tilde{z}' \tag{8.17}$$

where the limits have been inverted to eliminate the minus sign. We know that a state balance independent of time never exists. In the atmosphere density at a given point has a time dependence owing to diurnal variations of heating, evaporation, condensation, moisture, and so on, and in the ocean owing to temperature, salinity, or even variations of z' due to waves. Thus the change of pressure with respect to time can be connected to the conservation of mass when density is affected by these pressure changes:

$$\frac{\partial \tilde{p}}{\partial t} = g \int_z^\infty \frac{\partial \tilde{\rho}}{\partial t}\, dz'$$

According to the continuity equation (6.12) the time variation of the density can be replaced by $\nabla \cdot (\rho \boldsymbol{U})$ and

$$\frac{\partial \tilde{p}}{\partial t} = \int_\infty^z \left[\frac{\partial(\rho \tilde{u})}{\partial x} + \frac{\partial(\rho \tilde{v})}{\partial y} + \frac{\partial(\rho \tilde{w})}{\partial z'} \right] dz'$$

The limits of integration have been reversed again to omit the minus sign. The third term in this equation can be integrated in z' to give $(\rho \tilde{w})_z$ since $(\rho w)_\infty = 0$ because both ρ and $\tilde{w}$ go to zero at very high altitudes. Then

$$\frac{\partial \tilde{p}}{\partial t} = g \int_\infty^z \left[\frac{\partial(\rho \tilde{u})}{\partial x} + \frac{\partial(\rho \tilde{v})}{\partial y} \right] dz' + \rho g \tilde{w} \tag{8.18}$$

Using the turbulence decomposition equations (7.64) and (7.68) we have

$$\frac{\partial P}{\partial t} + \frac{\partial p}{\partial t} = g \int_\infty^z \left\{ \left[\frac{\partial(\bar{\rho} U)}{\partial x} + \frac{\partial(\bar{\rho} V)}{\partial y} \right] + \left[\frac{\partial(\rho u)}{\partial x} + \frac{\partial(\rho v)}{\partial y} \right] \right\} dz' + \bar{\rho} g W + \rho g w$$

When we take the time average of this equation *for a time period of the order of the time scale of turbulence*, and if the density–velocity correlations are zero, we have

$$\frac{\partial P}{\partial t} = g \int_\infty^z \left[\frac{\partial(\rho U)}{\partial x} + \frac{\partial(\rho V)}{\partial y} \right] dz' + \rho g W \tag{8.19}$$

and thus

$$\frac{\partial p}{\partial t} = g \int_\infty^z \left[\frac{\partial(\rho u)}{\partial x} + \frac{\partial(\rho v)}{\partial y} \right] dz' + \rho g w \tag{8.20}$$

Essentially these are two different continuity equations, one for the mean properties and one for the fluctuating properties.

8.4 HORIZONTAL WIND AND CURRENT WITH NEGLIGIBLE ACCELERATION

The simplest possible wind system in the free layer is the case of wind blowing in straight parallel horizontal lines. Strictly speaking this is not possible on a spherical earth and in view of the meandering of streamlines, especially in the mid-latitudes, which we discussed in Section 6.15. We know that curvature of streamlines must accompany centrifugal accelerations and spreading or contracting streamlines must accompany linear accelerations. However, in spite of the presence of some curvature, some spreading, and some contracting of the streamlines in the free layer of the environment, calculations show that in many cases the centrifugal and the linear accelerations do not account for more than 10–20% of the Coriolis or the pressure force (see Illustrative Example 8.2). Thus it is of practical value to consider this simplified geostrophic motion.

Then neglecting the whole acceleration term $d\boldsymbol{U}/dt = \partial \boldsymbol{U}/\partial t + (\boldsymbol{U} \cdot \nabla)\boldsymbol{U}$ in Eq. (8.1) we are left with

$$2\boldsymbol{\Omega} \times \boldsymbol{U} = -\frac{1}{\rho} \nabla P + \boldsymbol{g} \tag{8.21}$$

We shall proceed to the solution of this geostrophic equation, after some geometric considerations. Since we have agreed to consider horizontal motion only, Eq. (8.21) can be split into a horizontal and a vertical equation. Thus

$$\begin{aligned} 2\boldsymbol{k}\Omega_z \times (\boldsymbol{i}U + \boldsymbol{j}V) &= -\frac{1}{\rho} \nabla_{\mathrm{h}} P \\ g &= -\frac{1}{\rho}\frac{\partial P}{\partial z} \end{aligned} \tag{8.22}$$

We have seen in Section 7.3 that the horizontal components of the earth's rotation contribute little to the horizontal dynamics. The unit vectors $\boldsymbol{i}, \boldsymbol{j}$, and $\boldsymbol{k}$ are the conventional directions defined in the preceding chapter.

Vector multiplication of the horizontal equation gives

$$fV = \frac{1}{\rho}\frac{\partial P}{\partial x}, \qquad fU = -\frac{1}{\rho}\frac{\partial P}{\partial y}, \qquad g = -\frac{1}{\rho}\frac{\partial P}{\partial z} \tag{8.23}$$

where $f = 2\Omega_z$. The vector geometry of Eq. (8.22) implies that the velocity vector and the horizontal pressure gradient are perpendicular to each other. Since, by definition, the gradient of pressure is normal to the isobar at a given elevation, it follows that $\boldsymbol{U}$ must be tangent to the isobars and consequently that *isobars are also streamlines.* At first thought it may seem strange that the pressure difference is *not pushing* the environment in the direction of the motion. This is true; the role of the pressure is to maintain the motion constantly deflected away from the preceding direction to abide with the influence of the earth's rotation through the Coriolis. On second thought we already know that circular motions such as rigid body and free vortex motions also must have pressure gradients normal to the circular streamlines to balance the centrifugal force.

To obtain a solution for the horizontal velocity from Eq. (8.23) we need to orient our coordinates such that x, for instance, is in the direction of the resultant horizontal geostrophic velocity $U_{\mathrm{hg}} = (U^2 + V^2)^{1/2}$ and have one equation from which

$$U_{\mathrm{hg}} = -\frac{1}{\rho f}\frac{\partial P}{\partial y} \tag{8.24}$$

where y now is in the direction normal to U_{hg} and to isobars.

Since the horizontal velocity U_{hg} is defined positive toward the east (westerly), and since f and ρ are positive, $(\partial P/\partial y) < 0$, which implies that pressure must decrease toward the north or increase toward the south. Thus when one face the wind with his feet on the ground in the Northern Hemisphere the pressure field increases to the left. Figure 8.3 shows this geometry. The geometry just described carries the name of the *Buys Ballot law* after a Dutch meteorologist.

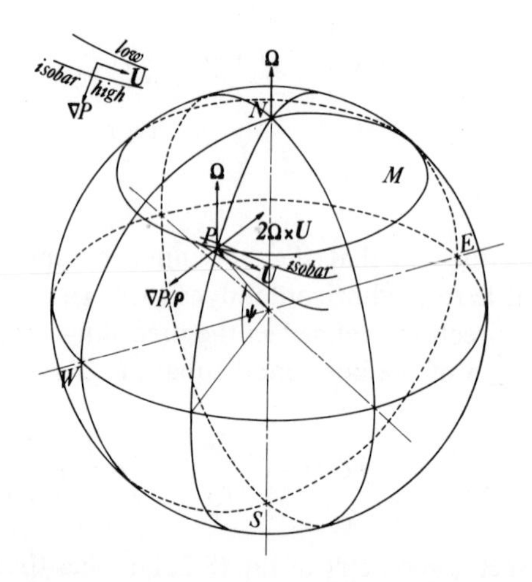

Fig. 8.3 Geostrophic wind and current.

One could conclude that from the conservation of mass standpoint, since the direction of flow is tangent to the isobars the magnitude of the wind speed will be inversely proportional to the spacing between isobars.

When the third z-component equation of hydrostatic equilibrium is added to the horizontal equilibrium, the direction of the pressure gradient is the resultant, and is slightly inclined to the vertical. In other words, the isobaric surfaces are slightly inclined from the horizontal because the magnitude of the vertical gradient is considerably larger than that of the horizontal. The angle of isobaric surface inclination α is

$$\tan \alpha = \frac{\partial P/\partial y}{\partial P/\partial z}$$

From Eqs. (8.24) and (8.23) we obtain

$$\tan \alpha = \alpha = \frac{f}{g} U_{\text{hg}} \tag{8.25}$$

In meteorology and oceanography, the inclination of the isobaric surfaces is measured in terms of *dynamic or geopotential topography* (see Fig. 8.4). This

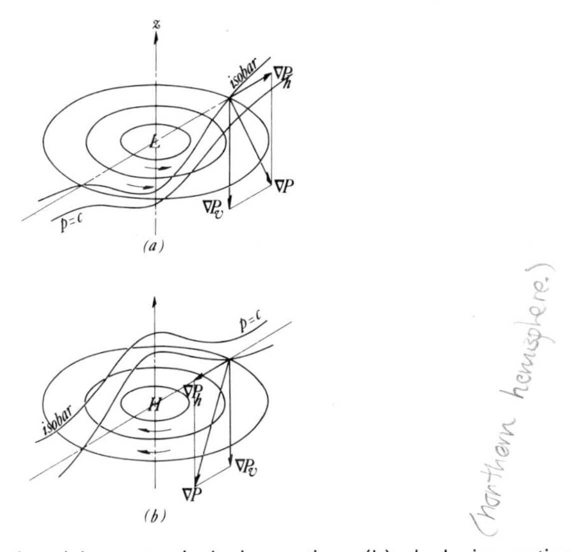

Fig. 8.4 Dynamic topography: (a) counterclockwise motion; (b) clockwise motion.

method consists in drawing heights or depths (as the case may be) from sea level of an isobaric surface of a given pressure value. The equation giving the topography of the isobars is obtained as follows: We form the dot product of the vector equation (8.21) with a generalized distance $d\boldsymbol{r}$, keeping in mind that the vertical component of the earth's rotation is the only influencing component and that the motion is horizontal as in Eq. (8.22). Thus

$$(\boldsymbol{k}f \times \boldsymbol{U}) \cdot d\boldsymbol{r} = -\frac{1}{\rho} \nabla P \cdot d\boldsymbol{r} + \boldsymbol{g} \cdot d\boldsymbol{r}$$

Decomposing this relation into its components, and taking the cross and dot products we obtain

$$f(U\,dy - V\,dx) = -\frac{1}{\rho}\,dP - g\,dz \tag{8.26}$$

From Eq. (6.19) the negative value of the terms in parentheses constitutes the differential of the stream function $d\Psi$, which is also equal to $U_\perp\,ds$, where s is the spacing coordinate between isobars, which are also streamlines (consult Fig. 6.6). Thus

$$f\,d\Psi = fU_\perp\,ds = \frac{1}{\rho}\,dP + g\,dz$$

For an isobaric surface $dP = 0$, and

$$\left.\frac{dz}{ds}\right|_P = \frac{fU_\perp}{g} \tag{8.27}$$

represents the slope of isobaric surfaces, as Eq. (8.25) did. Introducing the *geopotential height* defined in Section 2.7, $g\,dz = g_0\,dH = d\Phi_E$, we can also write

$$\frac{d\Phi_E}{ds} = fU_\perp \tag{8.28}$$

Since g is approximately 10 m/s^2, defining a *dynamic height* D, such that $d\Phi_E = 10dD$, then the height contour of a given isobar is given by

$$\left.\frac{dD}{ds}\right|_P = \frac{fU_\perp}{10} \tag{8.29}$$

Integrating and substituting Eq. (8.26)

$$D = \frac{1}{10}\int f\,U_\perp\,ds = \frac{1}{10}\int_0^P \frac{1}{\rho}\,dP + \frac{g}{10}\,z \tag{8.30}$$

From sea level the dynamic height is the geometric height plus a correction depending on the integral of $(1/\rho)\,dP$. The upper limit of the integral identifies the dynamic height for that isobar. Topographic charts showing lines of equal dynamic heights are called *dynamic isobaths*.

The problem suggested by Eq. (8.30) is whether from this equation, which is valid for the atmosphere and the ocean, it is possible to determine both geostrophic velocities U_g from direct measurements of pressure, and density from measurements of temperature, humidity, and salinity, as the case may be. The method is extensively used for the determination of circulation patterns.

Illustrative Example 8.2

The analysis developed in this section ignores the importance of the acceleration of the environment. Based on actual meteorological data, it is interesting to verify if this hypothesis is well supported. Figure 8.5 is a reproduction of a

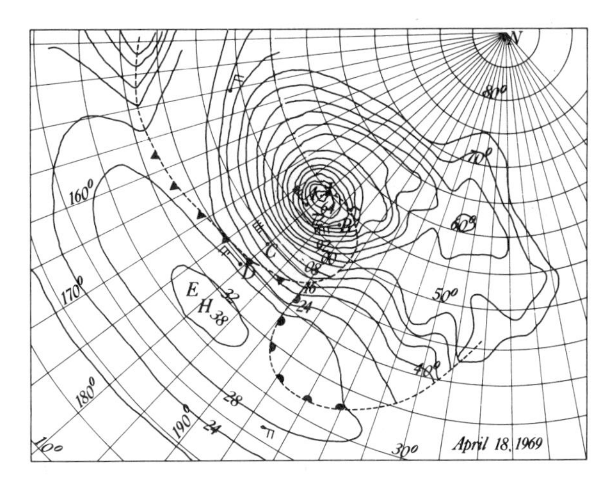

Fig. 8.5 Synoptic weather map over the Pacific Ocean.

synoptic weather map in a region of the Pacific Ocean. On this map are given the last two digits of the isobar values (in millibars) as well as some recorded wind velocities and directions represented by arrows at the tail of which a long cross line is valued at 10 knots and a short cross line at 5 knots. The sum of the values of the cross lines represents the wind speed to within 2.5 knots.

From the values given on the map it is possible to check the accuracy of Eq. (8.24) and consequently the validity of zero-acceleration geostrophic flows.

Conditions at Point B around the Low Pressure Center

The measured velocity of the wind at B is $U_h = 40$ knots, and since each knot represents approximately 1.85 km/h, $U_h = 74$ km/h. Notice that the direction of the velocity vector is in accord with being tangent to the isobar and that facing the wind the higher pressures are on the left of the observer. At B the isobars are fairly concentric so that the streamwise acceleration will be even smaller than the centrifugal acceleration.

To obtain the centrifugal acceleration it is necessary to establish the distance of point B to the curvature center A. In arc length on the surface of the earth this is equivalent to approximately 4°, which when multiplied by the radius of the earth gives a distance $r = 444$ km between A and B:

$$\text{centrifugal acceleration at } B = \frac{U_h^{\;2}}{r}$$
$$= 12.3 \quad \text{km/h}^2$$

The Coriolis acceleration can be computed at the latitude of point B which is approximately $\psi = 52°$. Thus

$$\begin{aligned}\text{Coriolis acc.} &= fU_h\\ &= (2\Omega \sin \psi)U_h\\ &= 2 \times 74 \times 7.3 \times 10^{-5} \times 3600 \times 0.79\\ &= 30.7 \quad \text{km/h}^2\end{aligned}$$

Thus the centrifugal acceleration is approximately 40% of the Coriolis and not small, as we have assumed in this section. This proves that there may be areas in the environment when the acceleration cannot be ignored. This sets the stage for improved analysis, as carried out in the next section. However, it is obvious to the reader that point B was purposely chosen to make this point clear since it must be emphasized that the atmospheric disturbance around A and B is of the magnitude of a strong storm. Thus the first conclusion we reach is that in the vicinities of strong disturbances the velocities are large, and since the centrifugal acceleration varies as the square of the velocity and the Coriolis as the velocity, the acceleration cannot be ignored under these circumstances.

The second question we need to answer is whether taking centrifugal acceleration into account, and omitting viscous forces, corresponds to a realistic model. To answer this we need to compute the pressure difference between A and B from the sum of the centrifugal acceleration and Coriolis and compare it with the data on Fig. 8.5. Judging from the isobar values from A to B on the figure an extrapolated value of the probable pressure at A is 956 mb. Thus $P_B - P_A = 16$ mb. The distance was already found to be 444 km. The sum of the centrifugal acceleration and the Coriolis is then 43 km/h^2. Addition is necessary for a low pressure area (see Fig. 8.7). Converting this into millibars per kilometer and multiplying by $r = 444$ km we obtain a pressure difference from the calculation of 15 mb which is in good agreement with the measured value of 16 mb. On the basis of this illustration alone, we can say that the viscous effects played a negligible role in the balance of forces in the free layer.

At the point C, 1130 km from the center A, the velocity is recorded as 35 knots or 65 km/h. The calculated centrifugal acceleration is 3.7 km/h^2 and the Coriolis is 24.3 km/h^2, making the ratio a little over 15%. The pressure gradient at this point C measured between adjacent isobars corresponds closely to the sum of the centrifugal acceleration and the Coriolis.

The point D, with a wind speed of 25 knots, gives a centrifugal acceleration of 3.2 km/h^2 when measured from the high pressure center E and a Coriolis of 24.2 km/h^2. The pressure difference computed from these two figures is about 11 mb, which agrees well with the measured values of 10 mb.

8.5 THE THERMAL WIND AND CURRENT

Illustrative Example 8.2 has shown that in a number of applications the centrifugal acceleration may not be neglected. However, before dealing with the effect of the centrifugal acceleration let us investigate the influence of elevation on the simple geostrophic motion. For the case of the atmosphere we shall assume that in addition to pressure, temperature and temperature variations with altitude do exert a significant influence on the motion.

To introduce the temperature we make use of the perfect gas law. We know that humidity may also have an influence on the density distribution but we shall assume that for the elevations involved in this analysis its role will be minor. Thus

$$\rho = \frac{P}{RT}$$

and the dynamical equations (8.23) become

$$fV = RT\frac{1}{P}\frac{\partial P}{\partial x}, \qquad fU = -RT\frac{1}{P}\frac{\partial P}{\partial y}, \qquad g = -RT\frac{1}{P}\frac{\partial P}{\partial z}$$

The three equations are equivalent to

$$\frac{fV}{T} = R\frac{\partial(\ln P)}{\partial x}, \qquad \frac{fU}{T} = -R\frac{\partial(\ln P)}{\partial y}, \qquad \frac{g}{T} = -R\frac{\partial(\ln P)}{\partial z} \tag{8.31}$$

In order to eliminate the pressure we differentiate the first two equations by z and substitute the derivative of $\ln P$ in the third equation. Therefore,

$$-f\frac{\partial}{\partial z}\left(\frac{V}{T}\right) = \frac{\partial}{\partial x}\left(\frac{g}{T}\right)$$

$$f\frac{\partial}{\partial z}\left(\frac{U}{T}\right) = \frac{\partial}{\partial y}\left(\frac{g}{T}\right)$$

Multiplying each of these equations by dz and integrating from z_0 to z we have

$$\frac{U}{T} = \frac{U_0}{T_0} - \frac{g}{f}\int_{z_0}^{z}\frac{1}{T^2}\frac{\partial T}{\partial y}\,dz$$

$$\frac{V}{T} = \frac{V_0}{T_0} + \frac{g}{f}\int_{z_0}^{z}\frac{1}{T^2}\frac{\partial T}{\partial x}\,dz \tag{8.32}$$

where the zero subscripts correspond to the lower bound location z_0. The integral part of these relations is called the *thermal wind* and as we see it subtracts from the westerly component as we increase in elevation and adds to the southerly component. For temperature variations along x and y with the same sign, one component of the geostrophic wind will increase while the other will decrease with altitude. This implies a change in magnitude as well as direction with altitude. The pressure gradients can also be introduced in the equations instead of f from Eq. (8.23)

$$\begin{aligned} \frac{U}{T} &= \frac{U_0}{T_0} + \frac{\rho g U}{\partial P/\partial y} \int_{z_0}^{z} \frac{1}{T^2} \frac{\partial T}{\partial y}\, dz \\ \frac{V}{T} &= \frac{V_0}{T_0} + \frac{\rho g V}{\partial P/\partial x} \int_{z_0}^{z} \frac{1}{T^2} \frac{\partial T}{\partial x}\, dz \end{aligned} \tag{8.33}$$

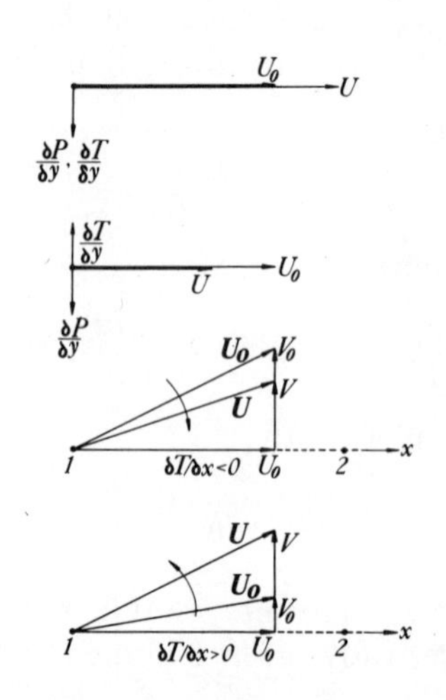

Fig. 8.6 The thermal wind effect.

Figure 8.6 explains the influence of the variation of horizontal temperature variations with altitude. Thus when $\partial P/\partial y$ and $\partial T/\partial y$ have the same sign then according to Eq. (8.33) the westerly component of the geostrophic velocity increases with altitude because of the thermal wind. The same is true with variations in the x direction. When the components of the gradients of pressure and temperature are both negative, the geostrophic velocity decreases with altitude. However, when the y gradients are of the same sign and the x gradients are of opposite sign, U increases and V decreases with altitude giving a clockwise rotation of the velocity with altitude. For instance, under such conditions a westerly at z_0 will veer south as it climbs in z. It must be emphasized also that $\partial T/\partial x$ like $\partial P/\partial x$ does not influence U. Consequently if a geostrophic wind blows toward cooler temperatures, say $\partial T/\partial x < 0$, then it will pick up a negative V and thus veer to the right looking in the direction of the original motion.

If the temperature does not vary with altitude as in the tropopause during certain seasons, then the integrand is constant with z, and Eq. (8.32) modifies to

$$
\begin{aligned}
U &= U_0 - \frac{gz}{fT_0}\frac{\partial T}{\partial y} \\
V &= V_0 + \frac{gz}{fT_0}\frac{\partial T}{\partial x}
\end{aligned}
\tag{8.34}
$$

In the case of the ocean, we are not able to use the perfect gas law. The density could vary with depth as a result of temperature and salinity gradients which in turn could be functions of depth. In Section 4.4.B we had developed in Eq. (4.25) an expression for the dependence of the density of the ocean on the temperature and the salinity. If ρ_* is the reference density of the water at a temperature T_* and salinity $\mathfrak{s}_*$ above which T and $\mathfrak{s}$ are measured, then

$$\rho = \rho_*(1 + \gamma\mathfrak{s} - \beta_1 T) \tag{8.35}$$

Beginning with Eq. (8.23)

$$
\begin{aligned}
f\frac{\partial}{\partial z}(\rho V) &= \frac{\partial}{\partial x}\left(\frac{\partial P}{\partial z}\right) \\
f\frac{\partial}{\partial z}(\rho U) &= -\frac{\partial}{\partial y}\left(\frac{\partial P}{\partial z}\right) \\
\rho g &= \frac{\partial P}{\partial z}
\end{aligned}
$$

The sign has changed in the third equation because in the ocean positive z is in the direction of depth. Substitution of the third equation yields

$$
\begin{aligned}
\frac{\partial}{\partial z}(\rho V) &= \frac{g}{f}\frac{\partial \rho}{\partial x} \\
\frac{\partial}{\partial z}(\rho U) &= -\frac{g}{f}\frac{\partial \rho}{\partial y}
\end{aligned}
$$

Integrating with respect to z between z_0 and z

$$
\begin{aligned}
\rho V &= \rho_0 V_0 + \frac{g}{f}\int_{z_0}^{z}\frac{\partial \rho}{\partial x}\,dz \\
\rho U &= \rho_0 U_0 - \frac{g}{f}\int_{z_0}^{z}\frac{\partial \rho}{\partial y}\,dz
\end{aligned}
\tag{8.36}
$$

From Eq. (8.35)

$$\frac{\partial \rho}{\partial x} = \rho_*\left(\gamma \frac{\partial s}{\partial x} - \beta_1 \frac{\partial T}{\partial x}\right)$$

$$\frac{\partial \rho}{\partial y} = \rho_*\left(\gamma \frac{\partial s}{\partial y} - \beta_1 \frac{\partial T}{\partial y}\right)$$

and finally,

$$\begin{aligned} \rho U &= \rho_0 U_0 - \frac{g\rho_*}{f} \int_{z_0}^{z} \left(\gamma \frac{\partial s}{\partial y} - \beta_1 \frac{\partial T}{\partial y}\right) dz \\ \rho V &= \rho_0 V_0 + \frac{g\rho_*}{f} \int_{z_0}^{z} \left(\gamma \frac{\partial s}{\partial x} - \beta_1 \frac{\partial T}{\partial x}\right) dz \end{aligned} \tag{8.37}$$

The arguments are the same as for the atmosphere.

Illustrative Example 8.3

During the winter months westerly geostrophic winds blow from the Pacific Ocean at a velocity of 15 m/s across the state of California and over the Rockies. In this distance, the temperature gradient $\partial T/\partial x$ could easily reach -0.02°K/km. The temperature gradient S–N, inland, can be of the order $\partial T/\partial y = -0.01$°K/km. Assuming that these gradients remain independent of z, and that the vertical temperature lapse rate is $\alpha = -\partial T/\partial z = 6$°K/km, compute the components of the wind velocity 10 km above, where the jet stream may be located. Also, what is the angle of the jet stream during these winter months relative to a westerly near the earth's surface?

From Eq. (8.32) we can write

$$\frac{U}{T} = \frac{U_0}{T_0} + \frac{g}{f} \int_0^{10} \frac{0.01}{(T_0 + \alpha z)^2} dz$$

$$\frac{V}{T} = \frac{V_0}{T_0} - \frac{g}{f} \int_0^{10} \frac{0.02}{(T_0 + \alpha z)^2} dz$$

Integrating and taking $T_0 = 278$°K we have $U = 46$ m/s and $V = -70$ m/s. The resultant speed at 10 km altitude is 84 m/s and the angle it makes with the westerly below is approximately 50°. The speed of the stream at that altitude is not unusual but the veering of 50° is excessive. Naturally, the assumption of horizontal temperature gradients constant with z is responsible for this error.

8.6 HORIZONTAL GEOSTROPHIC MOTION WITH CENTRIFUGAL ACCELERATION—GRADIENT WIND OR CURRENT

The streamlines in the free layer are seldom straight lines. They display curved patterns of many sorts. Figure 8.5 is a typical example of these patterns, especially in mid-latitudes. In Illustrative Example 8.2 we showed the need to consider the centrifugal acceleration resulting from the curvature of these streamlines. Neglecting transient effects and linear acceleration, according to Eq. (7.30) the centrifugal acceleration added to the Coriolis gives $(2\boldsymbol{\Omega} + \boldsymbol{\zeta}) \times \boldsymbol{U}$ and in the free layer the horizontal component of the equation of motion reads

$$(2\Omega_z + \zeta) \times \boldsymbol{U}_{\mathrm{h}} = -\frac{1}{\rho} \nabla_{\mathrm{h}} P \tag{8.38}$$

where the subscript z refers to the vertical component, and h to the horizontal component. Compared with Eq. (8.22) the centrifugal acceleration of the form $\boldsymbol{\omega} \times (\boldsymbol{\omega} \times \boldsymbol{r})$ is contained in the term $\boldsymbol{\zeta} \times \boldsymbol{U}_{\mathrm{h}}$. First, we ask whether the coincidence between the streamlines and isobars established in flows without acceleration (Section 8.4) is destroyed by the inclusion of the centrifugal acceleration. The geometry of Eq. (8.38) attests to the fact that $\boldsymbol{U}_{\mathrm{h}}$ is still parallel to $P =$ constant lines. *So for gradient winds, the streamlines are also isobars.* Now it remains to solve for U_{h} as in Eq. (8.24) for this case, taking into account the centrifugal acceleration.

This type of flow is best analyzed in cylindrical coordinates with the axis of the cylinder in the vertical direction. Since we assumed the acceleration in the direction of the motion to be negligible, in cylindrical coordinates it is like saying that axial symmetry exists. Then the only component of the horizontal motion is

$$\frac{V^2}{r} + fV = \frac{1}{\rho}\frac{dP}{dr} \tag{8.39}$$

where V is the velocity in θ, and r is the radius of curvature of the streamline or isobar. Although this equation is referred to as the *gradient wind equation*, it is simply the *radial equilibrium equation*.

This quadratic equation can be solved for V, and we have

$$V = \frac{fr}{2}\left[-1 \pm \left(1 + \frac{4}{\rho f^2 r}\frac{dP}{dr}\right)^{1/2}\right] \tag{8.40}$$

Thus, the gradient wind or current is computed from the geometry of the isobars, by calculating dP/dr at a given point where r is the radius of curvature of the isobar. To compare this relation with the pure geostrophic velocity of Eq. (8.23), $V_g = (1/\rho f)(dP/dr)$, we have

$$V_g = V + \frac{V^2}{fr} \tag{8.41}$$

In the Northern Hemisphere $f > 0$ and the gradient wind or current V is always less than the geostrophic wind V_g for $V > 0$ in counterclockwise cyclonic motion, as shown by Eq. (8.41). For anticyclonic motion the reverse is true.

Figure 8.7 sheds additional light onto the equilibrium of the gradient motion. In Section 8.4 we established a relationship between the direction of motion and the direction in which the pressure increased. This was called the Buys Ballot law. Since the centrifugal acceleration is almost always smaller than the Coriolis the law applies here as well. Facing the wind or current in the Northern Hemisphere produces an increase of pressure to the left of the observer. *For a counterclockwise cyclonic motion, since pressure increases with increasing r, the center of the cyclone is a low pressure.* Conversely, the center of an anticyclone is a *high pressure.*

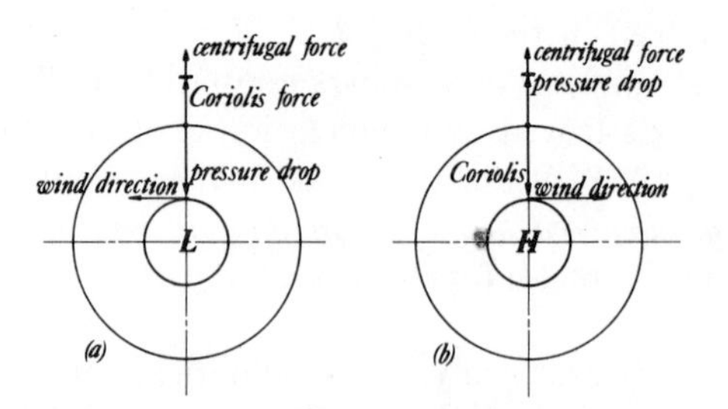

Fig. 8.7 Relationship between forces and wind direction in (a) cyclonic and (b) anticyclonic flows in the Northern Hemisphere.

There are four algebraic solutions to Eq. (8.40); however, two of these solutions seem to be possible in practice. For $dP/dr < 0$ (an *anticyclone*), $|dP/dr| < \rho f^2 r/4$ for a real solution, and naturally the value of the square root term in Eq. (8.40) will be less than 1; thus $V < 0$. However, both of the signs in front of the square root term would give $V < 0$. To find which sign is appropriate we reason as follows: Looking at Fig. 8.7b when $dP/dr \to 0$ the anticyclonic velocity does not have to go to zero for equilibrium, because for large enough V and small enough r the centrifugal acceleration could balance the Coriolis. Thus[3]

$$V = \frac{fr}{2}(-1 - \sqrt{1}) = -fr$$

[3] Dynamics does not exclude the positive dP/dr for anticyclonic motion but the centrifugal acceleration must be larger than the Coriolis, requiring very small radii of curvatures not usually encountered in the environment at the hemispherical scale.

The same cannot be said for the cyclonic flow where $dP/dr > 0$ and $V > 0$. Looking again at Fig. 8.7a, when $dP/dr \to 0$, the centrifugal acceleration and the Coriolis, being in the same direction, cannot possibly balance each other; thus V must tend to zero, and

$$V = \frac{fr}{2}\left[-1 + \left(1 + \frac{4}{\rho r f^2}\frac{dP}{dr}\right)^{1/2}\right]$$

We conclude that the pressure gradient and the velocity in the anticyclonic motion have a restriction on how high they can reach: $|dP/dr| < \rho f^2/4$. No such restriction exists in cyclonic motions except through viscous friction. This is an explanation of why velocities can get considerably higher in cyclonic flows than in anticyclonic flows.

If we substitute the value of V_g into Eq. (8.40) we have

$$V = \frac{fr}{2}\left[-1 \pm \left(1 + \frac{4V_g}{fr}\right)^{1/2}\right] \tag{8.42}$$

From this equation we can deduce that when the radius of curvature of the isobar tends to infinity the gradient wind V must tend toward the geostrophic value of V_g. Direct substitution of $r = \infty$ gives an indeterminate value. For large values of r the square root can be expanded in a Taylor series, giving $(1 + 2V_g/fr)$, and then

$$V = \frac{fr}{2}\left[-1 \pm \left(1 + \frac{2V_g}{fr}\right)\right]$$

For $V = V_g$ we must choose the positive sign in front of the square root. Thus, in the hemispherical environment we expect that

$$V = \frac{fr}{2}\left[-1 + \left(1 + \frac{4}{\rho f^2 r}\frac{\partial P}{dr}\right)^{1/2}\right] \tag{8.43}$$

Again, the solution for an anticyclone with a minus sign in front of the square root and $dP/dr > 0$, as a pathological case for the environment has already been discussed. Finally, it is interesting to note from Eqs. (8.42) and (8.43) that the square brackets take a value of -1 for $V_g = -fr/4$ or $dP/dr = -\rho f^2 r/4$ and a value of 0 for both $V_g = 0$ and $r = \infty$ corresponding to $dP/dr = 0$. Between these two limits the smallest value of V is $-fr/2$, which corresponds to an angular velocity of the environment equal but opposite in sign to that of the *vertical component* of the earth's rotation, $f/2 = \Omega \sin \psi$. This indicates that although an anticyclone rotates clockwise with reference to the earth, its absolute rotation is still positive when added to the earth's angular velocity, $\mathbf{\Omega}$.

8.7 MOTION IN THE CIRCLE OF INERTIA—INERTIA CURRENTS AND WINDS

In the preceding section we have questioned the possibility of negligible pressure gradients, thus creating a situation of the acceleration balancing the Coriolis. In the previous two cases the pressure gradient was considered important. Aside from the academic interest in this type of motion, experiments show that the oceans display this character (Ekman and Helland-Hansen, 1931; Gustafson and Kullenberg, 1936). In a homogeneous ocean where the density is constant, horizontal pressure gradients can only result from a slope on the sea surface, and during relatively short periods of time this can only be caused by changes in the wind field.

Let us assume then that on a calm sea the horizontal pressure gradients are negligible compared to the inertia dU_h/dt and the Coriolis $2\boldsymbol{\Omega}_z \times \boldsymbol{U}_h$. (The viscous forces on the free layers have been omitted in this chapter. The horizontal equations of motion are

$$\frac{dU}{dt} = fV, \qquad \frac{dV}{dt} = -fU \tag{8.44}$$

These are ordinary differential equations that can be solved. Since the dependent variables are mixed, and we need to separate them, differentiating once more with respect to t yields

$$\frac{dV}{dt} = \frac{1}{f}\frac{d^2U}{dt^2}, \qquad \frac{dU}{dt} = -\frac{1}{f}\frac{d^2V}{dt^2}$$

Replacing these in Eq. (8.44) we succeed in separating the variables:

$$\frac{d^2U}{dt^2} + f^2U = 0, \qquad \frac{d^2V}{dt^2} + f^2V = 0 \tag{8.45}$$

These are the harmonic vibrational equations with a natural frequency f; the solutions for initial conditions $U = U_0$ and $V = V_0$ at $t = 0$ are

$$\begin{aligned} U &= U_0 \cos ft + V_0 \sin ft \\ V &= V_0 \cos ft - U_0 \sin ft \end{aligned} \tag{8.46}$$

When we take the squares and add, this immediately suggests that

$$U^2 + V^2 = U_0^2 + V_0^2 = \text{constant}$$

implying that a particle which began its motion with a horizontal velocity $U_{h0} = (U_0^2 + V_0^2)^{1/2}$ must continue to move at the same horizontal speed $U_{h0} = (U^2 + V^2)^{1/2}$ (only the x and y decompositions of the velocity components change with time). A constant velocity motion with a natural frequency can only imply motion in a circle with frequency f, as shown in Fig. 8.8.

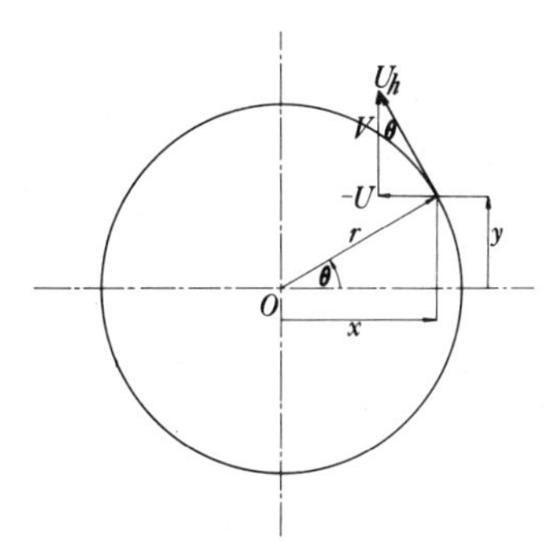

Fig. 8.8 Circular motion.

It remains for us to determine the spatial nature of this motion, that is, $U_h = F(r)$ where F is the functional relationship of the velocity with distance from the center of the *circle of inertia*. One could guess simply that since the motion is not irrotational and since f is the angular velocity of the motion that

$$U_h = fr \tag{8.47}$$

a rigid body motion, varying in intensity with latitude since $f = 2\Omega \sin \psi$. If we need to be convinced, by analysis, of this conclusion, Eqs. (8.44) can be written (Neumann, 1968, p. 151), using $V = dy/dt$ and $U = dx/dt$ and eliminating dt, as

$$dU = f\,dy \qquad \text{and} \qquad dV = -f\,dx$$

After integration

$$U = fy + m \qquad \text{and} \qquad V = -fx + n \tag{8.48}$$

where m and n are constants of integration.

Since we had discovered that $U^2 + V^2 = U_h^2$, squaring Eqs. (8.48) and adding we have

$$\left(x - \frac{n}{f}\right)^2 + \left(y + \frac{m}{f}\right)^2 = \left(\frac{U_h}{f}\right)^2$$

which is the equation of a circle with center at $(n/f, -m/f)$ and radius U_h/f. Thus Eq. (8.47) is justified. The period of this inertial rotation is then $T = 2\pi/f = 2\pi r/U_h$. Also substituting for $f = 2\Omega \sin \psi$

$$T = \frac{\pi}{\Omega \sin \psi} \tag{8.49}$$

The period depends only on the latitude, and is also called the *half-pendulum day* because it is one-half the sidereal day $2\pi/\Omega$. At the poles its value is approximately 12 h and at the equator it is infinite. At 30° latitude it is approximately 24 h. Furthermore, for a current speed of 10 cm/s and at 45° latitude the radius of the inertial circle is approximately 1 km.

Because of minor prevailing currents and because of the geometry of the shore basin, in actual conditions U_h will have additional noncircular components. In 1933 T. Gustafson and B. Kellenberg performed a famous experiment in the Baltic Sea between the coasts of Sweden and Gotland. This is known in the literature as the *Baltic Sea experiment*. They introduced a photographic recorder at a depth of 14 m below the surface of the sea and recorded the motion of the upper homogeneous layer. The experiment lasted for 162 h of continuous recording. The results traced in Fig. 8.9 are from this work. The continuous trace in

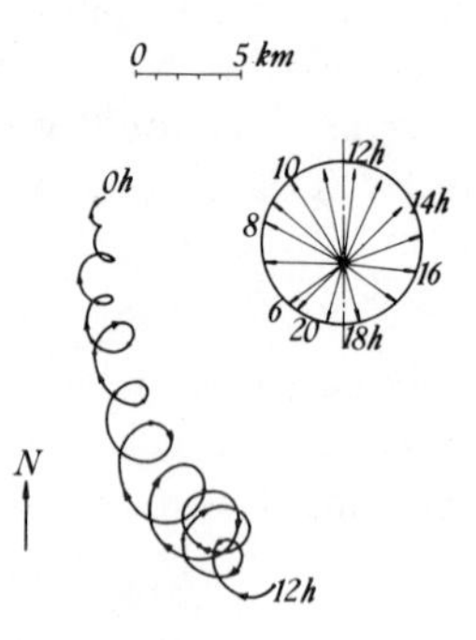

Fig. 8.9 Rotating current observed in the Baltic Sea. (After Gustafson and Kullen berg, 1936.)

the figure was drawn on the basis of velocities obtained from time traces of floating elements. Every twelfth hour, a short line is marked on the continuous trace. The superimposed drift motion was calculated to be first toward the northwest and then north. The direction of rotation is clockwise, as we indicated in the preceding section in our discussion of flows without pressure gradients. The vector diagram in Fig. 8.9 was constructed from the time traces. The endpoints of the velocity vectors fall nearly on a circle but the center is displaced due to the prevailing drifting motion.

The period of rotation calculated from Eq. (8.49) for this experiment was recorded to be 14 h, 8 min, while the period measured from the experiment was 14 h, which is close to the theoretical value.

Some measurements by Pochapsky (1966) and Nan'niti *et al.* (1964) with neutrally buoyant floats in the Atlantic and Pacific Oceans and at depths ranging to 2000 m show the same inertial current phenomena.

8.8 WAVE MOTION IN THE ENVIRONMENT—ZONAL CURRENTS

This subject is extremely vast in breadth and existing analysis. Among the most important phenomena in this major subject classification are stability of the stratified atmosphere and oceans subjected to small disturbances, frontal waves, tidal waves, wind-generated surface waves, gravitational waves, and so on. Fairly concise accounts of the treatment of a number of these problems are given by Stewart (1945, pp. 474–499) and Phillips (1969).

The sample treatment we discuss here deals with the analytical prediction of horizontal Rossby waves as discussed in Section 6.15.

The analysis that follows applies to the free layer with horizontal uniform motion of a barotropic environment. From Eq. (8.1)

$$\frac{\partial \tilde{\boldsymbol{U}}}{\partial t} + (\tilde{\boldsymbol{U}} \cdot \nabla)\tilde{\boldsymbol{U}} + 2\boldsymbol{\Omega} \times \tilde{\boldsymbol{U}} = -\frac{1}{\rho} \nabla \tilde{p} + \boldsymbol{g} \tag{8.50}$$

We shall assume that in the presence of wave motion in the horizontal plane, the perturbations with respect to time and space produced by these waves are small so that the nonlinear terms in $(\boldsymbol{U} \cdot \nabla)\boldsymbol{U}$ are negligible. This assumption becomes, when we substitute into Eq. (8.50) a mean value and a perturbed value for the velocity and the pressure, as in the case of turbulence in Eq. (7.68),

$$\tilde{\boldsymbol{U}} = \boldsymbol{U} + \boldsymbol{u}, \qquad \tilde{p} = P + p$$

Substituting into Eq. (8.50) and omitting the nonlinear perturbation $(\boldsymbol{u} \cdot \nabla)\boldsymbol{u}$ we obtain

$$\left[\frac{\partial \boldsymbol{U}}{\partial t} + (\boldsymbol{U} \cdot \nabla)\boldsymbol{U} + 2\boldsymbol{\Omega} \times \boldsymbol{U} + \frac{1}{\rho} \nabla P - \boldsymbol{g}\right]$$
$$+ \left[\frac{\partial \boldsymbol{u}}{\partial t} + (\boldsymbol{U} \cdot \nabla)\boldsymbol{u} + (\boldsymbol{u} \cdot \nabla)\boldsymbol{U} + 2\boldsymbol{\Omega} \times \boldsymbol{u} + \frac{1}{\rho} \nabla p\right] = 0$$

The terms in the first bracket are the mean terms and those in the second bracket are the perturbation terms. When a time average of this equation is taken, the second bracket vanishes because all fluctuating terms vanish to zero in the averaging process. Thus both brackets must be zero independently. We have worked with the mean equation in other applications; let us consider the properties of the perturbation equation:

$$\frac{\partial \boldsymbol{u}}{\partial t} + (\boldsymbol{U} \cdot \nabla)\boldsymbol{u} + (\boldsymbol{u} \cdot \nabla)U + 2\boldsymbol{\Omega} \times \boldsymbol{u} = -\frac{1}{\rho} \nabla p \tag{8.51}$$

This is a linearized equation for small perturbations. We have made the assumption, however, that we would consider a uniform mean velocity $\boldsymbol{U}$ in the x direction only. Thus $\boldsymbol{U} = \boldsymbol{i}U$. The perturbation velocity $\boldsymbol{u}$ is assumed not likely to be distributed in z so that in the horizontal plane it has two components u and v. Thus Eq. (8.51) in component form becomes[4]

$$\frac{\partial u}{\partial t} + U\frac{\partial u}{\partial x} - fv = -\frac{1}{\rho}\frac{\partial p}{\partial x}$$
$$\frac{\partial v}{\partial t} + U\frac{\partial v}{\partial x} + fu = -\frac{1}{\rho}\frac{\partial p}{\partial y} \tag{8.52}$$

and the continuity equation

$$\frac{\partial u}{\partial x} + \frac{\partial v}{\partial y} = 0 \tag{8.53}$$

The vertical component is the hydrostatic equation which will not interact in this analysis.

To eliminate the pressure in Eq. (8.52) we take the derivative of the first component with respect to y and the second with respect to x and equate. We must keep in mind that the size of these zonal perturbations are of the order of the earth's radius and that variations of f with respect to y must be important. We know that $df/dx = 0$. Keeping in mind also that $d/dt = \partial/\partial t + U\,\partial/\partial x$ the combined equations of motion become

$$\frac{d}{dt}\left(\frac{\partial v}{\partial x} - \frac{\partial u}{\partial y}\right) + v\beta = 0 \tag{8.54}$$

where $\beta = \partial f/\partial y$. Since we know that $f = 2\Omega \sin \psi$ and $\partial y = R\,\partial\psi$, then β is also equal to $2\Omega \cos \psi/R$. Equation (8.54) expresses the rate of change of vorticity. Since this equation must represent the wave phenomena for u, v, and p, we can assume a general form of wave solutions

$$u = \xi(y) \exp[i\lambda_x(x - ct)/2\pi]$$
$$v = \eta(y) \exp[i\lambda_x(x - ct)/2\pi] \tag{8.55}$$
$$p = \Pi(y) \exp[i\lambda_x(x - ct)/2\pi]$$

where $i = (-1)^{1/2}$, which yields sine and cosine solutions, $\lambda_x/2\pi$ is the wavelength in the x direction where the mean flow U is also directed, and c is the

[4] Had we considered gravity waves, we would have introduced the component w instead of v, and variations with z.

propagation velocity of the wave relative to an observer moving with the mean stream. Substituting Eq. (8.55) into Eqs. (8.52) and (8.53) we obtain the algebraic equations

$$\frac{2\pi i\xi(U-c)}{\lambda_x} - f\eta = -\frac{1}{\rho\lambda_x} i\Pi$$

$$\frac{2\pi i\eta(U-c)}{\lambda_x} + f\xi = -\frac{1}{\rho}\Pi'$$

$$\frac{2\pi i\xi}{\lambda_x} + \eta' = 0$$

where the primes denote derivatives with respect to y. Eliminating $\Pi(y)$ from the first two equations by taking the derivative of the first with respect to y and substituting for $\xi(y)$ in the third, we have

$$\eta'' + \left(\frac{\beta}{U-c} - \frac{4\pi^2}{\lambda_x^{\ 2}}\right)\eta = 0 \tag{8.56}$$

The solution of this second-order differential equation is

$$\eta = a\cos\left(\frac{\beta}{U-c} - \frac{4\pi^2}{\lambda_x^{\ 2}}\right)^{1/2} y + b\sin\left(\frac{\beta}{U-c} - \frac{4\pi^2}{\lambda_x^{\ 2}}\right)^{1/2} y \tag{8.57}$$

This is the solution of the wave equation. For boundary conditions which shall determine the coefficients a and b and a special relationship for the argument of the trigonometric functions we assume that v is maximum at $y = 0$ and that it is zero at a meridianal scale $y = \pm\lambda_y$. Substituting these conditions into Eq. (8.57) we get first that $b = 0$ because of the even behavior of the boundary conditions, and then that

$$\left(\frac{\beta}{U-c} - \frac{4\pi^2}{\lambda_x^{\ 2}}\right)^{1/2} \lambda_y = \frac{\pi}{2}$$

Solving for the speed of propagation c from the expression just given, we have

$$c = U - \frac{(\beta/4\pi^2)\lambda_x^{\ 2}}{1 + (\frac{1}{4}\lambda_x/\lambda_y)^2} \tag{8.58}$$

We see immediately from Figs. 6.22 and 8.5 that the two scales λ_x and λ_y are of the same order and that the terms in parentheses in the denominator of Eq. (8.58) are approximately 5%, and can be neglected. In that case the solution for λ_x is exactly what was given in Eq. (6.70):

$$\lambda_x = 2\pi\left(\frac{U-c}{\beta}\right)^{1/2} \tag{8.59}$$

TABLE 8.1

Wavelength of Zonal Patterns as a Function of Wind Speeds at Mid-Latitudes, Eq. (8.59)

Number of waves	2	3	4	5	6	7
Approximate wavelengths (km)	15,000	10,000	7500	6000	5000	4250
Approximate wind speeds at $\psi = 45°$ (m/s)	80	37	20	13	9	7
(miles/h)	180	83	45	29	21	16

This shows that the wavelength of this meandering westerly of velocity $(U - c)$ is inversely proportional to the square root of the cosine of the latitude and proportional to the square root of the particle speed. Normally we know that c is positive eastward and small so that in most cases it can be neglected.

Considering the mid-latitudes and letting $\psi = 45°$, it is easy to calculate the number of integer waves that will fit on a latitudinal circle, at a fixed wind velocity. Table 8.1 gives the relationship between the number of *integer* waves at $\psi = 45°$ corresponding to approximately 30,000 km around the earth and the wind speed, assuming that c is negligible.

The analysis that preceded, in Cartesian coordinates, was for a flat earth. If we take into account the curvature of the earth by considering the equations of motion in spherical coordinates and by repeating the same process (Haurwitz, 1940; see also Stewart, 1945), we obtain a relationship between the number of wavelengths n and the wind speed U, as follows:

$$n(n + 1) = 2\left(1 + \frac{\Omega R}{U \sin \psi}\right) \tag{8.60}$$

Table 8.2 gives the results of this equation, which show a significant difference when compared with the results of Table 8.1. In Eq. (8.60) R is the radius of the earth, Ω is the angular velocity of the earth, and ψ is the latitude angle. For comparison with the results of Table 8.1 we must take $\psi = 45°$.

TABLE 8.2

Wavelength of Zonal Patterns as a Function of Wind Speeds at Mid-Latitudes, Eq. (8.60)

Number of waves	2	3	4	5	6	7
Approximate wind speeds at $\psi = 45°$ (m/s)	164	65	36	23	16	12

8.9 THE RATE OF CHANGE OF CIRCULATION

Having already discussed the equations of motion, and having also discovered that the physics and the mathematics of these equations give rise to wave motions, it becomes of interest to reexamine Sections 6.14, 6.15, and 7.11 and Illustrative Examples 6.5 and 6.7.

By definition of Eq. (6.68) the *circulation* in any linear contour of length C and area $\boldsymbol{A}$ bound by the contour, is

$$\Gamma = \oint_C \boldsymbol{U} \cdot d\boldsymbol{c} \tag{8.61}$$

If C is a contour made up of the *same environmental parcels* as they move in time, this contour will move and deform in time and the rate of change of circulation, $d\Gamma/dt$, *following the parcels*, will be a combination of effects due to the rate of change of $\boldsymbol{U}$ and the rate of change of C. Thus

$$\begin{aligned} \frac{d\Gamma}{dt} &= \frac{d}{dt}\oint_C \boldsymbol{U} \cdot d\boldsymbol{c} \\ &= \oint_C \left(\frac{d\boldsymbol{U}}{dt} \cdot d\boldsymbol{c} + \boldsymbol{U} \cdot \frac{d}{dt}\, d\boldsymbol{c}\right) \end{aligned} \tag{8.62}$$

Let us first examine the second term due to the deformation of the contour C. Since we said that C is made up of the same parcels of the environment as they move downstream $d(d\boldsymbol{c}/dt) = d\boldsymbol{U}$, and thus

$$\boldsymbol{U} \cdot \frac{d}{dt}(d\boldsymbol{c}) = \boldsymbol{U} \cdot d\boldsymbol{U} = \tfrac{1}{2}\, dU^2$$

Although this differential is not necessarily zero, its integration over a complete cyclic integral is zero. This is because dU^2 is an exact differential and both limits in the cyclic integration are the same. Thus

$$\tfrac{1}{2}\oint_C dU^2 = 0$$

and

$$\frac{d\Gamma}{dt} = \oint_C \frac{d\boldsymbol{U}}{dt} \cdot d\boldsymbol{c} = \oint_C \boldsymbol{a} \cdot d\boldsymbol{c} \tag{8.63}$$

This is *Kelvin's theorem* which states first that the rate of change of circulation is equal to the integration of the acceleration vector $\boldsymbol{a}$ in the contour C where the circulation is defined; and second, that only in the case when the acceleration has a potential, will Γ not vary with time. We shall see that in an environment influenced by rotational forces such as the Coriolis, friction, and baroclinic conditions, it is not possible for the acceleration vector to have a potential and consequently the circulation will vary with time.

From Eq. (7.40) we can replace the acceleration by the forces per unit mass acting on the parcel. Thus

$$\boldsymbol{a} = \frac{\partial \boldsymbol{U}}{\partial t} + (\boldsymbol{U} \cdot \nabla)\boldsymbol{U}$$

$$= -\frac{1}{\rho} \nabla P + \boldsymbol{g} - 2\boldsymbol{\Omega} \times \boldsymbol{U} + \nu \nabla^2 \boldsymbol{U}$$

and

$$\frac{d\Gamma}{dt} = -\oint_C \frac{\nabla P}{\rho} \cdot d\boldsymbol{c} + \oint_C \boldsymbol{g} \cdot d\boldsymbol{c} - \oint_C (2\boldsymbol{\Omega} \times \boldsymbol{U}) \cdot d\boldsymbol{c} + \oint_C (\nu \nabla^2 \boldsymbol{U}) \cdot d\boldsymbol{c} \qquad (8.64)$$

This equation states that *circulation will change with time* if

(a) The *density* (*in the first integral*) *is not a unique function of pressure.* In other words, the fluid is baroclinic, thus making $(\nabla P/\rho) \cdot d\boldsymbol{c}$ an inexact differential. We have seen in Section 8.2.B that $(1/\rho)\, \nabla P = \nabla \int (1/\rho)\, dP$ and that only when $\rho = f(P)$ does the integrand of this first integral become an exact differential, and consequently the integral around a closed circuit will be zero. Baroclinicity is an integral part of most of the environment and therefore it will contribute significantly to the rate of change of circulation, as we shall see in forthcoming sections.

(b) The *gravitational acceleration* $\boldsymbol{g}$ poses no problems in Eq. (8.64). This is because we have asserted a number of times that it can be represented in terms of the gradient of a potential energy. Thus its integral in a closed circuit will be zero and so will its contribution to the change of circulation.

(c) The *Coriolis has an influence on the motion.* We have seen that the Coriolis is the major driving force in the free layer, and consequently this free layer will exhibit constant changes of circulation, with time.

(d) The *viscous forces have an influence on the motion.* Although viscous forces have negligible influence in the free layer, in the boundary layer circulation will be constantly changing as a result of the contribution of the last term in Eq. (8.64).

Thus in the *free layer* we may rewrite Eq. (8.64) as

$$\frac{d\Gamma}{dt} = -\oint_C \frac{\nabla P}{\rho} \cdot d\boldsymbol{c} - \oint_C (2\boldsymbol{\Omega} \times \boldsymbol{U}) \cdot d\boldsymbol{c} \qquad (8.65)$$

The second integral on the right-hand side can be developed further by using the triple scalar product relation (A.28)

$$(2\boldsymbol{\Omega} \times \boldsymbol{U}) \cdot d\boldsymbol{c} = (\boldsymbol{U} \times d\boldsymbol{c}) \cdot 2\boldsymbol{\Omega}$$

and since $\boldsymbol{\Omega}$ is a constant vector

$$-\oint_C (2\boldsymbol{\Omega} \times \boldsymbol{U}) \cdot d\boldsymbol{c} = -2\boldsymbol{\Omega} \cdot \oint_C (\boldsymbol{U} \times d\boldsymbol{c})$$

Considering a point on the contour moving at the velocity $\boldsymbol{U} = d\boldsymbol{r}/dt$ the cross product $d\boldsymbol{r} \times d\boldsymbol{c}$ will be the infinitesimal area $d\boldsymbol{A}$ bounded by the contour C and $(\boldsymbol{U} \times d\boldsymbol{c})$ the rate of change of this area as shown in Fig. 8.10. Finally Eq. (8.65) becomes

$$\frac{d\Gamma}{dt} = -\oint_C \frac{dP}{\rho} - 2\Omega \frac{dA}{dt} \tag{8.66}$$

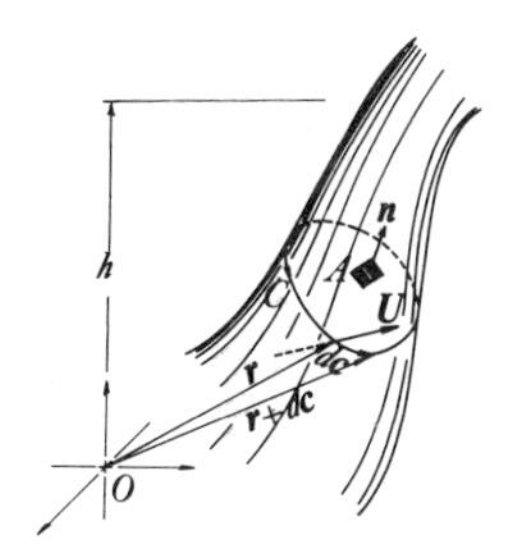

Fig. 8.10 Rate of change of circulation.

This final result is attributed to Bjerknes (1898). Equation (8.66) states that *in the absence of viscous forces, circulation in a circuit C moving with the fluid will change in time if the process is baroclinic and if the area of any contour moving with the fluid changes with time in the presence of the earth's rotation.*

Equation (8.66) can also be put in vorticity form since circulation is the integral of the vorticity in the area A. A simplified approach would be to consider the average vorticity in the area A so that $\Gamma = \zeta A$, and since $\boldsymbol{\Omega}$ is not a function of time we can add its rate of change to

$$\frac{d}{dt}(\zeta + 2\Omega) \cdot A = -\oint \frac{dP}{\rho}$$

$$= A \cdot \frac{d}{dt}(\zeta + 2\Omega) + (\zeta + 2\Omega) \cdot \frac{dA}{dt}$$

$$= A \cdot \left[\frac{d}{dt}\zeta_t + \zeta_t \frac{1}{A}\frac{dA}{dt}\right]$$

where $\zeta_t = (\zeta + 2\Omega)$ is the absolute vorticity. The quantity $(1/A)(dA/dt)$ is the divergence of the velocity since in two dimensions, for instance, $A = \Delta x \, \Delta y$, $dA = \Delta x \, d(\Delta y) + \Delta y \, d(\Delta x)$, and

$$\frac{1}{A}\frac{dA}{dt} = \frac{d(\Delta y)/dt}{\Delta y} + \frac{d(\Delta x)/dt}{\Delta x}$$

$$= \frac{\partial V}{\partial y} + \frac{\partial U}{\partial x}$$

Therefore,

$$\frac{d}{dt}(\zeta_t) = -\zeta_t \nabla \cdot \boldsymbol{U} - \frac{1}{A}\oint \frac{dP}{\rho} \tag{8.67}$$

where the integral is an average value in the area A.

For baroclinic situations the integral in Eq. (8.66), using Stokes's theorem in vector calculus, can be transformed into an area integral

$$\oint_C \frac{\nabla P}{\rho} \cdot d\boldsymbol{c} = \int_A \left[\nabla \times \left(\frac{\nabla P}{\rho}\right)\right] \cdot \boldsymbol{n} \, dA$$

The integrand can be expanded as follows:

$$\nabla \times \left(\frac{1}{\rho} \nabla P\right) = \nabla\left(\frac{1}{\rho}\right) \times \nabla P + \frac{1}{\rho} \nabla \times \nabla P$$

as in Section 7.11. Because the curl of the gradient is zero the pressure term becomes

$$\oint_C \frac{\nabla P}{\rho} \cdot d\boldsymbol{c} = \int_A \left[\nabla\left(\frac{1}{\rho}\right) \times \nabla P\right] \cdot \boldsymbol{n} \, dA$$

and Eq. (8.66) takes the form

$$\frac{d\Gamma}{dt} = \int_A \left[\nabla P \times \nabla\left(\frac{1}{\rho}\right)\right] \cdot \boldsymbol{n} \, dA - 2\Omega \frac{dA}{dt} \tag{8.68}$$

The sign of the integral has been changed by reversing the order of the cross product. For *barotropic processes* the integral becomes zero and the rate of change of circulation from Eq. (8.68) depends only on the rotationality of the earth's motion:

$$\frac{d\Gamma}{dt} = -2\Omega \frac{dA}{dt} \tag{8.69}$$

and from Eq. (8.67) the rate of change of vorticity is

$$\frac{d\zeta_t}{dt} + \zeta_t \nabla \cdot \boldsymbol{U} = 0 \tag{8.70}$$

Identifying by h the height of the vortical column in Fig. 8.10 the volume of air in the column is Ah which must be constant since we have been talking of circuits moving with the fluid. Thus

$$\frac{1}{A}\frac{dA}{dt} = -\frac{1}{h}\frac{dh}{dt} = \nabla \cdot \boldsymbol{U}$$

Replacing this in Eq. (8.70) we can finally write

$$\frac{d}{dt}\left(\frac{\zeta + 2\Omega}{h}\right) = 0 \tag{8.71}$$

This final statement specifies that in the absence of viscous and baroclinic effects the ratio of the total vorticity to the height of the column is conserved. This was discussed, in a less formal way, in Sections 6.14 and 6.15. Thus when the vortex column is stretched the vorticity also increases keeping the ratio constant.

When there is a *divergence*, or stretching of a vertical column (a) either the relative vorticity of the fluid increases locally keeping the vertical component of $\boldsymbol{\Omega}$ constant, or (b) the motion moves to larger latitudes where Ω_z is larger, or (c) both the relative vorticity increases and the motion moves to larger latitudes. We are considering in this argument that the vertical component of the absolute vorticity is nearly its total value.

The case of divergence implies tendency toward low pressure formation because of an increase in counterclockwise rotation relative to the earth as well as movement of the stream toward the north in the Northern Hemisphere. *Convergence* or compression of an environmental column follows the opposite argument, with a reduction of relative vorticity (high pressure tendency) and movement of the stream toward the south in the Northern Hemisphere. Also from Eq. (8.71) we expect that environmental columns will expand and contract and that a meandering tendency will prevail.

Looking at the equation of equilibrium for a gradient wind or current [Eq. (8.39)], at any point in the free layer

$$V = \frac{1}{f}\left(\frac{1}{\rho}\frac{dP}{dr} - \frac{V^2}{r}\right)$$

Given a meandering stream such as that shown in Fig. 8.11, let us look into the effects of curvature and those of changing latitude. Let us start with a unit velocity V_A at a point A and in balance with $dP/dr > 0$ for a given $f = f_A$ and $r = r_A$. Going from A to B the stream will have two principal effects, one of changing r and the other of changing f.

Consider first the effect of changing curvature. Since at B $(dP/dr)_B < 0$ (same sign as $-V^2/r$), the velocity there will be larger even if $r_A = r_B$. We designate by V_{Br} the velocity influenced by r alone. We conclude that between A and B,

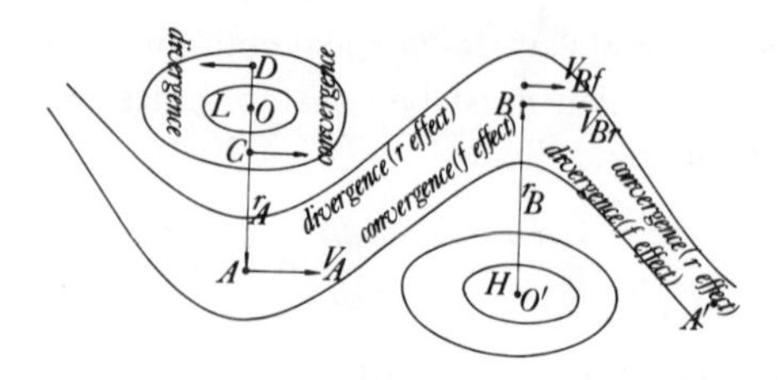

Fig. 8.11 Application of Bjerknes principle to conservation of absolute vorticity.

or between a low and a high, there must be a divergence owing to the increase in velocity. From B to A' it will be a convergence. In the curvature effect then, pressure must drop east of A since velocity has increased according to Bernoulli's principle. The pressure will rise east of B. This implies a movement eastward of the pressure system since the pressures outside the system push it from crest to trough.

In the preceding argument, we omitted the changes in latitude. Consider now the changes of f in the meandering flow. According to the gradient wind and current relation, when f increases V decreases, everything else remaining the same. Since $f_B > f_A$, $V_B < V_A$. This phenomenon implies a convergence in going from A to B, and following the same argument, the pressure system on the meandering motion will have a tendency to move it westward.

In the region of circular isobars where the effect of curvature is constant, in the low pressure area, $f_C < f_D$, therefore $V_D < V_C$ and we conclude that there must be divergence with pressure dropping west of O and convergence with pressure rising east of O.

Thus the influences of curvature and latitude changes oppose each other. We can argue that the combined effect is a function of the geometry of the pressure system. For long wavelengths and fixed amplitudes, the curvature effect in V^2/r may be small and the system will move westward. For short wavelengths and fixed amplitudes the curvature effect may dominate and the system will move eastward. Finally, for a given amplitude and wavelength there is a critical velocity which will keep the wave stationary. In reality, at high altitudes, these waves have a slow motion eastward, which implies that the curvature has a slight predominance.

We must remember that near the ground the land contour will also have an influence. For instance, the sudden appearance of a mountain range produces a compression of the column on the windward side of the mountain and therefore a reduction in vorticity which is accompanied by a high pressure tendency and a veering toward the south. On the lee side of the mountain a divergence occurs which will have a tendency to produce a low pressure and veering back toward the north.

8.10 APPLICATION OF KELVIN'S THEOREM TO A BAROCLINIC ENVIRONMENT

A The Atmospheric Circulation

The question often arises as to how winds are produced, intensify and decay if one begins with an atmosphere that is horizontally homogeneous. Consider such a stratified but horizontally homogeneous atmosphere whose constant-pressure, constant-density, and constant-temperature surfaces are parallel to the surface of the earth. Kelvin's theorem on the rate of change of circulation contains an integral in which the specific volume $1/\rho$ appears. We shall refer to the surfaces of constant specific volume as *isosteric surfaces*.

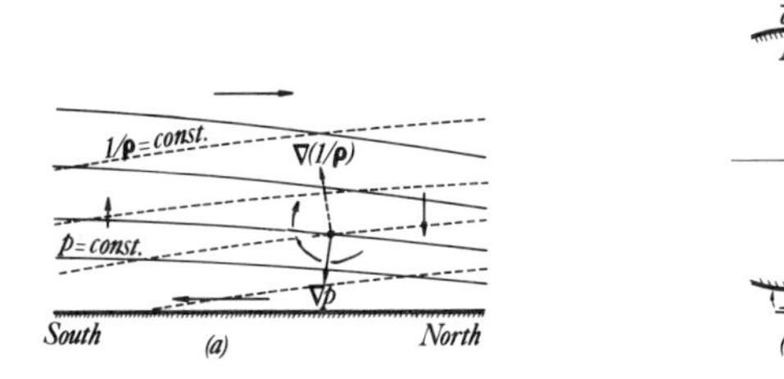

Fig. 8.12 The meridional circulation principle. Motion due to thermal gradients.

The earth's surface and the atmosphere are shown in Fig. 8.12. The southern latitudes are warmed by the sun more rapidly than the northern latitudes. The effect of this uneven heating (shown in Fig. 3.11) is to expand the atmosphere in the south and to contract it in the north. Since at the same elevation pressure is larger in the south because of higher temperatures, a particular isobar in the south is higher than the same isobar in the north. Thus isobars are raised in the warm regions and lowered in the cold regions. The pressure on the ground may remain unchanged because of the earth's temperature.

From the gas law, the specific volume depends on the pressure and temperature. Since the sun's heat is a major factor here, the atmosphere is no longer barotropic. The atmosphere, heated from the sun, will display rising temperatures along the isobar going from north to south. Thus the isobaric surfaces will no longer be parallel to the isosteric surfaces and the two surfaces will intersect. The specific volume along the isobar will increase from north to south.

From Kelvin's theorem, this baroclinic situation will generate a rate of change of circulation and vorticity proportional to the cross product of the pressure gradient and the gradient of the specific volume as given by Eq. (8.68). Looking at Fig. 8.12a the pressure gradient, in the process of uneven heating, has inclined toward the south from the vertical giving a component in that direction. The gradient of the specific volume has inclined toward the north. The cross product of the two gradients [appearing in Eq. (8.68)] is a vector perpendicular to the plane of the paper pointing down toward the west.

A meridional circulation will begin immediately in a clockwise direction in Fig. 8.12a corresponding to a counterclockwise direction in the representation of Fig. 8.12b, in the Northern Hemisphere, when looking at the east side of the earth. This rate of increase of circulation is in the direction of $\nabla P \times \nabla(1/\rho)$. Thus at high altitudes the wind will blow from south to north or from warm to cold. In the south this will entrain air from the ground up replacing the movement in the upper layer. Consequently the ground pressure will tend to drop in the south. There will be a surface movement of air from north to south bringing cold air to warmer areas. We shall see, shortly, that one circulation loop of the size shown in Fig. 8.12b is not stable and consequently the overall circulation pattern will eventually break into a multiloop system.

The effect just described should not be interpreted as being those of buoyancy as discussed in Sections 4.12 and 7.8. Buoyancy effects can occur even in barotropic conditions provided the temperature lapse rate is different than the adiabatic lapse rate. What we are considering here is a misalignment of the pressure and isosteric surfaces. This is not to say that baroclinic flows cannot influence the vertical stability, or vice versa.

Since the horizontal temperature gradients are the cause of this circulation, the stronger these gradients are, the stronger will be the circulation. Fortunately, these baroclinic motions have a tendency to *mix* the environment and make the gradients smaller. In time this mixing and friction will prevent the circulation from increasing continually without limits.

To evaluate the circulation on the earth we start with Eqs. (8.63) and (8.66)

$$\frac{d\Gamma}{dt} = \oint_C \boldsymbol{a} \cdot d\boldsymbol{c} = -\oint_C \frac{dP}{\rho} - 2\Omega \frac{dA}{dt}$$

Omitting the stretching term dA/dt, if we consider a closed path *abcd* in Fig. 8.12b and the integral $-\oint dP/\rho$ we can perform illustrative calculations for the order of the atmospheric circulation.

If segments *d* and *b* of the path are considered isobars, as a first approximation, the contribution to the integral along these segments will be zero, dP being zero along these segments. However, along *a* and *c* we can introduce the temperature and pressure instead of the density, such that

$$-\oint \frac{dP}{\rho} = -R \oint T \frac{dP}{P}$$

Then the rate of change of circulation is approximately

$$\frac{d\Gamma}{dt} = R(\overline{T}_a - \overline{T}_c) \ln \frac{P_d}{P_b}$$

where $\overline{T}_a$ and $\overline{T}_c$ are the average temperatures at the equator and north pole between sea level and the elevation b. Bjerknes (1933) has taken this temperature difference to be $\overline{T}_a - \overline{T}_c = 40°C$ and for an elevation of approximately 9 km giving a pressure ratio of $\frac{1}{3}$ he calculated the rate of change of circulation:

$$\frac{d\Gamma}{dt} = 138 \times 10^6 \quad \text{cm}^2/\text{s}^2$$

Thus both the velocity and the circulation will depend on time. After 6 h the average wind velocity becomes approximately 15 m/s, which is considerable, and in 24 h it reaches a hurricane level of 60 m/s. If it were not for the mixing that takes place and the accompanying reduction in temperature gradients, the winds would indeed quickly reach intense levels.

The influence of the Coriolis is to deflect the streams *abcd* and to produce a horizontal zonal component of circulation. The segment b will be displaced toward the east, and the southbound segment d is veered toward the west.

The meridional circulation just described in Fig. 8.12 breaks down into at least three separate cells on each hemisphere. A detailed account of this breakdown was given by Rossby (1941). Part of this account is given here:

A ring of air extending around the earth at the equator, at rest relative to the surface of the earth, spins around the polar axis with a speed equal to that of the earth itself at the equator. This is shown in Fig. 8.13a. If somehow this ring is pushed northward over the surface of the earth, its radius is correspondingly reduced; and it follows from the principle set forth that the absolute speed of the ring from west to east increases. Since the speed of the surface of the earth itself from west to east decreases northward, it follows that the moving ring, in addition to its northward velocity, must acquire a rapidly increasing speed from west to east relative to the earth's surface.

The principle itself may be illustrated very simply by an extreme and absurd case. The eastward speed of the earth itself at the equator is about 465 m/s. A ring of air displaced from this latitude to latitude 60, where the distance from the axis of the earth is only half that at the equator, would appear in its new position with double the original absolute velocity, or 930 m/s. Since the speed of the earth itself at this latitude is only half what it is at the equator, or 232 m/s, it follows that a ring of air thus displaced would move eastward over the surface of the earth with a relative speed of 698 m/s. Obviously such wind speeds never occur in the atmosphere, one reason being the effect of frictional forces, another the fact that large-scale atmospheric displacements never are symmetric around the earth's axis or as large as those indicated.

It is apparent, however, that this tendency toward the establishment of west winds in northward-moving rings of air and of east winds in southward-moving rings must modify the previously described meridional (north–south) circulation scheme considerably. The moment circulation begins, west winds (relative to the earth) would begin to develop in the upper atmosphere, with a slight component northward, and east winds in the lower atmosphere, with a slight component southward.

In this scheme, ground friction plays a basic role, since it prevents the development of excessive east winds in the surface layers. The upper atmosphere, in which west winds prevail,

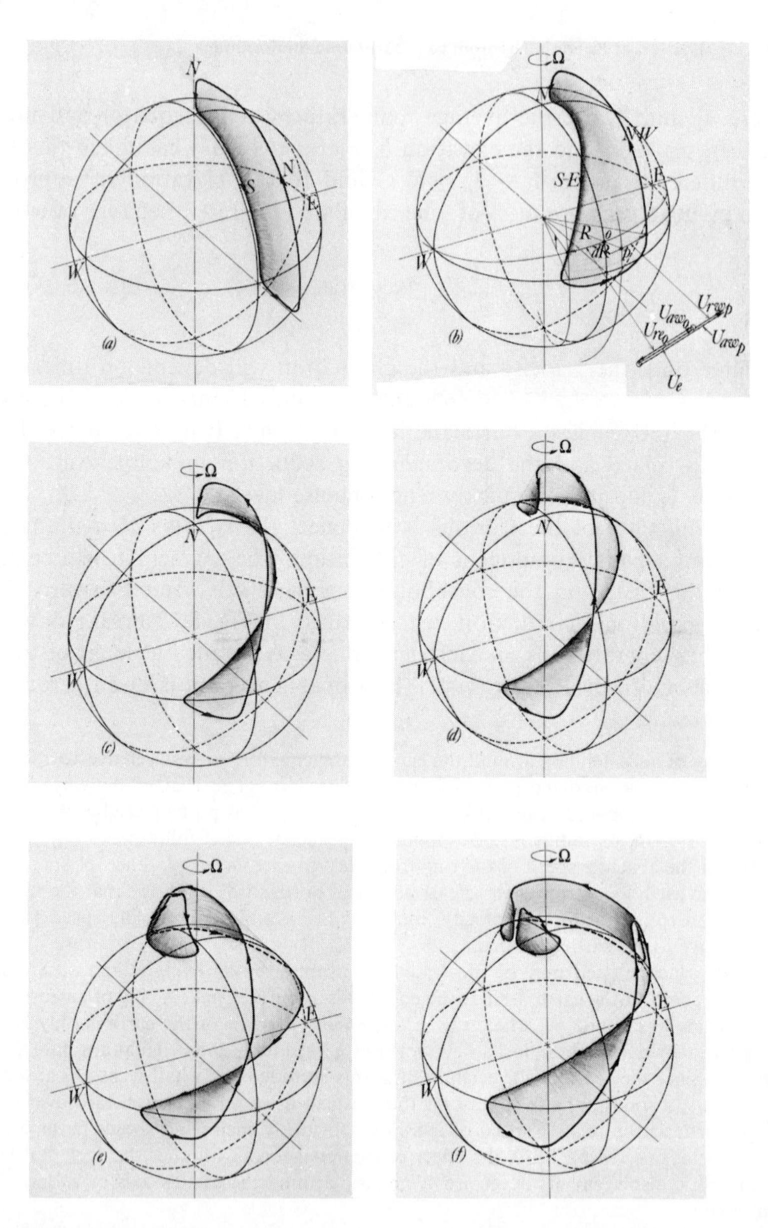

Fig. 8.13 How the earth's rotation leads to a breakdown into several cells of the simple meridional (N–S) circulation. (a) Model without earth's rotation. (b) Model with earth's rotation but not in radial equilibrium [a short time after circulation in (a) begins]. U_e earth's velocity; U_{awp} atmosphere west at zero; U_{re_0} relative velocity at zero; U_{awp} atmosphere west at p; U_{rwp} relative velocity at p. (c) Gradually the upper west winds are brought down to the ground near the poles and the east winds rise near the equator. (d) The west winds are retarded by friction and seek their way northward, but cooling and sinking continue next to the pole. (e) A two-cell circulation. (f) A complete three-cell circulation.

is not in direct contact with the earth's surface; however, mixing of air between the upper and lower strata must reduce the west winds aloft as well as the east winds below. Since the momentum of the east winds also is reduced from below, through the effect of ground friction, it is apparent that the mass of the west winds aloft would far exceed the mass of the easterlies near the surface. Figure 8.13b illustrates what the velocity distribution would be a short time after the rotation began.

Certain features of this picture agree well with observed conditions. Above 4 to 5 km westerly winds prevail in all latitudes. At sea level, easterly wind components are normally observed between latitude 30°N and 30°S. Other belts of easterly wind components are observed in the polar regions, north of latitude 60°N and south of latitude 60°S. Unexplained, however, is the fact that in each hemisphere westerly winds prevail also at sea level within a broad belt between latitude 30° and 60°, approximately.

It is fairly easy to see that the theoretical model in Fig. 8.13b characterized by east winds everywhere in the surface layers, is physically impossible as a steady state. If east winds prevailed in all latitudes, friction between the atmosphere and the solid earth would constantly tend to reduce the rotation of the earth. On the other hand, the atmosphere would constantly gain momentum from the earth. Sooner or later a state of equilibrium would be established in which the atmosphere would neither gain nor lose momentum through contact with the earth—an equilibrium that is known to prevail, since the rotation of the earth for practical purposes can be regarded as constant. Such an equilibrium requires that the retarding influence of the east winds must be offset by the accelerating influence of a belt or belts of west winds, also in the surface layers. It is clear, however, that this argument is incapable of determining the number, width, and strength of the required west-wind belts. It is the purpose of the five diagrams in Fig. 8.13 to explain why the initial meridional circulation, under the influence of the earth's rotation, necessarily must break down into at least three separate cells on each hemisphere.

In order to understand the successive stages of development indicated in this figure it is necessary to discuss, in some detail, the effect of the rotation of the earth on the relative motion of air over its surface. Wherever a ring of air parallel to a latitude circle is rotating more rapidly than the surface of the earth itself, it is acted upon by an excess of centrifugal force which tends to throw the ring away from the axis of the earth, which in the Northern Hemisphere means southward. If the ring rotates with the same speed as the earth (i.e., if it is at rest relative to the surface of the earth), this excess of centrifugal force vanishes. If this were not the case any object resting on the surface of the earth would be thrown towards the equator. A ring of air rotating more slowly than the earth itself, and hence appearing as an east wind relative to the earth, suffers a deficiency in centrifugal force and tends to move toward the axis of the earth, i.e., the Northern Hemisphere, towards the north. To keep a west-wind belt (which rotates more rapidly than the earth because it comes from southern latitudes) from being thrown southward, the atmospheric pressure must be higher to the south than to the north of the ring (in the Northern Hemisphere), thus producing a force directed northward and capable of balancing the excess centrifugal force. If no such pressure force (gradient) is available, the ring will be displaced slightly southward until enough air has piled up on its south side to bring about the required cross-current pressure rise to the south and equilibrium. The total displacement needed for this purpose is usually quite small as compared with the width of the current.

To keep an east-wind belt in equilibrium, the atmospheric pressure must be higher on the north side than on the south side (in the Northern Hemisphere), so that the resulting pressure force balances the deficiency in centrifugal force acting on the ring. It has already been brought out that, in the Northern Hemisphere, air moving northward tends to acquire a velocity eastward (westerly), while air moving southward tends to acquire a velocity westward (easterly).

To offset this tendency towards deflection eastward,[5] a northbound current of limited width piles up air to the east and creates higher pressure to the east than to the west, while the reverse applies to the southbound current.

All these results may be generalized so as to apply to any wind direction. It is thus found that in the Northern Hemisphere steady winds always blow in such a fashion that the air pressure drops from right to left across the current for an observer facing downstream. The stronger the current flows, the steeper the drop in cross-current pressure. If, in any horizontal plane, lines of constant air pressure (isobars) are drawn, it may be seen that the air follows the isobars and moves counterclockwise around regions of low pressure (cyclones) and clockwise around regions of high pressure (anticyclones). In the Southern Hemisphere, the direction of motion around highs and lows is reversed.

It is apparent from the preceding reasoning that the relationship between wind and horizontal pressure distribution is truly mutual; a prescribed pressure distribution will gradually set the air in motion in accordance with the law set forth; likewise, if somehow a system of horizontal currents has been set up in the atmosphere, the individual current branches will very quickly be displaced slightly to the right (in the Northern Hemisphere) until everywhere the proper cross-current pressure drop from right to left has been established. Owing to the ease with which the atmosphere thus builds up the cross-current pressure drop required for equilibrium flow, the reasoning just outlined merely helps in understanding why the pressure in the Northern Hemisphere always rises from left to right for an observer looking downstream but does not by itself indicate that one current pattern is more likely to be established than another. To establish the character of the current patterns, either the pressure distribution must be known, or additional physical principles must be utilized.

It is now possible to return to a discussion of the circulation development in Fig. 8.13. In an axially symmetric atmosphere, such as the one here discussed, the absolute angular momentum of individual parcels of air does not change except through the influence of frictional forces. Under these conditions it is evident that the meridional (north–south) movements indicated in Fig. 8.13b must gradually redistribute the absolute angular momentum so as to create west winds (easterlies) next to the ground in the polar regions, and east winds (westerlies) aloft over the equator. This is the state illustrated in Fig. 8.13c. If the meridional circulation is slow, the pressure distribution in the atmosphere must constantly adjust itself fairly closely to the prevailing zonal winds. Thus in Fig. 8.13c there would be a sea-level-pressure maximum at the transition point between the east winds in low latitudes and the west winds farther north. This latter belt of west winds can continue its southward displacement as long as it is acted upon by an excess of centrifugal force. However, part of the air in this west-wind belt must steadily lose momentum through frictional contact with the ground. Under the influence of the resulting deficiency in centrifugal force, this shallow portion of the belt next to the earth's surface must seek its way northward, as indicated in Fig. 8.13d. Since air continues to cool and sink next to the pole, it follows that the retarded west winds, for purely dynamic reasons, are forced to escape aloft some distance from the pole. Finally a cellular state develops, as indicated in Figs. 8.13e and 8.13f.

B Oceanic Circulation

Ocean currents, far from the shores which receive larger quantities of fresh water from runoff due to precipitation, depend on the depth of the layers.

[5] The method of compensation here described is obviously impossible for circumpolar rings of air. Hence rings of air displaced northward acquire west-wind tendencies, southbound rings east-wind tendencies, as brought out previously.

Unlike the atmosphere, the larger velocity and mass currents in the ocean occur at the surface as a result of wind stress as well as baroclinic conditions. In other words, the largest driving forces in the ocean are in the viscous layer and not in the free layer. We reserve the discussions pertaining to the environmental viscous boundary layer for the following chapter. Figure 8.14 shows the surface water circulation in the oceans of the earth during the winter months in the Northern Hemisphere.

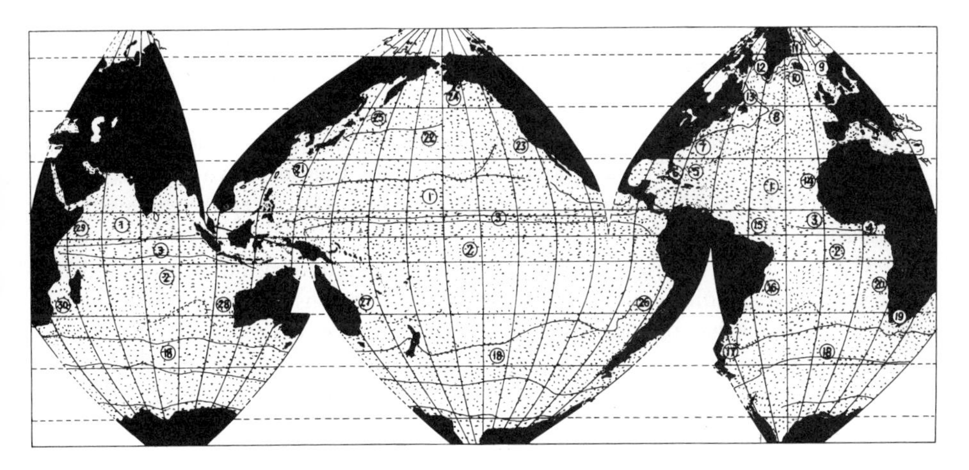

Fig. 8.14 World chart of ocean surface currents during Northern Hemisphere winter. (Neumann, 1968, "Ocean Currents"; courtesy of Elsevier Publishing Co., Amsterdam.) (1) North Equatorial Current; (2) South Equatorial Current; (3) Equatorial Countercurrent; (4) Guinea Current; (5) Antilles Current; (6) Florida Current; (7) Gulf Stream; (8) North Atlantic Current; (9) Norwegian Current; (10) Irminger Current; (11) East Greenland Current; (12) West Greenland Current; (13) Labrador Current; (14) Canary Current; (15) Guiana Current; (16) Brazil Current; (17) Falkland Current; (18) Antarctic Circumpolar Current; (19) Agulhas Current; (20) Benguela Current; (21) Kuroshio; (22) North Pacific Current; (23) California Current; (24) Aleutian Current; (25) Oya Shio: (26) Peru Current; (27) East Australian Current; (28) West Australian Current; (29) Somali Current; (30) Mozambique Current; (31) Monsoon Current.

Near the shores and particularly in river estuaries the baroclinic effects owing to strong salinity gradients become important. As in the case of Fig. 8.12a for the atmosphere, the isosteric surfaces in the estuaries are inclined away from the horizontal as shown in Fig. 8.15 in the form of wedges penetrating into the river and toward the river bottom. These inclined surfaces are called *salinity wedges*. The value of the specific volume and the *intrusion length* depend on the velocity of the fresh water, the density difference between the sea and the fresh water, the geometry of the estuary basin, and tides. These factors are discussed in more detail in Section 10.6. It is evident from the figure and on the basis of Eq. (8.68) that the cross product of the pressure gradient and the

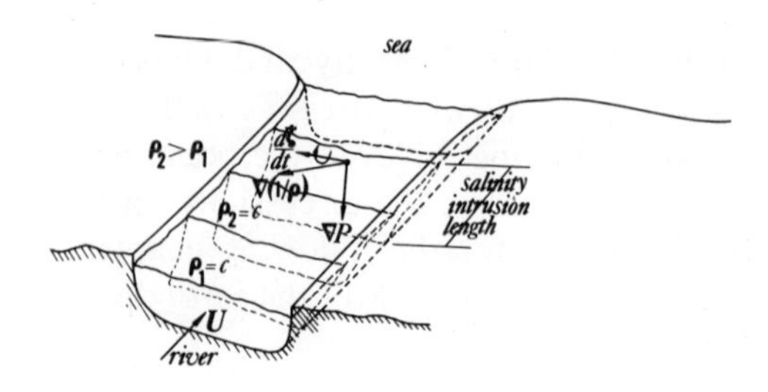

Fig. 8.15 Generation of vorticity at an estuary.

specific volume gradient will generate a rate of circulation and vorticity as shown in Fig. 8.15. These secondary flows in estuaries have been known, through the years, to produce erosion and shoaling that alter the geometry of the basin.

The problem of erosion and shoaling due to secondary flows produced by baroclinic conditions can also be found near regions of rivers, lakes, and oceans where large quantities of hot water are discharged from industrial cooling systems and power plants. Besides the biological implications of thermal pollution, the dynamical effects of pollution are explained through Kelvin's theorem.

9

The Frictional Shear Layer—The Boundary Layer

9.1 INTRODUCTION

In the general considerations of the dynamical equations of the environment (in Section 7.6), we dealt with frictional forces owing to the viscosity of the fluid and the velocity derivatives contained in the total rate of deformation of the same fluid. The product of the viscosity coefficient and the velocity derivatives gave rise to tangential frictional viscous stresses as stated in Eq. (7.39). We also explained that because the environment could be considered incompressible, the normal stresses were just due to pressure, and viscosity had no influence on them.

In Section 7.12 we also introduced another type of frictional stress, not due directly to viscosity at all, but to turbulent Reynolds stresses which add on to and often are larger in value than the viscous (laminar) stresses.

A characteristic of both viscous and turbulent stresses is that they both become significant in regions of fluid motion where the velocity field has considerable spacewise transverse variations. In the case of the atmosphere being free to move at speeds different than those on the surface of the earth near the ground, friction between the wind and the earth will ultimately develop a frictional layer in which the velocity field will build up smoothly from the value at the surface to the value of the environment in the free layer. When the free environment has a slower velocity than that of the earth (easterly) the earth will lose momentum to the environment, and vice versa.

Since the wind has a greater velocity than water, in the interaction between the atmosphere and the ocean at the surface, the wind will lose momentum to the ocean. Thus there develops near the surface an oceanic frictional layer in which the velocity at the surface is maximum but decreases smoothly to near zero value in the free layer below. Above the ocean surface, the atmosphere will also have a frictional layer in which the velocity of the atmosphere increases from that at the ocean surface to that in the atmospheric free layer. (Schematically, this was shown in Fig. 7.11.) We have already discovered that the uneven heating of the sun on the earth's surface and the Coriolis through the earth's rotation are the two primary motors, so to speak, of the atmospheric free layer.

In both the atmosphere and the oceans the thickness of the frictional layer is small compared to the remaining parts of the free layer. Nevertheless since man's activities are mostly within these layers, and since the evaporation processes which ultimately influence the dynamics and thermodynamics of the atmospheric free layer take place in the frictional layer, studies of these layers become very important.

9.2 THE BASIC DYNAMICAL EQUATIONS OF THE NEUTRAL BOUNDARY LAYER

We have referred to the environmental layer near the surface of the earth as the frictional layer. Because these planetary frictional layers are near the rigid boundary on the ground, they are often referred to as *boundary layers*. Although the boundary layer is a frictional shear layer, not all shear layers are boundary layers. As illustrative examples, the wake flow behind fixed immersed bodies is a shear flow without having the characteristics of a boundary layer, (a frictional layer near a rigid boundary). Similarly, the flow in the plume arising from a stack is a shear layer but does not have the characteristic mean and turbulent behavior of the boundary layer shear flow. We shall study the characteristic physical and mathematical behavior of the planetary boundary layer in the course of this chapter.

In Chapter 8 it was established that unless geostrophic motions of the free layer had the magnitude of a storm, the linear and radial acceleration components of these flows were a small part in the balance of forces. This is even truer in the boundary layers. We shall say that for most applications we shall consider the boundary layer equations as those of Eq. (7.70) without the linear and radial acceleration terms. Thus

$$
\begin{aligned}
fV &= \frac{1}{\rho}\frac{\partial P}{\partial x} - \frac{\partial}{\partial z}\left(\nu\frac{\partial U}{\partial z} - \overline{uw}\right) \\
fU &= -\frac{1}{\rho}\frac{\partial P}{\partial y} + \frac{\partial}{\partial z}\left(\nu\frac{\partial V}{\partial z} - \overline{vw}\right) \qquad (9.1) \\
g &= -\frac{1}{\rho}\frac{\partial P}{\partial z}
\end{aligned}
$$

constitute the basic dynamical equations for a *neutral* boundary layer. The buoyancy force having been omitted here in the z direction implies a state of neutral equilibrium with respect to buoyancy.

Looking at Fig. 9.1, the boundary layer is generally divided into two distinct layers: an *inner layer* of height z_0 of the order of centimeters in which the velocities relative to the earth's surface are so close to zero that the Coriolis has no

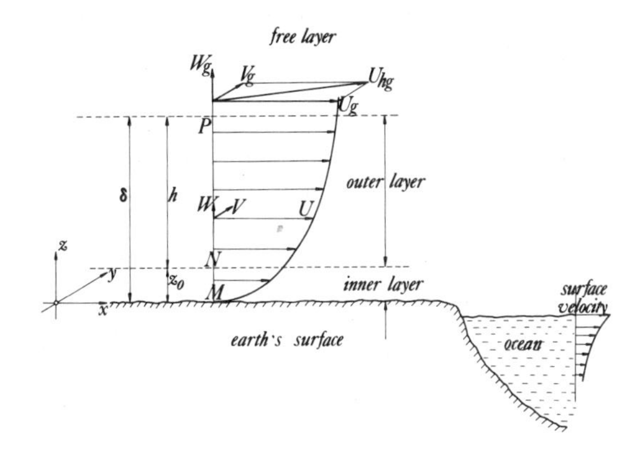

Fig. 9.1 The atmospheric and oceanic boundary layer.

effect on them and the viscous and Reynolds stresses dominate; and an *outer layer* of height h of the order of a kilometer where Coriolis, pressure, and Reynolds stresses dominate. Naturally, because of the relative importance of the various forces, the behavior of these two layers is different. One may wonder at the necessity of breaking the boundary layer into these two regions instead of solving Eqs. (9.1) directly for the entire boundary layer, subject to the boundary conditions on the ground and at the geostrophic layer. Because these partial differential equations have no known exact solutions in analytic form, except for computer-oriented numerical solutions, the subdivision of the boundary layer is for convenience.

The boundary layer near the surface of the ocean due to the action of the wind shear does not display the same characteristic insofar as the inner layer is concerned. There is no region in that oceanic layer where the Coriolis is not important and where the viscous and Reynolds stresses completely dominate the motion. The oceanic boundary layer is more like the outer layer of the atmosphere; it was first studied in detail by Ekman, after whom this layer is named.

9.3 THE EKMAN LAYER OF THE OCEAN—PURE DRIFT CURRENTS

V. W. Ekman (1905) was the first to develop this analysis. He considered an ocean infinite in extent and very deep compared to the thickness of the frictional layer generated by the steady blowing of wind over the surface. In reality the depth of the ocean need not exceed a few hundred meters for the motion induced by the wind shear to become negligible. Even though Fig. 1.5 showed that the density of the ocean varied near the surface, it can be shown that there is little to gain in an analysis maintaining variable density. Ekman considered constant density in his analysis, which now follows. The curvature of the earth is neglected. The surface of the sea is considered flat so that the isobars are horizontal, which also implies the absence of waves. We shall see later that such an assumption will lead to solutions whereby water is piled up in the same directions as long as the wind remains unchanged and that this will necessitate the free surface and the isobars to be inclined slightly away from the horizontal. We shall resolve this problem at the proper time.

Starting with Eqs. (9.1) the question is raised as to whether we are justified in omitting the turbulent Reynolds stresses in the frictional layer of the ocean, as Ekman did. Since the Reynolds number of the motion controls the effects of turbulence, we must investigate it. When comparing the atmospheric and oceanic frictional layers we find that the thickness of the frictional layer in the ocean is approximately 30 times smaller than that in the atmosphere, and the maximum velocity in the oceanic layer is at least one-tenth that of the atmospheric layer. This produces, for a kinematic viscosity of the water approximately one-tenth that of the air, a Reynolds number of the oceanic frictional layer one-thirtieth that of the atmosphere. These figures are approximate but they convey the idea of much lower Reynolds numbers in the ocean. Even though we cannot say that the ocean frictional layer is ever laminar, we can say, however, that the total stress in the parentheses of Eqs. (9.1) can be replaced by $\nu_T\, \partial U/\partial z$ and $\nu_T\, \partial V/\partial z$ where ν_T is an eddy viscosity depending on the turbulent state of the motion and larger than the kinematic molecular viscosity.

Thus we start with the equations of motion

$$
\begin{aligned}
fV &= \frac{1}{\rho}\frac{\partial P}{\partial x} - \nu_T \frac{\partial^2 U}{\partial z^2} \\
fU &= -\frac{1}{\rho}\frac{\partial P}{\partial y} + \nu_T \frac{\partial^2 V}{\partial z^2} \qquad (9.2) \\
g &= -\frac{1}{\rho}\frac{\partial P}{\partial z}
\end{aligned}
$$

and consider ν_T constant in the frictional layer. Ekman went through considerable explanations justifying why he should omit the horizontal pressure gradients as being those imposed by the atmospheric conditions and two small to be considered. This is not far from the truth. However, this was not necessary. The boundary conditions of the frictional layer state that at a depth $z = z_0$ the frictional layer produced by the wind stresses at the surface will terminate and beyond this point the motion of the ocean will be purely geostrophic in the absence of viscous forces. Below that depth z_0

$$fV_g = \frac{1}{\rho}\frac{\partial P}{\partial x}, \qquad fU_g = -\frac{1}{\rho}\frac{\partial P}{\partial y}, \qquad g = -\frac{1}{\rho}\frac{\partial P}{\partial z} \tag{9.3}$$

where U_g and V_g are the geostrophic velocities at $z > D$. Now it suffices to say that since $\partial P/\partial z$ depends on g only and is constant, $\partial P/\partial x$ and $\partial P/\partial y$ do not depend on z. Thus the horizontal pressure gradients in Eq. (9.3) are identical to those in Eq. (9.2), and after substitution we eliminate the pressure terms:

$$\begin{aligned} f(V - V_g) &= -\nu_T \frac{\partial^2 U}{\partial z^2} \\ f(U - U_g) &= \nu_T \frac{\partial^2 V}{\partial z^2} \end{aligned} \tag{9.4}$$

Since U_g and V_g vary little with z, we could have replaced $\partial^2 U/\partial z^2$ by $\partial^2(U - U_g)/\partial z^2$ and analogously for V. Thus we should have a pair of equations in terms of velocities in the frictional layer relative to the geostrophic flow outside the layer. Replacing $(V - V_g)$ by $\overline{V}$ and $(U - U_g)$ by $\overline{U}$, we have

$$\begin{aligned} f\overline{V} &= -\nu_T \frac{d^2\overline{U}}{\partial z^2} \\ f\overline{U} &= \nu_T \frac{d^2\overline{V}}{\partial z^2} \end{aligned} \tag{9.5}$$

where for small geostrophic velocities, these equations amount to having ignored the horizontal pressure gradients in the first place. The total derivatives in Eq. (9.5) replace the partial derivatives in Eq. (9.4) since $\overline{U}$ and $\overline{V}$ are only functions of z. Letting

$$a = (f/2\nu_T)^{1/2} \tag{9.6}$$

the equations of motion become

$$\frac{d^2\overline{U}}{dz^2} + 2a^2\overline{V} = 0, \qquad \frac{\partial^2\overline{V}}{\partial z^2} - 2a^2\overline{U} = 0 \tag{9.7}$$

where a is considered constant in the frictional layer. The general solutions of this pair of ordinary differential equations are

$$\begin{aligned} \overline{U} &= C_1 e^{az}\cos(az + c_1) + C_2 e^{-az}\cos(az + c_2) \\ \overline{V} &= C_1 e^{az}\sin(az + c_1) - C_2 e^{-az}\sin(az + c_2) \end{aligned} \tag{9.8}$$

and the pressure is determined from the hydrostatic equation:

$$P = \rho g z$$

Since it is necessary that $\overline{U}$ and $\overline{V}$ go to zero at a finite depth $z = D$, the first terms in e^{az} cannot be admissible in the solution. Therefore we must conclude that $C_1 = 0$, and

$$\begin{aligned} \overline{U} &= C_2 e^{-az}\cos(az + c_2) \\ \overline{V} &= -C_2 e^{-az}\sin(az + c_2) \end{aligned} \tag{9.9}$$

The coefficient C_2 can be evaluated by forming the resultant horizontal velocity $\overline{U}_h = (\overline{U}^2 + \overline{V}^2)^{1/2}$

$$\overline{U}_h = C_2 e^{-az} \tag{9.10}$$

Since at $z = 0$ the horizontal velocity must equal C_2, we let $\overline{U}_{h0} = C_2$ at the surface of the ocean. We shall see later how we can relate $\overline{U}_{h0}$ to the shear stress at the interface.

To compute the constant c_2 in the trigonometric argument in Eq. (9.9) we orient the coordinate system so that the wind and its stress τ_0 at the water surface are in the y direction, giving

$$\mu_T \left(\frac{d\overline{U}}{dz}\right)\bigg|_{z=0} = 0 \quad \text{and} \quad \tau_0 = -\mu_T \left(\frac{d\overline{V}}{dz}\right)\bigg|_{z=0} \tag{9.11}$$

From Eq. (9.9)

$$\frac{d\overline{U}}{dz} = -a\sqrt{2}\,\overline{U}_{h0} e^{-az}\sin(az + c_2 + 45^\circ)$$

$$\frac{d\overline{V}}{dz} = -a\sqrt{2}\,\overline{U}_{h0} e^{-az}\cos(az + c_2 + 45^\circ)$$

and we conclude that, for $d\overline{U}/dz = 0$ at $z = 0$, $c_2 = -45^\circ$. Thus the final form of the velocity distributions in depth is

$$\begin{aligned} \overline{U} &= \overline{U}_{h0} e^{-az}\cos(45^\circ - az) \\ \overline{V} &= \overline{U}_{h0} e^{-az}\sin(45^\circ - az) \end{aligned} \tag{9.12}$$

Through the shear stress in Eq. (9.11) we can find a relation between it and the surface velocity by taking the derivative of the second equation in (9.12). Thus we obtain

$$\overline{U}_{\mathrm{h0}} = \frac{\tau_0}{\mu_{\mathrm{T}} a \sqrt{2}} \tag{9.13}$$

implying that the surface current is proportional to the wind stress. The expressions in Eq. (9.12) represent the vertical distributions of the horizontal velocities in the *Ekman layer* of the ocean.

The following conclusions are derived from these solutions. First, due to the quantity e^{-az} the velocity decreases rapidly with z and the coefficient that determines the rate of decrease is a which is a function of latitude and eddy viscosity according to Eq. (9.6). Second, and most important, is that on the ocean surface $\overline{U} = \overline{U}_{\mathrm{h0}} \cos 45°$ and $\overline{V} = \overline{U}_{\mathrm{h0}} \sin 45°$ for a wind and its stress in the y direction. Looking at Fig. 9.2a if we place the wind velocity in the y direction, $\overline{U}$ and $\overline{V}$ at the surface, we see that the surface ocean current given by $\overline{U}_{\mathrm{h0}}$ is at 45° clockwise to the direction of the wind near the surface in

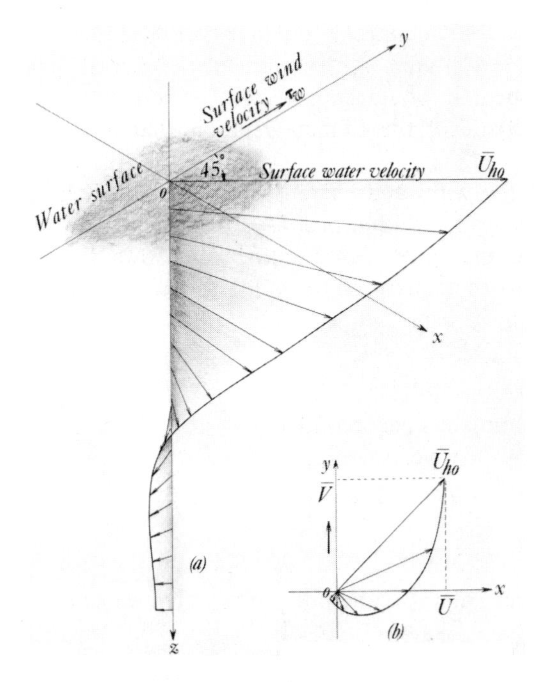

Fig. 9.2 (a) Direction and magnitude of ocean current generated by wind (Ekman spiral). (b) Projection on x–y plane.

the Northern Hemisphere. In the Southern Hemisphere it will be 45° to the left of the wind direction if we look in the direction of the arrow. This surprising shift in angle between the wind direction and the surface current direction was noticed as early as the turn of the century by F. Nansen of the Norwegian North Polar Expedition who observed that the drift of ice with respect to the wind was between 20 and 40° to the right of an observer looking in the wind direction. Third, with increasing z the magnitude of the current velocity decreases but at the same time it rotates clockwise because $\overline{U}$ increases with increasing z in the cosine and $\overline{V}$ decreases with increasing z in the sine. Figure 9.2b shows the resultant velocity of Eq. (9.12) plotted with z. A horizontal projection of the endpoints of these velocity vectors gives what is known as the *Ekman spiral.*

At a depth of $z = \pi/a$ the speed of the current is reduced to $e^{-\pi}$ which is $\frac{1}{23}$ of the value at the surface and the direction of the current is exactly opposite to that at the surface since the arguments in Eq. (9.12) have changed sign. For all practical purposes we can define the thickness of the Ekman layer as

$$z_0 = \pi/a = \pi(2\nu_T/f)^{1/2} \tag{9.14}$$

and we see immediately that this thickness is inversely proportional to the square root of the latitude parameter f and proportional to the square root of the eddy viscosity. In the extreme laminar case and at a latitude of $\psi = 45°$, with viscosity $\mu_T = 100$ g/cm·s, the depth of the Ekman layer is about 50 m. With turbulent shear it would be expected to increase to greater depths. The peculiar characteristic of the Ekman depth is that at the equator it reaches infinite depths.

The ratio of the wind and current speeds is called the *wind factor*. Some empirical approaches have been proposed to evaluate the wind factor and the thickness of the Ekman layer. Since in turbulent flows the friction can be taken as proportional to the square of the velocity, the wind stress in terms of the wind speed U_w can be expressed as

$$\tau_0 = 3.2 \times 10^{-6} U_w^{\ 2}$$

where U_w is in centimeters per second and τ_0 is in dynes per square centimeter. Then using Eq. (9.13) we can write

$$\frac{\overline{U}_{h0}}{U_w} = \frac{3.2 \times 10^{-6} U_w}{(f\mu_T\rho)^{1/2}}$$

Introducing typical values for μ_T and ρ for water we can write approximately that the *wind factor* is

$$\frac{\overline{U}_{h0}}{U_w} = \frac{2.65 \times 10^{-3} U_w}{(\sin\psi)^{1/2}}$$

However, as a result of a number of measurements (Sverdrup, 1942; Shuleikin, 1953), the ratio of the surface current to the wind velocity seems to obey a much simpler expression independent of U_w:

$$\frac{\overline{U}_{h0}}{U_w} = \frac{0.0127}{(\sin \psi)^{1/2}} \tag{9.15}$$

and from these two expressions we can establish that the eddy viscosity $\mu_T = cU_w^2$ in the Ekman layer, and with the help of Eq. (9.14) z_0 becomes

$$z_0 = 7.6 \frac{U_w}{(\sin \psi)^{1/2}} \tag{9.16}$$

where U_w is in meters per second and z_0 in meters. This is in fair agreement with observed values on the ocean.

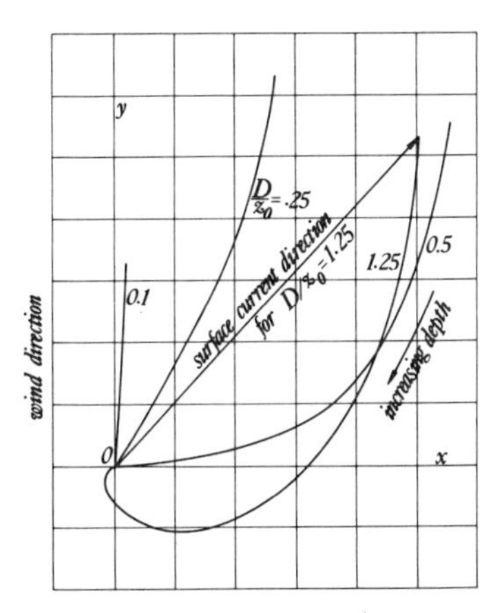

Fig. 9.3 Hodograph of current velocities projected on a horizontal plane.

Ekman also studied the time necessary for this pure drift to develop into a steady motion after the winds begins to blow. He discovered that at the start of the wind, the current velocity is in the direction of the wind velocity and that slowly, in a matter of hours, the current velocity drifts toward its 45° position (steady-state orientation), oscillating in time around 45° with ever decreasing angle increments. This implies that small overshoots beyond 45° are possible.

Ekman also studied the influence of the ocean depth on the thickness of the frictional layer z_0. Figure 9.3 shows the profile of Ekman spirals on a horizontal plane for various values of D/z_0, where D is the depth of the ocean floor. The model is still the same as that in Fig. 9.2 with the wind blowing in the y direction. On this figure the $D/z_0 = 1.25$ curve is already very close to the $D/z_0 = \infty$ line of Fig. 9.2b. Again the current velocity vector for any of the cases shown in Fig. 9.3 is obtained by drawing a line from the origin to any point on the curve corresponding to a specific depth. It is clear that the depth of the ocean has a significant influence on the spiral for depths less than $1.25z_0$.

The pure drift currents developed by Ekman's analysis lead to the conclusion that water is transported continuously in the horizontal direction. This suggests that there will be a piling up of water in regions of convergence and a reduction of level in regions of divergence. We conclude immediately that the isobars will be inclined to the horizontal surfaces and that the horizontal pressure gradients cannot be ignored. We shall return to this discussion on horizontal pressure gradients in the following sections. The influence of density variations with depth on the frictional layer was also considered by Ekman and is presented in his paper (Ekman, 1906). His results show that density stratification caused either by the sun's heating or by vertical salinity gradients produces significant departures from his classical model.

Illustrative Example 9.1

Physically, how can we explain the velocity vector's seemingly strange rotation with depth, namely, turning clockwise in direction as its magnitude decreases?

The explanation of this behavior can be followed with the help of Fig. 9.4. We start with the equilibrium of a geostrophic current which we studied in the preceding chapter. The important characteristic of the geostrophic motion was

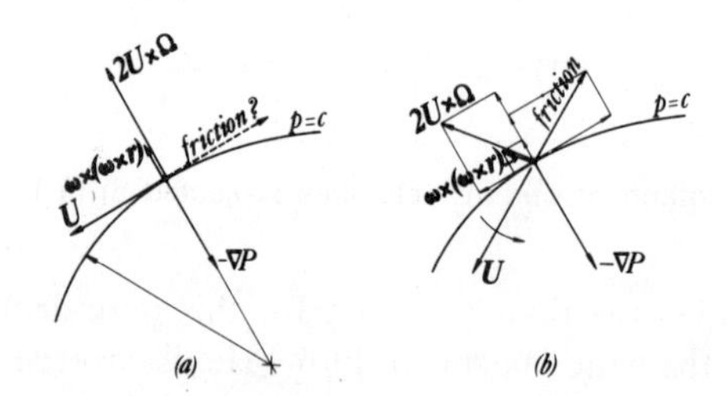

Fig. 9.4 The effect of friction on the rotation of the velocity vector.

that the motion was always tangential to the isobar, or in other words, the pressure gradient was always perpendicular to the velocity vector. This situation occurred in the absence of friction. Thus in the equilibrium of the geostrophic current *all* the forces, namely centrifugal, Coriolis, and pressure, are lined up along the radius of curvature $\boldsymbol{r}$ of the isobar.

Let us imagine that viscosity can be turned on, or more appropriately, that the current has penetrated a viscous region. We know that the effect of viscosity will be to reduce the momentum of the flow, and the viscous stress which we should add to our equilibrium must appear in the opposite sense as the motion given by $\boldsymbol{U}$. If we change nothing in the equilibrium diagram but only add a tangential shear, it is clear that there will be no other force to balance it for equilibrium.

This momentarily unbalanced situation must produce an effect on the environment to rectify the lack of equilibrium and to reestablish it. The only way to reestablish equilibrium is by veering the motion so as to produce a tangential force component to balance the friction, which will veer also. So, of the two choices of rotating the vector $\boldsymbol{U}$, only one is admissible. This is shown in Fig. 9.4b. Rotating $\boldsymbol{U}$ the right amount will produce a radial as well as a tangential equilibrium. Rotation of $\boldsymbol{U}$ into regions of larger radii of curvature of the isobar will make the situation worse.

9.4 THE ATMOSPHERIC BOUNDARY LAYER—VERTICALLY STABLE

In Section 9.2 we began to discuss the atmospheric boundary layer shown schematically in Fig. 9.1, and for a vertically stable layer we set up the equilibrium equations (9.1). We gave reasons for dividing the boundary layer into an *inner layer* and an *outer layer*.

We saw in the preceding section that the Reynolds number of the atmospheric boundary layer is at least 30 times larger than that of the ocean. Besides, the atmospheric boundary layer over land is significantly influenced by the terrain roughness, which we shall call k, as an average height of roughness. Thus, the internal structure and development of the atmospheric boundary layer are not only affected by internal shearing properties influenced by a larger Reynolds number but also by the roughness at the boundary. Since the atmospheric boundary layer will be strongly affected by turbulence, we shall define a characteristic turbulent velocity as $u_* = (\tau_0/\rho)^{1/2}$ where τ_0 is the shearing stress at the ground. This velocity is also called the *shearing velocity*. When we look into the turbulent behavior of these boundary layers we must be able to distinguish between effects due to the surface roughness and characterized by a *roughness Reynolds number* defined as $u_* k/\nu$, and effects due to dynamics in the interior of the boundary layer characterized by a *turbulent Reynolds number* defined as $u_* \delta/\nu$, where δ is a characteristic dimension in that layer. We shall solidify these concepts as we proceed.

Consider the atmospheric neutral boundary layer of Fig. 9.1 with the motion outside the layer being geostrophic. The subscript g will again designate the geostrophic state. In this free layer where viscosity has a negligible role, the horizontal equations of motion are the same as in Eq. (9.3). In the boundary layer since we could not neglect the Reynolds stress we must start with Eq. (9.1) and as in Section 9.3 after eliminating the pressure by substituting the geostrophic terms, we have

$$
\begin{aligned}
f(V - V_g) &= -\frac{d}{dz}\left(\nu \frac{dU}{dz} - \overline{uw}\right) \\
f(U - U_g) &= \frac{d}{dz}\left(\nu \frac{dV}{dz} - \overline{vw}\right)
\end{aligned}
\tag{9.17}
$$

This is similar to Eq. (9.4) except that it contains Reynolds stresses and thus there is no need to replace the molecular viscosity by the eddy viscosity. The arguments that the horizontal pressure gradients in a neutral atmosphere are not functions of altitude z were given in Section 9.3.

To solve Eqs. (9.17) we shall need the concept of the *inner layer*, of height z_0, in which the velocities are too small for the Coriolis to have a significant effect with respect to the total shear stress terms on the right-hand side of Eqs. (9.17). Thus in the inner layer, we conclude that the total shear is constant in the layer:

$$
\frac{d}{dz}\left(\nu \frac{dU}{dz} - \overline{uw}\right) = \frac{d}{dz}\left(\nu \frac{dV}{dz} - \overline{vw}\right) = 0
$$

and consequently

$$
\nu \frac{dU}{dz} - \overline{uw} = \text{constant} \qquad \text{and} \qquad \nu \frac{dV}{dz} - \overline{vw} = \text{constant} \tag{9.18}
$$

In the *outer layer* the molecular friction is too small compared with the Reynolds stress and the Coriolis so that

$$
f(V - V_g) = \frac{d}{dz}\overline{uw} \qquad \text{and} \qquad f(U - U_g) = -\frac{d}{dz}\overline{vw} \tag{9.19}
$$

Furthermore, in seeking solutions to Eqs. (9.18) and (9.19) we shall look for a special class of solutions appropriate to boundary layers that are *universal*, *affine*, or *self-preserving*. This implies that if we can find a velocity and a length scale characteristic of the flow in each layer, then the solutions for the velocity distribution in z, which are made dimensionless with these characteristic scales, will be universal for *all* planetary boundary layers. The question that needs to be answered is whether there exists such characteristic scales and if so, what are they? If there exists such characteristic scales to make the universal solutions good for all planetary boundary layers, then these scales must embody all internal and external influences on the layers.

The general study of self-preserving turbulent flows[1] can be found in a

[1] See, for instance, Townsend (1956).

number of treatises on turbulent shear flows. We shall assume, as experiments verify, that the shearing velocity is a characteristic velocity scale for these flows, in the inner and outer layers. However, the characteristic length will be different in the two layers.

A The Outer Layer

We shall treat the *outer layer* first, and for its characteristic length scale we shall take u_*/f which must be proportional to h, the thickness of the outer layer as shown in Fig. 9.1. Intuitively, we can justify this length scale as being characteristic because it contains u_*, which is a measure of the turbulent velocities, and f the Coriolis parameter, and both are influential in Eq. (9.19) for the outer layer. Thus, it is convenient to express the height of the outer layer $h = cu_*/f$ where c is a constant of the order of unity.

If we divide both sides of Eq. (9.19) by u_*^2, the characteristic velocity squared, we have

$$\frac{V - V_g}{u_*} = \frac{d}{d(zf/u_*)}\left(\frac{\overline{uw}}{u_*^2}\right)$$
$$\frac{U - U_g}{u_*} = -\frac{d}{d(zf/u_*)}\left(\frac{\overline{vw}}{u_*^2}\right) \tag{9.20}$$

Examining these equations, we conclude that their solution and particularly the solution for the mean velocities must depend on one independent dimensionless variable zf/u_*, or z/h, the reciprocal of the *friction Rossby number* according to the definition of Eq. (7.42). The frictional shearing velocity has been taken here for the characteristic velocity. Thus we can state, by letting $\eta = zf/u_*$,

$$\frac{U - U_g}{u_*} = F_1\left(\frac{zf}{u_*}\right) = F_1(\eta)$$
$$\frac{V - V_g}{u_*} = F_2\left(\frac{zf}{u_*}\right) = F_2(\eta) \tag{9.21}$$

Notice that the exact solutions of the preceding example, for the Ekman layer of the ocean given in Eq. (9.12), also obey the general law of the outer layer given in Eq. (9.21). The characteristic velocity in that case is obviously the surface current $\overline{U}_{h0}$ and from Eq. (9.13) we can see that a is inversely proportional to $\overline{U}_{h0}$ for a given wind condition. We conclude that the general mean velocity solutions of Eq. (9.21) are derived from Eq. (9.19) which are similar to Eq. (9.4) for the ocean except that in the atmospheric layer the Reynolds stress balances the Coriolis. It is easily concluded that if we would approximate the Reynolds stresses by an eddy viscosity times the derivative of the velocity with z, then the atmospheric outer layer will be an Ekman layer with velocity vectors rotating clockwise in a spiral going in the positive z direction, i.e., increasing elevation.

In the case of the ocean there was a reason to plot the Ekman spiral as in Fig. 9.3 because the wind velocity and the surface current velocity have different directions. However, in the case of the atmosphere it is customary to represent the same spiral by laying the geostrophic wind velocity at the end of the outer layer along the abscissa. Figure 9.5a represents the outer layer spiral in the (x, y) coordinate system and Fig. 9.5b shows the same spiral in the conventional atmospheric representation.

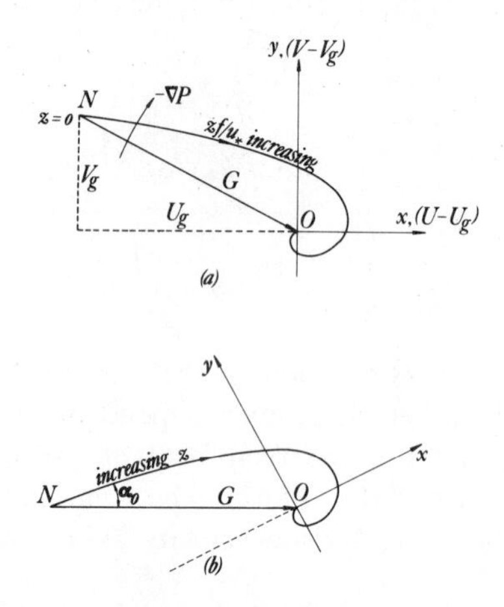

Fig. 9.5 Velocity hodograph in the outer layer of the boundary layer.

The point N in these figures is the end of the inner layer and the beginning of the outer layer. This point is also shown in Fig. 9.1 at $z = z_0$. The maximum value of z is at the edge of the boundary layer where $z = \delta$ and there, both $(U - U_g)$ and $(V - V_g)$ must go to zero, but the geostrophic velocity will be $G = (U_g^2 + V_g^2)^{1/2}$. The vectors drawn from N to any point on the spiral give the velocity vector of the wind at the altitude z corresponding to the endpoint on the spiral. Thus near the ground the velocity vector is tangent to the spiral at N, and the angle α_0 is the *geostrophic deviation* between the surface wind and the geostrophic wind. This angle is also called the *cross-isobar angle* of the outer layer, and it varies with ground roughness, latitude, and the vertical stability of the atmosphere. The dependence on ground roughness and latitude is not surprising because the independent variable in Eq. (9.21) is zf/u_*, where f is a function of latitude and u_* is a function of the ground friction. Table 9.1 gives typical cross-isobar angles as observed at three latitudes and for various ground roughnesses and vertical equilibria. Equation (9.37) and Fig. 9.7 provide further basic information on this angle.

TABLE 9.1

Observed Values[a] of Cross-Isobar Angles, α_0

Latitudes		Type of surface			
		Ocean	Smooth land	Average land	Rugged land
20°	Unstable	25	35	40	45
	Neutral	30	40	45	50
	Stable	40	50	55	60
45°	Unstable	15	25	30	35
	Neutral	20	30	35	40
	Stable	30	40	45	50
70°	Unstable	10	20	25	30
	Neutral	15	25	30	40
	Stable	25	35	40	45

[a] From "Dynamical and Physical Meterology" by G. J. Haltiner and F. L. Martin. Copyright 1957, McGraw-Hill. Used with permission of McGraw-Hill Book Co.

B The Inner Layer—Smooth Surface

The *inner layer*, as shown in Fig. 9.1, is in the region $z < z_0$. There are two essential differences in this layer when compared with the outer layer. First the characteristic length must be z_0 and not h or u_*/f. Second, the Coriolis has no influence on this layer. To show this we multiply both sides of Eq. (9.17) by z_0/u_*^2 to obtain

$$\begin{aligned} \frac{fz_0}{u_*}\left(\frac{V - V_g}{u_*}\right) &= -\frac{1}{u_*^2}\frac{d}{d(z/z_0)}\left(\nu\frac{dU}{dz} - \overline{uw}\right) \\ \frac{fz_0}{u_*}\left(\frac{U - U_g}{u_*}\right) &= \frac{1}{u_*^2}\frac{d}{d(z/z_0)}\left(\nu\frac{dV}{dz} - \overline{vw}\right) \end{aligned} \tag{9.22}$$

In this layer the left-hand side of these equations is at most of the order of $fz_0 U_g/u_*^2$. Since u_*/f is of the order of h, the thickness of the outer layer, the Coriolis terms in the inner layer are of the order of $z_0 U_g/hu_*$. Typically, in the atmospheric boundary layer h is approximately 1000 m while z_0 is of the order of 0.1 m. The ratio U_g/u_* is approximately 30 which makes the Coriolis terms in the inner layer of the order of 3×10^{-3}. Since $|\overline{uw}|$ is of the order of u_*^2 and z is of the order of z_0 the stress terms on the right-hand side of Eq. (9.22) are of the order of unity. Thus, in the inner layer the Coriolis is negligible, which means two separate things: the velocity vector in the inner layer does not change direction with altitude as in the Ekman layer, and the shearing stress in this layer is constant with altitude. *The inner layer is a constant stress layer.* Then we can write

$$\frac{d}{dz}\left(\nu \frac{dU}{dz} - \overline{uw}\right) = 0$$

Since there is no change of angle with z in this layer there is no need to keep two velocity components going. We can orient the coordinate to have only one velocity. Integrating,

$$\nu \frac{dU}{dz} - \overline{uw} = \text{constant} \tag{9.23}$$

Since one of the limits of the integration is on the ground where the stress is ρu_*^2, the constant of integration must be u_*^2. Thus

$$\nu \frac{dU}{dz} - \overline{uw} = u_*^2$$

In dimensionless terms,

$$\frac{d}{dz_*}\left(\frac{U}{u_*}\right) - \frac{\overline{uw}}{u_*^2} = 1 \tag{9.24}$$

where $z_* = zu_*/\nu$, a new dimensionless variable. So from Eq. (9.24) we see that there is only one independent variable in the ordinary differential equation and consequently the general solutions of the mean velocity and the Reynolds stress must be of the form

$$U = u_* F_3(z_*) \tag{9.25}$$

and

$$-\overline{uw} = u_*^2 G_1(z_*) \tag{9.26}$$

These general relations, together with the ones obtained for the outer layer in Eq. (9.21), are specified later.

C Inner Layer—Rough Surface

In most cases the roughness on the ground, such as buildings, forests, and mountains, is larger than the height of the smooth inner layer. We have indicated that normally z_0 is of the order of centimeters. When the *roughness parameter* k measured as a root mean square (rms) value of the land roughness, is of the order of z_0 or larger, then it is obvious that in the considerations which led to Eqs. (9.25) and (9.26) on the basis of the length scale ν/u_*, we must introduce a second length scale k. The ratio of these two length scales represents the *roughness Reynolds number*, $\mathscr{R}_k = ku_*/\nu$.

TABLE 9.2[a]

Type of earth surface	Roughness parameter, k (cm)	Eddy viscosity and diffusion coefficient 10^3 cm²/s	Shearing velocity, u_* (cm/s)
Exceptionally smooth (mud flats, ice, etc.)	1×10^{-3}	1.3	16
Smooth sea	2×10^{-2}	1.7	21
Level desert	3×10^{-2}	1.8	21
Lawn (grass height, 1 cm)	10^{-1}	2.2	27
Lawn (grass height 5 cm) also land without vegetation, winter	1–2	3.4	43
Long grass, 60 cm	4–9	4.8	60
Fully grown crops	14	5.6	70

[a] These values were published by Sutton (1949b) and adopted by Pasquill (1968).

Thus Eq. (9.25) must be replaced by

$$U/u_* = F_4(z_*, \mathscr{R}_k) \tag{9.27}$$

or

$$U/u_* = F_4(z/k, \mathscr{R}_k) \tag{9.28}$$

The outer layer is independent of k since, in most cases, $k/h \ll 1$.

Table 9.2 gives some typical values of roughness parameters, eddy viscosity and diffusion coefficients, and shearing velocities for various land roughnesses.

D Determination of the Functions F_1, F_2, F_3, F_4

The two separate regions of the boundary layer can now be identified as follows:

$$\text{Inner layer,} \qquad z_* \text{ finite} \quad \text{and} \quad zf/u_* \to 0$$
$$\text{Outer layer,} \qquad z_* \to \infty \quad \text{and} \quad 0 < zf/u_* < 1$$

In order to determine the character of these functions, we must proceed to match them and their derivatives at a point where the inner layer ends and the outer layer begins.

From Eq. (9.25) the velocity derivative in the *smooth inner layer* is

$$\frac{dU}{dz} = \frac{u_*^2}{\nu} \frac{dF_3}{dz_*}$$

and in the *outer layer*, from Eq. (9.21),

$$\frac{dU}{dz} = \frac{u_*}{h} \frac{dF_1}{d\eta}$$

If at a distance $z = z_0$, for $\eta \ll 1$ and $z_* \gg 1$, these derivatives must equal each other to give a continuity in the derivative of the velocity profile in z, we must set

$$\frac{u_*^2}{\nu} \frac{dF_3}{dz_*} = \frac{u_*}{h} \frac{dF_1}{d\eta}$$

Multiplying by z_*/u_* we have

$$z_* \frac{dF_3}{dz_*} = \eta \frac{dF_1}{d\eta} = \frac{1}{\kappa} \tag{9.29}$$

where κ is the *von Kármán* constant. Integration of Eq. (9.29) yields two special solutions: for the *smooth inner layer*

$$\frac{U}{u_*} = F_3(z_*) = \frac{1}{\kappa} \ln z_* + B \tag{9.30}$$

and for the *outer layer*

$$\frac{U - U_g}{u_*} = F_1(\eta) = \frac{1}{\kappa} \ln \eta + C$$

$$= \frac{1}{\kappa} \ln \frac{zf}{u_*} + C \tag{9.31}$$

where B and C are constants. For the outer layer, $(V - V_g)$ has a similar relation to Eq. (9.31) but this analysis cannot give us the relationship between

$(U - U_g)$ and $(V - V_g)$ to obtain the hodograph of the Ekman spiral. We discuss this point later.

For the *rough inner layer* we must consider Eq. (9.27) or (9.28) instead of (9.25). Since the character of the outer layer is assumed unchanged with roughness, the right-hand side of Eq. (9.29) must remain unchanged and the left-hand side must be modified by the additive dependence on $\mathscr{R}_k$. Thus the result will be

$$\frac{U}{u_*} = \frac{1}{\kappa} \ln z_* + \phi_1(\mathscr{R}_k)$$

$$= \frac{1}{\kappa} \ln \frac{z}{k} + \phi_2(\mathscr{R}_k) \tag{9.32}$$

We know from experience that, for a smooth surface $\mathscr{R}_k \to 0$, $\phi_1(\mathscr{R}_k) \to 5$, and that for $\mathscr{R}_k < 5$ the surface can be considered smooth because k is then of order much less than z_0 and ϕ_1 and ϕ_2 have no effect on Eq. (9.32). Coming back to Eq. (9.24) and multiplying it by k/u_*^2 we have with $\mathscr{R}_k = u_* k/\nu$

$$\frac{1}{\mathscr{R}_k} \frac{d(U/u_*)}{d(z/k)} - \frac{\overline{uw}}{u_*^2} = 1$$

Then for $\mathscr{R}_k \to \infty$ the viscous stress is much smaller than the Reynolds stress for z/k of the order of unity. From this we conclude that $\overline{uw}$ is nearly constant and that the velocity derivative is practically independent of $\mathscr{R}_k$ which according to Eq. (9.32) ϕ_2 must be constant. This occurs in practice for $\mathscr{R}_k > 30$. The physical interpretation of this is that the roughness elements with large $\mathscr{R}_k$ generate turbulent wakes which are responsible for a turbulent drag and not a viscous drag (Tennekes and Lumley, 1972). Therefore Eq. (9.32) becomes, for $\mathscr{R}_k > 30$,

$$\frac{U}{u_*} = \frac{1}{\kappa} \ln \frac{z}{k} + B' \tag{9.33}$$

Since for large roughness the location of $z = 0$ cannot be easily determined, the constant can be ignored and incorporated into the value of k in the logarithm of Eq. (9.33). Thus at $z = k$ the velocity is zero.

The constant B and C in Eqs. (9.30) and (9.31) cannot depend on the Reynolds number since the functions F_1, F_2, and F_3 do not depend on it. In matching the velocities at the point $z = z_0$ we obtain through Eqs. (9.30) and (9.31) for the *smooth surface*

$$\frac{U_g}{u_*} = \frac{1}{\kappa} \ln \frac{z_*}{\eta} + B - C$$

$$= \frac{1}{\kappa} \ln \mathscr{R}_h + B - C \tag{9.34}$$

where $\mathscr{R}_h = u_* h/\nu$, and for the *rough surface*

$$\frac{U_g}{u_*} = \frac{1}{\kappa} \ln\left(\frac{u_*}{fk}\right) + B' - C \tag{9.35}$$

At $z = 0$, because of the boundary $V = 0$ and from Eq. (9.21)

$$\frac{V_g}{u_*} = -F_2(0) = D \tag{9.36}$$

Thus these relations for rough surfaces are valid for large *friction Rossby numbers* u_*/zf. From Eqs. (9.35) and (9.36) we can show that $fkU_g/u_*{}^2 \to 0$ as $u_*/fk \to \infty$.

The cross-isobar angle α_0 can be computed by letting

$$\tan \alpha_0 = -\frac{V_g}{U_g} = \frac{Du_*}{U_g}$$

$$= \frac{D}{(1/\kappa)\ln(u_*/fk) + B' - C} \tag{9.37}$$

Figure 9.6 shows the validity of the analysis just given which led to the solution of Eq. (9.35). Furthermore, since Eq. (9.35) was obtained through a combination of inner and outer layer considerations, we conclude that the velocity distributions in the entire atmospheric boundary layer as given by Eqs. (9.30), (9.31), and (9.33) are representative of the observed data. Furthermore, Fig. 9.7, which is based on Blackadar's (1962) results which seem to fit experimental points from a number of sources, is also a good support for Eq. (9.37). From

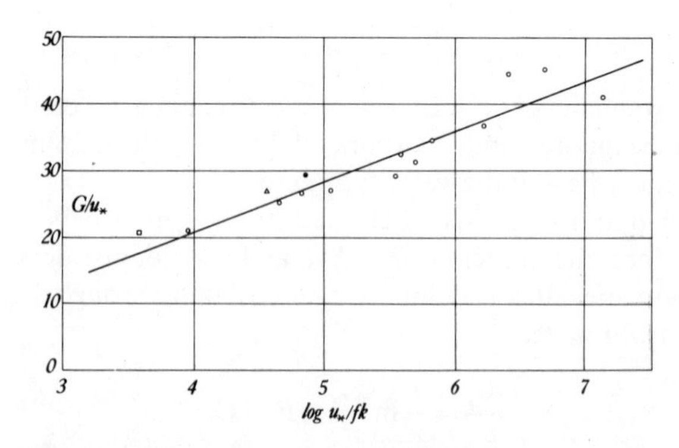

Fig. 9.6 Observed measurements in the outer layer to be compared with Eq. (9.35). [After data published by Blackadar (1962).] ○ Lettau; □ Brookhaven, 7, 1950; △ Brookhaven, 11, 1950; ● Hanford, 1959.

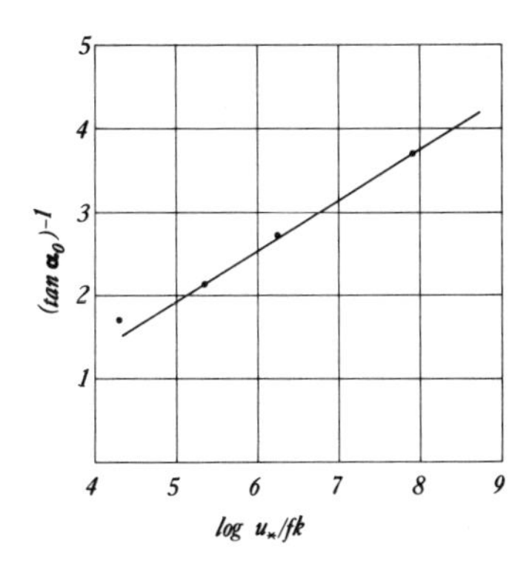

Fig. 9.7 Cross-isobar wind angle from predicted and observed results. (After Blackadar, 1962.)

experimental measurements the constant C is approximately 4 and the von Kármán constant $\kappa = 0.4$. It was suggested that the value of B' for rough surfaces could be made equal to zero if the value of k is adjusted to include that constant in Eq. (9.33).

With known values of k and one or two measurements of U, it is possible to compute the shearing velocity u_*, which is difficult to measure through the shear stress at the surface. From Fig. 9.7 the slope of the line when converted into a natural logarithm relation gives $D = 9.2$. The literature cites values for D as high as 12.

E The Turbulent Spiral in the Outer Layer

The relationship between U and V in the outer layer can only be established through solving Eqs. (9.20). A lack of knowledge of the distribution of the turbulent shear stress in z makes the exact solution unattainable. Because of this the Reynolds stresses are assumed to follow additional empirical relationships. There are many such approaches in the literature on turbulence. However, one of the most common methods employed is the introduction of an eddy viscosity ν_T, as we have done in Section 9.3, but allowing it to be variable in z. Blackadar obtained such a relation and solved for $(U - U_g)$ and $(V - V_g)$ as a function of z in the following manner.

As suggested in the text following Eq. (9.4) the equations of motion were taken as

$$\begin{aligned} f(V - V_g) + \frac{d}{dz}\left[\nu_T \frac{d}{dz}(U - U_g)\right] &= 0 \\ -f(U - U_g) + \frac{d}{dz}\left[\nu_T \frac{d}{dz}(V - V_g)\right] &= 0 \end{aligned} \tag{9.38}$$

Introducing Prandtl's *mixing length hypothesis* (Prandtl, 1932) relating the Reynolds stress to the mean velocity gradient

$$\nu_{\mathrm{T}} = \ell^2 \frac{dU}{dz} \tag{9.39}$$

Based on the experimental values of Lettau (1950), he let the mixing length ℓ be

$$\ell = \frac{\kappa z}{1 + \kappa z/a}$$

where κ is the von Kármán constant introduced in Section 9.4.D and a is the height at which the variation of ℓ with z levels off:

$$a = 27 \times 10^{-5} G/f$$

Then through a numerical method the solutions of Eq. (9.38) were obtained. Figure 9.8 gives the hodographs for a smooth surface corresponding to $k =$

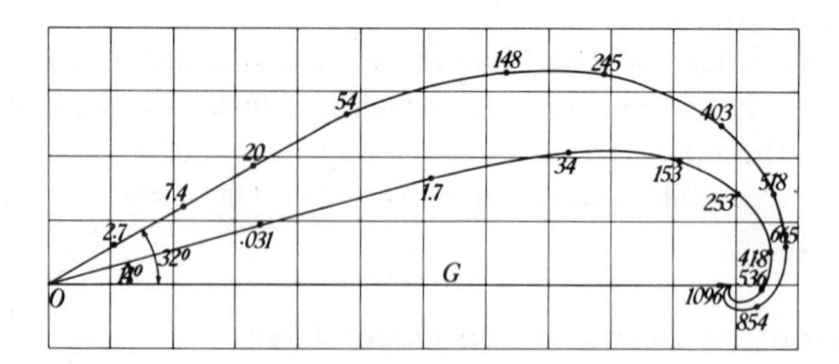

Fig. 9.8 Hodographs for smooth and rough surfaces. (After Blackadar, 1962.)

0.021 cm, and for a rough surface with $k = 106$ cm with the same geostrophic wind $G/f = 10^7$ cm. At the lower boundary the value of the roughness k will determine the value of τ_0 or u_* there. The numbers on the spiral correspond to the elevation from the ground, in meters. Drawing a line from the origin O to any point on the spiral gives the magnitude and direction of the velocity vector at that elevation. The cross-isobar wind direction is 14° for the smooth surface and 32° for the rough surface.

9.5 THE UNSTABLE BOUNDARY LAYER

When the environmental boundary layer is not stable, we must consider buoyancy along the vertical which will influence the horizontal velocity distribution in z as determined in the preceding section.

In Section 7.8 we discussed buoyancy effects at some length. We established that the Richardson number was a criterion for stability and that the Monin–Obukhov length scale was the characteristic vertical length scale of the buoyancy effects. These were

$$\mathscr{Ri} = \frac{g}{T_{\text{ad}}} \frac{\partial T}{\partial z} \bigg/ \left(\frac{\partial U}{\partial z}\right)^2$$

$$= -\frac{\kappa g H}{\rho u_*^3 T_{\text{ad}} c_p} z \tag{9.40}$$

and

$$L_{\text{M}} = -\frac{\rho c_p u_*^3 T_{\text{ad}}}{\kappa g H} \tag{9.41}$$

In the neutral case u_* was the characteristic velocity for the inner and outer layers, k (the roughness) was the characteristic length in the inner layer, and $h = cu_*/f$ was the characteristic length in the outer layer. Now, in the case of buoyancy L_{M} must be another characteristic length of the thermal effects, and we shall take t_* as being the characteristic temperature which must be defined the same way as u_*.

As u_* represents the value of the shearing velocity in the inner layer, i.e., $-\overline{uw} = u_*^2$, the turbulent heat transfer near the ground from the perturbed temperature equation in turbulent motion (Section 10.2.A) and analogously to the momentum, is $H = \rho c_p \overline{w\vartheta}$, where ϑ is the turbulent temperature fluctuation. This is approximately $H = \rho c_p u_* t_*$ and for the characteristic temperature we take

$$t_* = \frac{H}{\rho c_p u_*} \tag{9.42}$$

In the neutral boundary layer we have already computed the velocity derivative through the analysis leading to Eq. (9.29). Thus we can show that where the inner and outer layers meet and from Eq. (9.31), taking $dU/dz = (dU/d\eta)(d\eta/dz)$,

$$\frac{dU}{dz} = \frac{u_*}{\kappa}\left(\frac{u_*}{zf}\right)\frac{f}{u_*} = \frac{u_*}{\kappa z} \tag{9.43}$$

Whereas for the neutral layer $(\kappa z/u_*)(dU/dz)$ is constant with all z, in the thermally unstable layer it must be a function of the dimensionless altitude z/L_{M}, if buoyancy is to have an effect on it. Thus we write

$$\frac{\kappa z}{u_*}\frac{dU}{dz} = \varphi_1\left(\frac{z}{L_{\text{M}}}\right) \tag{9.44}$$

and by analogy

$$\frac{\kappa z}{t_*}\frac{dT}{dz} = \varphi_2\left(\frac{z}{L_{\text{M}}}\right) \tag{9.45}$$

We know from experience that the dimensionless functions φ_1 and φ_2 are the same, say φ. We shall show further on in this analysis that this corresponds to a constant Prandtl number, as defined in Section 7.7. On the basis of Eqs. (9.44) and (9.45) the Richardson number is

$$\mathcal{Ri} = \frac{g}{T}\frac{\partial T/\partial z}{(\partial U/\partial z)^2}$$

$$= \frac{\kappa^2 g}{T}\frac{t_* z}{u_*{}^2 \varphi(z/L_\mathrm{M})} \tag{9.46}$$

Substituting Eqs. (9.41) and (9.45)

$$\mathcal{Ri} = -\kappa \frac{z}{L_\mathrm{M}} \varphi^{-1}\left(\frac{z}{L_\mathrm{M}}\right) \tag{9.47}$$

Also the eddy viscosity and heat diffusion coefficients defined by

$$\nu_\mathrm{T} \frac{\partial U}{\partial z} = -\overline{uw} = u_*{}^2$$

$$\gamma_\mathrm{T} \frac{\partial T}{\partial z} = -\overline{\vartheta w} = u_* t_*$$

can be reduced by Eqs. (9.44) and (9.45) to

$$\nu_\mathrm{T} = \frac{\kappa u_* z}{\varphi_1(z/L_\mathrm{M})}$$
$$\gamma_\mathrm{T} = \frac{\kappa u_* z}{\varphi_2(z/L_\mathrm{M})} \tag{9.48}$$

This shows that when $\varphi_1 = \varphi_2 = \varphi$ we have the condition of the Prandtl number equal to a constant. Although this may not be strictly true in the atmosphere its assumption does not seem to affect the validity of the solutions. Bringing the Richardson number of Eq. (9.47) into the eddy coefficients we see that

$$\nu_\mathrm{T} = \gamma_\mathrm{T} = -\kappa u_* L_\mathrm{M} \mathcal{Ri} \tag{9.49}$$

In these deliberations we must remember that the signs of L_M and t_*, and all expressions related to them depend on the sign of H, the eddy heat flux in the vertical direction. When $H > 0$, implying unstable stratification, $L_\mathrm{M} < 0$ and $t_* > 0$. For stable stratification $H < 0$, $L_\mathrm{M} > 0$, while $t_* < 0$. For the neutral case the Monin–Obukhov length approaches infinity and the Richardson number approaches zero.

As in the preceding section the theory of dimensionless reasoning does not provide the form of $\varphi(z/L_\mathrm{M})$. It is generally assumed that this function can be expanded in a series of increasing positive powers since z/L_M is usually smaller than unity, closer to the ground. In the case of $z/L_M \ll 1$, which implies that either we stay in the vicinity of the ground or L_M is very large, implying heat transfer *near* neutral states, or both, we have

$$\varphi(z/L_M) = 1 + \beta(z/L_M) \tag{9.50}$$

where $\beta = 0.62$ from experimental data of Monin–Obukhov.[2] When we substitute this expression into Eqs. (9.44) and (9.45) we have

$$\frac{dU}{dz} = \frac{u_*}{\kappa z}\left[1 + \beta\left(\frac{z}{L_M}\right)\right]$$

$$\frac{dT}{dz} = \frac{t_*}{\kappa z}\left[1 + \beta\left(\frac{z}{L_M}\right)\right]$$

and in the *inner layer* when we integrate from $z = k$, the rms value of the roughness where $U = 0$ and $T = T_0$, to any point z we obtain

$$\begin{aligned} \frac{U}{u_*} &= \frac{1}{\kappa}\left(\ln\frac{z}{k} + \beta\,\frac{z-k}{L_M}\right) \\ \frac{T - T_0}{t_*} &= \frac{1}{\kappa}\left(\ln\frac{z}{k} + \beta\,\frac{z-k}{L_M}\right) \end{aligned} \tag{9.51}$$

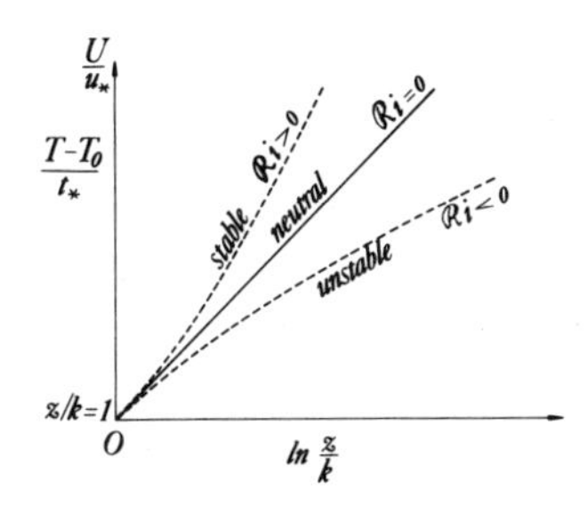

Fig. 9.9 Inner layer wind speed and temperature distributions in the vertical direction as influenced by the Richardson number.

These are the relations sought. It is clear that for the stable environment $L_M \to \infty$ and the correction term in $\beta(z - k)/L_M$ goes to zero, which brings us to the solution of the stable inner layer of Eq. (9.33). For the outer layer the integration must proceed from z to $h \simeq u_*/f$ and this provides a correction for Eq. (9.31).

Figure 9.9 compares the neutral layer vertical distribution of the wind

[2] For a discussion and added measurements of β not in accordance with Eq. (9.49) see Mcbean *et al.* (1971).

velocity and temperature with those of the stable and unstable boundary layers.

Another approach to the solution of U for various stability postures is through the so-called *power law*. Starting from Eqs. (9.44) and (9.45) we can let φ be a power expression of (z/k) in the inner layer

$$\frac{dU}{dz} = \frac{u_*}{k\kappa}\left(\frac{z}{k}\right)^{-m} \tag{9.52}$$

and determine the power m as a function of the Richardson number from experimental measurements. Such measurements were given by Deacon (1949), who showed that only in the case of $m = 1$ does the vertical distribution have a logarithmic profile.

Figure 9.10 shows the dependence of m on the Richardson number and Fig. 9.11 shows in a way similar to Fig. 9.9 the influence of the Richardson

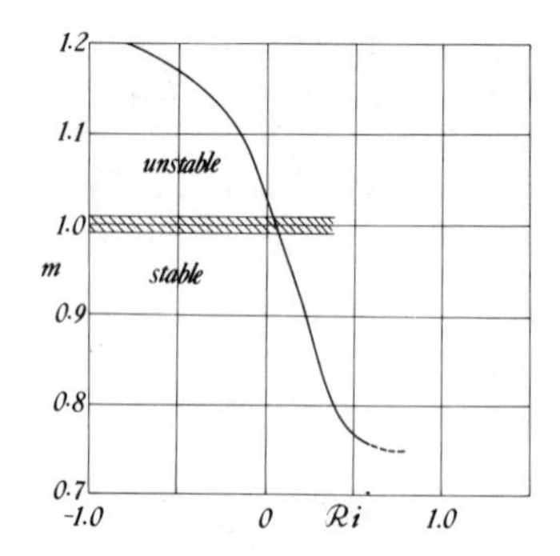

Fig. 9.10 Vertical stability from measured values in the atmosphere with ground surface roughness k approximately 0.25 cm. (After Deacon, 1949.)

number on the velocity profiles. The inner layer solutions are generally valid in the first 10 m from the ground with roughness comparable to 1 cm.

The results of Fig. 9.11 show at first, for an elevation of approximately 40 m, an opposite behavior to the logarithmic distribution in Fig. 9.9. For a stable layer the velocity increases slower with z than in the neutral layer, and at the crossover point P it comes into agreement with Fig. 9.9 insofar as relative increase with altitude is concerned. Brunt (1939) estimates the crossover point for the power law to be about 40 m, with $k = 1$ cm. The values of m in the atmosphere range between 0.8 and 1.2.

A third approach in dealing with vertically nonneutral layers is that of Swinbank (1964) and is called the *exponental law*. In his approach he postulates that instead of the independent variable z there exists a length χ such that Eq. (9.43), which is valid for the neutral layer, is valid for all stratified layers

$$\frac{dU}{d\chi} = \frac{u_*}{\kappa\chi} \tag{9.53}$$

and that the energy[3] in the turbulence owing to mechanical and thermal effects represented by

$$\frac{\tau}{\rho}\frac{dU}{dz} + \frac{gH}{\rho c_p T}$$

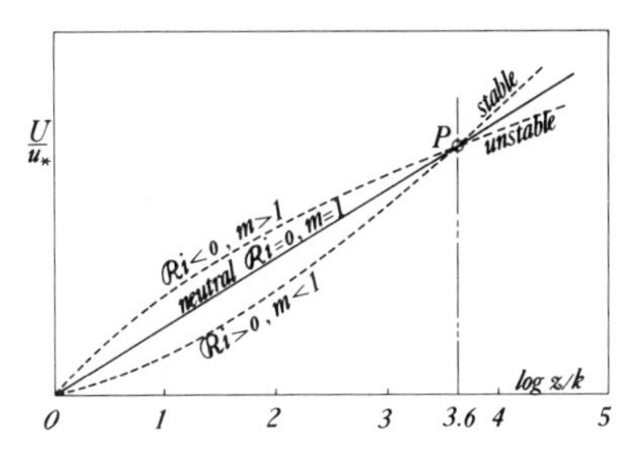

Fig. 9.11 Velocity profiles with the power law of Eq. (9.51) and values of m as in Fig. 9.10.

can be replaced by a single mechanical production term based on the new variable χ which is $(\tau/\rho)\, dU/d\chi$. Thus

$$\frac{\tau}{\rho}\frac{dU}{dz} + \frac{gH}{\rho c_p T} = \frac{\tau}{\rho}\frac{dU}{d\chi} \tag{9.54}$$

This is the differential equation with which we started. The first term on the left-hand side is the energy in the turbulence produced by mechanical means (shear), and the second term is the energy from buoyancy where H, the heat flux, was defined as $-\rho c_p\, dT/dz$ and when entered into Eq. (9.54) gives the numerator in the Richardson number.

[3] Strictly speaking the turbulent energy equation (which we do not develop in this book) should be used for a more formal approach in showing the mechanical and heat production terms.

Introducing the definition of the Monin–Obukhov length L_M in Eq. (7.58), dividing through by the first term, and replacing $(\tau/\rho) = u_*^2$ we obtain

$$1 - \frac{u_*}{\kappa L_M} \frac{dz}{dU} = \frac{dz}{d\chi} \tag{9.55}$$

The chain rule gives us

$$\frac{dz}{dU} = \frac{dz}{d\chi} \cdot \frac{d\chi}{dU}$$

Substitution into Eq. (9.55) and use of Eq. (9.53) gives

$$1 = \left(\frac{\chi}{L_M} + 1\right) \frac{dz}{d\chi}$$

or

$$\frac{d(\chi/L_M)}{d(z/L_M)} - \frac{\chi}{L_M} = 1 \tag{9.56}$$

Integrating we obtain the relationship sought between the buoyancy variable χ and the natural variable z:

$$\frac{\chi}{L_M} = \exp\left(\frac{z}{L_M}\right) - 1 \tag{9.57}$$

Returning to Eq. (9.53) and integrating we obtain the modified logarithmic law with density stratification:

$$\frac{U}{u_*} = \frac{1}{\kappa} \ln\left\{\frac{L_M[\exp(z/L_M) - 1]}{k}\right\} \tag{9.58}$$

As in the other methods the integration was performed for the *inner layer* from $z = k$ (where $U = 0$) to z.

mixing process. At the stack exit and shortly after, characterized as the *first phase*, since the specific momentum and enthalpy of the environment are relatively small compared to those of the plume, the turbulent dynamics and heat transfer of the plume[1] will be primarily controlled by its own properties. In other words, the wind will have little influence on the plume in the first phase. However, far from the stack exit, in the *second phase*, the plume, having slowed and cooled down considerably, will be dominated by the turbulent wind properties. The *intermediate phase*, as we shall see, will also play a major role in the restructuring of the internal dynamical properties of the plume.

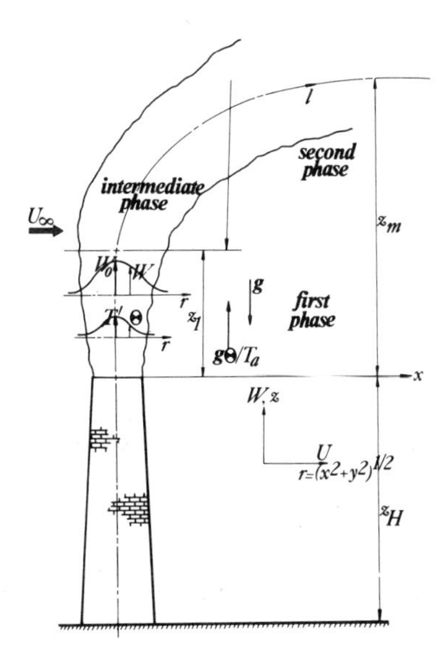

Fig. 10.1 Geometry of the plume.

A The First Phase

We shall treat this phase as if the outside environment were stationary, for reasons already explained. Figure 10.2 is an enlarged view of the development in this phase. All the properties of the plume will be considered turbulent and are subject to the decomposition in Eqs. (7.64) and (7.68), giving the temporal mean and fluctuating parts.

[1] In this analysis we shall consider the plume and the wind to be in turbulent states as there is hardly any practical application where laminar states exist.

Since most thermal plumes originate from stacks that are vertical, in this phase the main motion is along z with velocity $\tilde{w}$, and the spreading of the plume in the radial direction, $r = (x^2 + y^2)^{1/2}$, will contribute to a small but important radial velocity component U. The geometry of this motion and diffusion suggests cylindrical coordinates, and for obvious reasons we could assume axial symmetry. Let T be the temperature of a displaced plume parcel at any point in this field, and T_a the ambient environment temperature, assumed to be the adiabatic temperature. In other words, the environment itself is in neutral

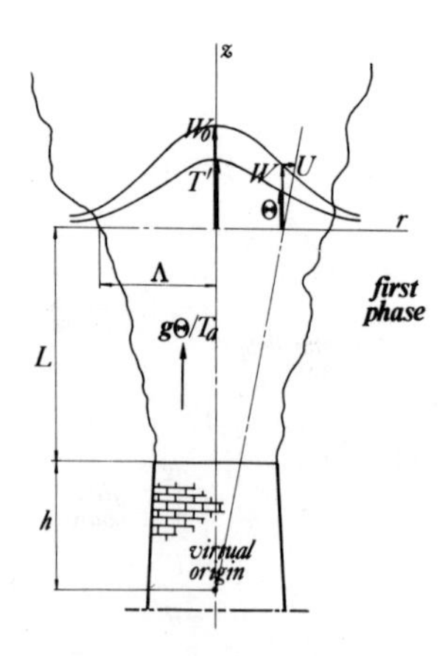

Fig. 10.2 Geometry and properties in the first phase: $\tilde{w} = W + w$; $\tilde{u} = U + u$; $\tilde{\vartheta} = \Theta + \vartheta$; $\tilde{p} = P + p$.

equilibrium. This is not always the case but it is preferable to wait for further developments before adding further complications. The buoyancy in Eq. (4.44) will soon be entered in the z direction in terms of the temperatures, as $g\tilde{\vartheta}/T_a$, where $\tilde{\vartheta} = T - T_a$, because in addition to dynamics we must consider heat diffusion where temperature is the principal variable.

The dynamical equations of motion in the axial and radial directions are respectively,

$$\tilde{u}\frac{\partial \tilde{w}}{\partial r} + \tilde{w}\frac{\partial \tilde{w}}{\partial z} = -\frac{1}{\rho}\frac{\partial \tilde{p}}{\partial z} + \frac{g}{T_a}\tilde{\vartheta} + \frac{\nu}{r}\frac{\partial}{\partial r}\left(r\frac{\partial \tilde{w}}{\partial r}\right) + \nu\frac{\partial^2 \tilde{w}}{\partial z^2}$$

$$\tilde{u}\frac{\partial \tilde{u}}{\partial r} + \tilde{w}\frac{\partial \tilde{w}}{\partial z} = -\frac{1}{\rho}\frac{\partial \tilde{p}}{\partial r} + \nu\left(\frac{\partial^2 \tilde{u}}{\partial r^2} - \frac{\tilde{u}}{r^2}\right) + \nu\frac{\partial^2 \tilde{u}}{\partial z^2} \tag{10.1}$$

As indicated in Section 7.12, the *tilde* above the properties signifies instantaneous values. Because of axisymmetry all terms in $\partial/\partial\theta$ have been omitted, θ being the angle on the plane normal to z.

As the plume moves, the scalar property[2] (the temperature in this case) will diffuse also according to the instantaneous diffusion equation

$$\tilde{u}\frac{\partial\tilde{\vartheta}}{\partial r} + \tilde{w}\frac{\partial\tilde{\vartheta}}{\partial z} = \frac{k}{\rho c_p}\left(\frac{\partial^2\tilde{\vartheta}}{\partial r^2} + \frac{1}{r}\frac{\partial\tilde{\vartheta}}{\partial r} + \frac{\partial^2\tilde{\vartheta}}{\partial z^2}\right) \tag{10.2}$$

where k is the molecular heat conductivity while c_p is the specific heat at constant pressure. The terms on the left-hand side are the convective terms while the ones on the right are the molecular diffusion terms. As one could guess at this point the latter will be shown to be unimportant in the turbulent state of the plume. We introduce again the turbulence decomposition of temporal mean and fluctuating values, while we have assumed all along that the temporal mean properties are not functions of time when the mean value is taken with reference to an averaging time of the order of the largest time scale in the turbulence:

$$\begin{aligned} \tilde{u} &= U + u, & \tilde{w} &= W + w \\ \tilde{p} &= P + p, & \tilde{\vartheta} &= \Theta + \vartheta \end{aligned} \tag{10.3}$$

The continuity equation will relate $\tilde{u}$ to $\tilde{w}$ for conservation of mass; from Eq. (6.15)

$$\frac{1}{r}\frac{\partial}{\partial r}(r\tilde{u}) + \frac{\partial\tilde{w}}{\partial z} = 0 \tag{10.4}$$

Substitution of the decompositions (10.3) into Eqs. (10.1), (10.2), and (10.4) followed by averaging all the equations yields the mean values of the equations of motion, diffusion, and conservation of mass:

Axial motion:

$$U\frac{\partial W}{\partial r} + W\frac{\partial W}{\partial z} + \frac{\partial}{\partial r}\overline{(uw)} + \frac{\partial}{\partial z}\overline{(w^2)} = -\frac{1}{\rho}\frac{\partial P}{\partial z} + \frac{g}{T_a}\Theta + \frac{\nu}{r}\frac{\partial}{\partial r}\left(r\frac{\partial W}{\partial r}\right) + \nu\frac{\partial^2 W}{\partial z^2} \tag{10.5}$$

Radial motion:

$$U\frac{\partial U}{\partial r} + W\frac{\partial U}{\partial z} + \frac{\partial}{\partial r}\overline{(u^2)} + \frac{\partial}{\partial z}\overline{(uw)} = -\frac{1}{\rho}\frac{\partial P}{\partial r} + \nu\left(\frac{\partial^2 U}{\partial r^2} - \frac{U}{r^2}\right) + \nu\frac{\partial^2 U}{\partial z^2} \tag{10.6}$$

Diffusion:

$$U\frac{\partial\Theta}{\partial r} + W\frac{\partial\Theta}{\partial z} + \frac{\partial}{\partial r}\overline{(\vartheta u)} + \frac{\partial}{\partial z}\overline{(\vartheta w)} = \frac{k}{\rho c_p}\left(\frac{\partial^2\Theta}{\partial r^2} + \frac{1}{r}\frac{\partial\Theta}{\partial r} + \frac{\partial^2\Theta}{\partial z^2}\right) \tag{10.7}$$

[2] All plumes do not necessarily diffuse heat only. The diffusion of any other scalar contaminant will obey the same type of equation and $\tilde{\upsilon}$ can be replaced, in that case, by $\tilde{\gamma}$, for instance, the concentration of another gas.

Conservation of mass:

$$\frac{1}{r}\frac{\partial}{\partial r}(rU) + \frac{\partial W}{\partial z} = \frac{1}{r}\frac{\partial}{\partial r}(ru) + \frac{\partial w}{\partial z} = 0 \qquad (10.8)$$

We shall immediately proceed to investigate relative orders of magnitude among the terms in these four equations, not only because there is no known method of solution when the equations are taken in the entire form but also because from a practical standpoint there is no need to keep insignificant terms that complicate matters unnecessarily.

To proceed with the order of magnitude evaluation[3] we must first choose characteristic length, velocity, and temperature scales of the process with which to compare all other such values. It is clear that there are two length scales in the first phase, namely a transverse length scale Λ, and a longitudinal length scale L. These two scales will eventually be related, as we shall find out from the analysis. We need two velocity scales: a mean axial characteristic velocity scale W_0 such that the axial velocity at the center of the plume, or a *densimetric buoyant velocity*, defined as $[\Lambda g(\rho_a - \rho)/\rho]^{1/2}$; and a turbulent velocity scale u', the rms value of the velocity fluctuation at the plume center. A transverse characteristic velocity scale will result from Eq. (10.8) in terms of the longitudinal velocity scale. Finally, we need to define two characteristic temperature scales: a mean temperature T' and t a rms value of the fluctuating temperature at the center of the plume, say.

Looking at the geometry of the mean streamline in Fig. 10.2 we conclude that all radial velocities must be *of the order of* $W_0 \Lambda/L$, written as $U = \mathcal{O}(W_0 \Lambda/L)$. Having established these characteristic dimensions, we proceed through Table 10.1 to evaluate the orders of magnitude of all the terms in Eqs. (10.5)–(10.8), making sure that all the relevant dimensions appear in each of the terms, for purposes of comparison. For this, in some cases, we need to multiply and divide by dimensions not appearing in them.

In evaluating the order of magnitude of the terms in Table 10.1 the ratios of differences were replaced by the differential terms. The pressure term was not evaluated because it is felt that it is important and therefore must be of the order of the largest term.

Looking at Table 10.1a the following conclusions can be drawn:

(1) Because mean convection is strong in the first phase and since turbulence plays an important role in the plume, the mean convective terms and the Reynolds stress must be kept in the axial equation. Both convection terms are of the same order. The second Reynolds stress in $\overline{w^2}$ is one order smaller because

[3] See Tennekes and Lumley (1972) for an excellent use of this method, which we need to employ in almost every turbulent process.

TABLE 10.1 Orders of Magnitude Comparison[a]

(a) Axial motion $[U = \mathcal{O}(W_0 \Lambda / L)]$

$$U \frac{\partial W}{\partial r} = \mathcal{O}\left(\frac{W_0 \Lambda}{L} \frac{W_0}{\Lambda}\right) = \mathcal{O}\left[\left(\frac{W_0^2}{u'^2} \frac{\Lambda}{L}\right) \frac{u'^2}{\Lambda}\right]$$

$$W \frac{\partial W}{\partial z} = \mathcal{O}\left(W_0 \frac{W_0}{L}\right) = \mathcal{O}\left[\left(\frac{W_0^2}{u'^2} \frac{\Lambda}{L}\right) \frac{u'^2}{\Lambda}\right]$$

$$\frac{\partial}{\partial r} \overline{(uw)} = \mathcal{O}\left(\frac{u'^2}{\Lambda}\right)$$

$$\frac{\partial}{\partial z} \overline{(w^2)} = \mathcal{O}\left[\left(\frac{u'^2}{\Lambda}\right) \frac{\Lambda}{L}\right]$$

$$\frac{g}{T_a} \Theta = \mathcal{O}\left(\frac{gT'}{T_a}\right) = \mathcal{O}\left[\left(\frac{g}{T_a} \frac{T' \Lambda}{u'^2}\right) \frac{u'^2}{\Lambda}\right]$$

$$\frac{\nu}{r} \frac{\partial}{\partial r}\left(r \frac{\partial W}{\partial r}\right) = \mathcal{O}\left(\frac{\nu}{\Lambda^2} \Lambda \frac{W_0}{\Lambda}\right) = \mathcal{O}\left[\left(\frac{1}{\mathcal{R}_\Lambda} \frac{W_0^2}{u'^2}\right) \frac{u'^2}{\Lambda}\right]$$

$$\nu \frac{\partial^2 W}{\partial z^2} = \mathcal{O}\left(\nu \frac{W_0}{L^2}\right) = \mathcal{O}\left[\left(\frac{1}{\mathcal{R}_\Lambda} \frac{W_0^2}{u'^2}\right)^2 \left(\frac{\Lambda}{L}\right)^2 \frac{u'^2}{\Lambda}\right]$$

where $\mathcal{R}_\Lambda = W_0 \Lambda / \nu$ (mean flow Reynolds number)

(b) Radial motion

$$U \frac{\partial U}{\partial r} = \mathcal{O}\left(\frac{W_0^2 \Lambda^2}{L^2} \frac{1}{\Lambda}\right) = \mathcal{O}\left[\left(\frac{W_0^2}{u'^2} \frac{\Lambda^2}{L^2}\right) \frac{u'^2}{\Lambda}\right]$$

$$W \frac{\partial U}{\partial z} = \mathcal{O}\left(\frac{W_0^2 \Lambda}{L^2}\right) = \mathcal{O}\left[\left(\frac{W_0^2}{u'^2} \frac{\Lambda^2}{L^2}\right) \frac{u'^2}{\Lambda}\right]$$

$$\frac{\partial}{\partial r} \overline{(u^2)} = \mathcal{O}\left(\frac{u'^2}{\Lambda}\right)$$

$$\frac{\partial}{\partial z} \overline{(uw)} = \mathcal{O}\left(\frac{u'^2}{L}\right) = \mathcal{O}\left(\frac{\Lambda}{L} \frac{u'^2}{\Lambda}\right)$$

$$\nu \frac{\partial^2 U}{\partial r^2} = \mathcal{O}\left(\nu \frac{W_0 \Lambda}{L} \frac{1}{\Lambda^2}\right) = \mathcal{O}\left[\left(\frac{1}{\mathcal{R}_\Lambda} \frac{W_0^2}{u'^2} \frac{\Lambda}{L}\right) \frac{u'^2}{\Lambda}\right]$$

$$\nu \frac{U}{r^2} = \mathcal{O}\left[\left(\frac{1}{\mathcal{R}_\Lambda} \frac{W_0^2}{u'^2} \frac{\Lambda}{L}\right) \frac{u'^2}{\Lambda}\right]$$

$$\nu \frac{\partial^2 U}{\partial z^2} = \mathcal{O}\left\{\left[\frac{1}{\mathcal{R}_\Lambda} \frac{W_0^2}{u'^2} \left(\frac{\Lambda}{L}\right)^3\right] \frac{u'^2}{\Lambda}\right\}$$

(c) Diffusion

$$U \frac{\partial \Theta}{\partial r} = \mathcal{O}\left(\frac{W_0 \Lambda}{L} \frac{T'}{\Lambda}\right) = \mathcal{O}\left[\left(\frac{W_0}{u'} \frac{T'}{t} \frac{\Lambda}{L}\right) \frac{tu'}{\Lambda}\right]$$

$$W \frac{\partial \Theta}{\partial z} = \mathcal{O}\left(W_0 \frac{T'}{L}\right) = \mathcal{O}\left[\left(\frac{W_0}{u'} \frac{T'}{t} \frac{\Lambda}{L}\right) \frac{tu'}{\Lambda}\right]$$

$$\frac{\partial}{\partial r} \overline{(\vartheta u)} = \mathcal{O}\left(\frac{tu'}{\Lambda}\right)$$

$$\frac{\partial}{\partial z} \overline{(\vartheta w)} = \mathcal{O}\left(\frac{tu'}{L}\right) = \mathcal{O}\left(\frac{\Lambda}{L} \frac{tu'}{\Lambda}\right)$$

$$\frac{k}{\rho c_p} \frac{\partial^2 \Theta}{\partial r^2} = \mathcal{O}\left(\frac{k}{\rho c_p} \frac{T'}{\Lambda^2}\right) = \mathcal{O}\left[\left(\frac{1}{\mathcal{P}r} \frac{1}{\mathcal{R}_\Lambda} \frac{W_0}{u'} \frac{T'}{t}\right) \frac{tu'}{\Lambda}\right]$$

$$\frac{k}{\rho c_p} \frac{1}{r} \frac{\partial \Theta}{\partial r} = \mathcal{O}\left[\left(\frac{1}{\mathcal{P}r} \frac{1}{\mathcal{R}_\Lambda} \frac{W_0}{u'} \frac{T'}{t}\right) \frac{tu'}{\Lambda}\right]$$

$$\frac{k}{\rho c_p} \frac{\partial^2 \Theta}{\partial z^2} = \mathcal{O}\left[\left(\frac{1}{\mathcal{P}r} \frac{1}{\mathcal{R}_\Lambda} \frac{\Lambda^2}{L^2} \frac{W_0}{u'} \frac{T'}{t}\right) \frac{tu'}{\Lambda}\right]$$

where $\mathcal{P}r = \rho c_p \nu / k$ [defined in Eq. (7.43)]

(d) Conservation of mass

$$\frac{1}{r} \frac{\partial}{\partial r}(rU) = \mathcal{O}\left(\frac{W_0}{L}\right) = \mathcal{O}\left[\left(\frac{W_0}{u'} \frac{\Lambda}{L}\right) \frac{u'}{\Lambda}\right]$$

$$\frac{\partial W}{\partial z} = \mathcal{O}\left(\frac{W_0}{L}\right) = \mathcal{O}\left[\left(\frac{W_0}{u'} \frac{\Lambda}{L}\right) \frac{u'}{\Lambda}\right]$$

[a] Length scales L; velocity scales W_0, u'; temperature scales T', t.

it is multiplied by Λ/L, which is less than 1. Thus, in order for the first three terms to be of the same order and be kept in the equation we conclude that

$$\frac{W_0{}^2}{u'^2}\frac{\Lambda}{L} = \mathcal{O}(1)$$

or that $u'/W_0 = \mathcal{O}(\Lambda/L)^{1/2}$. This is substantiated by experimental measurements.

(2) Since we postulated that buoyancy is important in the thermal plume it must be retained.

(3) Looking at the viscous terms in Table 10.1a we see that the last term is two orders smaller than the preceding one, which when compared with the convective term, is in itself $\Lambda\mathcal{R}_\Lambda/L$ smaller. We know from experience that $\mathcal{R}_\Lambda$ is a very large number for turbulent plumes, of the order of 10^4 or larger, or that $\mathcal{R}_\Lambda \gg L/\Lambda$. Thus we conclude that the viscous terms are negligible. In essence this implies that molecular shear is smaller than the turbulent shear given by the Reynolds stresses. Thus in the axial equation we shall keep only the mean convective terms, the largest Reynolds stress $\partial(\overline{uw})/\partial r$, the pressure, and the buoyancy.

(4) In the radial equation the Reynolds stress term $\partial(\overline{u^2})/\partial r$ is one order larger than the convective mean terms in the same equation. The frictional terms are even smaller than those in the axial equation. Thus in this equation we shall keep only the Reynolds stress term $\partial(\overline{u^2})/\partial r$ and the pressure.

(5) Looking at the terms in the temperature equation, in order for the mean convective terms to be as important as the turbulent convection, from Table 10.1c we must conclude that

$$\frac{W_0}{u'}\frac{T'}{t}\frac{\Lambda}{L} = \mathcal{O}(1)$$

Since we have already concluded that $u'/W_0 = \mathcal{O}(\Lambda/L)^{1/2}$, by substituting we obtain the same relation for $t/T' = \mathcal{O}(\Lambda/L)^{1/2}$, which is also substantiated by experience.

(6) The Reynolds term $\partial(\overline{\vartheta u})/\partial r$, being larger, will be the only one of the two kept in the simplified heat equation.

(7) The molecular diffusion terms on the right-hand side of Eq. (10.7), as evidenced by Table 10.1c, will be ignored for the same order of magnitude reasons given in (3). The Prandtl number appearing in those terms is of the order of unity. Thus in the temperature equation we shall keep the mean convective terms and the first Reynolds term.

(8) The mean conservation equation for mass yields both terms of the same magnitude and will be used as such.

Thus the simplified equations of motion, conservation of heat, and mass become

$$U\frac{\partial W}{\partial r} + W\frac{\partial W}{\partial z} + \frac{\partial}{\partial r}(\overline{uw}) = -\frac{1}{\rho}\frac{\partial P}{\partial z} + \frac{g}{T_a}\Theta \tag{10.9}$$

$$\frac{\partial \overline{u^2}}{\partial r} = -\frac{1}{\rho}\frac{\partial P}{\partial r} \tag{10.10}$$

$$U\frac{\partial \Theta}{\partial r} + W\frac{\partial \Theta}{\partial z} + \frac{\partial}{\partial r}(\overline{\vartheta u}) = 0 \tag{10.11}$$

$$\frac{1}{r}\frac{\partial}{\partial r}(rU) + \frac{\partial W}{\partial z} = 0 \tag{10.12}$$

These are the equations we must solve. In Section 9.4 we discussed the general idea of self-preserving solutions. For this problem, too, we shall look for this class of self-preserving solutions. We start with Eq. (10.10) and integrate in r from any radial distance to outside the plume, in the ambient environment. We shall designate by P_0 the ambient pressure and take $\overline{u_0{}^2}$ as the kinetic energy in the x component of the turbulence there. Thus

$$\rho\overline{u^2} - \rho\overline{u_0{}^2} = P_0 - P \tag{10.13}$$

Before we substitute P from Eq. (10.13) into the axial equation we must first take the derivative of Eq. (10.13) with respect to z and in the process assume that the turbulence level of the wind or ambient current does not depend on z. This gives

$$\rho\frac{\partial \overline{u^2}}{\partial z} = \frac{\partial P_0}{\partial z} - \frac{\partial P}{\partial z} \tag{10.14}$$

Since $\partial P_0/\partial z$ is the ambient vertical pressure gradient it is time we consider its state of equilibrium with respect to vertical stability and write that $\partial P_0/\partial z = \rho g \Theta_0/T_a$ where Θ_0 is the ambient temperature difference $T_0 - T_a$. Then finally

$$-\frac{1}{\rho}\frac{\partial P}{\partial z} = \frac{\partial \overline{u^2}}{\partial z} - g\frac{\Theta_0}{T_a} \tag{10.15}$$

Before substituting this into Eq. (10.9) we make two essential points. First, since $\partial\overline{u^2}/\partial z$ is an order smaller than $\partial(\overline{uw})/\partial r$ of Eq. (10.9), it will ultimately have to be dropped, to be consistent with the order of the terms kept. Second, the buoyancy term in Eq. (10.9) will become $g(\Theta - \Theta_0)/T_a$ when Eq. (10.15) is substituted and the pressure will not appear as such in the resulting axial equation. For the quantity $(\Theta - \Theta_0) = (T - T_a - T_0 + T_a) = (T - T_0)$ a new variable Θ' must be defined, and if T_0 or Θ_0 is considered constant with r and z, the temperature equation (10.11) can also be written in terms of the new variable Θ'. Thus the three remaining equations are

$$U\frac{\partial W}{\partial r} + W\frac{\partial W}{\partial z} + \frac{\partial}{\partial r}(\overline{uw}) = \frac{g}{T_a}\Theta' \tag{10.16}$$

$$\frac{1}{r}\frac{\partial}{\partial r}(rU) + \frac{\partial W}{\partial z} = 0 \tag{10.17}$$

$$U\frac{\partial \Theta'}{\partial r} + W\frac{\partial \Theta'}{\partial z} + \frac{\partial}{\partial r}(\overline{\vartheta u}) = 0 \tag{10.18}$$

These equations represent a set that must be integrated for solutions of velocity and temperature. It is clear that the set is not complete because there are five unknowns and only three equations. We shall look for the additional two independent equations relating the Reynolds terms to the mean terms, later in the analysis.

We know from analytical and experimental development of shear layers (except for the wake) that the characteristic width Λ spreads linearly with z. We shall see it formally in this development. This implies that Λ/L must be constant in the growth of the plume, and so are the ratios u'/W_0 and t/T' from conclusions (1) and (5) in the order of magnitude discussion. These results support the fact that W_0 and T', the center values of the axial velocity and temperature, make good characteristic scales for this problem.

If affine or self-preserving solutions exist for this problem, it then becomes necessary to reveal the special conditions that the characteristic dimensions in the plume must satisfy, in their development in z, for the existence of these types of solutions. To find these special relationships[4] for the characteristic scales we start with the postulation of self-preserving solutions of the general form

$$\frac{W}{W_0} = \phi_1(\eta) \tag{10.19}$$

$$-\frac{\overline{uw}}{W_0^2} = \phi_2(\eta) \tag{10.20}$$

$$\frac{\Theta'}{T'} = \phi_3(\eta) \tag{10.21}$$

$$-\frac{\overline{\vartheta u}}{T' W_0} = \phi_4(\eta) \tag{10.22}$$

[4] This implies that in all velocity and temperature solutions such as $W = W_0 f(r, z)$, there exists a scale $\Lambda(z)$ such that $\eta = r/\Lambda$ is independent of z.

where $\eta = r/\Lambda$. There is no need to define a similar relationship for U since it is related to W through Eq. (10.17), and indeed if we substitute Eq. (10.19) into Eq. (10.17), after a sequence of chain rule differentiation we have

$$U = -\frac{1}{r}\int_0^r r\frac{\partial W}{\partial z}\,dr$$

but

$$\frac{\partial W}{\partial z} = W_0\frac{\partial \phi_1}{\partial z} + \phi_1\frac{dW_0}{dz}$$

and

$$\frac{\partial \phi_1}{\partial z} = \frac{d\phi_1}{d\eta}\frac{d\eta}{dz}; \qquad \frac{\partial \eta}{\partial z} = -\frac{r}{\Lambda^2}\frac{d\Lambda}{dz}$$

Finally,

$$U = -\frac{\Lambda}{\eta}\int_0^\eta \eta\left(\phi_1\frac{dW_0}{dz} - \eta\phi_1'\frac{W_0}{\Lambda}\frac{d\Lambda}{dz}\right)d\eta \tag{10.23}$$

Partial derivatives have been replaced by total derivatives where applicable. Now we are ready to substitute Eqs. (10.19)–(10.23) into Eqs. (10.16) and (10.18). Let us also recall that the chain rule of differentiation with r reads $\partial/\partial r = (d/d\eta)(\partial\eta/\partial r) = (1/\Lambda)\,d/d\eta$. Then the dimensionless axial momentum and temperature equations become, after use has been made of the partial integration formula,

$$-\left(\frac{\Lambda}{W_0}\frac{dW_0}{dz} + \frac{2\,d\Lambda}{dz}\right)\frac{\phi_1'}{\eta}\int_0^\eta \eta\phi_1\,d\eta - \phi_2' = \frac{g}{T_a}\frac{T'\Lambda}{W_0^2}\phi_3 \tag{10.24}$$

$$-\left(\frac{\Lambda}{W_0}\frac{dW_0}{dz} + \frac{2\,d\Lambda}{dz}\right)\frac{\phi_3'}{\eta}\int_0^\eta \eta\phi_1\,d\eta - \phi_1\phi_3\frac{\Lambda}{T'}\frac{dT'}{dz} = \phi_4' \tag{10.25}$$

where the primes on the functions ϕ imply differentiation with η.

In order for these two dimensionless equations to be valid for *all* z in a plume and for *all* plumes (universality), *all* terms in these equations must be implicitly independent of r and z. Since we know this to be true for the functions ϕ and their derivatives, then the coefficients

$$\frac{\Lambda}{W_0}\frac{dW_0}{dz}, \qquad \frac{d\Lambda}{dz}, \qquad \frac{g}{T_a}\frac{T'\Lambda}{W_0^2}, \qquad \text{and} \qquad \frac{\Lambda}{T'}\frac{dT'}{dz}$$

must be constant.[5] Thus integration of the second coefficient yields that Λ

[5] Making the entire parentheses constant in Eqs. (10.24) and (10.25) is more restrictive because it requires a special relationship between W_0 and Λ.

increases linearly with z, as we anticipated earlier. Introducing this result into the first coefficient yields

$$\frac{z}{W_0}\frac{dW_0}{dz} = m \qquad \text{or} \qquad W_0 \sim z^m$$

In order to find the relationship of the temperature characteristic scale T' with z we use the two results thus obtained and introduce them in the third coefficient

$$\frac{g}{T_a}\frac{T'\Lambda}{W_0{}^2} \Rightarrow \frac{T'z}{z^{2m}} = c$$

$$T' \sim z^{2m-1}$$

The fourth coefficient is seen to be dependent on the other three and cannot yield the value of m which we seek. For this we must go to the conservation of total enthalpy relation[6] in the plume by integrating in the entire space of the plume the value of the enthalpy. Thus

$$H = 2\pi\rho c_p \int_0^{r_p} rW\Theta'\, dr = \text{constant} \tag{10.26}$$

Again replacing the expressions of Eqs. (10.19) and (10.21) we have

$$\frac{H}{2\pi\rho c_p} = \int_0^{r_p} \eta\Lambda^2 W_0 \phi_1 T'\phi_3\, d\eta = \text{constant}$$

For reasons already given $T'W_0\Lambda^2 = \text{constant}$, and substituting the relations in z for W_0 and Λ already found, we obtain

$$z^{2m-1}z^m z^2 = \text{constant}$$

or

$$3m + 1 = 0, \qquad m = -\tfrac{1}{3}$$

We conclude that in order for self-preserving solutions to exist the characteristic scales of the plume must obey the following proportional relations:

$$\Lambda \sim z, \qquad W_0 \sim z^{-1/3}, \qquad T' \sim z^{-5/3} \tag{10.27}$$

What is important in these results is that they were found without any specific knowledge of the functions ϕ_1, ϕ_2, ϕ_3, and ϕ_4. This implies that self-preservation itself fixes the z dependence of length, velocity, and temperature. What about the form of the solutions of Eqs. (10.19)–(10.22)? For this of course Eqs. (10.24) and (10.25) must be solved but we need two more independent equations because there are four functions to be determined. Essentially, what

[6] In the case of the cold plume in a stable atmosphere it is the momentum of the plume that is conserved.

this amounts to is the establishment of two additional expressions giving the relationships of the Reynolds terms to the mean properties. A number of semi-empirical methods are suggested in the literature, depending on the level of sophistication we need to have. However, one of the simplest is Prandtl's mixing length theory which, as discussed in Section 9.4.E, relates $-\overline{uw} = \nu_T \, dW/dr$ and $-\overline{vu} = \gamma_T \, d\Theta'/dr$. Using these additional relations and letting the Prandtl number equal unity ($\nu_T = \gamma_T$) we solve the following for the first phase of the plume:

$$\phi_1 = \phi_3 = \exp(-a\eta^2) \tag{10.28}$$

$$\phi_2 = \frac{1}{\mathscr{R}_\Lambda} \phi_1' = \phi_4 = -\frac{2a\eta}{\mathscr{R}_\Lambda} \exp(-a\eta^2) \tag{10.29}$$

and

$$\Lambda = \frac{z}{\mathscr{R}_\Lambda} \tag{10.30}$$

$$W_0^{\,3} = \frac{3Hg\mathscr{R}_\Lambda^{\,2}}{2\pi\rho c_p T_a}\left(\frac{1}{z} - \frac{h^2}{z^3}\right) + \frac{W_e^{\,3} h^3}{z^3} \tag{10.31}$$

$$T^3 = \frac{2H^2 T_a \mathscr{R}_\Lambda^{\,4}}{3(\pi\rho c_p)^2 z^5} \tag{10.32}$$

where $\mathscr{R}_\Lambda$ is the plume Reynolds number equal to $W_0 \Lambda/\nu_T$, h is the virtual origin[7] as shown in Fig. 10.2, and W_e is the velocity of the plume at the stack exit. When far from the exit h is ignored and we have Eq. (10.27):

$$W_0^{\,3} = \frac{3Hg\mathscr{R}_\Lambda^{\,2}}{2\pi\rho c_p T_a} \frac{1}{z} \tag{10.33}$$

Batchelor (1954) and Turner (1969) have arrived at the same results for buoyant plumes with an environment at rest, and Fay (1973) has extended Turner's work to a plume in a horizontal wind.

B The Intermediate Phase

There are a number of approaches used in the determination of the bending of a plume into a horizontal direction in the second phase. We know that ultimately, when the specific momentum of the plume becomes nearly that of the wind, it will reach a maximum height z_m and become horizontal. Essential internal changes take place in the plume during this transitional phase.

[7] In general Eq. (10.30) does not necessarily begin at the stack exit. At distances far from the stack the importance of h is negligible. See Pasquill (1968).

First consider a plume in a stationary environment, as shown in Fig. 10.3. The vorticity in the plume at the stack exit comes from the boundary layer on the stack walls, and at the discharge this vorticity is concentrated in vortex rings as shown in the figure. The baroclinic conditions in the plume owing to density gradients in the radial direction, interacting with the hydrostatic pressure gradient also generate secondary vorticity ζ_b to be added to the viscous boundary layer vorticity ζ_v such that the total vorticity $\zeta = \zeta_v + \zeta_b$. The vorticity due to baroclinicity was discussed in Section 8.9. The total vorticity vector is tangent to the ring in the direction shown in Fig. 10.3. We have seen in Eq. (7.29) that the acceleration of the steady plume can be expressed as

$$(\boldsymbol{U} \cdot \nabla)\boldsymbol{U} = \tfrac{1}{2}\nabla U^2 - \boldsymbol{U} \times \boldsymbol{\zeta} \tag{10.34}$$

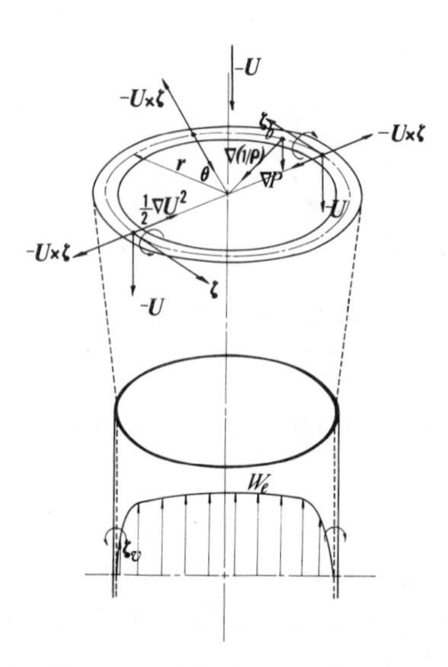

Fig. 10.3 Plume vorticity in a stationary environment.

Upon examination of Fig. 10.3, we see that $-\boldsymbol{U}$ is the velocity of the environment relative to the ring, which is oriented, in the case of a stationary environment, along the z axis and thus perpendicular to the plane of the ring. We conclude from the vector geometry that each of the terms in Eq. (10.34) is in a radial direction, with the Coriolis-like term $-\boldsymbol{U} \times \boldsymbol{\zeta}$ radial and positive, which can be compared to an outward lift on the ring due to the rotation and translation, and whose magnitude is generally larger than the gradient of the kinetic energy. Thus the plume opens, symmetrically, with downstream distance.

With cross wind $\boldsymbol{U}_{\infty}$, the resultant velocity vector $(\boldsymbol{U}_{\infty} - \boldsymbol{U})$ of the environment relative to the ring will make an angle β with the z axis of the ring as shown in Fig. 10.4. In reestimating the acceleration terms of Eq. (10.34) this angle β

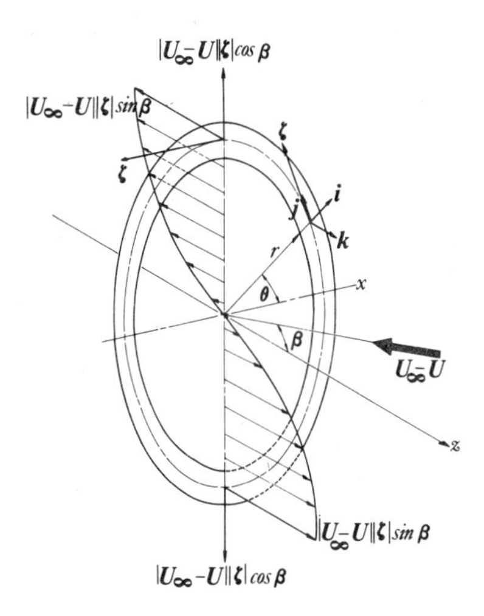

Fig. 10.4 Plume acceleration in a moving environment.

will give in the cross product a symmetric radial component equal to $|-\boldsymbol{U} + \boldsymbol{U}_{\infty}|$ $|\zeta| \cos \beta$ as well as an axial component $|-\boldsymbol{U} + \boldsymbol{U}_{\infty}|\,|\zeta| \sin \beta$ which is not symmetric because it depends on the sine of the angle. The radial component will spread the plume as before but the appearance of an axial component due to the angle β (which is zero at $\theta = 0$, maximum positive at the highest part of the ring, and maximum negative at the lowest part) creates an unsymmetry in the development. In time, these vortex rings will accelerate at the top and decelerate at the bottom and the plane of the ring will be inclined more and more toward the z axis, ultimately forming two counterrotating line vortices in the axial direction, as shown in Fig. 10.5. The presence of this axial vorticity far enough downstream of the stack exit has been reported experimentally (Pratte and Baines, 1967; Kamotani and Greber, 1972). Thus in this intermediate phase peripheral vorticity is oriented into axial vorticity.

Ultimately, when the vorticity has turned axial, the Coriolis-like term in the acceleration becomes zero and we obtain the so-called *Beltrami flow* which does not provide for transverse convection of vorticity. In other words, *the mean transverse convection in this motion going into the second phase will be nearly zero*. This is an important factor in the assumptions of the development in the second phase.

In order to determine the maximum height z_m to which the plume rises and consequently where the axis becomes nearly horizontal, a number of approaches have been used in the current literature. The determination of z_m is a major factor in the solution of the diffusion equation in the second phase. We shall discuss analytical and empirical methods for the evaluation of z_m.

Let us begin with the simplest method in which the plume is assumed to have become cold, or homogeneous with the environment, and that diffusion, from this point on, is confined to momentum and other mixed gases.

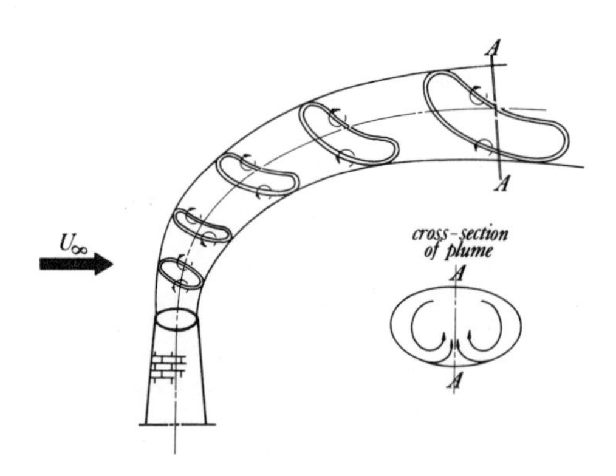

Fig. 10.5 Reorientation of vorticity.

Figure 10.6 shows a control volume around the plume in which momentum considerations will be applied.[8] Since the buoyancy is neglected here, the mean pressure on faces *AB* and *DC*, and *AD* and *BC* are essentially the same. Also along these faces there are no shear forces, especially if U_∞ is uniform in the dimensions of this control volume. Since there are no external forces on the control volume *ABCD*, the momentum entering the control volume must equal that leaving. To determine this momentum balance in the x and z directions, the momentum of the plume at the stack exit must be known. Assuming that the stack and plume are circular, the velocity distribution at any point in the plume and in the stack exit can be represented by

$$W_e = W_0 g_1\left(\frac{r}{r_s}\right), \qquad U_p = U_0 g_2\left(\frac{r}{r_p}\right) \tag{10.35}$$

where W_0 and U_0 are the outside center-line values of the velocities in the stack and in the plume, and r_s and r_p are the radii. There are a number of solutions

[8] Strictly speaking, momentum is not conserved in the plume because of pressure gradients due to vertical stratification. For more accurate estimates an energy balance must be taken, which of course brings into consideration quantities not easily measurable.

available from pipe-flow considerations that give the distribution of g_1. A solution of U_p was already given in the first phase.

The incoming x momentum across the face AB for a unit dimension b normal to the plane of the paper, is

$$\rho b(AB)U_\infty^2$$

The outgoing x momentum across the face CD is

$$\rho[(CD)b - \pi r_p^2]U_\infty^2 \cos\alpha + \int_0^{r_p} 2\pi\rho r U_p^2 \cos\alpha \, dr$$

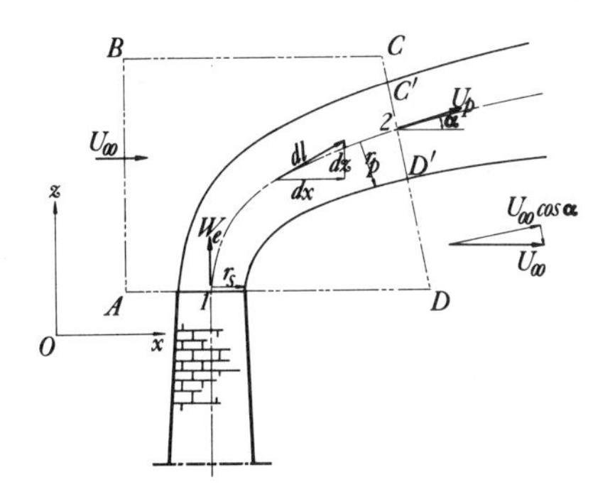

Fig. 10.6 Control volume $ABCD$ around the first and intermediate phases of the plume.

Equating the two x momenta because of conservation, and remembering that $b(CD)\cos\alpha = b(AB)$, the net result in this conservation of x momentum is

$$2\int_0^{r_p} U_p^2 r \, dr = r_p^2 U_\infty^2 \tag{10.36}$$

The same balance in the z momentum yields

$$\int_0^{r_s} W_e^2 \, r \, dr = \sin\alpha \int_0^{r_p} U_p^2 r \, dr = M_s = \text{constant} \tag{10.37}$$

Combining Eqs. (10.36) and (10.37), we have

$$r_p^2 \sin\alpha = 2M_s/U_\infty^2 \tag{10.38}$$

where M_s is the initial momentum of the stack divided by $2\pi\rho$ and is sometimes called the *bulk momentum*. Since M_s and U_∞ are constant for a given situation, the angle and size of the plume are related according to Eq. (10.38). Notice that the shape of the velocity distribution in the stack enters specifically in these results, through the value of M_s.

Let us define a differential length dl along the center line of the plume. Then

$$dz = dl \sin\alpha, \qquad dx = dl \cos\alpha$$

We have seen in the first phase that the radius of the plume, or any characteristic transverse dimension, grows linearly with distance along the axis of the plume. Thus we can write

$$r_p - r_s = \frac{1}{\mathscr{R}_\Lambda} l$$

where $1/\mathscr{R}_\Lambda$ is the tangent of the angle of spread and $\mathscr{R}_\Lambda$ is the plume Reynolds number. Substitution of this expression into Eq. (10.38) gives

$$\left(\frac{l}{\mathscr{R}_\Lambda} + r_s\right)^2 \sin\alpha = \frac{2M_s}{U_\infty{}^2}$$

and

$$l = \mathscr{R}_\Lambda \left(\frac{2M_s}{U_\infty{}^2 \sin\alpha}\right)^{1/2} - r_s \mathscr{R}_\Lambda$$

Then from the trigonometric relation

$$dz = -(2M_s/U_\infty{}^2)^{1/2} \mathscr{R}_\Lambda (\tfrac{1}{2} \sin^{-3/2}\alpha \cos\alpha) \sin\alpha$$

Integrating for z

$$z - z_1 = -\left(\frac{2\mathscr{R}_\Lambda{}^2 M_s}{U_\infty{}^2} \sin\alpha\right)\Bigg|_1^2$$

The lower limit has been chosen at the stack exit where $\alpha = \pi/2$ and after letting $z_1 = 0$, we have from the top of the stack

$$z = \left(\frac{2\mathscr{R}_\Lambda{}^2 M_s}{U_\infty{}^2}\right)(1 - \sin^{1/2}\alpha) \tag{10.39}$$

The *maximum plume rise* is obtained by letting $\alpha = 0$:

$$z_m = \frac{\mathscr{R}_\Lambda}{U_\infty} (2M_s)^{1/2} \tag{10.40}$$

To evaluate z_m we must know $\mathscr{R}_\Lambda$, U_∞, and especially the total bulk momentum at the stack exit.

As an example, the laminar flow in a stack has a nearly parabolic profile of velocity. Integrating a parabolic profile gives $M_s = 2\overline{W}^2 r_s{}^2/3$, where $\overline{W}$ is the spatial average velocity across the stack exist. Then

$$z_m/r_s = 1.155 \overline{W} \mathscr{R}_\Lambda / U_\infty$$

The flow in the stack is not likely to be laminar. The spatial average velocity in a turbulent flow is closer in value to the maximum velocity at the center; measurements show that the previous equation needs to be modified by an approximate factor $(2)^{1/2}$ for the *turbulent case*

$$z_m/r_s = 1.63\overline{W}\mathscr{R}_\Lambda/U_\infty \tag{10.41}$$

Another analytical method for determining the plume's maximum rise is that proposed by Laikhtman (1961). The length of the arc along the axis of the plume is

$$(dl)^2 = (dz)^2 + (dx)^2$$

and dividing through by $(dz)^2$ and substituting $U = dx/dt$ and $W_0 = dz/dt$, we have

$$\left(\frac{dl}{dz}\right)^2 = 1 + \left(\frac{U}{W_0}\right)^2 \tag{10.42}$$

We assume that the decay of W_0 in z, given by Eq. (10.33), is applicable in the intermediate phase, and from the linear relation of the plume spread for large values of $l = \mathscr{R}_\Lambda z$, we have

$$W_0 = \left(\frac{3Hg\mathscr{R}_\Lambda{}^3}{2\pi\rho c_p T_a}\right)^{1/3}\left(\frac{1}{l}\right)^{1/3} = W^*\left(\frac{1}{l}\right)^{1/3}$$

Then Eq. (10.42) becomes

$$\left(\frac{dl}{dz}\right)^2 = 1 + \left(\frac{U}{W^*}\right)^2 l^{2/3}$$

Integrating we obtain

$$z = \int \frac{dl}{[1 + (U/W^*)^2 l^{2/3}]^{1/2}} = \frac{3}{2}\left(\frac{W^*}{U}\right)^3 [\cot\alpha \csc\alpha - \ln(\cot\alpha + \csc\alpha)] \tag{10.43}$$

where we define α by

$$\cot^2\alpha = l\left(\frac{U}{W^*}\right)^3$$

which determines the angle of inclination of the plume trajectory with respect to the horizontal, at a given height z. The inconvenience of this simple solution is that z has an asymptotic dependence on the angle α. Laikhtman considers the maximum plume rise to occur at $\alpha = 10°$.

There are a number of empirical relations giving the maximum plume rise. We cannot treat them all here. Bosanquet *et al.* (1950) give a fairly involved empirical expression for z_m, taking into account the volume flow from the stack

Q in cubic meters per second and measured at the temperature T_1 in degrees Kelvin, when the density of the gas in the plume is reduced to that of the ambient air,

$$z_m = \frac{4.77}{1 + 0.43(U_\infty/\overline{W})} \frac{(Q\overline{W})^{1/2}}{U_\infty} + 6.37g \frac{Q\,\Delta T}{U_\infty{}^3 T_1}\left(\ln J^2 + \frac{2}{J} - 2\right) \tag{10.44}$$

where

$$J = \frac{U_\infty{}^2}{(Q\overline{W})^{1/2}}\left[0.43\left(\frac{T_1}{g\alpha_p}\right)^{1/2} - 0.28\frac{\overline{W}}{g}\frac{T_1}{\alpha_p}\right] + 1$$

and where $\overline{W}$ is the average speed of the gas at the stack exit, in meters per second; U_∞ is the average speed of the wind in meters per second; $\Delta T = T - T_1$ is the virtual temperature difference between that of the gas, T, and T_1 in degrees Kelvin; g is the gravitational constant; and α_p is the potential temperature lapse rate in degrees Celsius per meter.

The values of the temperature lapse rate can be computed according to Eq. (4.52). The following are a few suggested values; for $\alpha_{ad} = -0.01$°C/m (adiabatic conditions), $\alpha_p = 0$; for an isothermal atmosphere $\alpha_{ad} = 0$°C/m, $\alpha_p = 0.01$; for an inversion layer of $\alpha_{ad} = +0.02$, $\alpha_p = 0.03$; and for a weaker inversion of $\alpha_{ad} = 0.01$, $\alpha_p = 0.02$.

Lucas (1958) suggests a simpler formula based on experiments

$$z_m = 11.5H^{1/4}/U_\infty \tag{10.45}$$

where H is the heat content of the plume at the stack discharge in calories per second and U_∞ is the average wind speed in meters per second.

A Concawe publication (1966) has utilized experimental data and arrived at a slightly different formula:

$$z_m = 0.047H^{0.58}/U_\infty^{0.7} \tag{10.46}$$

where the symbols have already been defined.

Holland (1953) gives a simple relationship that is a slight modification of Eq. (10.41) derived on the basis of a cold plume:

$$z_m = (3\overline{W}r_s + 0.04H)/U_\infty \tag{10.47}$$

The symbols are the same except that the units of H are kilocalories per second.

Priestley (1956) proposes the following relation:

$$z_m - z_1 = \left(kW_1 + \frac{g\,\Delta T}{T_a}\right)\Big/\left(\frac{g}{T_a}\frac{\partial T_a}{\partial z} + k^2\right) \tag{10.48}$$

where k is the mixing rate given by $8\gamma_T/r_1{}^2$ in which γ_T is the eddy diffusivity and r_1 is the radius of the plume at a reference point (1) where z_1 and W_1 are also measured. These reference values can be obtained from Eq. (10.33) and the growth rate of the plume $r = r_s + l/\mathscr{R}_\Lambda$. The other quantities have already been defined.

We have said that these relations are very important in the determination of the emission point at a height $z_H + z_m$, where z_H is the stack height and z_m is the maximum plume rise from the stack exit. We shall soon see that this value of $(z_H + z_m)$ will be needed in Section 10.2.C, which deals with the development of the second phase.

The *turbulent entrainment model* based on the proportionality of the entrainment velocity into the plume to the centerline velocity of the plume has, since the original work of Batchelor (1954), generated a number of computational methods for flow properties and plume rise. The most recent of these relationships of plume rise is given, for very low Reynolds numbers, by Hewett *et al.* (1971) from best fits of measurements in a laboratory wind tunnel. This relationship is

$$\frac{z_m}{l_B S^{2/3}} = \left(\frac{6}{\beta^2}\right)^{1/3} \left(\frac{1}{1 - 2\beta\gamma}\right)^{1/3} \tag{10.49}$$

where l_B is the buoyancy scale defined by $l_B \equiv gH/\pi\rho c_p T_a U_\infty^3$ after Eq. (10.33) in which H, T_a, and U_∞ have been defined already, and ρc_p is the heat capacity of the stack effluent per unit volume. The quantity S is the stratification parameter U_∞/Nl_B, a form of the Strouhal number, where N is the Brunt–Väisälä frequency defined in Eqs. (7.51) and (7.54). The two quantities β and γ are the transverse entrainment coefficient and the coefficient of asymmetric entrainment, equal approximately to 0.71 and 0.20, respectively. There is still some disagreement between experiments and the last equation for values of $z_m/l_B S > 2.5$. The second parenthesis can be considered as a correction for asymmetry of the turbulent entrainment model, and is approximately equal to 1.12.

Illustrative Example 10.1

In order to test the amount of spread in the value of the maximum plume rise as calculated from Eqs. (10.41) and (10.43)–(10.46), we consider the data from a stack emptying in the atmosphere at the temperature $T_a = 20°C$ with its own temperature of $T = 100°C$. The volume flow rate is 30 m^3/s, converted at the temperature of the ambient air. The gas density at T_a is 1.1 kg/m^3. The exit diameter of the stack is 2 m and the exit mean velocity is $\overline{W} = 12$ m/s. The average wind speed at the top of the stack is $U_\infty = 5$ m/s. The stability of the atmosphere is given at $\alpha_p = 0.003°C/m$ on a day when the density of the air is 1.2 kg/m^3. The angle of spread of the plume width with distance along the axis of the plume is 30°.

The heat content, or enthalpy, of the gas relative to the temperature of the ambient air is given by $H = \dot{m}c_p(T - T_a)$, where $\dot{m}$ is the mass rate of flow and c_p is the specific heat of the gas at constant pressure which is nearly that of the air (0.24 cal/g · °K).

The mass rate is

$$\dot{m} = 30 \times 1.1 = 33 \times 10^3 \quad \text{g/s}$$

The enthalpy rate is

$$H = 33 \times 10^3 \times 0.24(100 - 20)$$
$$= 634 \times 10^3 \quad \text{cal/s}$$

For Bosanquet's formula we need to compute T_1, the temperature of the gas when the density of the gas equals that of the ambient air. Therefore,

$$T_1 = 1.1 \times 293/1.2 = 269°\text{K}$$

Then the virtual temperature difference is $\Delta T = 373 - 269 = 104°\text{K}$. In applying these data to Eq. (10.41) with a half-angle of spread of 15° and a turbulent plume, we have

$$z_m = 6r_s\overline{W}/U_\infty = 6 \times 12/5 = 14.4 \quad \text{m}$$

Applying the data of Eqs. (10.44)–(10.49) we obtain a comparison of the calculated maximum plume rise with the various formulas (Table 10.2). We return to the data in this problem and give further calculations in Illustrative Example 10.2. The large discrepancy in the results of Table 10.2 is a manifestation of the level of empiricism in all the analysis of plume rise.

TABLE 10.2

Comparison of the Maximum Plume Rise

Equation	z_m (m)
(10.41)	14.4
(10.44)	57
(10.45)	65
(10.46)	35
(10.47)	12
(10.49)	94

C The Second Phase

We have discovered that the problem of transition from the first to the second phase is not well resolved and, on the basis of Illustrative Example 10.1, that there are wide differences between the results of the various analyses given.

In this second phase since the dynamics of the plume are completely dependent on the dynamics of the turbulent environment, and since plumes normally carry contaminants, the practical aspect of this problem reduces to the solution of the diffusion equation in a turbulent environment starting from a point[9] at an altitude $(z_H + z_m)$.

A quantity of gaseous matter and its scalar properties will diffuse in the environment under the combined influence of molecular and turbulent motions. Since the turbulent diffusion is many orders of magnitude larger than the molecular diffusion, we must consider in this problem diffusion coefficients that are dependent on turbulence and position.

Since plumes carry heat as well as by-products of combustion, let us denote by Γ the concentration of any scalar constituent being diffused under the turbulent movement of the environment. This scalar quantity could be the percent composition of another gas, the temperature, the specific humidity, and so on.

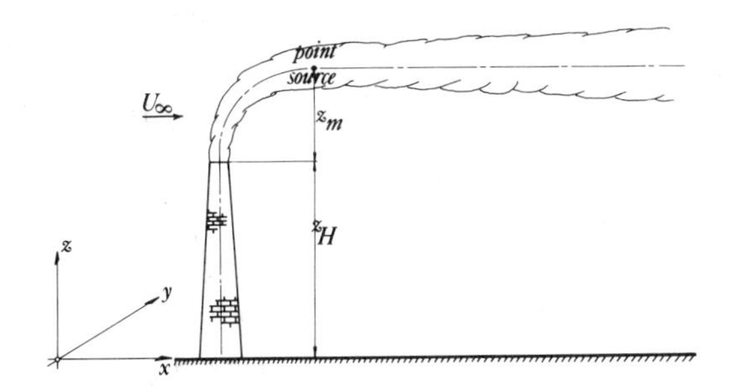

Fig. 10.7 Second phase of the plume. Point diffusion.

The mean steady-state diffusion equation in Cartesian coordinates, which is similar to the temperature equation (10.2), is given by

$$U\frac{\partial \Gamma}{\partial x} = \frac{\partial}{\partial y}\left(\gamma_y \frac{\partial \Gamma}{\partial y}\right) + \frac{\partial}{\partial z}\left(\gamma_z \frac{\partial \Gamma}{\partial z}\right) \tag{10.50}$$

Figure 10.7 gives the appropriate coordinate system of this equation. The simplified nature of this equation needs some explanation. In the intermediate phase we discovered that the orientation of the vorticity in the second phase was such that there was little transverse convection and for this reason the terms $V\,\partial\Gamma/\partial y$ and $W\,\partial\Gamma/\partial z$, which would normally appear in the diffusion

[9] Normally, the diameter of the plume at $(z_H + z_m)$ is small compared to this height, and therefore the emission region can be considered a point.

equation, were omitted, but the transverse diffusion terms on the right-hand side of the equation were retained. It is also true that the longitudinal convection $U\,\partial\Gamma/\partial x$ is estimated to be larger than the longitudinal diffusion, which has been neglected. The turbulent nature of this diffusion does not permit the assumption of constant diffusion coefficients γ. Ultimately they must be related to the turbulent field of the environment. There are many solutions of Eq. (10.50) with constant diffusion coefficients and a variety of boundary conditions, but such solutions of laminar plumes are not very useful in practice since the highly statistical nature of the turbulent environment plays a very important role. In the case of the atmosphere, the turbulent character depends essentially on two sources, namely the mechanically produced turbulence from the friction on the ground, and the thermally initiated turbulence from baroclinic and buoyant motions. These two influences on turbulence, which do not result from the same source, have different characters. Thus in any adequate solution of the diffusion equation we must take into consideration the effects of the roughness and geometry of the ground affecting the wind boundary layer and the stability posture due to stratification. To date there exists no method of solution of Eq. (10.50) for the turbulent diffusion, that fully takes into account the exact turbulent state of the wind.

Sutton (1949a,b) was among the first to consider eddy diffusion coefficients as more realistically related to the statistical nature of the wind motion, instead of the normal mixing-length analogies. Since then, Fleishman and Frankiel (1955) and others have developed similar methods for obtaining values of γ in a turbulent field from a statistical basis.

In order to derive a statistical concept of the eddy diffusion coefficient γ, we state as in molecular theory[10] that this coefficient represents the *mean rate of change of the variance of the statistical displacement of fluid parcels due to diffusion.* This eddy diffusion coefficient which is responsible for the random Lagrangian displacement χ_i of a parcel is proportional to the rate of increase of the mean square of this displacement, or the variance. Thus

$$\gamma_i = \frac{1}{2}\frac{d\overline{\chi_i^2}}{dt} \tag{10.51}$$

It must be clear that χ_i is a displacement vector and that γ_i is also a vector.

In a famous paper, Taylor (1921) related the variance of the parcel displacement $\overline{\chi_i^2}$ to the fluctuating eddy velocity autocorrelation as follows. For times much larger than the time necessary to move one mean free path, the eddy diffusion coefficient in Eq. (10.51) reaches an asymptotic constant value equal to $\gamma_i = \frac{1}{2}\overline{\chi_i^2}/t$.

[10] See, for instance, Hinze (1959, p. 42).

In a simple differentiation it can be shown that Eq. (10.51) is also

$$\gamma_i = \frac{1}{2}\frac{d\overline{\chi_i^2}}{dt} = \overline{\chi_i \frac{d\chi_i}{dt}} = \overline{\chi_i u_i} \tag{10.52}$$

where $u_i(t)$ is the instantaneous velocity fluctuation and $\chi_i(t)$ is the instantaneous parcel displacement. The displacement is related to the velocity after a short time τ as

$$\chi_i(t+\tau) = \int_0^t u_i(t+\tau)\, d\tau \tag{10.53}$$

Replacing this displacement due to diffusion in Eq. (10.52), we have that the eddy diffusion coefficient for a homogeneous turbulence is

$$\gamma_i = \int_0^t \overline{u_i(t)u_i(t+\tau)}\, d\tau \tag{10.54}$$

We should again recall that since velocity and displacement are vector quantities, Eq. (10.54) has three components in the three principal orthogonal directions. Thus, there are three such equations in the directions x, y, and z.

Introducing the *Lagrangian velocity correlation coefficient* $R_i(\tau) = \overline{u_i(t)u_i(t+\tau)}/u_i'^2$, where u_i' is the rms value of the turbulent velocity, we have

$$\gamma_i = u'^2 \int_0^t R_i(\tau)\, d\tau \tag{10.55}$$

In order to integrate this expression we must know the functional form of the correlation coefficient pertaining to the specific application. Universal relations for R_i can be obtained for very special isotropic turbulence which is certainly not that of the atmospheric boundary layer. The central problem here in atmospheric diffusion reduces to finding appropriate expressions for R_i and being able to locate z_m in the plume.

Kampé de Fériet (1939), following Taylor's analysis, has shown that the variance owing to turbulent diffusion can also be expressed as

$$\overline{\chi_i^2} = 2u'^2 \int_0^t (t-\tau) R_i(\tau)\, d\tau \tag{10.56}$$

Looking at the definition of $R_i(\tau)$ as being the average normalized product of the particle velocity at two different times t and $(t+\tau)$, we conclude that this function must be unity when $\tau = 0$, and zero when τ becomes larger than the characteristic Lagrangian time scale of the eddy, defined in Eq. (10.58). In other words, u correlates perfectly with itself at the same time, but has no correlation at a time larger than the life of the eddy, so to speak. There are many functions that take a maximum at the origin and become zero at infinity. The literature gives a number of these functional forms for R_i. Taylor had assumed one such relation as

$$R_i(\tau) = \exp(-\tau_i/\tau_L) \tag{10.57}$$

where τ_L is the Lagrangian time scale defined as

$$\tau_L = \int_0^\infty R(\tau)\, d\tau \tag{10.58}$$

Frankiel (1952) has taken the form

$$R_i(\tau) = \exp\left(-\frac{\pi\tau_i^2}{4\tau_L^2}\right) \tag{10.59}$$

for the correlation coefficient. Like that of Taylor's the form of R_i assumes a neutral boundary layer in which diffusion is taking place. Sutton (1949b) has postulated a relation for R_i which has general properties similar to those of Taylor and Frankiel but, in addition, which includes a variable parameter n loosely tied to the stability posture of the atmosphere because of stratification. This expression reads

$$R_i(\tau) = \left(\frac{\nu}{\nu + u_i'^2\tau}\right)^n \tag{10.60}$$

where ν is the kinematic viscosity[11] and n is a kind of stability parameter of the environment which also appears in the power law for the vertical distribution of the horizontal velocity in the boundary layer

$$U = U_0\left(\frac{z}{z_0}\right)^{n/2-n} \tag{10.61}$$

The subscript zero refers to a reference altitude.

Starting with Eq. (10.60), inserting it into Eq. (10.56), and integrating one obtains the variance of the Lagrangian displacement

$$\overline{\chi_i^2} = \tfrac{1}{2}C_i^2(Ut)^{2-n} \tag{10.62}$$

where the coefficients C_i^2 are given by

$$C_i^2 = \frac{4\nu^n}{(1-n)(2-n)U^n}\left(\frac{\overline{u_i^2}}{U^2}\right)^{1-n} \tag{10.63}$$

The x, y, and z components are identified with $i = 1, 2, 3$. Obtaining γ_i for insertion into Eq. (10.50) is now a simple matter, because of Eq. (10.51).

[11] The kinematic viscosity can be replaced, for improved analysis, by u_*k, as Sutton has also done, where u_* is the shearing velocity and k is the roughness (see Table 9.2).

For a point source with strength Q the solution of Eq. (10.50) yields

$$\Gamma(x, y, z) = \frac{Q}{\pi C_y C_z U x^{2-n}} \exp\left\{-\frac{1}{x^{2-n}}\left[\frac{y^2}{C_y^2} + \frac{(z - z_H)^2}{C_z^2}\right]\right\} \quad (10.64)$$

where C_y and C_z are the generalized eddy diffusion coefficients given in Eq. (10.63), z_H is the stack height, and Q is the strength of the contamination which can be evaluated at any plane normal to x in Fig. 10.7 by the integral

$$Q = \int_0^\infty U(z)\Gamma(x, y, z)\, dz \quad (10.65)$$

Equation (10.64) gives the general solution sought for the concentration at any point downstream of the source located at $(z_H + z_m)$. There are three undetermined constants in this solution, C_x, C_y, and n, which are all related to the turbulent and stability state of the environment in which the plume is diffusing.

In practical applications, what is of interest is the concentration on the ground which is normally populated. This is obtained by setting $z = 0$ in Eq. (10.64) and we have the concentration on the ground, Γ_g,

$$\Gamma_g = \frac{Q}{\pi C_y C_z U x^{2-n}} \exp\left\{-\frac{1}{x^{2-n}}\left[\frac{y^2}{C_y^2} + \frac{z_H^2}{C_z^2}\right]\right\} \quad (10.66)$$

From this distribution on the ground in x and y, the maximum values at any x will be found at $y = 0$. This distribution along the axis of the plume is given by

$$\Gamma_g = \frac{Q}{\pi C_y C_z U x^{2-n}} \exp\left(-\frac{1}{x^{2-n}} \frac{z_H^2}{C_z^2}\right) \quad (10.67)$$

Also, along the x axis, there is a point of optimum concentration or pollution. This point is very important in the design of plants and in particular in the determination of the stack height. This is found by differentiating Eq. (10.67) with respect to x and setting the result equal to zero. The position x_{max} and the maximum concentration at that position are found to be

$$x_{max} = \left(\frac{z_H}{C_z}\right)^{2/2-n} \quad (10.68)$$

and

$$\Gamma_{gmax} = \frac{Q}{\pi e U z_H^2}\left(\frac{C_z}{C_y}\right) \quad (10.69)$$

Thus the maximum concentration is inversely proportional to the square of the stack height and proportional to the ratio of the generalized eddy diffusion coefficients C_z/C_y. However, except for very near the ground, because of symmetry, this ratio C_z/C_y is nearly unity.

Sutton has provided values of eddy diffusion coefficients for various types of atmospheres. These values are given in Table 10.3.

For a more general approach, Eq. (10.64) is generalized to read

$$\Gamma(x, y, z) = A \exp[-(b|y|^r + c|z|^s)] \tag{10.70}$$

where the coefficients are to be found from experiments. If we define (see Pasquill, 1968) the rms deviation of the material particles

$$\sigma_y^2 = \int_0^\infty y^2 \Gamma \, dy \Big/ \int_0^y \Gamma \, dy$$

for each direction y and z, it can be shown for a continuous point source of strength Q that the concentration can be written as

$$\Gamma(x, y, z) = \frac{Q}{2\pi U \sigma_y \sigma_z} \exp\left[-\frac{1}{2}\left(\frac{y^2}{\sigma_y^2} + \frac{z^2}{\sigma_z^2}\right)\right] \tag{10.71}$$

In comparing this solution with that of Sutton [Eq. (10.64)] we see that $\sigma_y/x = C_y x^{-n/2}/(2)^{1/2}$ and $\sigma_z/x = C_z x^{-n/2}/(2)^{1/2}$. Other values of σ_y and σ_z have been proposed by Pasquill, Calder (1952), and Bosanquet and Pearson (1936).

TABLE 10.3

Generalized Eddy Diffusion Coefficients, $C_y = C_z$

Type of atmosphere	n	Altitude, z (m)			
		25	50	75	100
Large temperature gradient	0.20	0.21	0.17	0.16	0.12
Small temperature gradient	0.25	0.12	0.10	0.09	0.07
Weak inversion	0.33	0.08	0.06	0.05	0.04
Strong inversion	0.50	0.06	0.05	0.04	0.03

Above 25 m, $C_y = C_z$ in a neutral atmosphere

Illustrative Example 10.2

Consider the conditions of the plant of Illustrative Example 10.1 burning heavy fuel oil and discharging combustion gases at the rate of 33 kg/s. The plume contains 0.2% of SO_2 at discharge. The chimney height is 60 m and the wind has the velocity of 5 m/s, as already indicated.

Detrie (1970) chooses for this problem $C_y = C_z = 0.10$ and for n the value of 0.25. The plume rise is computed on the basis of Bosanquet's formula of $z_m = 57$ m, as shown in Illustrative Example 10.1, and the point source of pollution is located at $z_H + z_m = 60 + 57 = 117$ m.

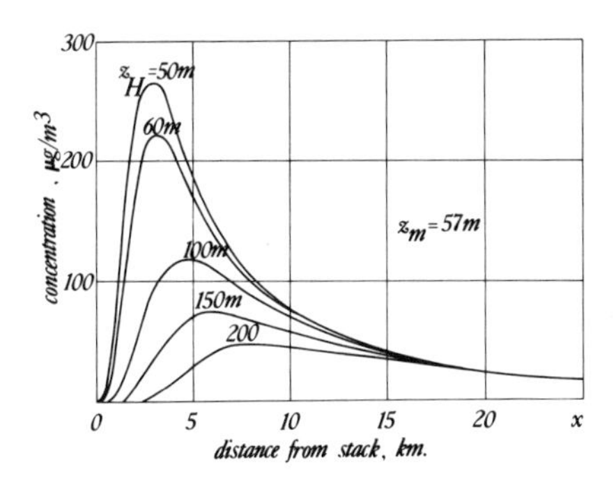

Fig. 10.8 Concentration of SO_2 on the ground for various chimney heights H corresponding to the data in Illustrative Examples 10.1 and 10.2. (After Detrie, 1970.)

The concentration distributions are plotted, from Eq. (10.67), in Fig. 10.8 for various stack heights z_H. Figure 10.9 shows the influence of the exponent n in the calculations, for the same stack height $z_H = 60$ m. The values of C_y and C_z are as given by Sutton (1949b, Table 10.2) for various values of n.

Although Sutton's analysis is among the best approaches available to evaluate diffusion of scalar contaminants in a plume, it fails to take into account, in a more precise way, the effects of the specific land contour and the stability posture in terms of stratification gradients. Furthermore, it depends on an *a priori* knowledge of the location of the maximum plume rise, whose estimation also depends on available analyses that seem to suffer from the same lack of completeness.

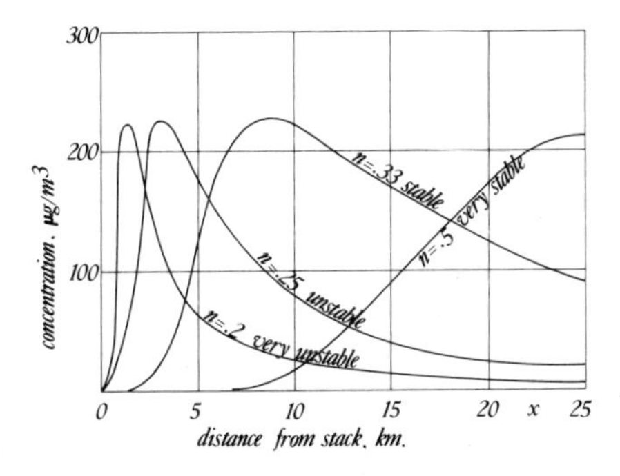

Fig. 10.9 Concentration of SO_2 on the ground for a chimney $z_H = 60$ m, but different meteorological conditions. (After Detrie, 1970.)

Figure 10.10 shows the general appearance of five different plume geometries, all of which are affected, to some degree, by the turbulence of the land contour and by the turbulence due to instabilities in the vertical stratification. The dynamical effects on the generation of turbulence are due to the production of eddies when the wind flows over ground roughness such as hills and buildings. The scales of the turbulent eddies near the ground are related to the shape and size of the roughness.

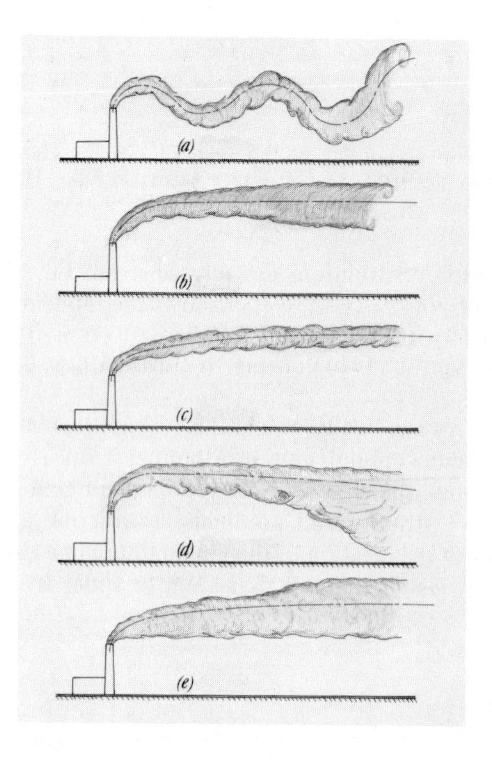

Fig. 10.10 Geometrical appearance of plumes: (a) looping; (b) coning; (c) fanning; (d) fumigating; (e) lofting.

The turbulence generated by the stratification of density is caused either by the production of secondary motion due to baroclinic conditions or simply by the vertical motion resulting from unbalanced vertical buoyant forces. We have seen in Chapter 4 that the vertical stability of environmental columns is measured on the basis of adiabatic vertical gradients of density or temperature. In the stable configuration any turbulence is suppressed. In a neutral environment the major part of turbulence will have to be from dynamical causes. In the unstable case both dynamical and thermal causes contribute to the production of turbulence.

In thermally generated turbulence, the intensity depends on the rate at which sunshine heats the ground and the inner layer of the boundary layer. Since this heating is principally from radiation, the wind or current plays a minor role. Only the transfer of the turbulent energy is affected by the wind speed. In slow winds, the size of the thermal eddies is larger. In strong winds, the dynamically generated eddies are more effective in transferring heat from the ground to the air.

It is apparent then that when the ground is warm due to a sustained sunshine, at slow winds the plume is spread out into greater widths than with strong winds. Conversely, when the ground is cold there is little spread and with strong winds the difference with varying ground temperatures becomes less noticeable.

Figure 10.10a, called the *looping plume*, occurs with high degree of turbulence, particularly with thermally convected turbulence, although large-scale dynamically generated eddies due to land topography are not excluded. This situation is typical on a clear day when the earth's surface receives the greatest amount of solar radiation thereby causing vertical temperature gradients larger than adiabatic. These large eddies can often grow to such sizes as to sweep the ground at some distance from the stack.

The *coning plume* (Fig. 10.10b) occurs when the environment is in a neutral state and when the dynamically produced turbulence is small in intensity and scale. When there is a heavy cloud cover during the day, which reduces the heat exchange between the sun and the earth, this condition is present. This case is the one most often treated in books because of its simplicity in analysis.

The third case (Fig. 10.10c), called the *fanning plume*, occurs when the environment is thermally stable and the vertical component of any dynamically generated turbulence is suppressed. Because motion in thermal stability is mainly restricted to the vertical direction, the horizontal component of turbulence is free to exhibit larger fluctuations than the vertical component under stable conditions. As a result, the plume is narrower along the vertical than any of the others treated. Often this plume will have a tendency to whip horizontally in a sinuous manner. Vertical stability is likely to be the highest at night when the earth's surface is coolest in reference to the air temperature.

When the thermal stability criteria vary significantly with elevation such that the atmospheric layer nearest the ground and below the plume is unstable while at the same time the layer above the plume is stable, the situation shown in Fig. 10.10d is observed. This is called the *fumigating plume*. It occurs normally during the morning hours, if stable conditions prevail during the preceding night. The *lofting plume* (Fig. 10.10e) is the opposite; the unstable layer is above the plume and greater diffusion is in that direction. This occurs in the late afternoon or evening when a stable layer develops below the plume.

10.3 THE FULLY DEVELOPED HURRICANE

Depending on the special interest of the observer, various models have been proposed for the large atmospheric disturbances called hurricanes. The model presented here is due to Carrier and his co-workers (1971). Hurricanes, sometimes called typhoons, are large cyclonic storms which develop in an area approximately 1000 km in radius and originate between 5 and 20° latitudes. Some observers believe that the surface temperature of the ocean is what determines the inception of hurricanes and not the latitudinal zone. For this they propose a surface water temperature of 25°C or higher.

The hurricane is a roughly circular cyclonic vortex with inflow at its lower end, due partly to the unbalance of radial forces in the boundary layer (as explained in Sections 6.13 and 9.3) but due mostly to the pumping of large quantities of wet air through its central core because of differences in buoyancy, and with outflow at its upper end from an annular region of intense precipitation, maximum winds, and strong updrafts.

As it is explained later, the center of the hurricane is much warmer, with extreme low pressure and lower humidity than the ambient conditions surrounding the center. This central column is called the *eye* of the hurricane. The winds in this planetary vortex may reach velocities as high as 200 knots or 100 m/s, tides could surge as high as 10 m above mean sea level, and the amount of rain could exceed 75 cm in 8 h. It should be pointed out that the height of the hurricane, and consequently the height of the atmosphere that is significantly affected by the hurricane process, is of the order of the height of the troposphere. These storms occur most frequently in the late summer months and early fall when the sun completes its second pass over the tropics and when ocean water temperatures are warmest.

During the summer and late fall there are hundreds of tropical disturbances around the globe that have the potential of becoming hurricanes. Despite the seemingly favorable oceanic and atmospheric conditions only about 50–60 disturbances a year, identified as hurricanes or typhoons, take place in the Northern Hemisphere and 15–20 in the Southern Hemisphere, as shown in

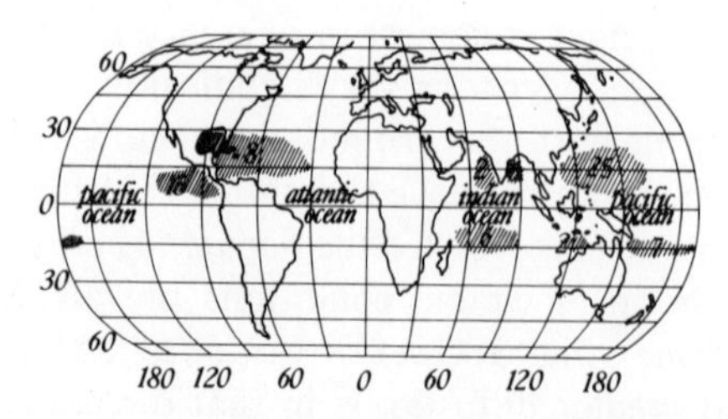

Fig. 10.11 Frequency of tropical storms around the world. Shaded areas are regions of tropical storms, the number in the area refers to the average number of storms in a year.

Fig. 10.11. Approximately 5–10 of these storms take place in the tropical Atlantic, between Western Africa and the Gulf of Mexico.

It is believed that a weak disturbance leading to the formation of a hurricane may start either from the updrafts generated from a cold front that penetrates a very large cloud mass in the tropical latitudes, or from an *easterly wave* in the streamline pattern around the equator. These waves in the streamline pattern moving from the east generate low pressure cells at the crest of the waves, setting up small cyclonic flows maintained by a radial pressure gradient. This wave phenomenon was discussed in Sections 8.8 and 8.9. Figure 10.12 shows these easterly waves[12] in the tropical Atlantic and the possible inception of cyclonic motions balanced by a positive pressure gradient

$$\frac{dP}{dr} = \rho \frac{V^2}{r} \tag{10.72}$$

around the crest of the wave, where V is the wind velocity, r is the radius of curvature of the easterly at the crest, and P is the local pressure.

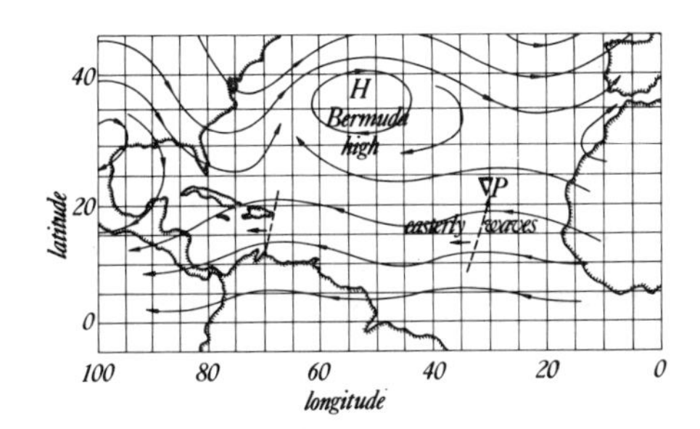

Fig. 10.12 Weather map showing tropical easterlies.

The stability of the air column in the center may be disturbed by the drop of pressure and result in induced updrafts with accompanying air intake, near saturation, from the surface of the ocean. We shall see later that the probability of intensification of such a disturbance born at the crest of an easterly, will depend on the occurrence of a sequence of events that will increase the energy of the air in the eye, thus creating stronger radial pressure gradients and faster swirls.

Beginning with the month of June and continuing through October, easterly waves move at the speed of approximately 10 knots from the coast of Africa toward the Caribbean. These waves move slowly during June and July and

[12] See also Burpee (1971).

intensify in August and September. Approximately one in ten to one in five of the waves that fold form strong swirls during this season. Satellite observations suggest that such low pressure disturbances may begin in the Sahara Desert, taking days to cross the Atlantic, and gaining strength until they penetrate the Caribbean.

In an average year (Howard *et al.*, 1972; see also Roberts, 1972) these cyclonic storms cause damages in the US alone amounting to hundreds of millions of dollars and kill 50–100 persons. In a bad year the property damage may exceed 1.5 billion dollars and the death toll may exceed 100 persons.

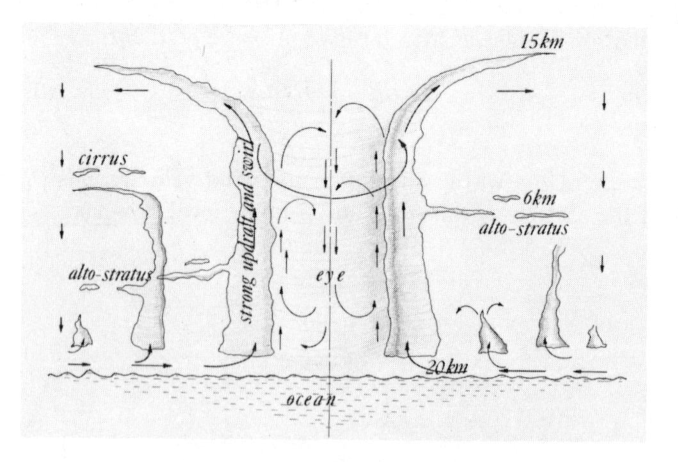

Fig. 10.13 Cross section of a hurricane.

The hurricane model shown in Figs. 10.13 and 10.14 is a vertical cross section of the hurricane which consists of four distinct regions. The warm, relatively dry, and quiescent *core* in which there is very small interaction with the major dynamical processes of the hurricane is the region marked IV (in Fig. 10.14) and called the *eye* of the hurricane. The fluid in this core is relatively motionless. Since it is warmer, at a given altitude it is less dense than the fluid in the outer regions, particularly the adjacent region III. This latter region is considerably more active and the one in which the most intense rainfall and accompanying latent heat release occurs. Although the ascending air in this region tends to preserve its angular momentum, at the same radial distance, this air has a deficit of total angular momentum compared to the air parcel in II because of the frictional losses in the boundary layer before climbing into region III. Region III is characterized by strong and tall convective updrafts which are fed by the swirling radial flow from region I. The positive gradient of density from IV to III to II is essential in the development of radial pressure gradients which balance the centripetal acceleration in the swirling motion.

Region II is characterized by fast rotational velocities with accompanying radial pressure gradients in the direction toward the center of the hurricane. There is almost no radial motion in region II. The air in this region replenishes the flow in region I going into III by settling down in the form of a downdraft mixed with a swirling motion. In fact the time of consumption of the energy of the hurricane is linked by Carrier *et al.* (1971) to the consumption of the volume of air in region II carrying latent heat.

Finally region I is characterized by a swirling boundary layer flow interacting through friction with the surface of the ocean. This movement supplies moist air to the updraft of region III. The radial pressure gradient in this region is nearly the same as that at the bottom of region II. Since friction reduces the swirl velocity in this region, there is a necessity for radial motion toward the center to create a momentum flux to balance the excess pressure gradient. This was discussed in Section 6.13.

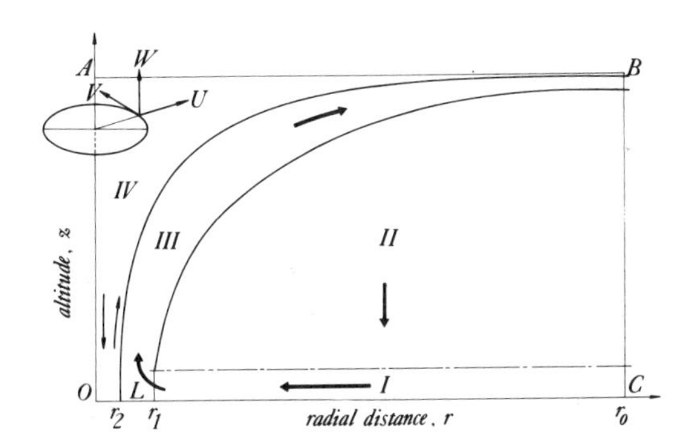

Fig. 10.14 Schematic cross section of a hurricane.

We have stated a number of times that in the scale of the phenomenon of the hurricane, dynamics alone cannot govern and be exclusively responsible for the entire process. We must consider thermodynamics to play an equal role.

A Thermodynamic Model

Figure 10.15 is a pressure–temperature diagram (discussed in Section 4.20) in which a number of the thermodynamic processes related to the storm are represented. In the troposphere, which is the range of this figure, and consequently the range of the hurricane, a standard atmosphere at the edge of the storm at $r = r_0$ has a nearly linear temperature drop with altitude (given by Fig. 2.6). On the coordinates of Fig. 10.15 the standard atmosphere at the edge of the storm is represented by the curve M starting with an ocean temperature of 30°C. Thus the pressure P, the temperature T, and the density ρ of the air particles at the edge of the storm are given by the thermodynamic states on this curve.

If a parcel of air were to be displaced up or down in the atmosphere without change of phase, it would follow the states of the dry adiabats (curves S) as already discussed in detail in Chapter 4. However, the same air parcel, saturated with vapor, and climbing in the atmosphere will condense its vapor constantly owing to the drop in temperature during the rise and will release the heat of condensation to increase the temperature of the rising air. Thus a wet

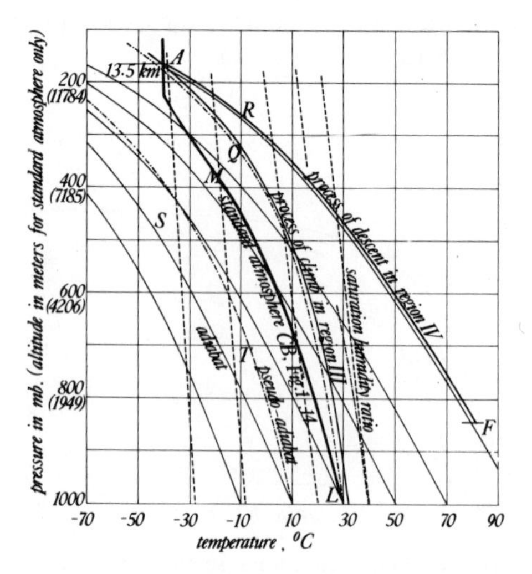

Fig. 10.15 Thermodynamic processes of various regions in a hurricane.

adiabat or pseudoadiabat process is represented by the dash-dot lines marked T. Near the ground these process lines have a steeper climb than the dry adiabats. In comparing a dry and saturated particle climbing the same height, we see that the saturated particle will be warmer at the same elevation and ambient pressure.

Figure 10.15 also shows the saturated humidity ratios (dashed lines) at various altitudes. From the manner in which the adiabats cross the humidity lines, it is apparent that air requires less and less vapor to be saturated at higher altitudes.

When air from the bottom of the hurricane in region I moves radially inward, it picks up sufficient moisture from the surface of the ocean to be saturated by the time it reaches r_1 (Fig. 10.14). As it starts to ascend in region III, beginning at the same temperature as the ocean (30°C), it will climb along the process line Q which is one in the family of the pseudoadiabats T, dropping moisture through intensive rainfall from a saturation humidity ratio of approximately 28 g/kg at the surface of the ocean to 0.4 g/kg at the top of the troposphere. A total of 27.6 grams of water per kilogram of air has been precipitated in the entire rise. Weather surveillance aircraft estimate this rainfall rate to reach as high as 15 cm/h, decreasing toward the edge of the hurricane to 3 mm/h. Knowing the mass rate of flow in the updraft of region III we could calculate the rate of precipitation and the accompanying release of the latent heat that is retained in the rising air. We can assume that at the altitude A in Fig. 10.15 and AB in Fig. 10.14 the strength of the storm is negligible. The reason for this is that the ambient air is likely to have the same conditions as that of the standard atmosphere at $r = r_0$ whose temperature above A is warmer than that of the air in region III if the air continued to climb beyond A (see top of Fig. 10.15). Thus the wet air beyond A would be colder (heavier) than that of the ambient air and consequently the updraft would terminate in that vicinity. This elevation is taken as 13.5 km and the pressure there is 180 mb. Because the updrafts are suppressed above this height we shall take this altitude as the upper boundary of the storm. Below this point the temperature T_{III} of the process in region III is higher than T_{II} in the region II and consequently $\rho_{\text{III}} < \rho_{\text{II}}$.

This part of the hurricane is often compared to a heat engine operating between a warm source (the tropical ocean) and a cold sink (the upper atmosphere). The warm air drawn from the ocean surface rises to release the energy in the moisture it contains. This released energy provides more buoyancy and the warmer air rises further to higher cold levels producing more pumping power for the air below.

At the interface between regions III and IV, small amounts of air from the eye are entrained upward but must be balanced by a small downdraft near the center along AF (Fig. 10.15). It is assumed in the model that this up and down recirculation in region IV is what maintains the equilibrium of the air in the eye. The rising air in the interface of the eye follows the process Q, and when it descends in the center of the eye in region IV it drops most of its moisture, making the descent from A to the surface of the ocean along a dry adiabat through process R to the point F. At F (sea level) the air is considerably warmer and drier than when it started at any point around r_1. Looking at Fig. 10.15 it may appear as an error to identify the point F as that at sea level in the eye. This is not so. The pressure–temperature values on that figure correspond to all processes; however, the values of the altitude in the parentheses of the ordinate correspond to the standard atmosphere of curve M only.

Starting at A with 180 mb, a dry adiabatic compression along R for a depth of 13.5 km would give a lower sea-level pressure at F than at L. This pressure at F can be computed from the adiabatic pressure–elevation relation of Eq.

(4.49) where $1/\kappa = \gamma/\gamma - 1$. Taking $\gamma = 1.4$ and $\alpha_{ad} = 9.66°K/km$ and beginning at A with the pressure 180 mb and the temperature 233°K we can compute P_F

$$P_F = P_A\left[1 + \frac{9.66}{233}(13.5)\right]^{3.5} = 853 \quad \text{mb}$$

Actually, this may be too low in reality. Perhaps 900 mb might be a more realistic figure for a strong hurricane. The temperature at the center of the eye is normally 15–20°C above ambient temperature. The calculations reflect a larger temperature difference.

Maintaining that the descending air in the center of the eye has a pressure of 853 mb and that at $r = r_0$ (edge of storm) the sea-level pressure is 1000 mb, the radial pressure difference is 147 mb. Based on this pressure differential we are ready to apply dynamical considerations to compute swirl velocities.

B Dynamical Model

Here, too, we need to make a few assumptions. For instance, we assume that the swirling motion is circular and thus axisymmetric. The photograph in Fig. 10.16 of Hurricane Gladys as seen from Apollo 7, at an altitude of 111 miles, substantiates this reasonable assumption. We also assume that there is no flow crossing the boundaries of the hurricane "box" *OABCO* in Fig. 10.14. Thus the storm is self-contained within those dimensions. Losses from the viscous region I into the updraft are compensated by the gain from the downdraft in region II. The velocities vanish at all boundaries of the box. We assume that viscosity effects, although of importance in this model, are everywhere negligible except in region I. This permits the maintenance of different angular momenta across the interface II–III. The negligible viscous and angular momentum coupling between these two regions is important in this model. Furthermore the flow in the fully developed hurricane is assumed to be steady (without necessarily excluding turbulence). This implies that the boundaries of the hurricane and the regions within it change very slowly with time, especially as the moist air in region II is used up and region III widens and decays. This assumption of steadiness can certainly be disputed.

C Equations of Motion

Although the density varies considerably in the processes of this problem, its variation is almost entirely dependent on the energy process and not on dynamics. Thus we can assume incompressible motion.

To justify steady flow in the process of consumption of the hurricane, we must admit that the consumption time is of an order much larger than the boundary layer formation and decay time in region I. The boundary layer

Fig. 10.16 Hurricane Gladys over the Gulf of Mexico, in 1968, as seen by Apollo 7 from an altitude of 111 miles.

development time, from dimensional reasoning, is of the order of $t^* \sim h^2/\nu_T$ where h is the boundary thickness in a swirl and ν_T is the eddy viscosity. From the study of the Ekman layer in Section 9.3 and Eq. (9.14) we discovered that this viscous layer has the dimensions of the order of $h \sim (2\nu_T/f)^{1/2} = (\nu_T/\Omega \sin \psi)^{1/2}$. Substituting this value of h into the time t^* we obtain a development time of the order of 5–10 h at tropical to mid-latitudes. The consumption time of the hurricane is often measured in days. However, it is clear that for regions close to the equator with smaller latitude angles and for consumption times of one to two days, the steadiness assumption begins to be doubtful. Some reconnaissance measurements seem to indicate that the strength of the storm increases with time, at first, owing to the conversion of thermal energy into mechanical energy. These measurements indicate that this growth could be as much as 50% in 12 h, in the initial period. Nevertheless, it is believed that the results obtained from the assumption of steadiness bring out revealing conclusions.

Let U, V, and W be the radial, circumferential (swirl), and vertical components of the mean velocity in the cylindrical coordinates r, θ, and z. In the boundary layer, the equations of motion and the continuity equation are

radial:

$$U\frac{\partial U}{\partial r} - \frac{V^2}{r} + W\frac{\partial U}{\partial z} - fV = -\frac{1}{\rho}\frac{\partial P}{\partial r} + \nu_{\mathrm{T}}\frac{\partial^2 U}{\partial z^2} \tag{10.73}$$

circumferential:

$$U\frac{\partial(rV)}{\partial r} + W\frac{\partial(rV)}{\partial z} + fU = \nu_{\mathrm{T}}\frac{\partial^2(rV)}{\partial z^2} \tag{10.74}$$

vertical:

$$\frac{\partial P}{\partial z} = -\rho g \tag{10.75}$$

continuity:

$$\frac{\partial(rU)}{\partial r} + \frac{\partial(rW)}{\partial z} = 0 \tag{10.76}$$

Because of axisymmetry, $\partial/\partial\theta = 0$ for any property, and in the boundary layer $\partial^2/\partial z^2 \gg \partial^2/\partial r^2$. We let the velocity $V(r, \infty)$ outside the boundary layer be the geostrophic velocity $V_g(r)$ and analogously for $U_g(r)$ and $W_g(r)$. For region II Eq. (10.73) becomes

$$\frac{1}{\rho}\frac{\partial P(r, \infty)}{\partial r} = \frac{V_g^{\,2}}{r} + fV_g \tag{10.77}$$

Since the change of pressure with altitude is dependent only on ρ and g according to the hydrostatic equation (10.75), the same dP/dr must hold inside and outside the boundary layer. We substitute Eq. (10.77) into Eq. (10.73) and we have

$$U\frac{\partial U}{\partial r} + W\frac{\partial U}{\partial z} + f(V_g - V) + \frac{V_g^{\,2} - V^2}{r} = \nu_{\mathrm{T}}\frac{\partial^2 U}{\partial z^2} \tag{10.78}$$

The boundary conditions are that at $z = 0$,

$$U = V = W = 0$$

For conservation of angular momentum in region II

$$U_g\left[\frac{\partial}{\partial r}(Vr) + fr\right] = 0$$

The free vortex with distribution $V_g \propto 1/r$ can be taken as a model of swirl velocity satisfying the conservation of momentum. But to make the swirl vanish at $r = r_0$ at the bounds of the hurricane box, we let

$$V_g = \begin{cases} C_1/r & \text{for } 0 < r \leq r' \\ C_2(r_0 - r) & \text{for } r' \leq r < r_0 \end{cases}$$

The distance r' can be fixed where the function $1/r$ can be approximated as a linear function going to zero at $r = r_0$.

D Maximum Velocity Estimates

Generalizing the vortex swirl velocity distribution as $V_g = C/r^n$ from Eq. (10.77) and in the region of maximum swirl velocities, the centrifugal acceleration dominating, we have

$$\frac{1}{\rho}\frac{dP}{dr} \simeq \frac{C^2}{r^{2n+1}} \tag{10.79}$$

The density varies little with radial position. Integrating from $r = r_1$ to $r = r_0$

$$P(r_0) - P(r_1) \simeq \frac{\rho_0}{2n}\left(\frac{C}{r^n}\right)^2 = \frac{\rho_0}{2n} V_g^{\,2}$$

Since $r_1 < r_0 - r_1$ and since $V_g(r)$ decreases rapidly with r

$$\frac{P(r_1) - P(r_2)}{P(r_0) - P(r_1)} \ll 1$$

so that

$$V_{\mathrm{gmax}} = \left\{\frac{2n}{\rho_0}[P(r_0) - P(r_2)]\right\}^{1/2} \tag{10.80}$$

The pressure difference $P(r_0) - P(r_2)$ was calculated in the thermodynamic model to be 147 mb. Picking a value for n that represents a type of swirl velocity distribution, we can see that the maximum value of V_g can be calculated. For $n = \frac{1}{2}$ we compute from Eq. (10.80) a maximum swirl of approximately 120 m/s or 270 miles/h. For $n = 1$, the free vortex, the maximum swirl is approximately 170 m/s. Thus this analysis shows that the thermodynamic–dynamic model we assume can support swirls of this order. Of course, we must realize that the thermodynamic process chosen gives excessive swirl velocities. Also it should be pointed out that Eq. (10.80) was obtained by reducing the differential equation (10.79) to a difference equation and allowing the pressure drop to take place linearly in the whole extent of the hurricane. Nevertheless, peak swirls in strong hurricanes can approach 100 m/s.

E Numerical Solution of the Boundary Layer

By a sequence of steps linearizing Eqs. (10.73)–(10.78) and by use of the fast converging series expansion method with the Galerkin technique, Carrier and his co-workers arrived at the results represented in the following set of graphs depicting the velocity field in the hurricane. First, for a prescribed swirl outside the boundary layer that satisfies $V_g = 0$ at $r = r_0$ in region II, continuity of velocity and stresses gives for the stream function

$$\Psi = (rV_g/\Psi_0)\begin{cases} 1 & \text{for } 0 < x \le \frac{1}{4}x_0 \\ 4[(x/x_0)^{1/2} - x/x_0] & \text{for } \frac{1}{4}x_0 < x \le x_0 \end{cases} \tag{10.81}$$

where Ψ_0 is the maximum value of rV_g, or the strength of the storm, while the dimensionless radial distance $x = r^2\Omega'/\Psi_0$, and x_0 is the outer edge of the storm where the swirl vanishes. Notice that the stream function is made to go to zero there. The corresponding swirl velocity distribution outside the boundary layer is

$$V = \begin{cases} C_1/r & \text{for } 0 < r \le r' \\ C_2(r_0 - r) & \text{for } r' \le r < r_0 \end{cases} \tag{10.82}$$

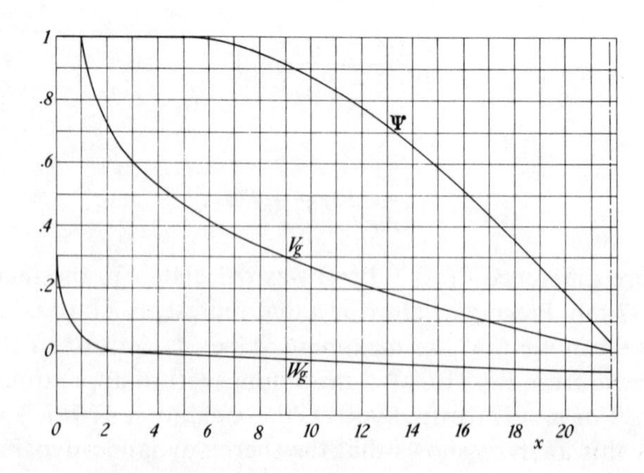

Fig. 10.17 Radial distribution of Ψ, V_g, and W_g for a prescribed swirl.

where r' is the radius where the changeover occurs from a free vortex distribution to a rigid body motion. As can be judged from Fig. 10.17 this transition distance r' is quite far from the center of the storm. The figure also contains the solution for the downdraft velocity W_g outside the boundary layer. It is interesting to note that for $x < 2$ there are strong updrafts. This corresponds to a region approximately 100 km in radius, and in the remainder of the hurricane the sign of W_g is negative, implying downdrafts which, as we shall see, replenish the flow going up region III. The values of the abscissa were determined by taking a typical case where $V_{gmax} = 240$ km/h, with $\nu_T = 10^5$ cm²/s, $r_1 = 30$ km where the maximum swirl occurs, and Ω', the vertical component of the earth's rotation at 15° latitude, was taken as $\frac{2}{3}$ rad/day. The maximum downdraft at the edge of the storm is calculated to be $-W = 0.266$ cm/s.

So Fig. 10.17 gives the velocities outside the boundary layer. For the solutions inside the boundary layer we need a dimensionless vertical variable $z' = z/(\nu_T/2\Omega')^{1/2}$ where again Ω' is the local vertical component of the earth's rotation. Figures 10.18–10.20 show the radial, swirl, and vertical velocity

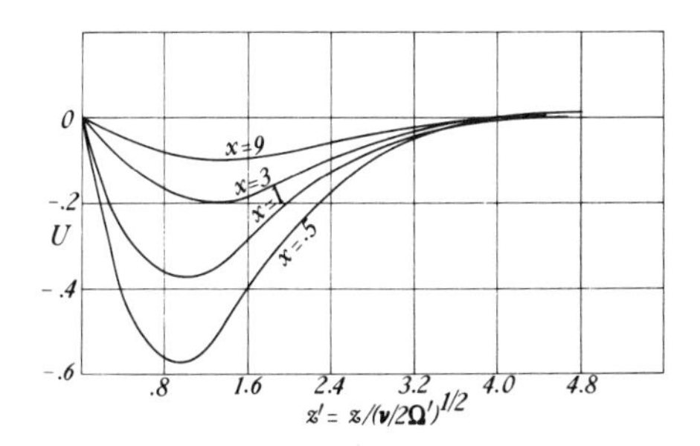

Fig. 10.18 Radial velocity distribution in the boundary layer of region I.

distributions in the boundary layer of the hurricane extending in z' to the geostrophic conditions. In each figure the family of curves belongs to different radial distances x. All values of these velocities have been normalized with V_{gmax} at $x = 1.0$ in Fig. 10.17. Looking at Fig. 10.18, there exists inflow in the boundary layer, for all x, for as high an altitude as $z' = 4.0$, which corresponds to an altitude of approximately 1 km above sea level if we substitute the value of ν_T and Ω' in the definition of z'. Figure 10.19 shows the swirl velocity distribution in the boundary layer region I. Comparing Figs. 10.18 and 10.19, the radial velocity is about one-third of the swirl velocity at each radial distance. These solutions are not applicable for very small radial distances $r < r_1$ because of the strong updrafts there, and consequently Eq. (10.75), which is used for these solutions for the vertical equilibrium, does not contain inertia terms.

Figure 10.20 shows that the vertical velocity is everywhere downward in region I and that its maximum value is at the edge of the boundary layer. Only for $x = 3$ does the solution give a slight indication of an updraft at the edge of the boundary layer.

Because the distributions of U, V, and W depend on the prescribed swirl $\Psi(r)$ in Fig. 10.17, Carrier and his co-workers attempted to determine the sensitivity of the results to different swirl distributions. Their conclusion is that the downdraft velocities are not sensitive to the details of the swirl distribution. Since the amount of pumping is directly related to the downdraft velocities in the major part of the hurricane, the reliability of this velocity becomes important.

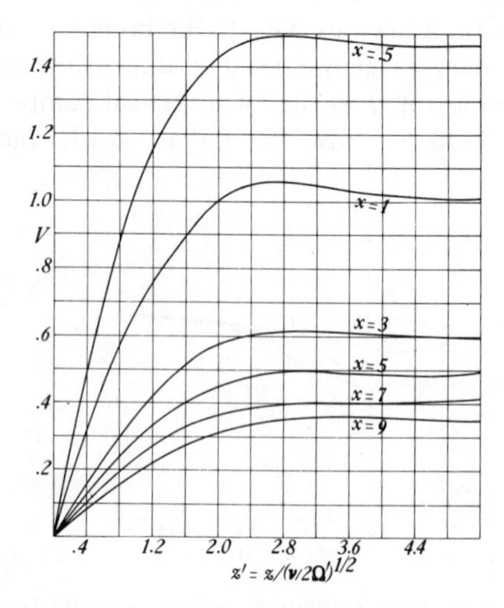

Fig. 10.19 Swirl velocity with altitude at a given radial distance.

With the help of the known velocity field, Carrier proceeds to calculate the energy transfer in the hurricane with the purpose of assessing the main source of energy of the hurricane. It is first concluded that radial convection is smaller than vertical convection, and second that radiation and conduction from the sea surface ("oceanic heating") are small compared to convection. When the largest convection term in the vertical direction is studied for the entire vertical extent of the hurricane, it is concluded that the vertical enthalpy distribution inside the storm is not much different than the distribution outside the storm, which seems to indicate that the main energy process of the storm does not come from direct energy transfer from the surface of the ocean owing to temperature differences. It is concluded that the main energy source of the hurricane is from the release of internal energy in the form of latent heat in the updraft region III.

The work also concludes that the life of the hurricane depends on the time it takes to deplete the volume of moist air in region II, a region 10 km high and 1000 km in radius, and to replace it by the dry and warmer air pumped through region III and filling region II from the top. Since the downdraft velocity in region II was found to be of the order of 0.3 cm/s, a hurricane could last a time given by

$$t \simeq \frac{\pi r^2 h}{\pi r^2 W} \simeq \frac{h}{W}$$

where h is the height of the hurricane (approximately 10 km) or the height of the troposphere, W is the downdraft mean velocity (taken here as 0.3 cm/s), and r is the radius of the storm (taken as 1000 km). With these figures in mind, the time t is of the order of one month. Naturally, this simple estimate excludes energy going into frictional losses and, of course, the entire volume of the storm will have to go through the process of climbing through the funnel of region III. A more realistic life of a hurricane can be of the order of 15 days.

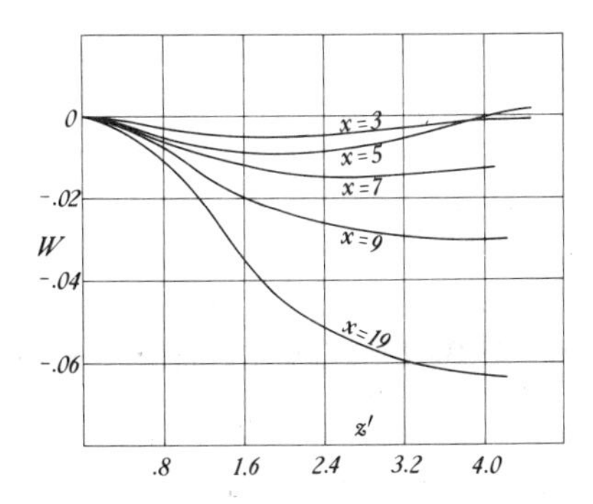

Fig. 10.20 Vertical component of the velocity in the boundary layer.

The dynamical model of the tornado is essentially the same as that of the hurricane just described, except that the inception cannot be assigned to easterly waves, and furthermore the earth's surface cannot provide the unlimited source of latent heat energy as in the case of the ocean for the hurricane. Thus the life of a tornado is considerably shorter.

There have been significant studies made on seeding hurricanes with silver iodide crystals over regions III and II, a few hours before a hurricane enters US coastal areas. Silver iodide crystals greatly increase sublimation (Edwards and Evans, 1968) of the large quantities of vapor in the air, and thus consume the available latent heat in the confines of the hurricane at a faster rate, and weaken the storm faster before it reaches the mainland. Experiments show that this reduction in strength varies between 15 and 30% of the hurricane's strength at the time of seeding; but in almost all cases, the strength of the storm when it reaches the mainland is higher than the strength at the time of seeding. This is illustrated in Fig. 10.21 when seeding takes place 12 h before a hurricane reaches land. Because of this and because states A and B in Fig. 10.21 have variances within a probability function that are higher or lower in value depending on the type of hurricane and on the seeding mission, the decision of whether or not to seed is not an easy matter (Howard *et al.*, 1972). In other words, even with a good knowledge of C, the state A is not absolutely known after seeding has taken place. It could have been considerably higher than B or slightly higher than B, but in all cases B is higher than C.

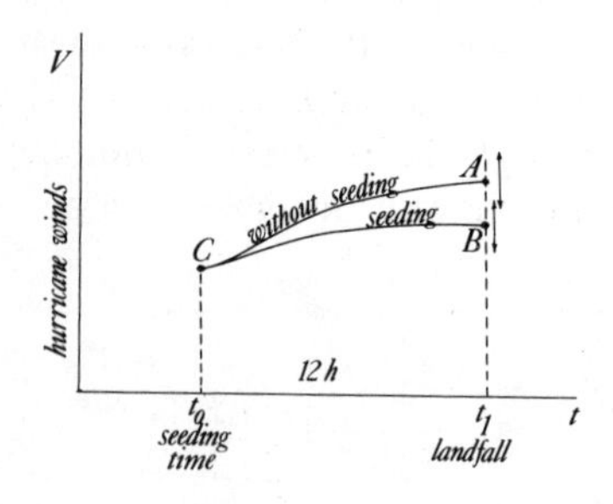

Fig. 10.21 Maximum hurricane winds over time.

10.4 DYNAMICS OF A BALLOON IN A HURRICANE OR A TORNADO

We have seen in the preceding section that large cyclonic storms maintain a radial and a vertical pressure gradient to balance the centrifugal and gravitational accelerations. When a closed finite surface such as that of a balloon is introduced into this pressure field, there will be two components of a generalized force, namely buoyant and centrifugal, which will act on this surface. The net pressure force in the vertical direction will make the balloon rise when the density of the fluid in the balloon is smaller than that of the surrounding air. There will also be a radial (horizontal) net force owing to the difference in pressure on the horizontal projection of the balloon surface which will move the balloon toward the center of the storm since the pressure gradient of a cyclonic storm is positive.

It becomes of interest, after this background information, to investigate the trajectory and the time necessary for the balloon to reach the center of the storm after it has been released, with instruments, at a given location on the ground. (Grant, 1971) gives the following analysis.

Let us define the position of the balloon, at any time, in cylindrical coordinates r, θ, and z as indicated in Fig. 10.22. Let the velocity of the balloon be $\boldsymbol{U}_b$ with components U, V, and W in the three directions, respectively, Let the wind velocity $\boldsymbol{V}_w$ be determined by the velocity of a free vortex

$$|V_w| = \frac{\Gamma}{2\pi}\frac{1}{r} \tag{10.83}$$

We assume throughout this simple analysis that V is everywhere in the θ direction. The density of the air is assumed to vary with elevation only, and is represented, from a dynamical point of view, in the hydrostatic equation. As we shall see, the trajectory of the balloon is not affected by the density variation with altitude. The quantity $\Gamma/2\pi$ is the strength or circulation of the storm.

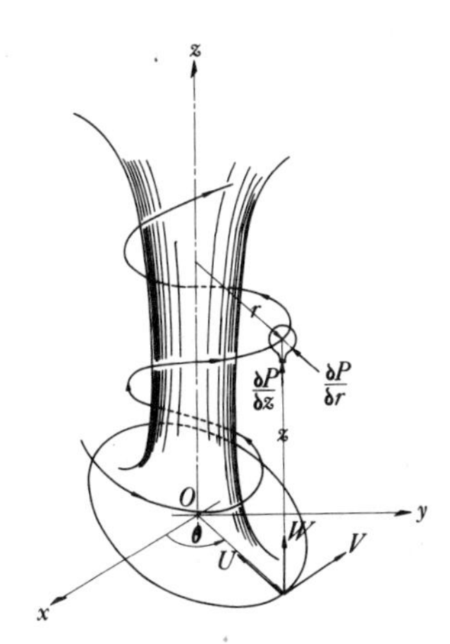

Fig. 10.22 Equilibrium of a balloon in a cyclonic storm.

The forces acting on the balloon are surface forces in the form of pressure and friction, and body forces in the form of gravity and inertia. Over the entire volume $\mathscr{V}$ of the balloon and instruments, this balance of forces is represented by

$$\int_{\mathscr{V}} \rho_b \boldsymbol{a}\, d\mathscr{V} = \int_{\mathscr{V}} \rho_b \boldsymbol{g}\, d\mathscr{V} - \oint_S \boldsymbol{n}P\, dS - C_D \rho \frac{A}{2} |\boldsymbol{U}_b - \boldsymbol{V}_w|(\boldsymbol{U}_b - \boldsymbol{V}_w) \tag{10.84}$$

Here the properties of the balloon are ρ_b, the density of the entire balloon and instruments; $\boldsymbol{a}$ its acceleration; A, the cross-sectional surface of $\mathscr{V}$ normal to the velocity of the balloon relative to the storm; S, the outer surface of $\mathscr{V}$, and $\boldsymbol{U}_b$, the velocity of the balloon. The quantity $(\boldsymbol{U}_b - \boldsymbol{V}_w)$ is the velocity of the balloon relative to the storm. Outside the core of the storm the air has negligible radial and vertical velocity components. The atmospheric properties are P for the pressure, ρ for the density, and V_w for the wind velocity. Thus

$$(\boldsymbol{U}_b - \boldsymbol{V}_w) = \boldsymbol{i}U + \boldsymbol{j}(V - V_w) + \boldsymbol{k}W$$

The quantity C_d is the drag coefficient of the balloon, a dimensionless parameter which is function of the shape of the balloon and the Reynolds number of the motion. We shall consider it constant for a particular application. The frictional force [the last term in Eq. (10.84)] is proportional to the square of the relative velocity, the density of the air, and the surface of the balloon.

Gauss's theorem can be applied to Eq. (10.84), and

$$\int_{\mathscr{V}} [\rho_b(\boldsymbol{a} \cdot \boldsymbol{g}) + \nabla P] \, d\mathscr{V} = -C_d \rho \frac{A}{2} |\boldsymbol{U}_b - \boldsymbol{V}_w| (\boldsymbol{U}_b - \boldsymbol{V}_w)$$

Within the size of the balloon, we can assume that the properties in the volume integral do not vary. Then

$$\rho_b \mathscr{V} \boldsymbol{a} = \rho_b \mathscr{V} \boldsymbol{g} - \mathscr{V} \nabla P - C_d \rho \frac{A}{2} |\boldsymbol{U}_b - \boldsymbol{V}_w| (\boldsymbol{U}_b - \boldsymbol{V}_w) \tag{10.85}$$

or

$$m_b \boldsymbol{a} = \rho_b \mathscr{V} \boldsymbol{g} - m_b \frac{\nabla P}{\rho_b} - C_d \rho \frac{A}{2} |\boldsymbol{U}_b - \boldsymbol{V}_w| (\boldsymbol{U}_b - \boldsymbol{V}_w)$$

where m_b is the total mass of the balloon and instruments. In component form the gravitational force is only in the vertical direction while the pressure force has vertical and radial components. The acceleration of the balloon, in cylindrical coordinates, excluding the Coriolis force (too small for this scale of motion) is

$$\boldsymbol{a} = \boldsymbol{i}\left(\frac{\partial U}{\partial t} - \frac{V^2}{r}\right) + \boldsymbol{j}\left(\frac{\partial V}{\partial t} + \frac{UV}{r}\right) + \boldsymbol{k}\frac{\partial W}{\partial t}$$

This is obtained by expanding the vectorial terms in Eq. (7.23) in cylindrical coordinates [see also Eq. (A.46)]. The three components of the equations of motion of the balloon are

radial:

$$m_b\left(\frac{\partial U}{\partial t} - \frac{V^2}{r}\right) = -\frac{m_b}{\rho_b}\frac{\partial P}{\partial r} - C_d \rho \frac{A}{2} U U_r \tag{10.86}$$

peripheral:

$$m_b\left(\frac{\partial V}{\partial t} + \frac{UV}{r}\right) = -C_d \rho \frac{A}{2} (V - V_w) U_r \tag{10.87}$$

altitude:

$$m_b\left(\frac{\partial W}{\partial t}\right) = -m_b g - \frac{m_b}{\rho_b}\frac{\partial P}{\partial z} - C_d\,\rho\,\frac{A}{2}\,WU_r \tag{10.88}$$

where U_r, the relative velocity, has been substituted for $|\boldsymbol{U}_b - \boldsymbol{V}_w|$. These are the equations to be solved in order to obtain the trajectory of the balloon. The following additional relations are also true:

$$m_b = \rho_b\frac{\pi D^3}{6}, \qquad A = \frac{\pi D^2}{4}, \qquad \frac{\partial P}{\partial z} = -\rho g, \qquad \frac{\partial P}{\partial r} = \rho\frac{V^2}{r} \tag{10.89}$$

where D is the diameter of the balloon.

After the initial period of release of the balloon, there develops a terminal velocity in z, r, and θ such that the accelerations $\partial W/\partial t$, $(\partial V/\partial t + UV/r)$, and $\partial U/\partial t$ are practically negligible. Beyond the initial period the component equations (10.86)–(10.88) reduce to

$$\begin{aligned} -m_b\frac{V^2}{r} &= -\frac{m_b}{\rho_b}\frac{\partial P}{\partial r} - C_d\,\rho\,\frac{A}{2}\,UU_r \\ V &= V_w \\ m_b g &= -\frac{m_b}{\rho_b}\frac{\partial P}{\partial z} - C_d\,\rho\,\frac{A}{2}\,WU_r \end{aligned} \tag{10.90}$$

Now substituting expressions (10.89) we have

$$\begin{aligned} -(V_w^2/r)(\rho - \rho_b)D &= \tfrac{3}{4}C_d\,\rho\,UU_r \\ (\rho - \rho_b)Dg &= \tfrac{3}{4}C_d\,\rho\,WU_r \end{aligned} \tag{10.91}$$

Dividing these two equations we have

$$-\frac{V_w^2}{rg} = \frac{U}{W} = \frac{dr}{dz} \tag{10.92}$$

When we substitute Eq. (10.83) we obtain

$$dz = -g\left(\frac{2\pi}{\Gamma}\right)^2 r^3\,dr$$

which integrates to

$$z_2 - z_1 = \frac{g}{4}\left(\frac{2\pi}{\Gamma}\right)^2 (r_1^4 - r_2^4) \tag{10.93}$$

where the subscripts 1 and 2 refer to points on the trajectory of the balloon. If we pick $z_1 = 0$ on the earth's surface and r_1 as the radius from the center of the storm at that point of release, then z_2 is the elevation to which the balloon will rise. For $r_2 = 0$ we obtain an expression for the elevation $z_2 = z_c$ the height of the balloon when it reaches the center:

$$z_c = \frac{g}{4}\left(\frac{2\pi}{\Gamma}\right)^2 r_1^{\ 4} \tag{10.94}$$

This is an interesting result since it states that the altitude at which the balloon will reach the center of the storm is independent of the properties of the balloon. This altitude is a function of the strength of the storm and the radius at which the balloon is released on the surface of the earth. This technique could also be used, through Eq. (10.94), to determine the strength of the storm from the known values of r_1 and z_c measured independently.

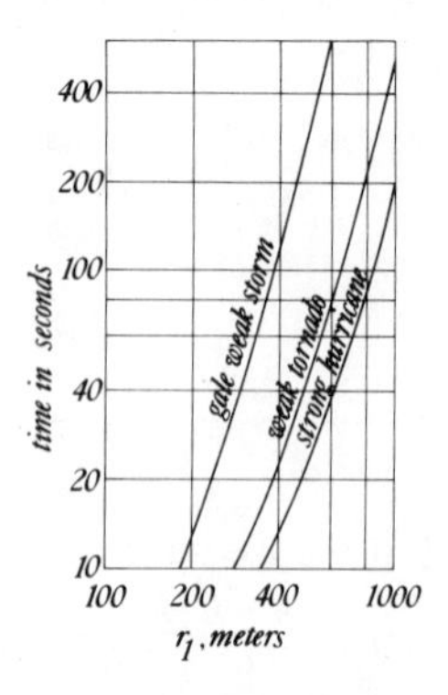

Fig. 10.23 Time for a balloon to reach the center of a storm. (After Grant, 1973.)

Grant computed the time it takes the balloon to reach the center of a gale, an average tornado, and a strong hurricane. He took a balloon 2 m in diameter filled with a gas of density 0.5 kg/cm^3 and computed the time as a function of launch radius r_1. The drag coefficient was taken as $C_d = 0.4$. The results are shown in Fig. 10.23.

10.5 MUNK'S GENERALIZED APPROACH TO WIND-DRIVEN WATER CIRCULATION

In Section 9.3 we discussed Ekman's simplified approach to wind-driven ocean circulation which consisted in ignoring the horizontal pressure gradients, maintaining that the Coriolis parameter f is constant in the confines of the motion, and that the viscous stresses are limited to gradients of velocity with depth, in the z direction. Munk's analysis (1963) begins with the general equation of motion in the horizontal plane, the vertical motion being negligible, and essentially letting the horizontal equilibrium be governed by the hydrostatic equation. Assuming steady motion of the water generated by a steady wind stress at the interface and no other acceleration except the Coriolis, the horizontal dynamical equation is

$$\boldsymbol{k}f \times (\rho \boldsymbol{U}_{\mathrm{h}}) = -\nabla_{\mathrm{h}} P + \nabla \cdot (\mu \nabla \boldsymbol{U}_{\mathrm{h}}) \tag{10.95}$$

For horizontal dynamics only the vertical component of the angular velocity of the earth interacts with the horizontal velocity $\boldsymbol{U}_{\mathrm{h}} = \boldsymbol{i}U + \boldsymbol{j}V$ to give a horizontal Coriolis force. Similarly the horizontal pressure force must include only $\nabla_{\mathrm{h}} = \boldsymbol{i}\nabla_x + \boldsymbol{j}\nabla_y$. The viscosity coefficient is assumed to be constant in the horizontal plane. Variations of viscosity with depth do not alter the results of this analysis.

As Sverdrup (1942) had done earlier, Munk defines a pressure transport and mass transport function as follows:

$$P_0 = \int_0^d P\, dz \qquad \text{and} \qquad \boldsymbol{Q} = \int_0^d \rho \boldsymbol{U}_{\mathrm{h}}\, dz \tag{10.96}$$

where d is the depth beneath which the motion and the horizontal pressure gradients vanish.

When we integrate all of Eq. (10.95) from 0 to d, the quantity $-\int_0^d \nabla_{\mathrm{h}} P\, dz = -\nabla_{\mathrm{h}} P_0$ where at $z = d$ the value of $\nabla_{\mathrm{h}} P$ is zero, and at $z = 0$ (the surface), it is finite.[13] The divergence of the viscous stress tensor $\nabla \cdot \mathscr{S} = \nabla \cdot (\mu \nabla \boldsymbol{U}_{\mathrm{h}})$ in Eq. (10.95) can be expanded as

$$\nabla \cdot (\mu \nabla \boldsymbol{U}_{\mathrm{h}}) = \frac{\partial}{\partial x}\left(\mu \frac{\partial \boldsymbol{U}_{\mathrm{h}}}{\partial x}\right) + \frac{\partial}{\partial y}\left(\mu \frac{\partial \boldsymbol{U}_{\mathrm{h}}}{\partial y}\right) + \frac{\partial}{\partial z}\left(\mu \frac{\partial \boldsymbol{U}_{\mathrm{h}}}{\partial z}\right)$$

For a homogeneous stress field we can break the component expansion just given into the horizontal derivatives

$$\mu \nabla_{\mathrm{h}}^2 \boldsymbol{U}_{\mathrm{h}} = \frac{\partial}{\partial x}\left(\mu \frac{\partial \boldsymbol{U}_{\mathrm{h}}}{\partial x}\right) + \frac{\partial}{\partial y}\left(\mu \frac{\partial \boldsymbol{U}_{\mathrm{h}}}{\partial y}\right)$$

which when integrated in z give

$$\int_0^d \nu \nabla_{\mathrm{h}}^2 (\rho \boldsymbol{U}_{\mathrm{h}})\, dz = -\nu \nabla_{\mathrm{h}}^2 \boldsymbol{Q} \tag{10.97}$$

[13] It might be argued that the ocean surface, when rough, contributes to the value of the integral of the pressure gradient owing to the variations of z at the surface. Although it is expected that the value of P at any horizontal surface will be affected by the surface contour, the effect on ∇P_0 will be small.

and the third component of the stress field when integrated will be

$$\int_0^d \frac{\partial}{\partial z}\left(\mu \frac{\partial \boldsymbol{U}_{\mathrm{h}}}{\partial z}\right) dz = \mu\left(\frac{\partial \boldsymbol{U}_{\mathrm{h}}}{\partial z}\right)_{\mathrm{d}} - \mu\left(\frac{\partial \boldsymbol{U}_{\mathrm{h}}}{\partial z}\right)_0$$
$$= -\boldsymbol{\tau}_{\mathrm{w}} \tag{10.98}$$

We have assumed all along that there is no flow at $z = d$; thus $\boldsymbol{Q}_{\mathrm{d}} = 0$. The minus signs in the last two expressions come from the fact that the upper limit of these integrals is zero. The quantity $\boldsymbol{\tau}_{\mathrm{w}}$ is the stress vector at the ocean surface due to the wind shear, with components τ_{zx} and τ_{zy}. The quantity $\boldsymbol{U}_{\mathrm{h}}$ when integrated between 0 and d gives, as in Eq. (10.96), $\boldsymbol{Q}$ which is the mass transport per unit horizontal distance. Consequently, the term on the left-hand side of Eq. (10.95) becomes, after integration,

$$\boldsymbol{k} \times \int_0^d f\rho \boldsymbol{U}_{\mathrm{h}}\, dz = -f\boldsymbol{k} \times \boldsymbol{Q} \tag{10.99}$$

Again the minus sign is because the upper limit is zero. Now the complete integrated form of Eq. (10.95) is

$$f\boldsymbol{k} \times \boldsymbol{Q} = -\nabla_{\mathrm{h}} P_0 + \nu\, \nabla_{\mathrm{h}}^2 \boldsymbol{Q} + \boldsymbol{\tau}_{\mathrm{w}} \tag{10.100}$$

Introducing the stream function defined in Section 6.5, but applied to the mass transport function $\boldsymbol{Q}$, we obtain

$$Q_x = \int_0^d \rho U\, dz = -\frac{\partial \Psi}{\partial y}; \qquad Q_y = \int_0^d \rho V\, dz = \frac{\partial \Psi}{\partial x}$$

This can be done because the vector $\boldsymbol{Q}$ also obeys the conservation law

$$\frac{\partial Q_x}{\partial x} + \frac{\partial Q_y}{\partial y} = 0$$

It is clear then[14]

$$\boldsymbol{Q} = \boldsymbol{k} \times \nabla \Psi \tag{10.101}$$

In attempting to reduce the dynamical equation to a simpler form, we take the curl of Eq. (10.100)

$$\nabla \times (f\boldsymbol{k} \times \boldsymbol{Q}) = -\nabla \times (\nabla_{\mathrm{h}} P_0) + \nu\, \nabla \times (\nabla_{\mathrm{h}}^2 \boldsymbol{Q}) + \nabla \times \boldsymbol{\tau}_{\mathrm{w}} \tag{10.102}$$

Let us develop the left-hand side of this equation. Using the triple vector product of Eq. (A.59e), we have

$$\nabla \times (f\boldsymbol{k} \times \boldsymbol{Q}) = f\boldsymbol{k}(\nabla \cdot \boldsymbol{Q}) + (\boldsymbol{Q} \cdot \nabla) f\boldsymbol{k} - \boldsymbol{Q}(\nabla \cdot f\boldsymbol{k}) - (f\boldsymbol{k} \cdot \nabla)\boldsymbol{Q}$$

[14] This is the same as Eq. (6.23) but integrated in z. The definition of Ψ here is mass per unit time, while in Section 6.5 it is mass per unit time and length.

Every term in this expression is zero except $(\boldsymbol{Q} \cdot \nabla) f \boldsymbol{k}$, the reason being that $\nabla \cdot \boldsymbol{Q} = 0$ owing to conservation of mass of an isochoric substance like water. Then since $\boldsymbol{k}$ is a constant unit vector $\nabla \cdot \boldsymbol{k} = 0$, $\boldsymbol{k} \cdot \nabla = 0$ because the nabla has significance only in the horizontal plane in this problem. Thus

$$(\boldsymbol{Q} \cdot \nabla)\boldsymbol{k} f = \boldsymbol{k}\left(Q_x \frac{\partial f}{\partial x} + Q_y \frac{\partial f}{\partial y}\right)$$

$$= \boldsymbol{k} Q_y \frac{\partial f}{\partial y} = \boldsymbol{k} \frac{\partial f}{\partial y} \frac{\partial \Psi}{\partial x} \tag{10.103}$$

We know that the curl of $\nabla_{\mathrm{h}} P_0$ is zero, and the curl of the Laplacian in Eq. (10.102) can be replaced by the Laplacian of the curl. In addition, f is not a function of longitude. Thus

$$\nu \nabla \times (\nabla_{\mathrm{h}}^2 \boldsymbol{Q}) = \nu \nabla_{\mathrm{h}}^2 (\nabla \times \boldsymbol{Q})$$

From Eq. (10.101)

$$\nabla \times \boldsymbol{Q} = \nabla \times (\boldsymbol{k} \times \nabla \Psi)$$

Again using the triple vector product

$$\begin{aligned} \nabla \times \boldsymbol{Q} &= \nabla \times (\boldsymbol{k} \times \nabla \Psi) \\ &= \boldsymbol{k}(\nabla \cdot \nabla \Psi) + (\nabla \Psi \cdot \nabla)\boldsymbol{k} - \nabla \Psi (\nabla \cdot \boldsymbol{k}) - (\boldsymbol{k} \cdot \nabla)\, \nabla \Psi \\ &= \boldsymbol{k}\, \nabla^2 \Psi \end{aligned}$$

Only the first term in the expansion is nonzero for reasons already given. Thus the viscous term in Eq. (10.102) becomes

$$\nu \nabla \times (\nabla_{\mathrm{h}}^2 \boldsymbol{Q}) = \boldsymbol{k} \nu\, \nabla^4 \Psi$$

Finally Eq. (10.102) can be rewritten by omitting the subscript h on the gradients:

$$\left(\nu \nabla^4 - \frac{\partial f}{\partial y} \frac{\partial}{\partial x}\right) \Psi = -(\nabla \times \boldsymbol{\tau}_{\mathrm{w}})_z \tag{10.104}$$

The subscript z in the curl of the wind stress implies the z component of the curl only, and contains terms in the x–y plane. Thus Eq. (10.104) is a two-dimensional equation. Thus this final equation is the governing equation for wind-driven ocean currents. The biharmonic operator

$$\nabla^4 = \frac{\partial^4}{\partial x^4} + \frac{2\, \partial^4}{\partial x^2\, \partial y^2} + \frac{\partial^4}{\partial y^4}$$

Equation (10.104) expresses the circulation in terms of the stream function. Since it was obtained by taking the curl of Eq. (10.100), it can be considered as the vorticity equation for ocean circulation. It represents the balance between the rate of change of vorticity due to the horizontal viscous stresses in the biharmonic operator, and the rate of vorticity generated by the wind, and also by $Q_y\,\partial f/\partial y$, which Ekman called the planetary vorticity due to the variation of the vertical component of the angular velocity of the earth.

At high latitudes, $\partial f/\partial y$ is very small and can be neglected. Thus the vorticity equation of the ocean circulation takes a special form:

$$\nu\,\nabla^4\Psi = -\left(\frac{\partial\tau_{zy}}{\partial x} - \frac{\partial\tau_{zy}}{\partial y}\right) \tag{10.105}$$

This equation is analogous to the equation for the deflection Ψ of a plate of uniform flexural rigidity ν, clamped at its boundaries, and subjected to a load equal to $-\nabla \times \boldsymbol{\tau}_w$.

Munk has presented a general solution of Eq. (10.104) subject to a number of applicable boundary conditions. The general solution for the stream function is taken as

$$\Psi = -b\chi\beta^{-1}\left(\frac{\partial\tau_{zy}}{\partial x} - \frac{\partial\tau_{zx}}{\partial y}\right) \tag{10.106}$$

where b is the extent of the ocean in the x direction, χ is a function obeying a fourth-order differential equation and having a general solution

$$\chi = -Ke^{-1/2kx}\cos\left(\frac{\sqrt{3}}{2}kx + \frac{\sqrt{3}}{2kb} - \frac{\pi}{6}\right) + 1 - \frac{1}{kb}\left[kx - e^{-k(b-x)} - 1\right]$$

We shall investigate a special case of this solution. The coefficient β is $\partial f/\partial y$. The quantity k is the Coriolis–friction wave number which is equal to $(\beta/\nu)^{1/3}$, and K is $(2/\sqrt{3}) - (\sqrt{3}/kb)$.

In mid-ocean, outside eastern and western boundaries, the function χ reduces to a simpler form:

$$\chi = 1 - \frac{x}{b} \tag{10.107}$$

For this special case let us examine the ocean circulation generated by a strong cyclonic windstorm. Let the circumferential velocity of the cyclone be represented by a general expression

$$V = \frac{\Gamma}{a}\left(\frac{r}{a}\right)\exp\left(-\frac{r^2}{a^2}\right) \tag{10.108}$$

This swirl distribution is more realistic for a cyclone than for a free vortex, as assumed in Section 10.3, because it gives a finite shear in the center of the storm. The constant a has the units of distance which can be established by the fact that the velocity of the vortex is maximum at $r = a\sqrt{2}/2$. The coefficient Γ is related to the strength of the vortex.

Since the radial velocity $U = 0$ and we assumed axisymmetry, the only component of the shear stress is $\tau_{r\theta}$ which is expressed as

$$\tau_{r\theta} = \mu\left[r\frac{d}{dr}\left(\frac{V}{r}\right)\right]$$

With Eq. (10.108) the shear stress for this vortex becomes

$$\begin{aligned}\tau_{r\theta} &= \mu\left\{r\frac{d}{dr}\left[\frac{\Gamma}{ar}\left(\frac{r}{a}\right)\exp\left(-\frac{r^2}{a^2}\right)\right]\right\}\\ &= -\frac{2\Gamma\mu}{a^2}\frac{r^2}{a^2}\exp\left(-\frac{r^2}{a^2}\right)\end{aligned} \qquad (10.109)$$

Equations (10.108) and (10.109) are shown in Fig. 10.24.

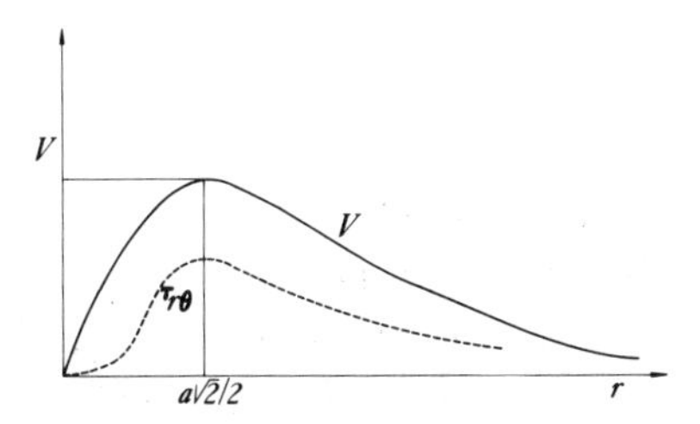

Fig. 10.24 Velocity distribution of an atmospheric cyclone.

We define the average curl of the shear along a latitudinal segment as

$$\overline{\nabla_h \times \tau_w} = \frac{1}{b-x}\int_x^b (\nabla_h \times \tau_w)\, dx$$

and this value of the average is replaced in the stream function solution of Eq. (10.106)

$$\begin{aligned}\Psi &= -b\left(1-\frac{x}{b}\right)\beta^{-1}\frac{1}{b-x}\int_x^b (\nabla_h \times \tau_w)\, dx\\ &= \beta^{-1}\int_x^{\infty} (\nabla_h \times \tau_w)\, dx\end{aligned}$$

We have taken b, the size of the storm, as infinite. We know that for the two-dimensional circulation we have adopted all along, the curl of the stress has one component only. That is,

$$\nabla_h \times \tau_w = \frac{1}{r}\frac{d}{dr}(r\tau_{r\theta})$$

$$= -\frac{2\Gamma\mu}{a^3}\left[3\left(\frac{r}{a}\right) - 2\left(\frac{r}{a}\right)^3\right]\exp\left(-\frac{r^2}{a^2}\right)$$

Then the solution of this problem in terms of the stream function is

$$\Psi = \frac{2\Gamma\mu}{a^3}\beta^{-1}\int_x^\infty \left[3\left(\frac{r}{a}\right) - 2\left(\frac{r}{a}\right)^3\right]\exp\left(-\frac{r^2}{a^2}\right)dx \tag{10.110}$$

Munk, in his analysis, started with wind stress distribution as given by Eq. (10.108) instead of the velocity distribution and obtained the following results. He assumed the motion to be anticyclonic and obtained

$$\Psi = \tfrac{1}{2}\sqrt{\pi}\,\Gamma\beta^{-1}\{[1 - 2(y/a)^2][1 - I] + \tfrac{1}{2}I''\}$$

where

$$I\left(\frac{r}{a}\right) = \frac{2}{\pi}\int_0^{r/a}\exp\left(-\frac{r^2}{a^2}\right)d\left(\frac{r}{a}\right)$$

is the probability integral and $I'' = d^2I/d(r/a)^2$. Figure 10.25 is the result of his analysis, where the ocean mass transport streamlines are represented by the lines $\Psi\beta/\Gamma$. The anticyclonic vortex is located at $(-a, 0)$, and a trough with a saddle point occurs at a, 0. The center of the oceanic vortex is displaced west-

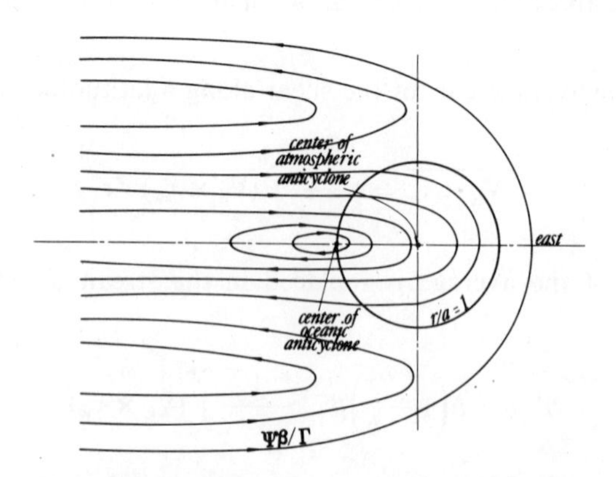

Fig. 10.25 Oceanographic circulation induced by atmospheric anticyclone. (After Munk, 1963.)

ward from the center of the atmospheric vortex by $\sqrt{2}$ times the radius of the maximum stress circle. Munk claims that data in the Eastern Pacific show a westward displacement of the oceanic anticyclone, relative to the Pacific high pressure area with which it is associated. The magnitude of the displacement is approximately 1000 km.

For large anticyclonic atmospheric motions in mid-ocean, one expects a tail of zonal currents west of the wind system feeding water into the system north of the wind axis and expelling it south of the wind axis (vice versa for cyclones) with relative weak compensation currents farther to the north and south.

10.6 BAROCLINIC SECONDARY FLOW IN ESTUARIES

In Section 8.10.B we introduced the problem of ocean circulation owing to the secondary motion resulting from the misalignment of the isobaric and isosteric surfaces. This production of circulation owing to baroclinicity is contained in Kelvin's theorem and expressed in Eq. (8.64). The term that produces circulation because of the baroclinic conditions in the estuaries is the first integral to the right of the equality sign. The analysis that follows is equally valid for discharges of industrial liquid wastes into the ocean, lakes, and rivers provided baroclinic conditions exist.

Consulting Fig. 8.15 again, baroclinicity takes place in the estuary because of the diffusion of salt from the sea into the freshwater river. The density of the estuary waters depends more on the diffusion process than on the pressure distribution. The isobaric surfaces, even with horizontal motion of the river, are very nearly horizontal surfaces, whereas the isosteric surfaces are inclined from the vertical, toward upstream of the river in the form of wedges as shown in the schematic diagram of Fig. 10.26.

The state of mixing of salt water and fresh water may depend considerably on the magnitude and frequency of tidal motion in the sea. For this reason, estuaries are called *well mixed, partially mixed, or stratified* on the basis of the level of density stratification. When the tide has little or no influence on the

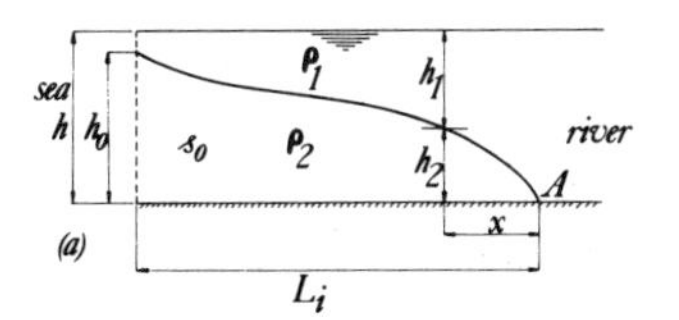

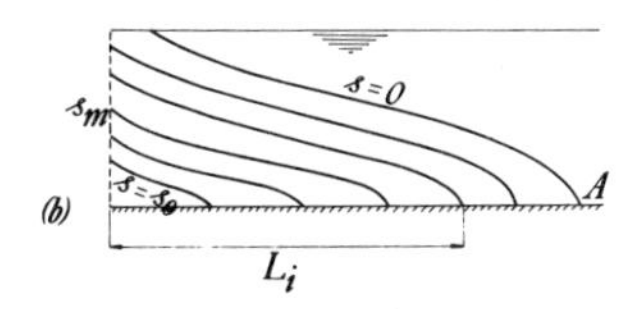

Fig. 10.26 (a) Saline arrested wedge. (b) Well-mixed estuary.

motion of the waters in the estuary, the salinity stratification from seawater salinity to fresh water is abrupt along an *arrested wedge* having maximum seawater salinity on one side and no salinity on the other, as shown in Fig. 10.26a. In locations where tidal motion produces considerable mixing, the level of salinity in the estuary is distributed and the constant-salinity surfaces varying in value from sea-level salinity to zero are parallel surfaces as shown in Fig. 10.26b.

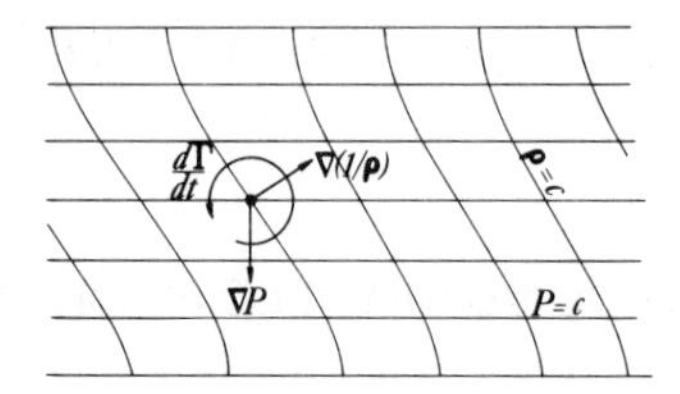

Fig. 10.27 Estuarine baroclinicity.

Regardless of the mixing level, the production of circulation due to the conditions shown in Fig. 10.27, and with the help of the development given in Eq. (8.68),

$$\frac{d\Gamma}{dt} = \int_A \left[\nabla P \times \nabla\left(\frac{1}{\rho}\right)\right] \cdot \mathbf{n}\, dA \tag{10.111}$$

and the rate of increase of circulation will be counterclockwise on the plane of the coordinates of Fig. 10.27. This implies that this secondary motion will tend to accelerate the river flow on layers near the surface of the river and retard flow on layers at the bottom. Figure 10.28 shows the development of the through-flow velocity distribution in the estuary as it progresses toward the ocean. It is apparent that the boundary layer vorticity due to friction is added on to the vorticity generated by baroclinicity, as predicted by Eq. (8.64).

How does one estimate this secondary circulation which, in some estuary conditions, is crucial for navigation, and for estimation of erosion and ensuing shoaling? The most basic but involved approach is to solve the combined dynamical and salt diffusion equations subject to the boundary conditions of the estuary. Outside laboratory models, this can be quite involved and even unsatisfying owing to many other real life effects, mainly in boundary conditions such as three-dimensionality, roughness of river bottom, runoff from land, etc. Instead, a similarity or self-preservation method such as the one outlined for the thermal plume in Section 10.2 can be more direct and satisfying, provided that the motion and the diffusion process lend themselves to this approach. In other words, the physical processes in the estuary, at any location, must be dominantly dependent on local conditions and not on the history of development. At any rate, if we assume that self-preserving solutions exist in the form of Eqs. (10.19)–(10.22) for velocities and salinity, substitution of these relations in the equations of motion and diffusion ought to give the conditions we must impose on the characteristic length, velocity, and salinity for such solutions to exist. It is well known that boundary layer flows admit self-preserving solutions, although there is not a single unique expression such as Eq. (10.19) to represent the entire boundary layer. Nevertheless, by dividing the boundary layer into an inner layer and an outer layer logarithmic expressions for the velocity profiles in the inner and outer layers satisfy the dynamics of the motion.[15] With this in mind the salinity diffusion equation, similar to Eq. (10.18) after orders of magnitude have been compared, is

$$U\frac{\partial s}{\partial x} + V\frac{\partial s}{\partial y} = -\frac{\partial \overline{vs'}}{\partial y} - \frac{\partial \overline{us'}}{\partial x} \tag{10.112}$$

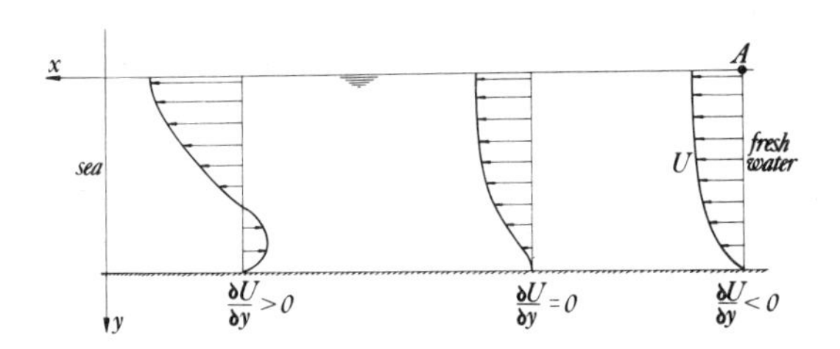

Fig. 10.28 Velocity profiles in an estuary.

According to the coordinate directions in Fig. 10.28, the x momentum of the stream decreases in the x direction, however, the salinity increases in that direction. Thus the motion of the stream is in the opposite sense of the diffusion. For this reason, as in the case of self-preservation of wake flows, we must take in Eq. (10.112) $s - s_m$ as the quantity decreasing in the direction of x, where s_m is the maximum salinity and s' is the fluctuation. Thus we may now say that the salinity difference

$$\frac{s - s_m}{s_m} = \phi_1(\eta) \tag{10.113}$$

$$\frac{-\overline{vs'}}{u_* s_m} = \phi_2(\eta) \tag{10.114}$$

$$\frac{-\overline{us'}}{u_* s_m} = \phi_3(\eta) \tag{10.115}$$

[15] Consult Section 9.4 and, for instance, Tennekes and Lumley (1972) or Townsend (1956).

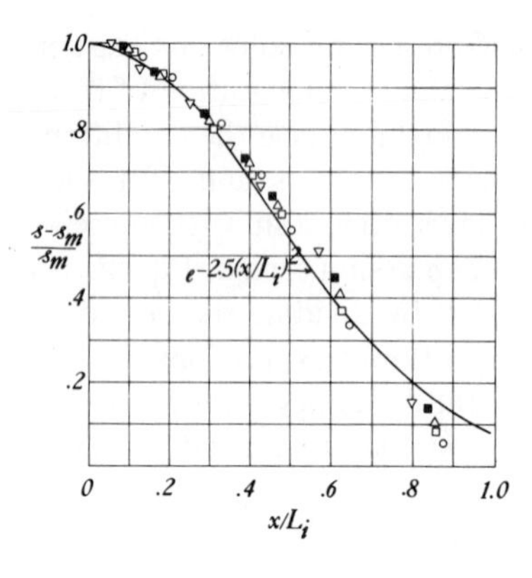

Fig. 10.29 Form of salinity distribution in a well-mixed estuary. Values of y/h are plotted as follows: ○ 0.1; □ 0.3; △ 0.5; ■ 0.7; ▽ 0.9.

where u_* is the characteristic velocity of the motion which can be taken as the *shearing velocity* and the *densimetric velocity* in the inner and outer layers, respectively. The densimetric velocity is defined as $(hg\,\Delta\rho/\rho)^{1/2}$, where h is the depth of the river, g is the gravitational constant, and $\Delta\rho$ is the difference be- ween the seawater density and the density of fresh water ρ.

Silverman (1973) has shown from Harleman and Ippen's data (1967) that Eq. (10.113) seems to be justified from experimental data in estuaries that are well mixed, partially mixed, and tending to become stratified. Denoting $\eta = x/L_i$, where L_i is the salinity intrusion length, Figs. 10.29–10.31 show that for a given intensity of stratification and at all depths y/h the salinity deficiency can be expressed in the general form

$$\frac{s - s_m}{s_m} = \exp\left[-K\left(\frac{x}{L_i}\right)^2\right] \tag{10.116}$$

where K is dependent on the level of stratification, which from the figures is shown to vary between 2.5 and 5.0. Since x is measured, according to Fig.

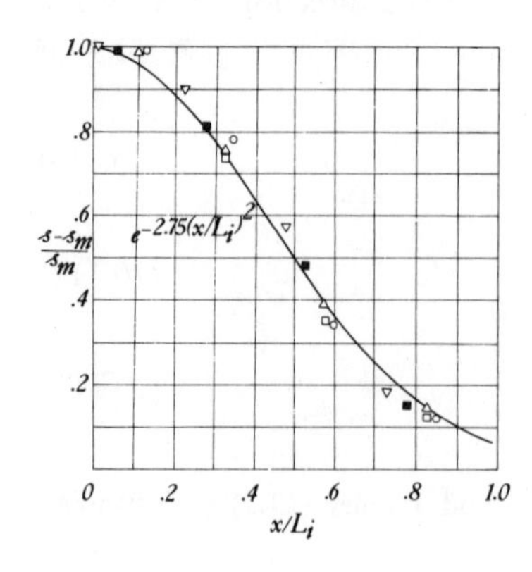

Fig. 10.30 Form of salinity distribution in a partially mixed estuary. See legend of Fig. 10.29 for y/h values.

10.26, from the point A where the salinity everywhere in the river is zero, then at the mouth of the estuary, $y = h$ and $x = L_i$, the salinity is that of the sea, or $s = s_m = s_0$. At any other y, $s_m < s_0$.

Assuming that Eq. (10.116) holds true for a great number of estuaries, knowing $s_m = f(y)$, K, and L_i one could determine the salinity at any point in the estuary. The density of the water can be expressed in terms of salinity in the form of Eq. (4.25) and the gradients of $(1/\rho)$ appearing in Eq. (10.111) can be computed. Although there is motion in the x direction, the pressure field in the estuary is predominantly governed by the hydrostatic equation, and the gradient of the pressure is given by $\rho \boldsymbol{g}$. Thus the production of circulation can be computed on this basis.

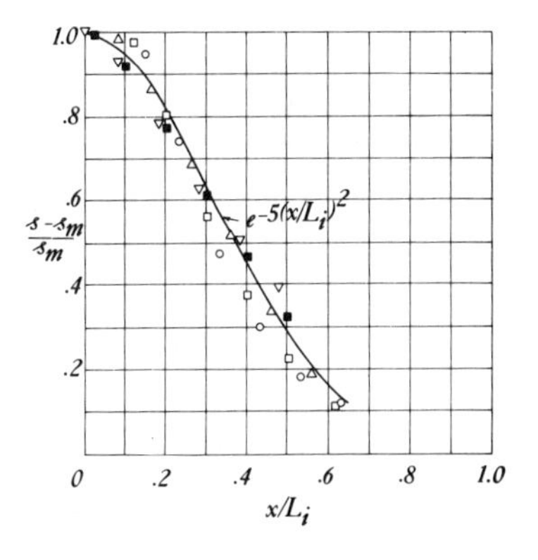

Fig. 10.31 Form of salinity distribution in a partially mixed (becoming stratified) estuary. See legend of Fig. 10.29 for y/h values.

As in the case of the discussion in Section 8.10.A, the circulation actually does not grow without bounds, as Eq. (10.111) indicates. Viscosity effects and mixing owing to this secondary motion maintain an upper bound in the production of circulation. If strong viscous forces impose a time limit on growth, then the characteristic time to final growth is determined from boundary layer considerations to be proportional to h^2/ν, where h is the river depth and ν is the kinematic viscosity of the water. If mixing due to stratification controls the time of growth of circulation, then the buoyancy development time $(h/g\ \Delta\rho/\rho)^{1/2}$ may control the process of development.

When Γ_{max} is obtained from the integration of Eq. (10.111), then induced velocities can be obtained from this result.

APPENDIX A

Basic Concepts of Vector Analysis

A.1 SCALARS AND VECTORS

When physical quantities are to be described, it is noticed that some require a number or magnitude with the proper units while others require in addition a direction relative to a given frame of reference. This distinction in the inherent representation of physical quantities brings interesting and specialized analysis in mathematics of considerable importance to the engineer and scientist.

When temperature is described at a given point in space, it is sufficient to give a single number a accompanied by the proper units—degrees centigrade (°C) or degrees Fahrenheit (°F). The single, real number a does not have the same value followed by different units, but the fact still remains that a physical quantity such as the temperature requires only one number for its description and consequently is called a *scalar quantity*. For identification, symbols representing a scalar quantity will be italicized. Other scalar quantities include the density, the energy, and most thermodynamic properties. We can say that in the absence of motion thermodynamics is an exclusive science of scalar properties.

Other physical quantities such as the velocity or the force require for their description, besides the magnitude and its proper units, a direction of action. This is extremely important for without their direction the position of a moving particle or the effect of a force cannot be determined. These physical quantities are then called *vectors* because they are directed quantities. To establish the direction of a vector, a frame of reference may be chosen and its choice is just as arbitrary as the units associated with the magnitude.

Graphically a vector can be represented completely by a pointed straight line originating at a point in space. Its length is representative of the magnitude and its position in space establishes its *sense of action*. The arrowhead at the other end from the *origin* represents its direction. A *unit vector* is a vector whose magnitude is unity.

Two vectors are said to be equal when they have the same length and direction. For this to be true the two vectors need not coincide but they must be parallel and of equal length. If the origin of a vector may be chosen arbitrarily the vector is said to be *free* or *axial.*[1] A vector is said to be *bound* or *polar* when its origin is fixed in space. The origin of the vector then describes the point where the vector applies. For instance a force is a polar vector since when moved to different positions it causes different moments although its magnitude and direction are kept the same. A vector symbol will be distinguished by italic, *boldface* type.

Since a vector needs a direction in order to be identified and since its direction can only be identified relative to a frame of reference, it is easy to see that one scalar number a is no longer sufficient for its description. For instance if a reference system of three mutually perpendicular axes is chosen, it will be necessary to specify at least three components of this vector (a_1, a_2, and a_3) on the respective axes, or its magnitude and two angles in the same coordinate system. Now it becomes clear that, unlike the scalar, a vector needs three independent scalar quantities for its description. Since two vectors are equal only when their magnitude and direction are the same, this implies that their respective components in a given coordinate system are equal.

A.2 VECTORS IN ORTHOGONAL COORDINATE AXES

In order to represent a point in space it is customary to use an orthogonal coordinate system. For this we choose three mutually perpendicular lines as coordinate axes, say x, y, and z, and to each point in space a value of x, y, and z is assigned. In setting up the coordinate system, for reasons of consistency it is necessary to pay attention to the orientation of the coordinates.

Rotation is the rate of displacement around an axis. Its positive or negative sense is merely a question of definition. Since rotation is often described on a plane with its axis of rotation normal to it, the sense of rotation and the direction of the axis normal to the plane of rotation must conventionally be established. Since a screw turns around its axis and advances at the same time it is used as the symbol for this correspondence. A *right-handed* screw is one which, when turned clockwise (looking at its head), advances in the direction of its pointed end. A *left-handed* screw will do the opposite. Now a coordinate system is called right-handed when, looking at the succession *xyzxyzx*..., any axis rotated toward the following coordinate gives the direction of advancement of a right-handed screw coinciding with the direction of the third coordinate. This is shown in Fig. A.1. A left-handed coordinate system will have one axis reversed.

By convention the right-handed coordinate system is used in science.

[1] The reason for the choice of this name is that vectors representing rigid body rotation, such as the rotation of the earth, are of this type. The position of the rotation vector is of no importance as long as it maintains its direction and magnitude.

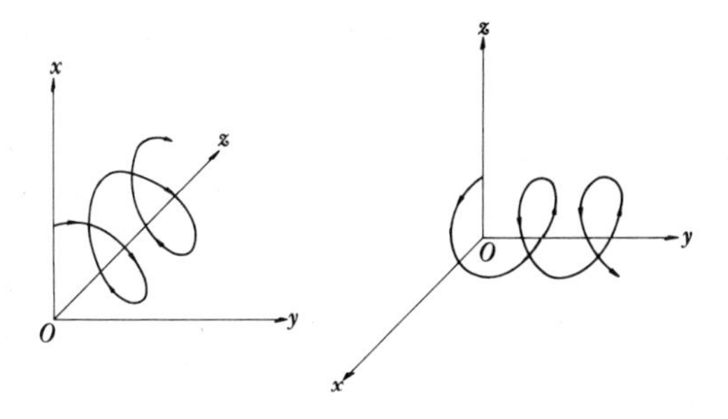

Fig. A.1 Right-handed coordinate system.

Now it is important to find out the relationship of the vector to its arbitrary sets of components. For the purpose of illustration consider a point P whose position is specified by the *position vector* $\overrightarrow{OP} = \boldsymbol{r}$ in terms of the three Cartesian coordinates in a three-dimensional Euclidean space with reference to an origin O.

In Fig. A.2, γ_1, γ_2, and γ_3 are the angles the position vector makes with the orthogonal axes, and, if m_1, m_2, and m_3 are the directional cosines,

$$m_1 = \cos \gamma_1$$

$$m_2 = \cos \gamma_2$$

$$m_3 = \cos \gamma_3$$

or in general we can say that

$$m_i = \cos \gamma_i$$

where i is an index varying from 1 to 3, in this case, showing the angle to the appropriate coordinate. Also we can say that

$$x_i = rm_i \tag{A.1}$$

represents three scalar equations. If the square of each of these equations is added, we obtain the square of the magnitude of $\boldsymbol{r}$:

$$x_1^{\,2} + x_2^{\,2} + x_3^{\,2} = r^2(m_1^{\,2} + m_2^{\,2} + m_3^{\,2})$$

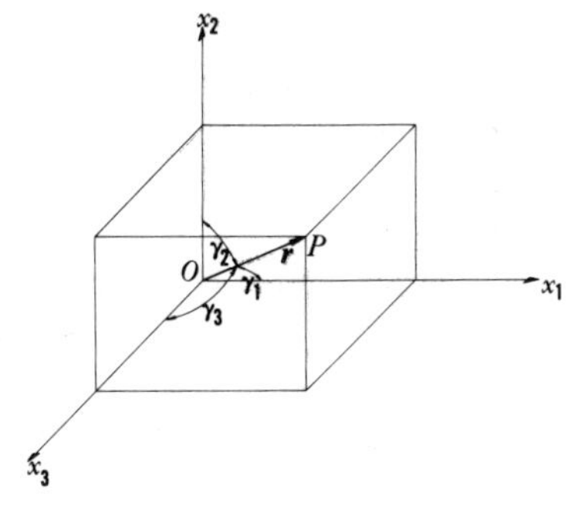

Fig. A.2 A vector in a Cartesian reference frame.

With the use of the Pythagorean theorem we can conclude that

$$m_1^2 + m_2^2 + m_3^2 = 1$$

Knowing the three components x_i, with the sum of the squares the magnitude of $\boldsymbol{r}$ is found. Its direction is established through Eq. (A.1) by solving for m_i. Therefore to specify a vector completely through the use of an orthogonal coordinate system we need to know the three equations x_i instead of one as in the case of a scalar. Since these relations are valid for any Cartesian coordinate system, although the values of x_i and m_i will vary from one coordinate system to the other with O as its origin, the vector $\boldsymbol{r}$ will not depend on the choice of coordinate. This leads us to believe that if vectors and vector operations are treated without any use of coordinates, the representation would be unique and independent on the coordinates used.

Since vector quantities have a magnitude as well as a direction, in other words since three scalars are necessary to describe them, algebra and calculus must be extended for vector operations. We must regard at all times *a vector as an entity that has different components in different coordinate systems and yet the resultant of any set of coordinates gives the same vector*. This is a property of vectors that is very important in vector operations.

A.3 ADDITION AND SUBTRACTION OF VECTORS

The rules of operations with vectors are analogous to those of operations with ordinary scalar numbers; namely, that the *result of the operation must be an entity and therefore not dependent on the frame of reference used to describe it.*

Let us define the sum of two vectors $\boldsymbol{a}$ and $\boldsymbol{b}$ to be a third vector $\boldsymbol{c}$ whose components are the sums of the components of $\boldsymbol{a}$ and $\boldsymbol{b}$ on the same axis. First it is necessary to see if the new components from the result of the additions

$$\begin{aligned} a_x + b_x &= c_x \\ a_y + b_y &= c_y \\ a_z + b_z &= c_z \end{aligned} \tag{A.2}$$

are independent of the orientation of the coordinate system. Every arbitrary operation may not necessarily give a result that is independent of the choice of coordinates. To illustrate that the sum as defined gives an entity, consider Fig. A.3 where two coordinate systems are chosen, both on the plane formed by $\boldsymbol{a}$ and $\boldsymbol{b}$. Since in doing so the z components of the vectors are zero, some generality in the proof is lost. However, this loss of generality was made for the purpose of clarity. These results are just as valid if the third coordinate was considered.

If the origin of $\boldsymbol{b}$ is brought to the end of $\boldsymbol{a}$, as shown, then from the projections on Oxy and $Ox'y'$ it is seen that

$$a_x + b_x = c_x$$
$$a_y + b_y = c_y$$

conforms with the definition. It is also seen that the *commutative law* of addition is satisfied,

$$\boldsymbol{a} + \boldsymbol{b} = \boldsymbol{b} + \boldsymbol{a}$$

and thus to form an addition of two vectors we need to form a parallelogram with the vectors $\boldsymbol{a}$ and $\boldsymbol{b}$ following in direction, and the diagonal extending from the origin of the first to the end of the other determines the addition vector. The law of vector addition is often called the *parallelogram law*.

The rule for subtraction is obtained readily from the figure since $\boldsymbol{c} - \boldsymbol{b} = \boldsymbol{a}$. The vector $\boldsymbol{b}$ is opposed to $\boldsymbol{c}$ in that their arrows oppose at Q and thus $\boldsymbol{a}$ starting at the origin of $\boldsymbol{c}$ becomes the result of the subtraction.

The *associative law* can be shown to hold such that

$$\boldsymbol{a} + (\boldsymbol{b} + \boldsymbol{c}) = (\boldsymbol{a} + \boldsymbol{b}) + \boldsymbol{c} = \boldsymbol{a} + \boldsymbol{b} + \boldsymbol{c} \tag{A.3}$$

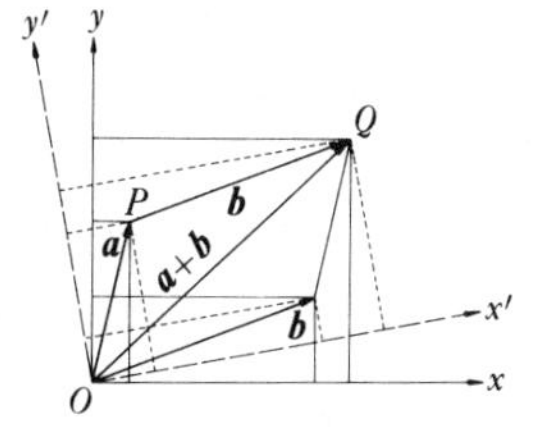

Fig. A.3 Addition of two vectors.

Finally, from Eq. (A.2) we see that a *vector equation* is equivalent to three scalar equations and that the result of a summation of two vectors is *coplanar* with the two vectors.

A.4 MULTIPLICATION OF A VECTOR WITH A SCALAR

At least for the case when the scalar is an integer it is easy to see that $m\boldsymbol{a}$ implies m additions of the same vector which results in a new vector in the same direction but with a magnitude m times larger. The same is true for the case when m is not an integer. The following relations are then true:

$$\begin{aligned} (m + n)\boldsymbol{a} &= m\boldsymbol{a} + n\boldsymbol{a} \\ m(n\boldsymbol{a}) = n(m\boldsymbol{a}) &= (mn)\boldsymbol{a} = mn\boldsymbol{a} \end{aligned} \tag{A.4}$$

A.5 LINEAR RELATION OF COPLANAR VECTORS

Given two coplanar, nonparallel, nonzero vectors $\boldsymbol{a}$ and $\boldsymbol{b}$, any third coplanar nonparallel vector $\boldsymbol{c}$ can be expressed as a linear combination of $\boldsymbol{a}$ and $\boldsymbol{b}$. In other words

$$\boldsymbol{c} = m\boldsymbol{a} + n\boldsymbol{b} \tag{A.5}$$

This can be shown readily by geometrical construction (see Fig. A.4). Let $\boldsymbol{a}$, $\boldsymbol{b}$,

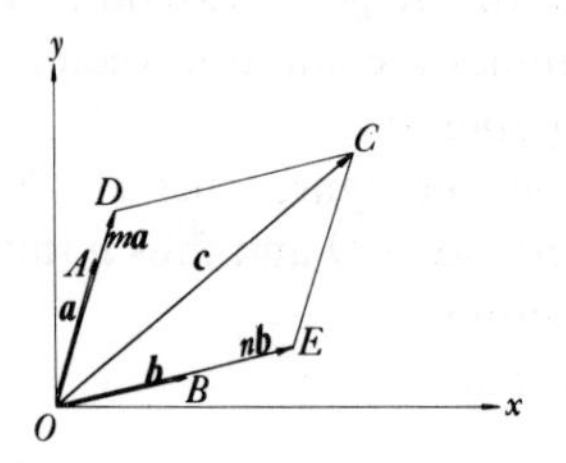

Fig. A.4 Linear relation of coplanar vectors.

and $\boldsymbol{c}$ in Fig. A.4 be three coplanar vectors; by translating them their origin can be made to coincide at O. If from $\boldsymbol{c}$ two parallel lines are drawn, one to $\boldsymbol{a}$ and the other to $\boldsymbol{b}$, the extensions of OA and OB will meet them at D and E. If $m = OD/OA$ and $n = OE/OB$, then from the rule of vector addition Eq. (A.5) is satisfied. It can be shown readily that the three component equations are

$$ma_x + nb_x = c_x$$

$$ma_y + nb_y = c_y$$

$$ma_z + nb_z = c_z$$

and therefore the determinant is

$$\begin{vmatrix} a_x & b_x & c_x \\ a_y & b_y & c_y \\ a_z & b_z & c_z \end{vmatrix} = 0$$

A.6 UNIT VECTOR AND VECTOR ADDITION OF COMPONENTS

Let a vector $\boldsymbol{a}$ be represented in a coordinate system as shown in Fig. A.5. Let $\boldsymbol{i}, \boldsymbol{j}$, and $\boldsymbol{k}$ be three unit vectors along the coordinate axis defined to be positive in the increasing direction of the coordinate. These unit vectors will have as components

$$\boldsymbol{i}(1, 0, 0)$$

$$\boldsymbol{j}(0, 1, 0)$$

$$\boldsymbol{k}(0, 0, 1)$$

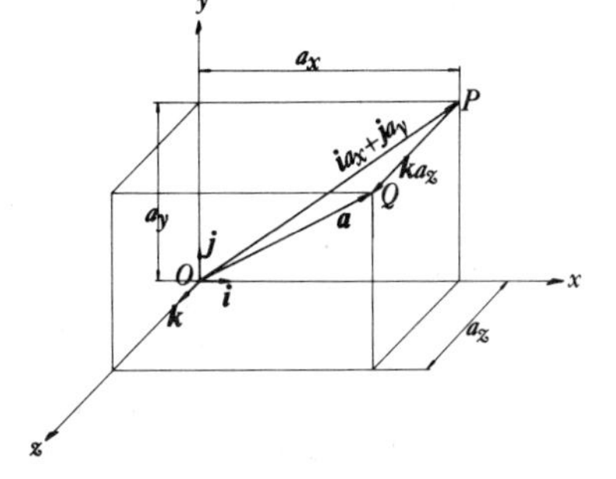

Fig. A.5 Vector and components.

Now according to Section A.5 the quantities $\boldsymbol{i}a_x$, $\boldsymbol{j}a_y$, and $\boldsymbol{k}a_z$ represent vectors of magnitude a_x, a_y, and a_z in the three corresponding directions. According to the rule of addition $\boldsymbol{i}a_x + \boldsymbol{j}a_y$ is the vector $\overrightarrow{OP}$ and

$$\boldsymbol{i}a_x + \boldsymbol{j}a_y + \boldsymbol{k}a_z = \boldsymbol{a} \tag{A.6}$$

represents the vector $\boldsymbol{a}$ in terms of its components a_x, a_y, and a_z and the unit vectors. Equation (A.6) will be the basis for representing in an equation form any vector in terms of its components. Now looking again at Fig. A.2,

$$\cos \gamma_1 = \frac{a_x}{|\boldsymbol{a}|}$$

$$\cos \gamma_2 = \frac{a_y}{|\boldsymbol{a}|}$$

$$\cos \gamma_3 = \frac{a_z}{|\boldsymbol{a}|}$$

where $|\boldsymbol{a}|$ is the absolute value or magnitude of a equal to

$$|\boldsymbol{a}| = a = (a_x^2 + a_y^2 + a_z^2)^{1/2} \tag{A.7}$$

From Fig. A.3, the projection of the sum of two vectors on a coordinate is seen to equal the sum of the projections. Therefore if

$$\boldsymbol{c} = \boldsymbol{a} + \boldsymbol{b}$$

then

$$c_x = a_x + b_x$$

$$c_y = a_y + b_y$$

$$c_z = a_z + b_z$$

$$\begin{aligned} \boldsymbol{c} &= \boldsymbol{i}c_x + \boldsymbol{j}c_y + \boldsymbol{k}c_z \\ &= \boldsymbol{i}(a_x + b_x) + \boldsymbol{j}(a_y + b_y) + \boldsymbol{k}(a_z + b_z) \end{aligned} \tag{A.8a}$$

and

$$c^2 = c_x{}^2 + c_y{}^2 + c_z{}^2 = (a_x + b_x)^2 + (a_y + b_y)^2 + (a_z + b_z)^2 \tag{A.8b}$$

When vector quantities are added the resultant quantity is no longer the sum of the magnitude as in the case of scalars but a vector whose components are the sum of the components of each vector on the same axis. Equation (A.8a) shows this relationship.

A.7 THE PRODUCT OF VECTORS

A product of vectors must be defined such that the resultant is not a function of coordinate rotation. The idea is to find which combination of coordinate multiplication gives a resultant, scalar or vector, that is independent of coordinate rotation. Obviously not any arbitrary combination of coordinate products will satisfy this requirement.

For instance, let a vector $\boldsymbol{a}$ make an angle θ with a vector $\boldsymbol{b}$ as shown in Fig. A.6. Let $\boldsymbol{e}_1$ and $\boldsymbol{e}_2$ be the unit vectors along $\boldsymbol{a}$ and $\boldsymbol{b}$ such that $\boldsymbol{e}_1 = \boldsymbol{a}/|\boldsymbol{a}|$ and $\boldsymbol{e}_2 = \boldsymbol{b}/|\boldsymbol{b}|$. The triangle $OA'B'$ is isosceles and furthermore according to Eq. (A.8b)

$$(\overrightarrow{A'B'})^2 = (e_{1x} - e_{2x})^2 + (e_{1y} - e_{2y})^2 + (e_{1z} - e_{2z})^2$$

But since $e_{1x}^2 + e_{1y}^2 + e_{1z}^2 = e_{2x}^2 + e_{2y}^2 + e_{2z}^2 = 1$,

$$(\overrightarrow{A'B'})^2 = 2 - 2(e_{1x}e_{2x} + e_{1y}e_{2y} + e_{1z}e_{2z}) \tag{A.9a}$$

Now from triangle $OA'B'$ with $A'M$ perpendicular to OB',

$$\begin{aligned} (\overrightarrow{A'B'})^2 &= (\overrightarrow{A'M})^2 + (\overrightarrow{MB'})^2 \\ &= \sin^2\theta + (1 - \cos\theta)^2 \\ &= 2 - 2\cos\theta \end{aligned} \tag{A.9b}$$

From Eqs. (A.9a) and (A.9b)

$$\cos\theta = e_{1x}e_{2x} + e_{1y}e_{2y} + e_{1z}e_{2z}$$

Substituting for values of $\boldsymbol{e}_1$ and $\boldsymbol{e}_2$ the terms $\boldsymbol{a}/|\boldsymbol{a}|$ and $\boldsymbol{b}/|\boldsymbol{b}|$,

$$\cos\theta = \frac{a_x b_x + a_y b_y + a_z b_z}{|\boldsymbol{a}|\,|\boldsymbol{b}|} \tag{A.10}$$

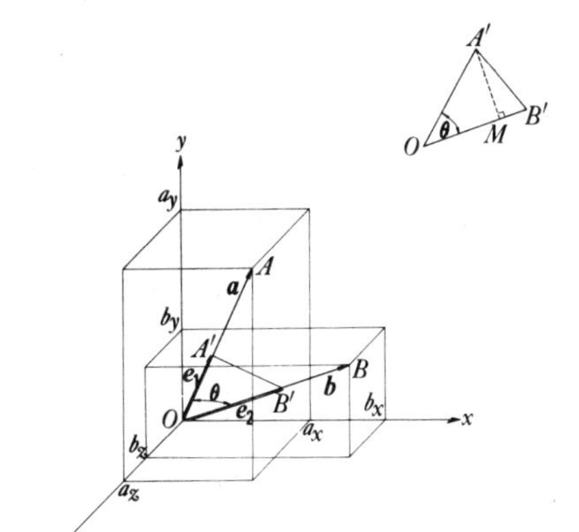

Fig. A.6 Product of vectors.

Now when two vectors are given, the angle θ between them is fixed and so is $\cos\theta$. Therefore the right-hand side of Eq. (A.10) is independent of which orthogonal coordinate system is used at the origin O, and consequently the scalar quantity $a_x b_x + a_y b_y + a_z b_z$ is always invariant with coordinate rotation provided $\boldsymbol{a}$ and $\boldsymbol{b}$ are fixed.[2]

A.8 THE SCALAR PRODUCT

On the basis of the results found in the previous section the *scalar product* or the *inner product* or the *dot product* is then defined as

$$\begin{aligned}\boldsymbol{a}\cdot\boldsymbol{b} &= a_x b_x + a_y b_y + a_z b_z \\ &= |a|\,|b|\cos\theta \end{aligned} \tag{A.11}$$

and is read as "$\boldsymbol{a}$ dot $\boldsymbol{b}$."

[2] For another proof showing the independence of these results of coordinate rotation see, for instance, R. Aris, "Vectors, Tensors, and the Basic Equations of Fluid Mechanics" p. 15. Prentice-Hall, Englewood Cliffs, New Jersey, 1962.

This definition of the dot product shows that it is equal to the product of the absolute value of $\boldsymbol{b}$ and the projection of $\boldsymbol{a}$ on $\boldsymbol{b}$. As long as this projection remains the same, as shown in Fig. A.7, $\boldsymbol{a} \cdot \boldsymbol{b} = \boldsymbol{c} \cdot \boldsymbol{b}$ and $\boldsymbol{a} \neq \boldsymbol{c}$. From Eq. (A.11)

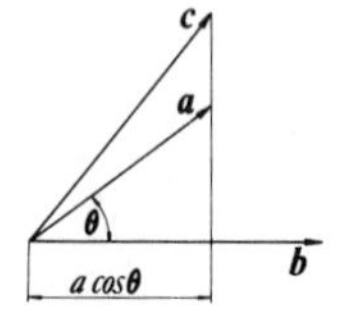

Fig. A.7 Representation of the scalar product.

we verify that since $\cos\theta$ is an even function the order of the product $\boldsymbol{a}$ to $\boldsymbol{b}$ or $\boldsymbol{b}$ to $\boldsymbol{a}$, or positive and negative θ, gives the same result,

$$\boldsymbol{a} \cdot \boldsymbol{b} = \boldsymbol{b} \cdot \boldsymbol{a}$$

and that

$$\boldsymbol{a} \cdot \boldsymbol{a} = |a|^2 = \boldsymbol{a}^2 = a^2 \tag{A.12}$$

If $\boldsymbol{a}$ is perpendicular to $\boldsymbol{b}$ then

$$\boldsymbol{a} \cdot \boldsymbol{b} = 0$$

The dot products of the unit vectors are then

$$\begin{aligned} \boldsymbol{i} \cdot \boldsymbol{i} = \boldsymbol{j} \cdot \boldsymbol{j} = \boldsymbol{k} \cdot \boldsymbol{k} = 1 \\ \boldsymbol{i} \cdot \boldsymbol{j} = \boldsymbol{j} \cdot \boldsymbol{k} = \boldsymbol{k} \cdot \boldsymbol{i} = 0 \end{aligned} \tag{A.13}$$

One can show easily that the *distributive law* applies to the dot product:

$$\boldsymbol{a} \cdot (\boldsymbol{b} + \boldsymbol{c}) = \boldsymbol{a} \cdot \boldsymbol{b} + \boldsymbol{a} \cdot \boldsymbol{c}$$

An application of the dot product in mechanics is seen when it is necessary to find the energy involved in the movement of the point of application of a force. This energy known as *work* is given by

$$E = \boldsymbol{F} \cdot \boldsymbol{r} \tag{A.14}$$

This shows that the maximum energy involved in the movement of a force $\boldsymbol{F}$ along a distance $\boldsymbol{r}$ is when the distance is along $\boldsymbol{F}$.

A.9 THE VECTOR PRODUCT

The scalar product is a linear combination of three component products of the total nine possible. It will become evident that the other six can be combined in pairs to form a so-called *vector product*. To show which combination is independent of rotation of coordinates the following method is used: Let the resultant of the vector product of $\boldsymbol{a}$ and $\boldsymbol{b}$, making an angle θ, be a vector $\boldsymbol{c}$ perpendicular to the plane formed by $\boldsymbol{a}$ and $\boldsymbol{b}$. Then it follows that $\boldsymbol{a} \cdot \boldsymbol{c} = \boldsymbol{b} \cdot \boldsymbol{c} = 0$ or

$$a_x c_x + a_y c_y + a_z c_z = 0, \quad b_x c_x + b_y c_y + b_z c_z = 0 \tag{A.15}$$

If c_x and c_y are solved in terms of c_z,

$$\frac{c_x}{a_y b_z - a_z b_y} = \frac{c_y}{a_z b_x - a_x b_z} = \frac{c_z}{a_x b_y - a_y b_x} = m \tag{A.16}$$

To preserve the symmetry of the dependence of the component of $\boldsymbol{c}$ in terms of those of $\boldsymbol{a}$ and $\boldsymbol{b}$ the quantity m must be a constant. Solving for c_x, c_y, and c_z,

$$\begin{aligned} c_x &= m(a_y b_z - a_z b_y) \\ c_y &= m(a_z b_x - a_x b_z) \\ c_z &= m(a_x b_y - a_y b_x) \end{aligned} \tag{A.17}$$

The absolute value of $\boldsymbol{c}$ is

$$\begin{aligned} c^2 &= c_x^{\,2} + c_y^{\,2} + c_z^{\,2} \\ &= m^2[a_x^{\,2}(b_y^{\,2} + b_z^{\,2}) + a_y^{\,2}(b_z^{\,2} + b_x^{\,2}) + a_z^{\,2}(b_x^{\,2} + b_y^{\,2}) \\ &\qquad - 2(a_y b_y a_z b_z + a_z b_z a_x b_x + a_x b_x a_y b_y)] \\ &= m^2[a_x^{\,2}(b^2 - b_x^{\,2}) + a_y^{\,2}(b^2 - b_y^{\,2}) + a_z^{\,2}(b^2 - b_z^{\,2}) \\ &\qquad - 2(a_y b_y a_z b_z + a_z b_z a_x b_x + a_x b_x a_y b_y)] \\ &= m^2[a^2 b^2 - (a_x b_x + a_y b_y + a_z b_z)^2] \end{aligned}$$

According to Eq. (A.11) the preceding equation becomes

$$c^2 = m^2 a^2 b^2 (1 - \cos^2\theta) = m^2 a^2 b^2 \sin^2\theta \tag{A.18}$$

This equation shows that the definition of the vector product gives a vector $\boldsymbol{c}$ whose magnitude is dependent on the magnitudes of $\boldsymbol{a}$ and $\boldsymbol{b}$ and the angle between them and consequently independent of the orthogonal coordinate system chosen. To evaluate m we can choose $\boldsymbol{a}$, $\boldsymbol{b}$, and $\boldsymbol{c}$ to be three mutually perpendicular unit vectors, say $\boldsymbol{i}$, $\boldsymbol{j}$, and $\boldsymbol{k}$, along x, y, and z. Then $a_x = b_y = c_z = 1$ and $a_y = a_z = b_x = b_z = c_x = c_y = 0$. From Eqs. (A.17) $m = 1$ and in general

$$\begin{aligned} c_x &= a_y b_z - a_z b_y \\ c_y &= a_z b_x - a_x b_z \\ c_z &= a_x b_y - a_y b_x \end{aligned} \tag{A.19}$$

Thus the *vector product* of two vectors $\boldsymbol{a}$ and $\boldsymbol{b}$ (see Fig. A.8) with an angle θ measured positive from $\boldsymbol{a}$ to $\boldsymbol{b}$ is a third vector $\boldsymbol{c}$ perpendicular to the plane formed by $\boldsymbol{a}$ and $\boldsymbol{b}$ and in the direction prescribed by the third axis of a conventional coordinate system (the right-handed system described in Section A.2).

The *vector product* or *cross product* or *outer product* is then given by

$$\begin{aligned}\boldsymbol{a} \times \boldsymbol{b} &= \boldsymbol{c} \\ &= \boldsymbol{i}(a_y b_z - a_z b_y) + \boldsymbol{j}(a_z b_x - a_x b_z) + \boldsymbol{k}(a_x b_y - a_y b_x) \\ &= \boldsymbol{e} ab \sin \theta \end{aligned} \tag{A.20}$$

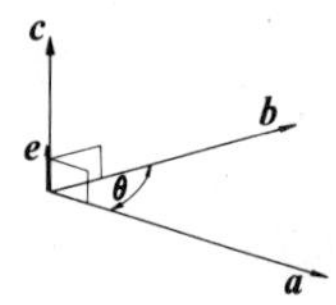

Fig. A.8 Vector product.

where $\boldsymbol{e}$ is a unit vector normal to the plane of $\boldsymbol{a}$ and $\boldsymbol{b}$ in the direction of the right-hand rule. Since $\sin \theta$ is an odd function it shows that

$$\boldsymbol{a} \times \boldsymbol{b} = -\boldsymbol{b} \times \boldsymbol{a}$$

and that

$$\begin{aligned}&\boldsymbol{i} \times \boldsymbol{j} = \boldsymbol{k}; \qquad \boldsymbol{j} \times \boldsymbol{k} = \boldsymbol{i}; \qquad \boldsymbol{k} \times \boldsymbol{i} = \boldsymbol{j} \\ &\boldsymbol{i} \times \boldsymbol{i} = \boldsymbol{j} \times \boldsymbol{j} = \boldsymbol{k} \times \boldsymbol{k} = 0\end{aligned} \tag{A.21}$$

The cross product of two parallel vectors is always zero.

From Eq. (A.20)

$$\boldsymbol{a} \times \boldsymbol{b} = \begin{vmatrix} \boldsymbol{i} & \boldsymbol{j} & \boldsymbol{k} \\ a_x & a_y & a_z \\ b_x & b_y & b_z \end{vmatrix} \tag{A.22}$$

If m and n are scalar numbers then

$$m\boldsymbol{a} \times n\boldsymbol{b} = mn\boldsymbol{a} \times \boldsymbol{b} = mn(\boldsymbol{a} \times \boldsymbol{b}) \tag{A.23}$$

Physically the cross product is involved for instance in the evaluation of the peripheral velocity of a point P around an axis OO' as shown in Fig. A.9. If $\boldsymbol{\omega}$ is the angular velocity of the point P and $\boldsymbol{r}$ the distance of P to an origin O on the axis of rotation then the peripheral velocity[3] of P is

$$\boldsymbol{U} = \boldsymbol{\omega} \times \boldsymbol{r} = \boldsymbol{\omega} \times \boldsymbol{r}' \tag{A.24}$$

[3] This is proved in Section A.13.

This last equation does not imply that $\boldsymbol{r}$ is equal to $\boldsymbol{r}'$ but rather that any vector extending from P to any point on the axis, including $\boldsymbol{r}'$ that is perpendicular to the axis of rotation, will give the same cross product with $\boldsymbol{\omega}$

$$|\boldsymbol{U}| = \omega(OP)\sin\theta$$

$$= \omega(O'P)$$

The direction of the vector product is along $\boldsymbol{U}$ perpendicular to the plane of $\boldsymbol{\omega}$ and $\boldsymbol{r}$ or $\boldsymbol{r}'$.

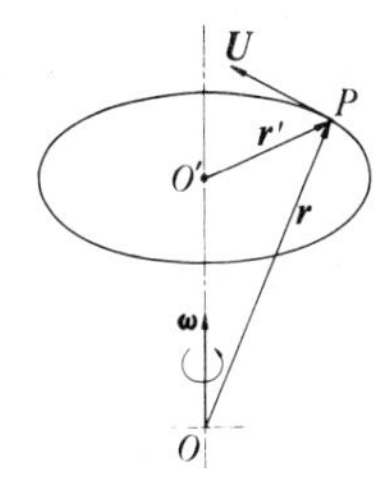

Fig. A.9 Peripheral velocity.

Another application is in the evaluation of the moment or the torque produced by a force $\boldsymbol{F}$ at a distance $\boldsymbol{r}$ from an axis:

$$\boldsymbol{T} = \boldsymbol{r} \times \boldsymbol{F}$$

If $\boldsymbol{F}$ instead of $\boldsymbol{U}$ is put into Fig. A.9, then $\boldsymbol{T}$ will be in the direction of $\boldsymbol{\omega}$.

Finally, the cross product of two vectors represents the area of the parallelogram formed by the two vectors.

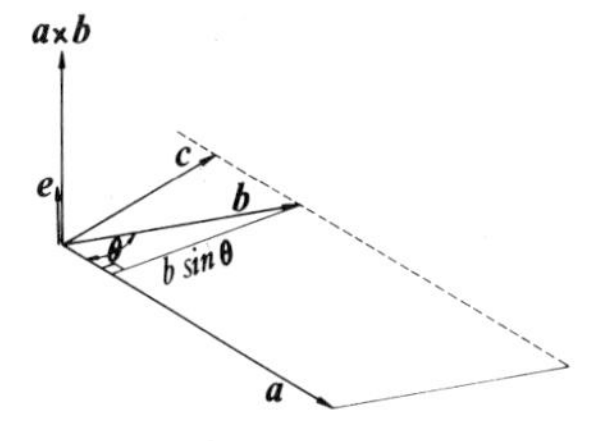

Fig. A.10 The area vector.

Here in Fig. A.10 it can be seen that

$$\boldsymbol{a} \times \boldsymbol{b} = \boldsymbol{e}|a|\,|b|\sin\theta \tag{A.25}$$

where $\boldsymbol{e}$ is a unit vector normal to the plane of $\boldsymbol{a}$ and $\boldsymbol{b}$. The value $ab \sin\theta$ is the area of the parallelogram and thus $\boldsymbol{a} \times \boldsymbol{b}$ represents the area vector. As long as the projection of a third vector $\boldsymbol{c}$ is $|\boldsymbol{b}|\sin\theta$, $\boldsymbol{a} \times \boldsymbol{b} = \boldsymbol{a} \times \boldsymbol{c}$ while $\boldsymbol{b} \neq \boldsymbol{c}$, knowing

$\boldsymbol{a} \times \boldsymbol{b}$ and $\boldsymbol{a}$ does not uniquely determine $\boldsymbol{b}$. The same was true for the dot product.

The distribution law applies to cross products and to show this it is necessary to prove that

$$\boldsymbol{a} \times (\boldsymbol{b} + \boldsymbol{c}) = \boldsymbol{a} \times \boldsymbol{b} + \boldsymbol{a} \times \boldsymbol{c} \tag{A.26}$$

From Eq. (A.20) it is easily seen[4] that the x component, for instance, of this product is

$$[a_y(b_z + c_z) - a_z(b_y + c_y)] = (a_y b_z - a_z b_y) + (a_y c_z - a_z c_y)$$

which is the sum of the x-components of $\boldsymbol{a} \times \boldsymbol{b}$ and $\boldsymbol{a} \times \boldsymbol{c}$. Repeating the same for the other two components proves the distributive law for a cross product given in Eq. (A.26).

A.10 THE TRIPLE SCALAR PRODUCT

The combined vector-scalar products $(\boldsymbol{a} \times \boldsymbol{b}) \cdot \boldsymbol{c}$ of three vectors $\boldsymbol{a}$, $\boldsymbol{b}$, and $\boldsymbol{c}$ results in a scalar quantity. This is immediately verified by the fact that $(\boldsymbol{a} \times \boldsymbol{b})$ is a vector, and when its dot product is taken with another vector $\boldsymbol{c}$ the result is a scalar. *The parentheses in these triple products are very meaningful. It implies that the operation enclosed in parentheses must be completed before carrying out the rest of the operations.* From Eqs. (A.20) and (A.11) it is immediately verified that

$$\begin{aligned}(\boldsymbol{a} \times \boldsymbol{b}) \cdot \boldsymbol{c} &= \boldsymbol{c} \cdot (\boldsymbol{a} \times \boldsymbol{b}) \\ &= (a_x b_y c_z + a_y b_z c_x + a_z b_x c_y) - (a_x b_z c_y + a_y b_x c_z + a_z b_y c_x) \\ &= \begin{vmatrix} c_x & c_y & c_z \\ a_x & a_y & a_z \\ b_x & b_y & b_z \end{vmatrix}\end{aligned} \tag{A.27}$$

From the determinant in Eq. (A.27) it can be seen that as long as the sequence of the rows is maintained its value remains unchanged. That is,

$$\begin{aligned}\boldsymbol{c} \cdot (\boldsymbol{a} \times \boldsymbol{b}) &= \boldsymbol{a} \cdot (\boldsymbol{b} \times \boldsymbol{c}) = \boldsymbol{b} \cdot (\boldsymbol{c} \times \boldsymbol{a}) \\ &= -\boldsymbol{c} \cdot (\boldsymbol{b} \times \boldsymbol{a}) = -\boldsymbol{a} \cdot (\boldsymbol{c} \times \boldsymbol{b}) = -\boldsymbol{b} \cdot (\boldsymbol{a} \times \boldsymbol{c})\end{aligned} \tag{A.28}$$

This is called the *cyclic rule* of *the triple scalar product.*

This product has a physical significance[5] in that it represents the volume of a parallelepiped with the three vectors for sides. From Fig. A.11 the product $\boldsymbol{a} \times \boldsymbol{b}$

[4] For an interesting geometrical proof of the distribution law, see R. Courant, "Differential and Integral Calculus," Vol. II, pp. 15–16. Wiley (Interscience), New York, 1936.

[5] For further physical significance of this product consult H. Yeh and J. I. Abrams, "Principles of Mechanics of Solids and Fluids," Vol. 1, p. 42. McGraw-Hill, New York, 1960.

represents a vector normal to the plane ab and with magnitude $ab \sin \theta$, the area of that plane. The triple scalar product then is equal to

$$\boldsymbol{c} \cdot (\boldsymbol{a} \times \boldsymbol{b}) = (c \cos \beta)(ab \sin \theta)$$

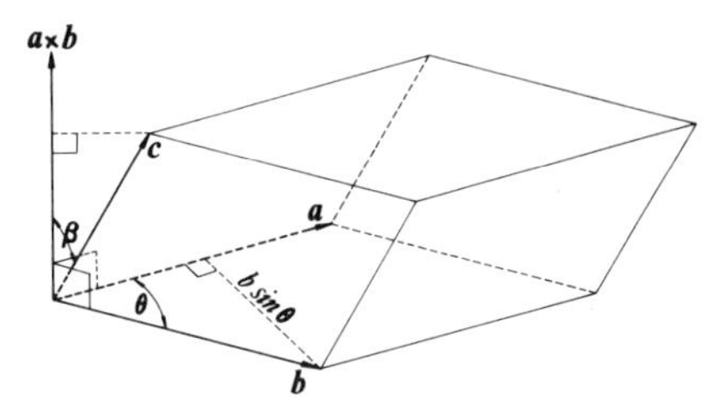

Fig. A.11
The triple scalar product.

which is the volume of the parallelepiped. The following relations are now true:

$$\boldsymbol{i} \cdot (\boldsymbol{j} \times \boldsymbol{k}) = \boldsymbol{j} \cdot (\boldsymbol{k} \times \boldsymbol{i}) = \boldsymbol{k} \cdot (\boldsymbol{i} \times \boldsymbol{j}) = 1 \tag{A.29}$$

A.11 THE TRIPLE VECTOR PRODUCT

The *triple vector product* is the cross product of three vectors which results in a fourth vector. The triple vector product

$$\begin{aligned} \boldsymbol{a} \times (\boldsymbol{b} \times \boldsymbol{c}) &= (\boldsymbol{a} \cdot \boldsymbol{c})\boldsymbol{b} - (\boldsymbol{a} \cdot \boldsymbol{b})\boldsymbol{c} \\ &= (\boldsymbol{c} \times \boldsymbol{b}) \times \boldsymbol{a} \end{aligned} \tag{A.30}$$

will be shown to be true in the following geometrical proof which has purposely been chosen to give physical significance to vectorial operations.[6] This multiple product will occur often in this text. In the first line the products in parentheses on the right-hand side of Eq. (A.30) are scalars, thus the triple vector product is on the plane formed by $\boldsymbol{b}$ and $\boldsymbol{c}$ and consequently is expressed in a linear combination of them.

Let $\boldsymbol{a}$, $\boldsymbol{b}$, and $\boldsymbol{c}$ be three vectors in space as shown in Fig. A.12. The fact that the three vectors are shown to have the same origin should not be considered a lack of generality. The vector $\boldsymbol{b} \times \boldsymbol{c}$ is perpendicular to the plane of $\boldsymbol{b}$ and $\boldsymbol{c}$.

(1) Now since $\boldsymbol{a} \times (\boldsymbol{b} \times \boldsymbol{c})$ is perpendicular to $(\boldsymbol{b} \times \boldsymbol{c})$ and $\boldsymbol{a}$ it must lie on the plane of $\boldsymbol{b}$ and $\boldsymbol{c}$. In fact it is for this reason that it could be expressed in a linear combination of these two vectors as shown in Eq. (A.30) (see Section A.5). Therefore,

$$\boldsymbol{a} \times (\boldsymbol{b} \times \boldsymbol{c}) = m\boldsymbol{b} - n\boldsymbol{c} \tag{A.31}$$

The object is to evaluate the scalar quantities m and n.

[6] A shorter algebraic verification of Eq. (A.30) is, for instance, in G. E. Hay, "Vector and Tensor Analysis," p. 17. Dover Publications, New York, 1953.

(2) The vector $\boldsymbol{a} \times (\boldsymbol{b} \times \boldsymbol{c})$ is perpendicular to $\boldsymbol{a}$ and for this reason

$$\boldsymbol{a} \cdot [\boldsymbol{a} \times (\boldsymbol{b} \times \boldsymbol{c})] = 0$$

or from Eq. (A.31)

$$m\boldsymbol{a} \cdot \boldsymbol{b} - n\boldsymbol{a} \cdot \boldsymbol{c} = 0$$

$$\frac{m}{n} = \frac{\boldsymbol{a} \cdot \boldsymbol{c}}{\boldsymbol{a} \cdot \boldsymbol{b}}$$

$\boldsymbol{a} \times (\boldsymbol{b} \times \boldsymbol{c})$

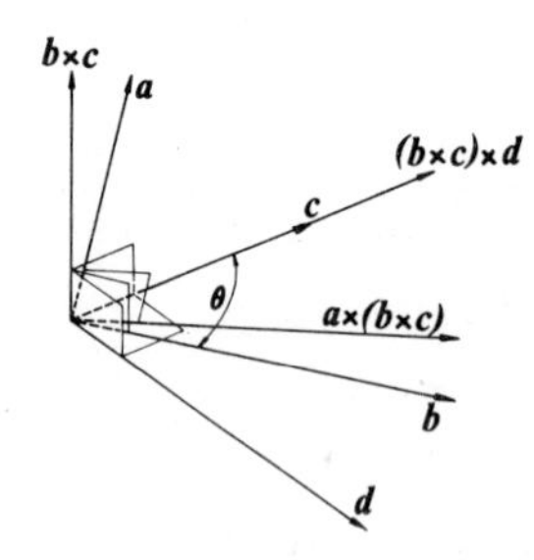

Fig. A.12 The triple vector product.

where at the most $m = p\boldsymbol{a} \cdot \boldsymbol{c}$ and $n = p\boldsymbol{a} \cdot \boldsymbol{b}$, p being a constant. Then

$$\boldsymbol{a} \times (\boldsymbol{b} \times \boldsymbol{c}) = p(\boldsymbol{a} \cdot \boldsymbol{c})\boldsymbol{b} - p(\boldsymbol{a} \cdot \boldsymbol{b})\boldsymbol{c} \tag{A.32}$$

(3) To determine the constant scalar p we choose a vector $\boldsymbol{d}$ coplanar with $\boldsymbol{b}$ and $\boldsymbol{c}$ and perpendicular to $\boldsymbol{c}$. Then

$$\boldsymbol{d} \cdot [\boldsymbol{a} \times (\boldsymbol{b} \times \boldsymbol{c})] = p\boldsymbol{d} \cdot (\boldsymbol{a} \cdot \boldsymbol{c})\boldsymbol{b} - p\boldsymbol{d} \cdot (\boldsymbol{a} \cdot \boldsymbol{b})\boldsymbol{c} \tag{A.33}$$

The second term on the right is zero since $\boldsymbol{d}$ is perpendicular to $\boldsymbol{c}$.

(4) From the rules of triple scalar product of Eq. (A.28) applied to $\boldsymbol{d}$, $\boldsymbol{a}$, and $(\boldsymbol{b} \times \boldsymbol{c})$,

$$\boldsymbol{d} \cdot [\boldsymbol{a} \times (\boldsymbol{b} \times \boldsymbol{c})] = \boldsymbol{a} \cdot [(\boldsymbol{b} \times \boldsymbol{c}) \times \boldsymbol{d}] \tag{A.34}$$

(5) Since the three vectors $\boldsymbol{d}$, $\boldsymbol{c}$, and $(\boldsymbol{b} \times \boldsymbol{c})$ are mutually perpendicular the vector $(\boldsymbol{b} \times \boldsymbol{c}) \times \boldsymbol{d}$ is along $\boldsymbol{c}$. Its magnitude is $|(\boldsymbol{b} \times \boldsymbol{c})|\,|\boldsymbol{d}| \sin 90° = |(\boldsymbol{b} \times \boldsymbol{c})|\,|\boldsymbol{d}| = (bc \sin\theta)d$ or $[bd\cos(90° - \theta)]c$. But from Fig. A.12 the quantity $(\boldsymbol{b} \cdot \boldsymbol{d})\boldsymbol{c}$ has also the same magnitude and direction as $(\boldsymbol{b} \times \boldsymbol{c}) \times \boldsymbol{d}$ and consequently they must be equal. Therefore,

$$(\boldsymbol{b} \times \boldsymbol{c}) \times \boldsymbol{d} = (\boldsymbol{b} \cdot \boldsymbol{d})\boldsymbol{c} \tag{A.35}$$

From the combination of Eqs. (A.33)–(A.35)

$$p(\boldsymbol{d} \cdot \boldsymbol{b})(\boldsymbol{a} \cdot \boldsymbol{c}) = (\boldsymbol{a} \cdot \boldsymbol{c})(\boldsymbol{b} \cdot \boldsymbol{d})$$

which means that $p = 1$. Therefore, in conclusion Eq. (A.33) becomes Eq. (A.30), the proof of the theorem.

A.12 VECTOR FUNCTION OF A SCALAR

So far we have considered vectors as quantities with magnitude and direction that were fixed in space, time, or any other variable. When dealing with most statics problems in mechanics of rigid bodies, the knowledge of vector algebra discussed so far is sufficient. In kinematics or dynamics of a continuous medium, the medium properties, scalar or vector, vary across the whole *field* to be studied. This concept of a *field of vectors* immediately requires differential or integral calculus in order to evaluate the changes of vectors in the field. Consider for instance a velocity field or rather, since the subject of this book concerns fluids, a fluid velocity field $\boldsymbol{U}$, a function of time and space. This velocity in terms of its Cartesian components can be written as

$$\boldsymbol{U} = \boldsymbol{i}u(x, y, z, t) + \boldsymbol{j}v(x, y, z, t) + \boldsymbol{k}w(x, y, z, t) \tag{A.36}$$

when u, v, and w are the x, y, and z scalar components of the velocity vector whose values depend on the position of a point $P(x, y, z)$ considered and the time t.

The value $\boldsymbol{U}$ can vary in time at the same point or at the same time at various points. When dealing with vector fields it is important to determine the change of this vector with space and time. In calculus the student has learned to find the change of a single scalar function of x, y, z, and t. The question now arises as to whether the same method can be applied to vectors. Since $\boldsymbol{U}$ changes only if the components change, the total change of the vector is the vector sum of the change of the components:

$$d\boldsymbol{U} = d(\boldsymbol{i}u) + d(\boldsymbol{j}v) + d(\boldsymbol{k}w)$$

If the description of these quantities is viewed with respect to a coordinate system that does not change in orientation, then the change of the unit vectors $\boldsymbol{i}$, $\boldsymbol{j}$, and $\boldsymbol{k}$ in space is zero and

$$d\boldsymbol{u} = \boldsymbol{i}\,du + \boldsymbol{j}\,dv + \boldsymbol{k}\,dw \tag{A.37}$$

If this was not the case then the change of the unit vector must be taken into account. This is described in sections dealing with cylindrical and spherical coordinates. Since the scalar components of $\boldsymbol{U}$ are functions of x, y, z, and t then this total change involves partial differentiation with respect to all four variables. A very useful application of Eq. (A.37) is encountered in the description of the change of position of a particle $P(x, y, z)$ in Euclidean space.

If $\boldsymbol{r} = \boldsymbol{i}x + \boldsymbol{j}y + \boldsymbol{k}z$ is its position at time $t = t_0$, in an infinitesimal time change the position change will be

$$d\boldsymbol{r} = \boldsymbol{i}\,dx + \boldsymbol{j}\,dy + \boldsymbol{k}\,dz \tag{A.38}$$

A.13 DIFFERENTIATION WITH RESPECT TO A SCALAR VARIABLE

Let a vector $\boldsymbol{a}$ be a continuous function of a scalar σ. By convention this is expressed in the form $\boldsymbol{a}(\sigma)$. For a particular value of σ, $\boldsymbol{a}(\sigma)$ has a value shown in Fig. A.13 by the vector $\overrightarrow{OM}$.

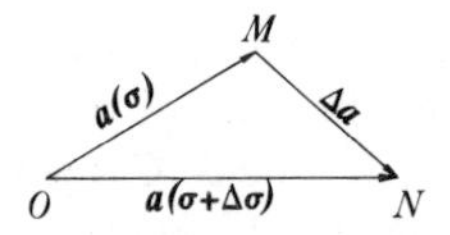

Fig. A.13 Change of a vector with one variable.

If the value of the independent quantity is increased by $\Delta\sigma$ then the vector corresponding to this new value $(\sigma + \Delta\sigma)$ will be $\boldsymbol{a}(\sigma + \Delta\sigma)$, shown in Fig. A.13 by ON. Because of the change $\Delta\sigma$, the resulting change in $\boldsymbol{a}$ is

$$\Delta\boldsymbol{a} = \boldsymbol{a}(\sigma + \Delta\sigma) - \boldsymbol{a}(\sigma)$$

This is shown by the vector $\overrightarrow{MN}$. The average change of the vector $\boldsymbol{a}$ per unit change in σ is $\Delta\boldsymbol{a}/\Delta\sigma$. If $\Delta\sigma > 0$ then $\Delta\boldsymbol{a}/\Delta\sigma$ is a vector in the direction of $\Delta\boldsymbol{a}$ but $1/\Delta\sigma$ as long. If $\Delta\sigma < 0$, then $\Delta\boldsymbol{a}/\Delta\sigma$ has the direction of $-\Delta\boldsymbol{a}$. If now $\Delta\sigma$ is made to approach zero without really reaching it, in other words $\Delta\sigma$ is made very small and hence $\Delta\boldsymbol{a}$ is very small compared to $\boldsymbol{a}$, then

$$\frac{d\boldsymbol{a}}{d\sigma} = \lim_{\Delta\sigma \to 0} \frac{\boldsymbol{a}(\sigma + \Delta\sigma) - \boldsymbol{a}(\sigma)}{\Delta\sigma} \tag{A.39a}$$

is called the *derivative* of $\boldsymbol{a}(\sigma)$ with respect to σ.

If we consider a point P moving on a circle of radius r with constant angular speed $d\theta/dt$ as shown in Fig. A.14, then

$$\boldsymbol{r} = \boldsymbol{i}r\cos\theta + \boldsymbol{j}r\sin\theta$$

Then

$$\frac{d\boldsymbol{r}}{dt} = \boldsymbol{U} = \boldsymbol{i}r\frac{d}{dt}\cos\theta + \boldsymbol{j}r\frac{d}{dt}\sin\theta$$

Here the coordinate system has been fixed at O and thus $\boldsymbol{i}$ and $\boldsymbol{j}$ are constant with the motion of P. Then

$$\boldsymbol{U} = (-\boldsymbol{i}r\sin\theta + \boldsymbol{j}r\cos\theta)\frac{d\theta}{dt} \tag{A.39b}$$

The rotation vector $\boldsymbol{\omega}$ is perpendicular to the paper and points towards the reader and therefore is equal to $\boldsymbol{\omega} = \boldsymbol{k}\, d\theta/dt$. It is easily verified then that Eq. (A.24) is satisfied,

$$\begin{aligned} \boldsymbol{U} &= \boldsymbol{\omega} \times \boldsymbol{r} \\ &= \left(\boldsymbol{k}\,\frac{d\theta}{dt}\right) \times (\boldsymbol{i}r\cos\theta + \boldsymbol{j}r\sin\theta) \end{aligned}$$

and that it is equal to Eq. (A.39b).

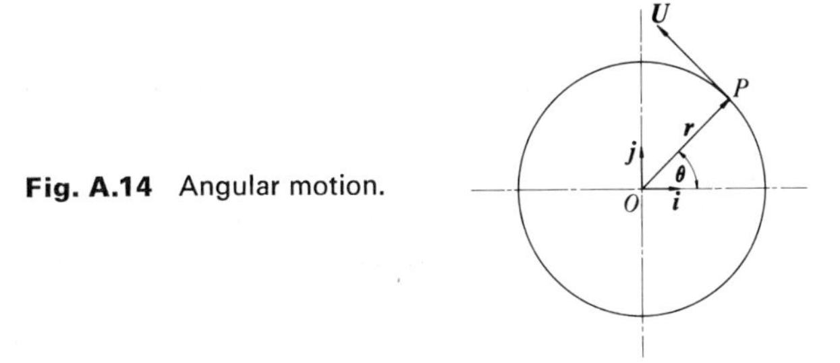

Fig. A.14 Angular motion.

The acceleration which is a rate of change of velocity is nonzero because the direction of the velocity is changing even though its magnitude remains constant. Then

$$\begin{aligned} \boldsymbol{a} &= \frac{d\boldsymbol{U}}{dt} = \frac{d^2\boldsymbol{r}}{dt^2} \\ &= (-\boldsymbol{i}r\cos\theta - \boldsymbol{j}r\sin\theta)\left(\frac{d\theta}{dt}\right)^2 \\ &= -\omega^2\boldsymbol{r} \end{aligned} \tag{A.40}$$

This is the *centripetal acceleration* directed towards the center of rotation.

A.14 DIFFERENTIATION RULES

Following the reasoning in the previous section it can be shown easily that if $\boldsymbol{a}$, $\boldsymbol{b}$ and m are functions of σ

$$\begin{aligned} &\text{(a)} \quad \frac{d}{d\sigma}(m\boldsymbol{a}) = m\,\frac{d\boldsymbol{a}}{d\sigma} + \frac{dm}{d\sigma}\,\boldsymbol{a} \\ &\text{(b)} \quad \frac{d}{d\sigma}(\boldsymbol{a} + \boldsymbol{b}) = \frac{d\boldsymbol{a}}{d\sigma} + \frac{d\boldsymbol{b}}{d\sigma} \\ &\text{(c)} \quad \frac{d}{d\sigma}(\boldsymbol{a} \cdot \boldsymbol{b}) = \boldsymbol{a} \cdot \frac{d\boldsymbol{b}}{d\sigma} + \frac{d\boldsymbol{a}}{d\sigma} \cdot \boldsymbol{b} \\ &\text{(d)} \quad \frac{d}{d\sigma}(\boldsymbol{a} \times \boldsymbol{b}) = \boldsymbol{a} \times \frac{d\boldsymbol{b}}{d\sigma} + \frac{d\boldsymbol{a}}{d\sigma} \times \boldsymbol{b} \end{aligned} \tag{A.41}$$

An important result appears from the third relation of Eq. (A.41). For a vector of constant magnitude $\boldsymbol{a} \cdot \boldsymbol{a}$ is constant. Then

$$\boldsymbol{a} \cdot \frac{d\boldsymbol{a}}{d\sigma} + \frac{d\boldsymbol{a}}{d\sigma} \cdot \boldsymbol{a} = 0$$

which means that $d\boldsymbol{a}/d\sigma$ is perpendicular to $\boldsymbol{a}$. Since $d\boldsymbol{a}/d\sigma$ has the same direction as $d\boldsymbol{a}$, $d\boldsymbol{a}$ is perpendicular to $\boldsymbol{a}$.

A.15 THE GRADIENT OF A SCALAR FUNCTION

Let f, a scalar function of position such as the temperature, be represented in three-dimensional Cartesian coordinates. The values of f for all x, y, z represent the scalar field. In Fig. A.15, given a point P a distance $\boldsymbol{r}$ from O, a

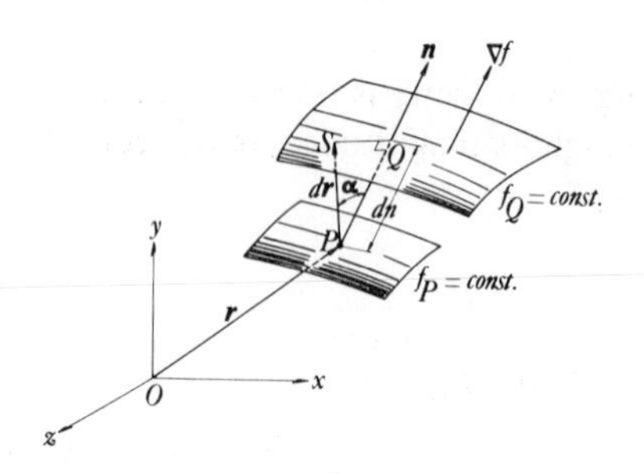

Fig. A.15 Gradient of a scalar function.

surface $f_P = c_1$ passes through P where everywhere on this surface the value of the scalar is a constant. Let f and its change in the neighborhood of P be continuous. If we draw a normal to f_P at P in the direction of the unit normal vector $\boldsymbol{n}$, this normal intersects another surface $f_Q = c_2$ at the point Q. For differential changes, since the limiting case of these two surfaces approaching each other must be taken, let the difference $f_Q - f_P$ be a differential change df in a distance dn. Then for small changes by Taylor's expansion along PQ

$$f_Q - f_P = PQ(df/dn)$$

Let S be any other neighboring point of P such that $f_S = f_Q$; then

$$f_S = f_Q = f_P + PQ(df/dn)$$

$$f_S - f_P = PQ(df/dn)$$

$$= PS \cos \alpha \, (df/dn) \tag{A.42}$$

Since f is a scalar, the total change of f is also a scalar. However, it is seen that this scalar change is the scalar product of two vectors $\overline{PS} = \boldsymbol{dr}$ and the directed unit change $\boldsymbol{n}\, df/dn$ of the scalar function f.

Therefore,

$$df = \boldsymbol{n}\,\frac{df}{dn} \cdot \boldsymbol{dr}$$

It is seen from Eq. (A.42) that for no matter what the displacement $\boldsymbol{dr}$ from P, it is its projection $\boldsymbol{n} \cdot \boldsymbol{dr} = dn$ along the normal that counts. The vector $\boldsymbol{n}\, df/dn$ is called the *gradient of the scalar function f* often expressed as grad f or ∇f and points in the direction of maximum rate of change of f. The symbol ∇ is called "nabla" or "del" and is analogous to the scalar operator $D = d/dx$. From the previous equation it is seen that the gradient can be symbolically expressed as $df/\boldsymbol{dr}$:

$$\begin{aligned} df &= \frac{df}{\boldsymbol{dr}} \cdot \boldsymbol{dr} \\ &= \nabla f \cdot \boldsymbol{dr} \\ &= \boldsymbol{dr} \cdot \nabla f \\ &= (\boldsymbol{dr} \cdot \nabla) f \end{aligned} \tag{A.43}$$

Since $\boldsymbol{r}$ is generally used to represent the radius vector, in order to avoid confusion any small directed displacement $\boldsymbol{ds}$ can be used instead of $\boldsymbol{dr}$, and this will establish the total change of f along $\boldsymbol{ds}$:

$$df = \nabla f \cdot \boldsymbol{ds} \tag{A.44}$$

In *Cartesian coordinates* $\boldsymbol{ds} = \boldsymbol{i}\,dx + \boldsymbol{j}\,dy + \boldsymbol{k}\,dz$ and since $f = f(x, y, z)$ its total change is also

$$df = \frac{\partial f}{\partial x}\,dx + \frac{\partial f}{\partial y}\,dy + \frac{\partial f}{\partial z}\,dz$$

Since ∇f is a vector, in terms of its Cartesian components

$$\nabla f = \boldsymbol{i}(\nabla f)_x + \boldsymbol{j}(\nabla f)_y + \boldsymbol{k}(\nabla f)_z$$

and according to Eq. (A.44)

$$df = (\nabla f)_x\,dx + (\nabla f)_y\,dy + (\nabla f)_z\,dz$$

Comparing this with the sum of the differential changes the components of the gradient in Cartesian coordinates are

$$(\nabla f)_x = \partial f/\partial x; \qquad (\nabla f)_y = \partial f/\partial y; \qquad (\nabla f)_z = \partial f/\partial z$$

and the vector operator

$$\nabla \equiv \boldsymbol{i}\frac{\partial}{\partial x} + \boldsymbol{j}\frac{\partial}{\partial y} + \boldsymbol{k}\frac{\partial}{\partial z} \tag{A.45}$$

In *cylindrical coordinates* (see Fig. A.16) the components of any linear element ds are dr, $r\,d\theta$, and dz:

$$d\boldsymbol{s} = \boldsymbol{i}\;dr + \boldsymbol{j}r\;d\theta + \boldsymbol{k}\;dz$$

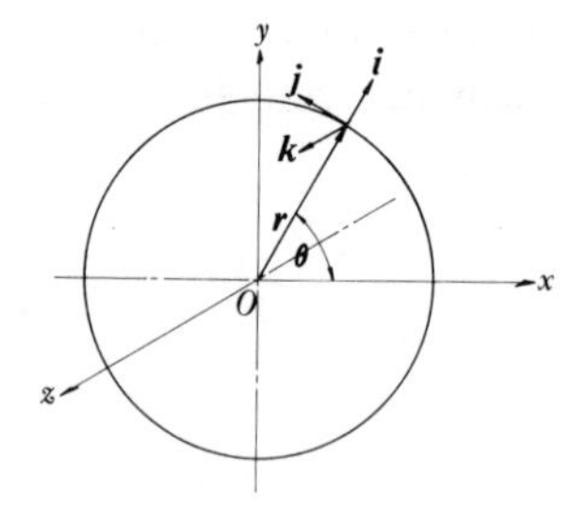

Fig. A.16 Cylindrical unit coordinates.

The gradient is also expressed by

$$\nabla f = \boldsymbol{i}(\nabla f)_r + \boldsymbol{j}(\nabla f)_\theta + \boldsymbol{k}(\nabla f)_z$$

From Eq. (A.44)

$$df = (\nabla f)_r\,dr + (\nabla f)_\theta r\,d_\theta + (\nabla f)_z\,dz$$

Since $f = f(r, \theta, z)$ then also

$$df = \frac{\partial f}{\partial r}\,dr + \frac{\partial f}{\partial \theta}\,d\theta + \frac{\partial f}{\partial z}\,dz$$

Equating these two expressions the components of the gradient in cylindrical coordinates are

$$(\nabla f)_r = \frac{\partial f}{\partial r}; \qquad (\nabla f)_\theta = \frac{1}{r}\frac{\partial f}{\partial \theta}; \qquad (\nabla f)_z = \frac{\partial f}{\partial z}$$

and the vector operator in cylindrical coordinates is

$$\nabla \equiv \boldsymbol{i}\frac{\partial}{\partial r} + \boldsymbol{j}\frac{1}{r}\frac{\partial}{\partial \theta} + \boldsymbol{k}\frac{\partial}{\partial z} \tag{A.46}$$

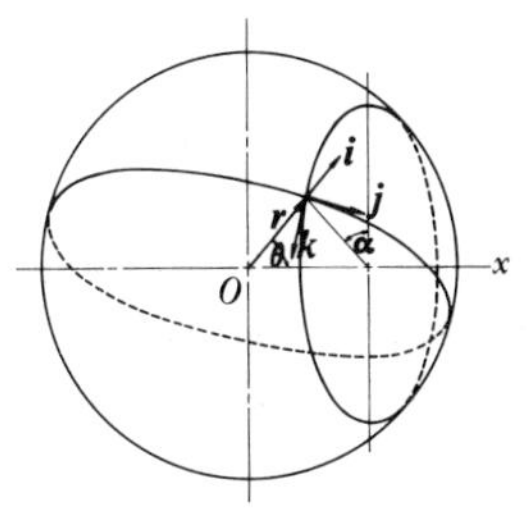

Fig. A.17
Spherical unit coordinates.

In the *polar* or *spherical coordinates* (see Fig. A.17) $ds = \boldsymbol{i}\,dr + \boldsymbol{j}r\,d\theta + \boldsymbol{k}r\sin\theta\,d\alpha$ for the three variables r, θ, and α:

$$\nabla f = \boldsymbol{i}(\nabla f)_r + \boldsymbol{j}(\nabla f)_\theta + \boldsymbol{k}(\nabla f)_\alpha$$

and

$$df = \frac{\partial f}{\partial r}\,dr + \frac{\partial f}{\partial \theta}\,d\theta + \frac{\partial f}{\partial \alpha}\,d\alpha$$

From Eq. (A.44)

$$(\nabla f)_r = \frac{\partial f}{\partial r}; \qquad (\nabla f)_\theta = \frac{1}{r}\frac{\partial f}{\partial \theta}; \qquad (\nabla f)_\alpha = \frac{1}{r\sin\theta}\frac{\partial f}{\partial \alpha}$$

and

$$\nabla \equiv \boldsymbol{i}\frac{\partial}{\partial r} + \boldsymbol{j}\frac{1}{r}\frac{\partial}{\partial \theta} + \boldsymbol{k}\frac{1}{r\sin\theta}\frac{\partial}{\partial \alpha} \tag{A.47}$$

With the background in this appendix the following relations can be shown to be true:

(a) $\nabla f = \dfrac{df}{d\varphi}\nabla\varphi = f'\nabla\varphi$ if $\varphi = \varphi\,(x, y, z)$ and $f = f(\varphi)$;

(b) $\nabla(\varphi + \psi) = \nabla\varphi + \nabla\psi$; (A.48)

(c) $\nabla(\varphi\psi) = \varphi\nabla\psi + \psi\nabla\varphi$.

If, for instance, φ is a constant then

$$\nabla(\varphi\psi) = \varphi\nabla\psi$$

A.16 THE DOT PRODUCT OF A VECTOR WITH THE OPERATOR ∇

The scalar product of a vector with the *vector operator* ∇ has already appeared in Eq. (A.43). The result is by the rules of scalar multiplication or *scalar operator*. For instance, in Cartesian coordinates if $\boldsymbol{a} = \boldsymbol{i}a_x + \boldsymbol{j}a_y + \boldsymbol{k}a_z$

then

$$(\boldsymbol{a} \cdot \nabla) = a_x \frac{\partial}{\partial x} + a_y \frac{\partial}{\partial y} + a_z \frac{\partial}{\partial z} \tag{A.49}$$

This operator is very important in fluid mechanics since it appears in the Eulerean representations of the rate of change of a quantity (see Section 7.4). As an example, if $\varphi = \varphi(x, y, z, t)$ then the total rate of change of φ, $d\varphi/dt$, can be shown to be

$$d\varphi = \frac{\partial \varphi}{\partial t} dt + \frac{\partial \varphi}{\partial \boldsymbol{r}} \cdot d\boldsymbol{r}$$

$$\frac{d\varphi}{dt} = \frac{\partial \varphi}{\partial t} + \left(\frac{d\boldsymbol{r}}{dt} \cdot \nabla\right)\varphi \tag{A.50}$$

where $\boldsymbol{r} = \boldsymbol{i}x + \boldsymbol{j}y + \boldsymbol{k}z$, while $\partial\varphi/\partial\boldsymbol{r} = \nabla\varphi$ as shown in Section A.15. The partial differentiation is necessary since properties are functions of time and space here.

Since the operation in the parentheses of Eq. (A.50) applies to a scalar function φ then the parentheses can be removed and the result will be the same:

$$\left(\frac{d\boldsymbol{r}}{dt} \cdot \nabla\right)\varphi = \frac{d\boldsymbol{r}}{dt} \cdot \nabla\varphi \tag{A.51}$$

Caution must be exercised in placing parentheses in vector operations since their free use does not always give the same result.

A.17 THE CROSS PRODUCT OF A VECTOR AND THE OPERATOR ∇

According to Eqs. (A.20) and (A.22)

$$\boldsymbol{a} \times \nabla = \begin{vmatrix} \boldsymbol{i} & \boldsymbol{j} & \boldsymbol{k} \\ a_x & a_y & a_z \\ \dfrac{\partial}{\partial x} & \dfrac{\partial}{\partial y} & \dfrac{\partial}{\partial z} \end{vmatrix}$$

$$= \boldsymbol{i}\left(a_y \frac{\partial}{\partial z} - a_z \frac{\partial}{\partial y}\right) + \boldsymbol{j}\left(a_z \frac{\partial}{\partial x} - a_x \frac{\partial}{\partial z}\right) + \boldsymbol{k}\left(a_x \frac{\partial}{\partial y} - a_y \frac{\partial}{\partial x}\right) \tag{A.52}$$

This resultant quantity is a vector operator. When applied to a scalar function φ it is also observed that

$$(\boldsymbol{a} \times \nabla)\varphi = \boldsymbol{a} \times \nabla\varphi \tag{A.53}$$

Since ∇ is a *vector operator*, the order of the product, scalar or vector, with another vector cannot be interchanged.

A.18 THE DIVERGENCE OF A VECTOR

In vector calculus, so far, we have considered the vector sum of the changes of a scalar function in the three coordinate directions. This gave rise to the concept of the gradient. This concept could be applied to vectors as well but we can see right away that, since the gradient has three components and if it operates on a vector that also has three components, the complete operation involves nine derivatives to describe this concept. This then brings us to quantities more complex in their description called tensor quantities, which will be discussed in Section A.27.

However, the scalar sum of the change of each vector component in its own direction has a real physical significance and is called the *divergence*. It represents the extent to which the vector field diverges away from a point. It is seen immediately that the divergence of a vector is just the dot product of the operator del with the vector:

$$\nabla \cdot \boldsymbol{a} = \left(\boldsymbol{i}\frac{\partial}{\partial x} + \boldsymbol{j}\frac{\partial}{\partial y} + \boldsymbol{k}\frac{\partial}{\partial z}\right) \cdot (\boldsymbol{i}a_x + \boldsymbol{j}a_y + \boldsymbol{k}a_z)$$

$$= \frac{\partial a_x}{\partial x} + \frac{\partial a_y}{\partial y} + \frac{\partial a_z}{\partial z} \tag{A.54}$$

Here, in the Cartesian coordinate system, the unit vectors do not change their orientation with x, y, z. This fact is not true in cylindrical or spherical coordinate systems.

For instance, from Eq. (A.46) in *cylindrical coordinates*

$$\nabla \equiv \boldsymbol{i}\frac{\partial}{\partial r} + \boldsymbol{j}\frac{1}{r}\frac{\partial}{\partial \theta} + \boldsymbol{k}\frac{\partial}{\partial z}$$

Now if a vector $\boldsymbol{a}$ is given in terms of its cylindrical components a_r, a_θ, and a_z then the divergence is

$$\nabla \cdot \boldsymbol{a} = \left(\boldsymbol{i}\frac{\partial}{\partial r} + \boldsymbol{j}\frac{1}{r}\frac{\partial}{\partial \theta} + \boldsymbol{k}\frac{\partial}{\partial z}\right) \cdot (\boldsymbol{i}a_r + \boldsymbol{j}a_\theta + \boldsymbol{k}a_z)$$

$$= \boldsymbol{i} \cdot \left(\boldsymbol{i}\frac{\partial a_r}{\partial r} + a_r\frac{\partial \boldsymbol{i}}{\partial r} + \boldsymbol{j}\frac{\partial a_\theta}{\partial r} + a_\theta\frac{\partial \boldsymbol{j}}{\partial r} + \boldsymbol{k}\frac{\partial a_z}{\partial r} + a_z\frac{\partial \boldsymbol{k}}{\partial r}\right)$$

$$+ \boldsymbol{j}\frac{1}{r} \cdot \left(\boldsymbol{i}\frac{\partial a_r}{\partial \theta} + a_r\frac{\partial \boldsymbol{i}}{\partial \theta} + \boldsymbol{j}\frac{\partial a_\theta}{\partial \theta} + a_\theta\frac{\partial \boldsymbol{j}}{\partial \theta} + k\frac{\partial a_z}{\partial \theta} + a_z\frac{\partial \boldsymbol{k}}{\partial \theta}\right)$$

$$+ \boldsymbol{k} \cdot \left(\boldsymbol{i}\frac{\partial a_r}{\partial z} + a_r\frac{\partial \boldsymbol{i}}{\partial z} + \boldsymbol{j}\frac{\partial a_\theta}{z} + a_\theta\frac{\partial \boldsymbol{j}}{\partial z} + \boldsymbol{k}\frac{\partial a_z}{\partial z} + a_z\frac{\partial \boldsymbol{k}}{\partial z}\right)$$

In this relation six terms drop out immediately. The terms $\partial \boldsymbol{i}/\partial r = \partial \boldsymbol{j}/\partial r = \partial \boldsymbol{k}/\partial r = 0$ because referring to Fig. A.18 it is seen that the unit vectors do not

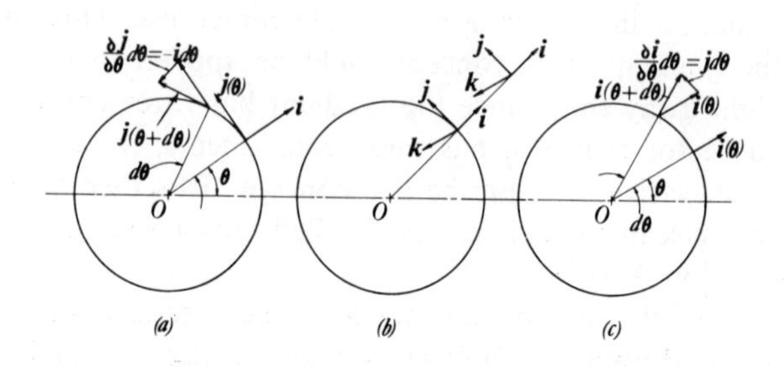

Fig. A.18 Change of unit vector in a cylindrical frame.

change with r. The same is true for $\partial \boldsymbol{k}/\partial z = \partial \boldsymbol{j}/\partial z = \partial \boldsymbol{i}/\partial z = 0$; this eliminates six more terms. However, $\partial \boldsymbol{j}/\partial \theta \neq 0$ because from Fig. A.18a using Taylor's expansion $\boldsymbol{j}(\theta + d\theta) = \boldsymbol{j}(\theta) + (\partial \boldsymbol{j}/\partial \theta)\, d\theta$ and consequently $\boldsymbol{j}(\theta + d\theta) - \boldsymbol{j}(\theta) = (\partial \boldsymbol{j}/\partial \theta)\, d\theta$. This resultant vector is in the direction of $-\boldsymbol{i}$ and its magnitude is the sine of the angle or the angle itself. Thus $\partial \boldsymbol{j}/\partial \theta = -\boldsymbol{i}$. From Fig. A.18c, $(\partial \boldsymbol{i}/\partial \theta)\, d\theta = \boldsymbol{j}\, d\theta$, giving $\partial \boldsymbol{i}/\partial \theta = \boldsymbol{j}$. The final quantity $\partial \boldsymbol{k}/\partial \theta = 0$. Now the divergence in *cylindrical coordinates* becomes

$$\nabla \cdot \boldsymbol{a} = \frac{\partial a_r}{\partial r} + \frac{a_r}{r} + \frac{1}{r}\frac{\partial a_\theta}{\partial \theta} + \frac{\partial a_z}{\partial z}$$

$$= \frac{1}{r}\frac{\partial}{\partial r}(r a_r) + \frac{1}{r}\frac{\partial a_\theta}{\partial \theta} + \frac{\partial a_z}{\partial z} \qquad \text{(A.55)}$$

In *spherical coordinates* the reader should show from Fig. A.19 that

$$\frac{\partial}{\partial r}(\boldsymbol{i}, \boldsymbol{j}, \boldsymbol{k}) = 0$$

$$\frac{\partial \boldsymbol{i}}{\partial \theta} = \boldsymbol{j}; \qquad \frac{\partial \boldsymbol{j}}{\partial \theta} = -\boldsymbol{i}; \qquad \frac{\partial \boldsymbol{k}}{\partial \theta} = 0$$

$$\frac{\partial \boldsymbol{i}}{\partial \alpha} = \boldsymbol{k} \sin \theta; \qquad \frac{\partial \boldsymbol{j}}{\partial \alpha} = \boldsymbol{k} \cos \theta$$

$$\partial \boldsymbol{k}/\partial \alpha = -(\boldsymbol{i} \sin \theta + \boldsymbol{j} \cos \theta)$$

and thus

$$\nabla \cdot \boldsymbol{a} = \frac{1}{r^2}\frac{\partial}{\partial r}(r^2 a_r) + \frac{1}{r \sin\theta}\frac{\partial}{\partial \theta}(a_\theta \sin\theta) + \frac{1}{r \sin\theta}\frac{\partial a_\alpha}{\partial \alpha} = 0 \quad \text{(A.56)}$$

We can easily show that the divergence of the sum of two vectors is

$$\nabla \cdot (\boldsymbol{a} + \boldsymbol{b}) = \nabla \cdot \boldsymbol{a} + \nabla \cdot \boldsymbol{b} \quad \text{(A.57a)}$$

The divergence of the product of a scalar and a vector is

$$\nabla \cdot (\varphi \boldsymbol{a}) = \varphi(\nabla \cdot \boldsymbol{a}) + (\nabla\varphi) \cdot \boldsymbol{a} \quad \text{(A.57b)}$$

The divergence of a gradient of a scalar or the *Laplacian* of a scalar is

$$\nabla \cdot (\nabla\varphi) = \nabla^2\varphi = \frac{\partial^2\varphi}{\partial x^2} + \frac{\partial^2\varphi}{\partial y^2} + \frac{\partial^2\varphi}{\partial z^2} \quad \text{(A.57c)}$$

where

$$(\nabla \cdot \nabla) = \nabla^2 \quad \text{(A.57d)}$$

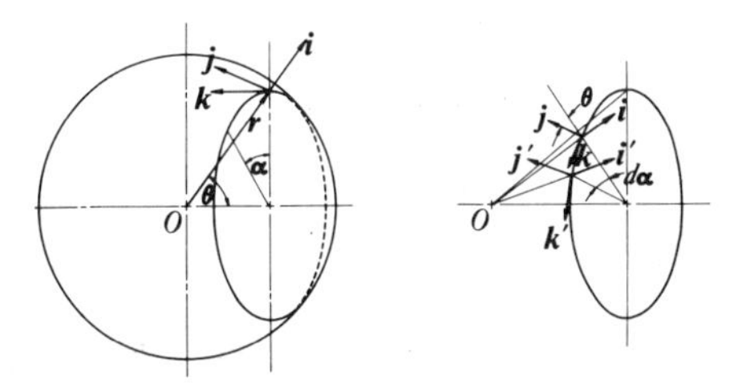

Fig. A.19 Change of unit vector in spherical frame.

A vector whose divergence is equal to zero is called a *solenoidal vector*. If the velocity vector is such a vector, then from Eq. (A.54) we can see that the divergence will represent the sum of the changes of the velocity components away from a point which physically implies that the *net flux* of the substance is zero. More physical significance of the divergence is given in Section 6.4.

A.19 THE CURL OF A VECTOR

The vector operator ∇ has been used so far as a differential operation on a function giving the gradient and as a differential vector dotted to another vector giving a scalar quantity called the divergence. The cross product of ∇

and a vector gives a vector quantity called the *curl* which has a physical significance in mechanics. From Eq. (A.45)

$$\begin{aligned}
\nabla \times \boldsymbol{a} &= \text{curl}\ \boldsymbol{a} \\
&= \left(\boldsymbol{i}\frac{\partial}{\partial x} + \boldsymbol{j}\frac{\partial}{\partial y} + \boldsymbol{k}\frac{\partial}{\partial z}\right) \times (\boldsymbol{i}a_x + \boldsymbol{j}a_y + \boldsymbol{k}a_z) \\
&= \boldsymbol{i}\left(\frac{\partial a_z}{\partial y} - \frac{\partial a_y}{\partial z}\right) + \boldsymbol{j}\left(\frac{\partial a_x}{\partial z} - \frac{\partial a_z}{\partial x}\right) + \boldsymbol{k}\left(\frac{\partial a_y}{\partial x} - \frac{\partial a_x}{\partial y}\right) \\
&= \begin{vmatrix} \boldsymbol{i} & \boldsymbol{j} & \boldsymbol{k} \\ \dfrac{\partial}{\partial x} & \dfrac{\partial}{\partial y} & \dfrac{\partial}{\partial z} \\ a_x & a_y & a_z \end{vmatrix} = \boldsymbol{i} \times \frac{\partial \boldsymbol{a}}{\partial x} + \boldsymbol{j} \times \frac{\partial \boldsymbol{a}}{\partial y} + \boldsymbol{k} \times \frac{\partial \boldsymbol{a}}{\partial z}
\end{aligned} \tag{A.58}$$

If $\boldsymbol{a}$ is the velocity vector then the curl represents twice the angular velocity. This will be easily verified shortly after introducing the triple vector product involving ∇. First it can easily be shown when differentiating by parts and using the triple scalar and vector products that

$$\begin{aligned}
&\text{(a)} \quad \nabla \times (\boldsymbol{a} + \boldsymbol{b}) = \nabla \times \boldsymbol{a} + \nabla \times \boldsymbol{b} \\
&\text{(b)} \quad \nabla \times (\boldsymbol{a}\varphi) = \varphi(\nabla \times \boldsymbol{a}) + (\nabla\varphi) \times \boldsymbol{a} \\
&\text{(c)} \quad (\boldsymbol{a} \times \nabla) \cdot \boldsymbol{b} = \boldsymbol{a} \cdot (\nabla \times \boldsymbol{b}) \\
&\text{(d)} \quad \nabla \cdot (\boldsymbol{a} \times \boldsymbol{b}) = \boldsymbol{b} \cdot (\nabla \times \boldsymbol{a}) - \boldsymbol{a} \cdot (\nabla \times \boldsymbol{b}) \\
&\text{(e)} \quad \nabla \times (\boldsymbol{a} \times \boldsymbol{b}) = \boldsymbol{a}(\nabla \cdot \boldsymbol{b}) + (\boldsymbol{b} \cdot \nabla)\boldsymbol{a} - \boldsymbol{b}(\nabla \cdot \boldsymbol{a}) - (\boldsymbol{a} \cdot \nabla)\boldsymbol{b} \\
&\text{(f)} \quad \nabla \times (\nabla\varphi) = 0 \\
&\text{(g)} \quad \nabla \cdot (\nabla \times \boldsymbol{a}) = 0 \\
&\text{(h)} \quad \nabla \times (\nabla \times \boldsymbol{a}) = \nabla(\nabla \cdot \boldsymbol{a}) - \nabla^2 \boldsymbol{a}
\end{aligned} \tag{A.59}$$

It would be interesting to incorporate proofs of all relations in Eqs. (A.59). With the help of the principles of vectors treated so far the student should be able to establish all these relations.[7]

To show the physical significance of the curl of the velocity we choose a solid body rotating with a velocity according to Eq. (A.24):

$$\boldsymbol{U} = \boldsymbol{\omega} \times \boldsymbol{r}$$

[7] Detailed proofs of these relations may be found in most vector calculus books: for example, see H. Lass, "Vector and Tensor Analysis." McGraw-Hill, New York, 1950.

Consider the curl of $\boldsymbol{U}$:

$$\nabla \times \boldsymbol{U} = \nabla \times (\boldsymbol{\omega} \times \boldsymbol{r})$$

Using Eq. (A.59e)

$$\nabla \times \boldsymbol{U} = \boldsymbol{\omega}(\nabla \cdot \boldsymbol{r}) + (\boldsymbol{r} \cdot \nabla)\boldsymbol{\omega} - \boldsymbol{r}(\nabla \cdot \boldsymbol{\omega}) - (\boldsymbol{\omega} \cdot \nabla)\boldsymbol{r}$$

The first term on the right-hand side is easily seen to be $3\boldsymbol{\omega}$. The second and third terms are zero because there are no space variations of $\boldsymbol{\omega}$ on a rigid body. The last term is equal to $-\boldsymbol{\omega}$. Then

$$\nabla \times \boldsymbol{U} = 2\boldsymbol{\omega} \tag{A.60}$$

This expression could have been obtained just as easily from the last expression of Eq. (A.58). Then the curl of the velocity is twice the angular velocity. It was shown in Section 6.10 that this concept is extended with more generality to fluid rotation as well.

A.20 INVARIANCE OF THE VECTOR OPERATOR

At the beginning of this appendix it was pointed out that a vector is a quantity whose magnitude and direction are independent of orientation of the coordinate system. The operator

$$\nabla \equiv \boldsymbol{i}\frac{\partial}{\partial x} + \boldsymbol{j}\frac{\partial}{\partial y} + \boldsymbol{k}\frac{\partial}{\partial z}$$

is a vector operator in the x, y, z Cartesian coordinate system. If another coordinate system x', y', z' is considered such that its orientation is different from the first, a new del operator can be defined:

$$\nabla' \equiv \boldsymbol{i}'\frac{\partial}{\partial x'} + \boldsymbol{j}'\frac{\partial}{\partial y'} + \boldsymbol{k}'\frac{\partial}{\partial z'}$$

where $\boldsymbol{i}'$, $\boldsymbol{j}'$, and $\boldsymbol{k}'$ are the new unit vectors in the new coordinates. Now if a scalar function φ is expressed either in the x, y, z system as $\varphi(x, y, z)$ or in the x', y', z' system as $\varphi'(x', y, z')$, then it can be shown[8] that

$$\nabla\varphi = \nabla'\varphi'$$

The same is found to be true for the divergence and the curl. If a vector $\boldsymbol{a}$ is given then

$$\nabla \cdot \boldsymbol{a} = \nabla' \cdot \boldsymbol{a} \qquad \text{and} \qquad \nabla \times \boldsymbol{a} = \nabla' \times \boldsymbol{a}$$

These are very important useful relations in transformation of coordinates or in expressing motion relative to a moving and rotating coordinate system.

[8] See, for instance, G. E. Hay, "Vector and Tensor Analysis," p. 111. Dover Publications, New York, 1953.

A.21 CONDITION FOR A VECTOR TO BE THE GRADIENT OF A SCALAR FUNCTION

Even though the gradient of a scalar function is a vector, not all vector fields can be represented as the gradient of a scalar field.

From Eq. (A.59f) for a vector $\boldsymbol{U}$ to be expressed as the gradient of φ it must follow that $\nabla \times \boldsymbol{U} = 0$. So conversely it can be said that *if the curl of a vector is zero, then*, because of the consequences of Eq. (A.59f), *the vector can be expressed as the gradient of some function* φ. This is proved formally in Section A.23.

A.22 THE LINE INTEGRAL

Let $\boldsymbol{f}$ be a continuous vector field along a path AB represented by $\boldsymbol{r} = \boldsymbol{r}(s)$ in Fig. A.20, where $\boldsymbol{r}$ is the radius vector from O to any point on the path and s the

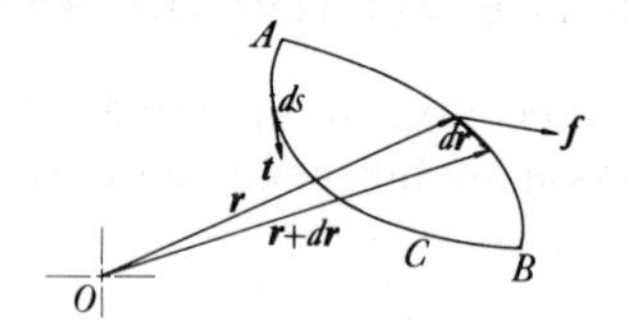

Fig. A.20 The line integral.

path length from a given origin. The infinitesimal arc length is $d\boldsymbol{r} = \boldsymbol{t}\, ds$, where $\boldsymbol{t}$ is the unit vector tangent to the path. The line integral is defined as

$$\int_C \boldsymbol{f} \cdot d\boldsymbol{r} = \int_A^B \left(\boldsymbol{f} \cdot \frac{d\boldsymbol{r}}{ds} \right) ds = \int_A^B f_x\, dx + f_y\, dy + f_z\, dz \tag{A.61}$$

This integral appears, for instance, in mechanics when a fluid or solid particle moves along a path C under a force $\boldsymbol{f}$. The work done is then given by Eq. (A.61). In general this integral is a function of the path chosen from A to B; in other words the work of all particles going from A to B following different paths may not be the same. There is, however, one condition for which this integral is independent of the path provided the end points are kept the same. Mathematically speaking, the product $\boldsymbol{f} \cdot d\boldsymbol{r}$ must be an exact differential. Looking at Eq. (A.43), we see that if $\boldsymbol{f}$ is the gradient of a scalar φ, then Eq. (A.61) can be written

$$\int_C \boldsymbol{f} \cdot d\boldsymbol{r} = \int_C \nabla\varphi \cdot d\boldsymbol{r} = \int_C d\varphi$$

$$= \varphi_B - \varphi_A$$

The result then is uniquely dependent on the values of φ at A and B and not on the path. If f is a force and can be expressed as $\nabla\varphi$ then the force field is called *conservative*. The previous section already has shown under what condition f can be expressed as the gradient of a scalar function. Since the integral from A to B for two different paths is the same for a conservative field, then the integration around a closed contour is

$$\Gamma = \oint \boldsymbol{f} \cdot d\boldsymbol{r} = 0 \tag{A.62}$$

This integral sign indicates integration around a closed contour. This contour integration is called the *circulation* Γ and in general the circulation need not be zero.

A.23 ORIENTED SURFACE

We have seen in Section A.9 that a surface is oriented and that the orientation depends on the order of the cross product of its sides. The *surface vector* is in the direction normal to the surface, along the positive unit normal $\boldsymbol{n}$, and its magnitude is the value of the area. The infinitesimal surface vector is then $\boldsymbol{n}\,dS$ or $d\boldsymbol{S}$. This is shown in Fig. A.21. The positive unit normal is defined as the outer normal in the direction of the radius of curvature.

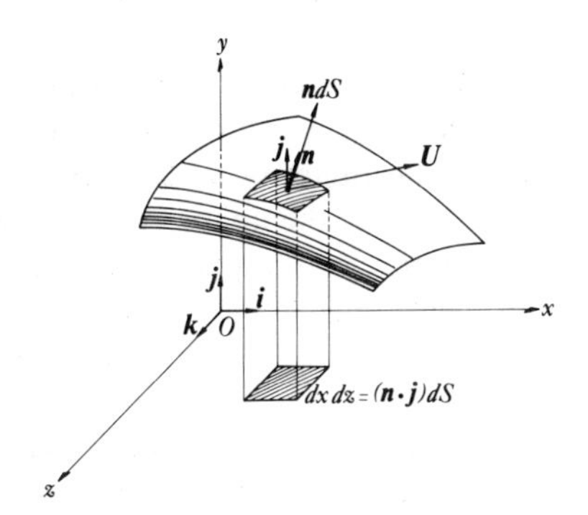

Fig. A.21 Vector orientation of a surface.

If at the point P a vector, say the velocity $\boldsymbol{U}$, is defined then the scalar quantity $\boldsymbol{n} \cdot \boldsymbol{U}\,dS$, by the scalar product rule, represents the component of $\boldsymbol{U}$ on $\boldsymbol{n}$ multiplied by the infinitesimal area or the projection of $\boldsymbol{n}\,dS$ normal to $\boldsymbol{U}$ multiplied by the magnitude of $\boldsymbol{U}$. Physically, this is the volume rate of flow crossing the area dS.

By the same token $(\boldsymbol{n}\,dS) \cdot \boldsymbol{j}$ is the projection of the area dS normal to $\boldsymbol{j}$. Since $\boldsymbol{n} \cdot \boldsymbol{j} = |n|\,|j| \cos\theta = \cos\theta$, then $(\boldsymbol{n}\,dS) \cdot \boldsymbol{j} = dS \cos\theta = dx\,dz$.

A.24 STOKES'S THEOREM—LINE AND SURFACE INTEGRALS

Let S be a surface bounded by a closed curve C. Let $\boldsymbol{n}$ be the unit normal on the surface[9] defined in the outward direction as shown in Fig. A.22. If a

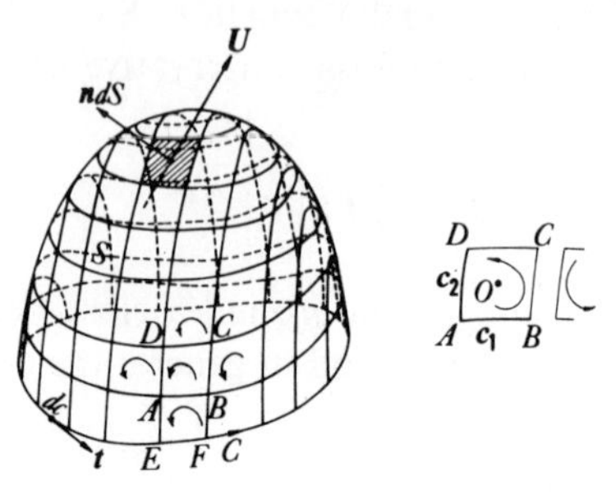

Fig. A.22 Surface and contour in Stokes's theorem.

vector field $\boldsymbol{U}$ and its derivatives are continuous and defined on the surface S then Stokes's theorem states that

$$\oint_C \boldsymbol{U} \cdot \boldsymbol{t}\, dc = \int_S \boldsymbol{n} \cdot (\nabla \times \boldsymbol{U})\, dS$$
$$\oint_C \boldsymbol{U} \cdot d\boldsymbol{c} = \int_S (\nabla \times \boldsymbol{U}) \cdot d\boldsymbol{S} \qquad \text{(A.63)}$$

This is a relationship of the line integral of $\boldsymbol{U}$ around the closed contour C of infinitesimal length $\boldsymbol{t}\, dc = d\boldsymbol{c}$ in terms of the curl of $\boldsymbol{U}$ on any open surface S that ends on the contour C. The quantity $\boldsymbol{n}\, dS = d\boldsymbol{S}$ is the infinitesimal area vector normal to the surface while $\boldsymbol{t}$ is the unit vector tangent to C. This is a very important theorem in mathematics that applies to any vector field that is continuous and has continuous derivatives. Its use is very broad in engineering and physics. To prove the theorem we proceed as follows with the following geometrical proof.

Dividing the surface S into small meshes as shown and considering one such mesh $ABCD$, if the vector at the center of $ABCD$ is $\boldsymbol{U}$ then by Taylor's expansion the corresponding quantities at all sides are

$$(\boldsymbol{U})_{AB} = \boldsymbol{U} - \tfrac{1}{2}(\boldsymbol{c}_2 \cdot \nabla)\boldsymbol{U}$$
$$(\boldsymbol{U})_{DC} = \boldsymbol{U} + \tfrac{1}{2}(\boldsymbol{c}_2 \cdot \nabla)\boldsymbol{U}$$
$$(\boldsymbol{U})_{AD} = \boldsymbol{U} - \tfrac{1}{2}(\boldsymbol{c}_1 \cdot \nabla)\boldsymbol{U}$$
$$(\boldsymbol{U})_{BC} = \boldsymbol{U} + \tfrac{1}{2}(\boldsymbol{c}_1 \cdot \nabla)\boldsymbol{U}$$

[9] This surface is arbitrary provided it is *simply connected*. For this definition see, for instance, Harry F. Davis, "Introduction to Vector Analysis," p. 130. Allyn and Bacon, Boston, 1961.

The circulation Γ according to Eq. (A.62) around this small mesh is

$$\oint_{ABCD} \boldsymbol{U} \cdot d\boldsymbol{c} = (\boldsymbol{U})_{AB} \cdot \boldsymbol{c}_1 + (\boldsymbol{U})_{BC} \cdot \boldsymbol{c}_2 - (\boldsymbol{U})_{DC} \cdot \boldsymbol{c}_1 - (\boldsymbol{U})_{AD} \cdot \boldsymbol{c}_2$$

The negative signs on the last two terms appear because if the integration is performed counterclockwise, along CD and DA, the length vectors are negative. Evaluating this integral by substitution of the expanded vectors $\boldsymbol{U}$ at each edge,

$$\oint_{ABCD} \boldsymbol{U} \cdot d\boldsymbol{c} = \boldsymbol{c}_2 \cdot [(\boldsymbol{c}_1 \cdot \nabla)\boldsymbol{U}] - \boldsymbol{c}_1 \cdot [(\boldsymbol{c}_2 \cdot \nabla)\boldsymbol{U}]$$

From the rules of the triple vector product, Eq. (A.30), the previous equation for the circulation around $ABCD$ becomes

$$\oint_{ABCD} \boldsymbol{U} \cdot d\boldsymbol{c} = [(\boldsymbol{c}_1 \times \boldsymbol{c}_2) \times \nabla]\boldsymbol{U}$$

Now according to permutations of the triple scalar product of vectors $(\boldsymbol{c}_1 \times \boldsymbol{c}_2)$, ∇, and $\boldsymbol{U}$,

$$\begin{aligned}\oint_{ABCD} \boldsymbol{U} \cdot d\boldsymbol{c} &= (\boldsymbol{c}_1 \times \boldsymbol{c}_2) \cdot (\nabla \times \boldsymbol{U}) \\ &= (\nabla \times \boldsymbol{U}) \cdot d\boldsymbol{S}\end{aligned} \tag{A.64}$$

This is very similar to Eq. (A.63). It applies just to the small mesh. Now if a summation is performed one mesh after another, each edge, as AB, will be integrated twice, once in each direction and therefore canceling, except for the edges EF along C which will not be common to two meshes. However the right-hand side of Eq. (A.64) depends on the area of the mesh and in the summation these contributions add throughout the surface S. Thus

$$\oint_C \boldsymbol{U} \cdot d\boldsymbol{c} = \int_S (\nabla \times \boldsymbol{U}) \cdot d\boldsymbol{S}$$

which is the proof of the theorem. From the triple scalar product $\boldsymbol{n} \cdot (\nabla \times \boldsymbol{U}) = (\boldsymbol{n} \times \nabla) \cdot \boldsymbol{U}$, consequently Stokes's equation (A.63) can be written as

$$\oint_C \boldsymbol{U} \cdot d\boldsymbol{c} = \int_S (\boldsymbol{n} \times \nabla) \cdot \boldsymbol{U}\, dS \tag{A.65}$$

This theorem can be shown to extend also for scalar fields φ instead of vector field $\boldsymbol{U}$. Then

$$\oint_C \varphi\, d\boldsymbol{c} = \int_S (\boldsymbol{n} \times \nabla)\varphi\, dS \tag{A.66}$$

Also it could be shown to apply to the cross product

$$-\oint_C \boldsymbol{U} \times d\boldsymbol{c} = \int_S (\boldsymbol{n} \times \nabla) \times \boldsymbol{U}\, dS \tag{A.67}$$

The condition for a vector field U to be represented as the gradient of a scalar field φ discussed in Section A.21 can be obtained formally here. It is seen immediately from Eq. (A.63) that, if $U = \nabla\varphi$, the left-hand integral vanishes because of Eq. (A.62). Then since for any S whose boundary is C the right-hand integral is zero; then if the integrand is continuous in S it must be zero: $\nabla \times U = 0$. So if in a region $\nabla \times U = 0$ then U can be represented in that region as the gradient of a scalar function.

A.25 THE DIVERGENCE THEOREM—GAUSS'S THEOREM

This is one of the most important theorems[10] in vector analysis. It relates the vector field in a closed arbitrary surface[11] to the vector field in the volume enclosed by the surface. Given a continuously differential vector U in a volume $\mathscr{V}$ bounded by a surface S, in which $\boldsymbol{n}$ is the unit outward normal,

$$\oint_S \boldsymbol{n} \cdot \boldsymbol{U}\, dS = \int_{\mathscr{V}} (\nabla \cdot \boldsymbol{U})\, d\mathscr{V} \tag{A.68}$$

The integral on the left is the so-called net flux integral over the entire surface of the vector U crossing the surface S which is by Eq. (A.68) equal to the summation of the divergence inside the volume. If a physical significance is attached to U, say the velocity, then Eq. (A.68) states physically that, in order to conserve mass, the net flux of mass across S, if not zero, must be equal to the summation of sources and sinks[12] in the volume. To prove this we proceed as follows with the guide of Fig. A.23.

The integral

$$\oint_S \boldsymbol{n} \cdot \boldsymbol{U}\, dS = \oint_S [(\boldsymbol{n} \cdot \boldsymbol{i})u + (\boldsymbol{n} \cdot \boldsymbol{j})v + (\boldsymbol{n} \cdot \boldsymbol{k})w]\, dS \tag{A.69}$$

since $\boldsymbol{U} = \boldsymbol{i}u + \boldsymbol{j}v + \boldsymbol{k}w$. Out of the arbitrary volume $\mathscr{V}$ consider an infinitesimal volume intersecting the surface S at dS' and dS''. Consider the second term in the last integral. The quantity $(\boldsymbol{n} \cdot \boldsymbol{j})\, dS = \boldsymbol{j} \cdot \boldsymbol{n}\, dS = \boldsymbol{j} \cdot d\boldsymbol{S} = dx\, dz$ by Section A.23. Then the contribution of the integral in the differential strip is $(\boldsymbol{n}'' \cdot \boldsymbol{j})\, dS''\, v'' - (\boldsymbol{n}' \cdot \boldsymbol{j})\, dS'\, v'$ or $(v'' - v')\, dx\, dz$. By calculus

$$v'' - v' = \int_{y'}^{y''} \frac{\partial v}{\partial y}\, dy$$

[10] Often the Stokes theorem of Section A.23 is shown to be deduced from this Gauss theorem. For instance, see I. J. Sokolnikoff and R. M. Redheffer, "Mathematics of Physics and Modern Engineering," p. 391. McGraw-Hill, New York, 1958.

[11] Surface of a simply connected region (see footnote 9).

[12] The physical analogy here pertains to a fluid motion with constant density.

Now if all such strips are integrated in the whole surface the contribution to the second term on the right of Eq. (A.69) is

$$\oint_S \int_{y'}^{y''} \left(\frac{\partial v}{\partial y} \, dy \right) dx \, dz$$

where the first integration is taken over the xz projection of S and the second integration is taken from y'' to y'. This is exactly the volume integration of $\partial v / \partial y$:

$$\int_{\mathcal{V}} \frac{\partial v}{\partial y} \, d\mathcal{V}$$

The same reasoning can be carried out for the other two quantities in Eq. (A.69) and thus shows that

$$\oint_S \boldsymbol{n} \cdot \boldsymbol{U} \, dS = \int_{\mathcal{V}} \left(\frac{\partial u}{\partial x} + \frac{\partial v}{\partial y} + \frac{\partial w}{\partial z} \right) d\mathcal{V}$$

$$= \int_{\mathcal{V}} (\nabla \cdot \boldsymbol{U}) \, d\mathcal{V}$$

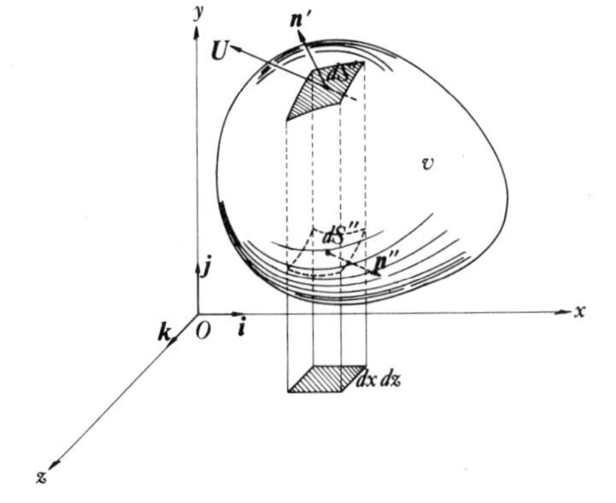

Fig. A.23 Geometric representation for Gauss's theorem.

the proof of the theorem. From Gauss's theorem we can see that another alternate definition of the divergence[13] is

$$\nabla \cdot \boldsymbol{U} = \lim_{\Delta \mathcal{V} \to 0} \frac{1}{\Delta_{\mathcal{V}}} \oint_S (\boldsymbol{n} \cdot \boldsymbol{U}) \, dS \tag{A.70}$$

[13] This is done by defining an average $\nabla \cdot \boldsymbol{U}$ over a volume $\mathcal{V}$,

$$(\nabla \cdot \boldsymbol{U})_{\text{av}} = \frac{1}{\mathcal{V}} \int (\nabla \cdot \boldsymbol{U}) \, d\mathcal{V}$$

and in the limit as $\mathcal{V} \to 0$, $(\nabla \cdot \boldsymbol{U})_{\text{av}} = \nabla \cdot \boldsymbol{U}$. For the integral form of the operations with "del" see, for instance, L. M. Thomson, "Theoretical Hydrodynamics," p. 39. Macmillan, New York, 1960.

From Eq. (A.70) it is seen that the divergence of a vector $\boldsymbol{U}$ has for a physical meaning the net flux of $\boldsymbol{U}$ across the closed surface S when the volume $\mathscr{V}$ shrinks to a point.

A.26 CONSEQUENCES OF GAUSS'S THEOREM

It can be shown easily that the generalized Gauss theorem applies not only to the divergence but also to the gradient, the curl, and the Laplacian:

$$\text{(a)} \quad \oint_S \boldsymbol{n} \times \boldsymbol{U}\, dS = \int_{\mathscr{V}} \nabla \times \boldsymbol{U}\, d\mathscr{V}$$

$$\text{(b)} \quad \oint_S \boldsymbol{n}\varphi\, dS = \int_{\mathscr{V}} \nabla\varphi\, d\mathscr{V}$$

$$\text{(c)} \quad \oint_S (\boldsymbol{n} \cdot \nabla)\varphi\, dS = \int_S \frac{\partial \varphi}{\partial n}\, dS = \int_{\mathscr{V}} \nabla^2\varphi\, d\mathscr{V} \qquad \text{(A.71)}$$

$$\text{(d)} \quad \oint_S (\boldsymbol{n} \cdot \nabla)\boldsymbol{U}\, dS = \int_{\mathscr{V}} \nabla^2\boldsymbol{U}\, d\mathscr{V}$$

$$\text{(e)} \quad \oint_S \boldsymbol{a}(\boldsymbol{n} \cdot \boldsymbol{U})\, dS = \int_{\mathscr{V}} [\boldsymbol{a}(\nabla \cdot \boldsymbol{U}) + (\boldsymbol{U} \cdot \nabla)\boldsymbol{a}]\, d\mathscr{V}$$

Now letting $\boldsymbol{a} = u\nabla v$ and $b = v\nabla u$, applying Eq. (A.68),

$$\oint_S \boldsymbol{n} \cdot (u\nabla v)\, dS = \int_{\mathscr{V}} \nabla \cdot (u\nabla v)\, d\mathscr{V} = \int_{\mathscr{V}} (u\nabla^2 v + \nabla u \cdot \nabla v)\, d\mathscr{V}$$

$$\oint_S \boldsymbol{n} \cdot (v\nabla u)\, dS = \int_{\mathscr{V}} \nabla \cdot (v\nabla u)\, d\mathscr{V} = \int_{\mathscr{V}} (v\nabla^2 u + \nabla v \cdot \nabla u)\, d\mathscr{V}$$

After subtraction the well-known *Green's theorem* is obtained:

$$\oint_S \boldsymbol{n} \cdot (u\nabla v - v\nabla u)\, dS = \int_{\mathscr{V}} (u\nabla^2 v - v\nabla^2 u)\, d\mathscr{V} \qquad \text{(A.72)}$$

according to Section A.15 since $\nabla \equiv \boldsymbol{n}\, \partial/\partial n$. Then Green's formula can be written in a form usually recognized:

$$\oint_S \left(u\frac{\partial v}{\partial n} - v\frac{\partial u}{\partial n} \right) dS = \int_{\mathscr{V}} (u\nabla^2 v - v\nabla^2 u)\, d\mathscr{V} \qquad \text{(A.73)}$$

Finally, Stokes's theorem, Eqs. (A.63), and Gauss's theorem, Eq. (A.68), could be combined to show that Eq. (A.59g) states that the divergence of the curl is always zero. To show this, if a lower half surface were drawn in Fig. A.21 it would also be equal to the left-hand side of Eq. (A.63). Therefore the complete closed integral

$$\oint_S \boldsymbol{n} \cdot (\nabla \times \boldsymbol{U})\, dS = 0$$

Now in Eq. (A.68) for another vector $\boldsymbol{a}$ in the region

$$\oint_S \boldsymbol{n} \cdot \boldsymbol{a}\, dS = \int_{\mathscr{V}} (\nabla \cdot \boldsymbol{a})\, d\mathscr{V}$$

For $\boldsymbol{a} = \nabla \times \boldsymbol{U}$ the divergence of $\boldsymbol{a}$ must be equal to zero since for all $\mathscr{V}$ the answer would be the same. This then verifies that the divergence of the curl is always zero provided the vector fields and their derivatives are continuous in the volume $\mathscr{V}$.

Another consequence of Gauss's theorem and its limit as shown in Eq. (A.70) for the divergence is

$$\nabla \varphi = \lim_{\mathscr{V} \to 0} \frac{1}{\mathscr{V}} \oint_S \boldsymbol{n} \varphi\, dS \qquad \nabla \times \boldsymbol{U} = \lim_{\mathscr{V} \to 0} \frac{1}{\mathscr{V}} \oint_S \boldsymbol{n} \times \boldsymbol{U}\, dS \tag{A.74}$$

A.27 THE INDEFINITE OR DYADIC PRODUCT

In Sections A.1 and A.2 we showed the essential differences in the description of scalars and vectors. Their main difference was that, besides the units necessary in the description of both scalars and vectors, a scalar needed one number and a vector three numbers for their complete description. It was deduced then that the vector was a more complex quantity than the scalar.

However, some quantities are even more complex than vectors and need nine numbers for their description. Such quantities were encountered in the triple vector product of Section A.11 but were not fully developed. For instance, $(\boldsymbol{a} \cdot \boldsymbol{b})\boldsymbol{c}$ is described in terms of nine independent quantities. Expanding,

$$\begin{aligned}(\boldsymbol{a} \cdot \boldsymbol{b})\boldsymbol{c} &= \boldsymbol{i} a_x b_x c_x + \boldsymbol{i} a_y b_y c_x + \boldsymbol{i} a_z b_z c_x \\ &\quad + \boldsymbol{j} a_x b_x c_y + \boldsymbol{j} a_y b_y c_y + \boldsymbol{j} a_z b_z c_y \\ &\quad + \boldsymbol{k} a_x b_x c_z + \boldsymbol{k} a_y b_y c_z + \boldsymbol{k} a_z b_z c_z \end{aligned} \tag{A.75}$$

This indeed is the sum of three independent vectors when added by columns with coefficients:

$$\begin{pmatrix} \lambda_{xx} & \lambda_{xy} & \lambda_{xz} \\ \lambda_{yx} & \lambda_{yy} & \lambda_{yz} \\ \lambda_{zx} & \lambda_{zy} & \lambda_{zz} \end{pmatrix} \tag{A.76a}$$

where $\lambda_{xx} = a_x b_x c_x$, $\lambda_{xy} = a_x b_x c_y$, etc. The same would have been accomplished if a new product were defined. Looking over Eq. (A.75), the relationship between

[14] This product was first introduced by J. Willard Gibbs.

$\boldsymbol{b}$ and $\boldsymbol{c}$ is one neither of a dot product nor of a cross product. We shall define this new product as the *dyadic* or *indefinite* product[14] and symbolically write it as $\boldsymbol{bc}$ without any sign of multiplication. From Eq. (A.75) it is seen that a dyad is a linear combination of vectors at a point, or a linear vector function of vectors. Then $(\boldsymbol{bc})$ is a *dyad* and if $\boldsymbol{a} \cdot (\boldsymbol{bc})$ must be equal to $(\boldsymbol{a} \cdot \boldsymbol{b})\boldsymbol{c}$ then the rules of the dyadic multiplication consist of putting the unit vectors and components together without any sign of multiplication *in the order of the product*:

$$\begin{aligned}\boldsymbol{bc} = \boldsymbol{ii}b_xc_x + \boldsymbol{ij}b_xc_y + \boldsymbol{ik}b_xc_z + \boldsymbol{ji}b_yc_x + \boldsymbol{jj}b_yc_y + \boldsymbol{jk}b_yc_z \\ + \boldsymbol{ki}b_zc_x + \boldsymbol{kj}b_zc_y + \boldsymbol{kk}b_zc_z\end{aligned} \tag{A.76b}$$

Dropping the nine basic dyads, $\boldsymbol{ii}$, $\boldsymbol{ij}$, $\boldsymbol{ik}$, etc., in a 3×3 matrix form the dyad $(\boldsymbol{bc})$ can be written

$$(\boldsymbol{bc}) = \begin{pmatrix} b_xc_x & b_xc_y & b_xc_z \\ b_yc_x & b_yc_y & b_yc_z \\ b_zc_x & b_zc_y & b_zc_z \end{pmatrix} \tag{A.76c}$$

This has no geometrical significance but when it operates on a vector or a vector operates on it, it gives a definite geometrical quantity: another vector. Then if $\boldsymbol{a}$ is dotted to these nine quantities, remembering that $(\boldsymbol{i} \cdot \boldsymbol{i})\boldsymbol{i} = (\boldsymbol{j} \cdot \boldsymbol{j})\boldsymbol{i} = (\boldsymbol{k} \cdot \boldsymbol{k})\boldsymbol{i} = \boldsymbol{i}$, $(\boldsymbol{i} \cdot \boldsymbol{j})\boldsymbol{i} = (\boldsymbol{j} \cdot \boldsymbol{k})\boldsymbol{i} = (\boldsymbol{k} \cdot \boldsymbol{i})\boldsymbol{j} = 0$, etc., then it is easily verified[15] that $\boldsymbol{a} \cdot (\boldsymbol{bc}) = (\boldsymbol{a} \cdot \boldsymbol{b})\boldsymbol{c}$. This same situation occurred in Eq. (A.50) in the Eulerian rate of change[16] $(\boldsymbol{U} \cdot \nabla)\boldsymbol{U}$, which could be written as $\boldsymbol{U} \cdot (\nabla \boldsymbol{U})$ where the gradient of the velocity vector $\nabla \boldsymbol{U}$ is a dyad with nine coefficients[17]:

$$\begin{aligned}\nabla \boldsymbol{U} = \boldsymbol{ii}\frac{\partial u}{\partial x} + \boldsymbol{ij}\frac{\partial v}{\partial x} + \boldsymbol{ik}\frac{\partial w}{\partial x} + \boldsymbol{ji}\frac{\partial u}{\partial y} + \boldsymbol{jj}\frac{\partial v}{\partial y} + \boldsymbol{jk}\frac{\partial w}{\partial y} \\ + \boldsymbol{ki}\frac{\partial u}{\partial z} + \boldsymbol{kj}\frac{\partial v}{\partial z} + \boldsymbol{kk}\frac{\partial w}{\partial z}\end{aligned} \tag{A.77}$$

If we drop the nine basic dyads the gradient of a vector can be represented in the matrix form

$$\nabla \boldsymbol{U} = \begin{pmatrix} \dfrac{\partial u}{\partial x} & \dfrac{\partial v}{\partial x} & \dfrac{\partial w}{\partial x} \\[2ex] \dfrac{\partial u}{\partial y} & \dfrac{\partial v}{\partial y} & \dfrac{\partial w}{\partial y} \\[2ex] \dfrac{\partial u}{\partial z} & \dfrac{\partial v}{\partial z} & \dfrac{\partial w}{\partial z} \end{pmatrix} \tag{A.78}$$

[15] The reader should verify this identity.

[16] This should be developed in component form as in Eq. (A.75).

[17] This was developed in Section 7.6.

APPENDIX B

Standard Environmental Data

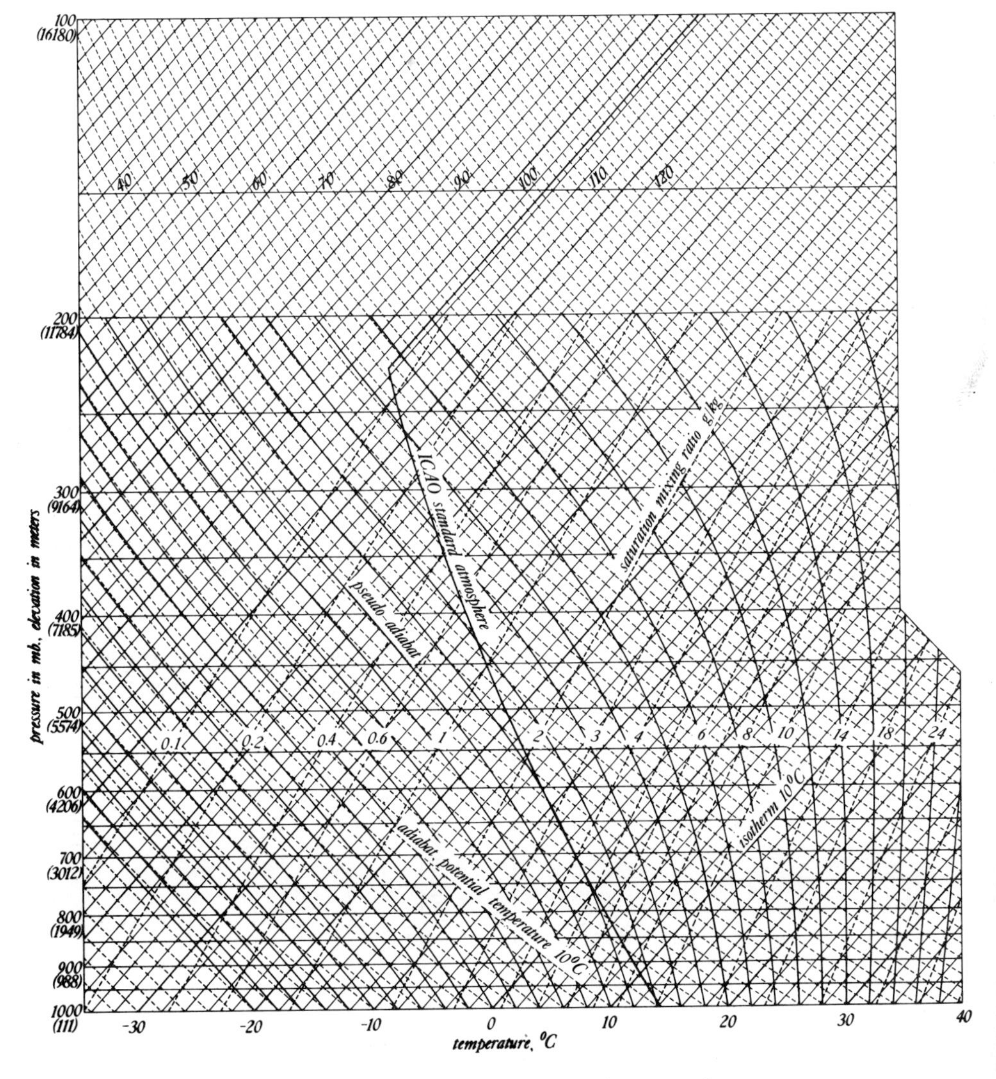

Fig. B.1 Skew T–log p chart for the atmosphere.

TABLE B.1

Atmospheric Properties Mid-Latitude Spring–Fall[a]

Altitude (km)							
H	z	Temperature (°K)	Pressure (mb)	Density (kg/m³)	Speed of sound (m/s)	Viscosity, $\mu \times 10^5$ (kg/m·s)	Thermal conductivity, $k \times 10^6$ (kcal/m·s·°K)
0	0	288.15	1.01325×10^3	1.225	340.3	1.789	6.053
1	1.000	281.65	8.987×10^2	1.112	336.4	1.758	5.931
2	2.001	275.15	7.950	1.006	332.5	1.726	5.807
3	3.001	268.65	7.011	9.091×10^{-1}	328.6	1.694	5.683
4	4.003	262.15	6.164	8.191	324.6	1.661	5.558
5	5.004	255.65	5.402	7.361	320.5	1.628	5.433
6	6.006	249.15	4.718	6.597	316.4	1.595	5.306
7	7.008	242.65	4.106	5.895	312.3	1.561	5.179
8	8.010	236.15	3.560	5.252	308.1	1.527	5.051
9	9.013	229.65	3.074	4.663	303.8	1.492	4.992
10	10.016	223.15	2.644	4.127	299.5	1.457	4.792
11	11.019	216.65	2.263	3.639	295.1	1.422	4.662
12	12.023	216.65	1.933	3.108	295.1	1.422	4.662

13	13.027	216.65	1.651	2.655	295.1	1.422	4.662
14	14.031	216.65	1.410	2.268	295.1	1.422	4.662
15	15.035	216.65	1.204	1.937	295.1	1.422	4.662
20	20.063	216.65	5.475×10^{1}	8.804×10^{-2}	295.1	1.422	4.662
25	25.099	221.65	2.511	3.947	298.5	1.449	4.762
30	20.142	226.65	1.172	1.801	301.8	1.476	4.862
35	35.194	237.05	5.589×10^{0}	8.214×10^{-3}	308.6	1.532	5.069
40	40.253	251.05	2.775	3.851	317.6	1.605	5.343
45	45.321	265.05	1.431	1.881	326.4	1.676	5.614
50	50.396	270.65	7.594×10^{-1}	9.775×10^{-4}	329.8	1.704	5.721
55	55.480	264.65	4.023	5.296	326.1	1.674	5.606
60	60.572	254.65	2.084	2.850	319.9	1.623	5.413
65	65.672	236.65	1.041	1.533	308.4	1.529	5.061
70	70.779	217.65	4.901×10^{-2}	7.844×10^{-5}	295.7	1.427	4.682
75	75.896	202.65	2.173	3.736	285.4	1.344	4.378
80	81.020	190.64	9.066×10^{-3}	1.657	276.8	1.275	4.132
85	86.152	190.57	3.701	6.762×10^{-6}	276.8	1.275	4.132
90	91.293	190.23	1.511	2.760	276.8	1.275	4.132
96	97.472	199.59	5.325×10^{-4}	9.155×10^{-7}	285.4	1.344	4.378
100	101.598	204.72	2.749	4.546	291.0	1.388	4.541
110	111.937	240.18	6.295×10^{-5}	8.625×10^{-8}	319.7	1.621	5.405

[a] After US Standard Atmosphere Supplements (1966).

TABLE B.2

Atmospheric Properties 30°N Latitude[a]

Altitude (km)		Temperature (°K)		Pressure (mb)		Density, ρ (kg/m³)		Speed of (sound m/s)	Viscosity $\mu \times 10^5$ (kg/m·s)	Thermal conductivity $k \times 10^6$ (kcal/m·s·°K)
H	z	January	July	January	July	January	July	July		
0	0	288.52	304.58	1.021×10^3	1.013×10^3	1.233	1.159	349.9	1.868	6.359
1	1.002	285.24	295.58	9.064×10^2	9.044×10^2	1.107	1.066	344.7	1.825	6.192
2	2.003	281.66	289.54	8.035	8.048	9.931×10^{-1}	9.683×10^{-1}	341.1	1.796	6.079
3	3.006	275.10	283.72	7.107	7.143	9.000	8.771	337.7	1.768	5.970
4	4.008	268.39	277.82	6.268	6.325	8.135	7.931	334.1	1.739	5.858
5	5.011	261.81	272.13	5.510	5.586	7.331	7.151	330.7	1.711	5.750
6	6.014	255.24	266.44	4.828	4.920	6.589	6.433	327.2	1.683	5.641
7	7.017	248.71	259.36	4.216	4.321	5.905	5.804	322.8	1.647	5.504
8	8.021	242.18	252.27	3.668	3.781	5.276	5.221	318.4	1.611	5.367
9	9.025	235.67	245.22	3.179	3.295	4.699	4.681	313.9	1.574	5.229
10	10.030	229.16	238.18	2.744	2.861	4.172	4.185	309.4	1.537	5.091

15	15.056	208.35	203.15	1.246	1.317	2.083	2.258	285.7	1.346	4.388
20	20.091	208.15	211.95	5.145×10^{1}	5.763×10^{1}	9.068×10^{-2}	9.471×10^{-2}	291.9	1.396	4.567
25	25.133	219.15	222.15	2.438	2.623	3.876	4.114	298.8	1.452	4.772
30	30.184	229.15	232.15	1.138	1.237	1.730	1.856	305.4	1.506	4.971
35	35.243	240.35	243.35	5.491×10^{0}	6.023×10^{0}	7.958×10^{-3}	8.622×10^{-3}	312.7	1.565	5.193
40	40.309	252.35	255.35	2.744	3.036	3.789	4.141	320.3	1.627	5.427
45	45.384	264.35	267.35	1.417	1.579	1.867	2.057	327.8	1.687	5.658
50	50.467	269.15	272.15	7.492×10^{-1}	8.410×10^{-1}	9.697×10^{-4}	1.076	330.7	1.711	5.750
55	55.558	261.15	264.15	3.941	4.456	5.258	5.876×10^{-4}	325.8	1.671	5.597
60	60.656	250.05	252.35	2.022	2.304	2.818	3.180	318.5	1.611	5.368
65	65.764	234.55	233.35	9.990×10^{-2}	1.140	1.484	1.701	306.2	1.512	4.995
70	70.879	219.05	214.35	4.703	5.311×10^{-2}	7.479×10^{-5}	8.632×10^{-5}	293.5	1.409	4.615
75	76.002	203.55	195.35	2.095	2.306	3.585	4.112	280.2	1.302	4.229
80	81.134	191.15	176.34	8.764×10^{-3}	9.189×10^{-3}	1.597	1.815	266.2	1.191	3.836
85	86.274	191.07	172.47	3.586	3.422	6.535×10^{-6}	6.909×10^{-6}	263.3	1.168	3.756
90	91.422	197.72	174.50	1.477	1.273	2.596	2.535	265.2	1.182	3.807
96	97.610	216.04	186.16	5.524×10^{-4}	4.130×10^{-4}	8.781×10^{-7}	7.598×10^{-7}	275.9	1.267	4.105
100	101.742	227.07	192.85	3.018	2.043	4.510	3.577	282.8	1.323	4.302
110	112.096	276.22	266.40	8.150×10^{-5}	4.880×10^{-5}	9.761×10^{-8}	6.002×10^{-8}	337.4	1.766	5.961

[a] After US Standard Atmosphere Supplements (1966).

TABLE B.3

Conversion Table

Quantity	Units	Equivalents	
Energy	1 joule	10^7 ergs	0.239 cal
	1 erg	9.48×10^{-11} Btu	1 dyne-cm
Enthalpy	1 kJ/kg	0.239 kcal/kg	0.43 Btu/lb
	1 kcal/kg	4.19 kJ/kg	1.8 Btu/lb
	1 Btu/lb	2.33 kJ/kg	0.56 kcal/kg
Gravitational acceleration		980.66 cm/s^2	32.174 ft/s^2
Length	1 m	3.281 ft	
	1 ft	0.3048 m	
	1 mile	1.609 km	
	1 km	0.621 mile (statute)	
Mass	1 kg	2.205 lb	
	1 lb	0.454 kg	
Pressure	1 bar	1.0197 atm	14.504 $lb/in.^2$
	1 atm	0.980 bar	14.223 $lb/in.^2$
	1 $lb/in.^2$	0.0689 bar	0.070 atm
	1 bar	10^6 $dynes/cm^2$	10^5 N/m^2
Specific heat	1 kJ/kg·°K	0.239 kcal/kg·°K	0.239 Btu/lb·°R
	1 Btu/lb·°K	4.19 kJ/kg·°K	1 kcal/kg·°K
Specific volume	1 m^3/kg	16.019 ft^3/lb	
	1 ft^3/lb	0.0625 m^3/kg	
Speed	1 knot	1.152 miles/h	1.853 km/h
Temperature	$\frac{°C}{100} = \frac{°F - 32}{180}$	$°K = °C + 273.16$	$°R = °F + 459.69$
Thermal conductivity	1 kcal/m·h·°K	0.672 Btu/ft·h·°F	
	1 Btu/ft·h·°F	1.488 kcal/m·h·°K	
Viscosity	1 kg/m·s	101.97 $P·s/m^2$	0.672 lb/ft·s
	1 $P·s/m^2$	0.0098 kg/m·s	0.00659 lb/ft·s
	1 lb/ft·s	1.488 kg/m·s	151.75 $P·s/m^2$
	1 poise (P) = 1 g/cm·s		
Volume	1 liter	0.0353 ft^3	

References

Batchelor, G. K. (1954). *Quart. J. Royal Meterol. Soc.* **80**, 339.

Belinskii, V. A. (1948). "Dynamic Meteorology." Gostekhizdat, Moscow.

Belinskii, V. A. (1967). "Dynamic Meteorology," "Gidrometeorologicheskoe Izdatel'stov," translated. Israel Program for Scientific Translations, Jerusalem.

Bjerknes, V. (1898). "Über die Bildung von Circulationsbewegung Videskabs." Skrifter, Oslo.

Bjerknes, V. (1933). "Physikalische Hydrodynamik." Springer-Verlag, Berlin and New York.

Blackadar, A. K. (1962). The Vertical Distribution of Wind and Turbulent Exchange in a Neutral Atmosphere, *J. Geophys. Res.* **67**, 8.

Blaton, J. (1938). *Bull. Soc. Geophys.* (*Warsaw*).

Bosanquet, C., Carey, W. F., and Halton, E. M. (1950). Dust Deposition from Chimney Stacks, *Proc. Inst. Mech. Engrs.* **162**, 355.

Bosanquet, C., and Pearson, J. L. (1936). The Spread of Smoke and Gases from Chimneys, *Trans. Faraday Soc.* **32**, 1249.

Brooks, F. A. (1959). "Introduction to Physical Microclimatology." Associated Student Stores, Univ. of California, Davis.

Brunt, D. (1939). "Physical and Dynamical Meteorology." Cambridge Univ. Press, London and New York.

Burpee, R. (1971). Birth of a Hurricane, *Sail* November, 40–44.

Cadle, R. C. (1966). "Particles in the Atmosphere and Space." Van Nostrand-Reinhold, Princeton, New Jersey.

Calder, K. L. (1952). Some Recent British Work on the Problem of Diffusion in the Lower Atmosphere. "Proceedings of a US Technical Conference on Air Pollution." McGraw-Hill, New York.

Carrier, G. F., Hammond, A. L., and George, O. D. (1971). A Model of the Mature Hurricane, *J. Fluid Mech.* **47**, 145–170.

Concawe Publication (1966). "The Calculation of Atmospheric Dispersion from a Stack." Le Hague.

Coriolis, G. (1835). Mémoire sur les Equations du Mouvement Relatif des Systémes des Corps, *J. Ecole Recherche Polytech.* **15**, 142.

Deacon, E. L. (1949). Vertical Diffusion in the Lowest Layers of the Amosphere, *Quart. J. Royal Meteorol. Soc.* **75**, 89.

Detrie, (1970). "La Pollution Atmospherique." Dunod, Paris.

Dufour, L., and Defay, R. (1963). "Thermodynamics of Clouds," Academic Press, New York.

Earth Photographs from Gemini III, IV, and V (1967). National Aeronautics and Space Administration, SP-129. Washington, D.C.

Edwards, G. R., and Evans, L. F. (1968). Ice Nucleation by Silver Iodide: III. The Nature of the Nucleating Site, *J. Atmos. Sci.* **25**, 249–256.

Ekman, V. W. (1905). On the Influence of the Earth's Rotation on Ocean Currents, *Ark. Mat. Astronom. Och Fysik.* **2**, 53.

Ekman, V. W. (1906). Beiträge zur Theorie der Meeresströmungen, *Ann. Hydrograph-Maritimen Meteorol.* **34**, 423.

Ekman, V. W., and Helland-Hansen, B. (1931). Measurements of Ocean Currents, *Kgl. Fysiograf. Sällskap. Lund, Forh.* **1**, 7.

Elsasser, W. M. (1942). Heat Transfer by Infrared Radiation in the Atmosphere, *Harvard Meteorol. Studies* **6**, 277–300.

Encyclopedia Brittanica (1965). Vol. 15. Univ. of Chicago Press, Chicago, Illinois.

Eskinazi, S. (1967). "Vector Mechanics of Fluids and Magnetofluids," Academic Press, New York.

Eskinazi, S. (1968). "Principles of Fluid Mechanics," 2nd ed. Allyn & Bacon, Rockleigh, New Jersey.

Eskinazi, S. (1973). Mean Particle Paths and Streaklines from Synoptic Weather Maps, *Int. J. Environ. Studies* **5**, 49–57.

Faire, A. C., and Champion, K. S. W. (1968). Recent Density, Temperature, and Pressure Results Obtained at White Sands Missile Range Compared with IQSY Results, *Space Res.* **8**, 845–854.

Faire, A. C., and Champion, K. S. W. (1969). *Space Res.* **9**.

Fay, J. A. (1973). Buoyant Plumes and Wakes, *Ann. Rev. Fluid Mechanics.*

Fischback, F. F., Graves, M. E., and Jones, L. M. (1969). Microwave Refractions as a Technique for Satellite Meteorology, *Space Res.* **9**.

Fleagle, R. G., and Businger, J. A. (1963). "An Introduction to Atmospheric Physics." Academic Press, New York.

Fleishman, B. A., and Frankiel, F. N. (1955). *J. Meteorol.* **12**, 141.

Frankiel, F. N. (1952). "On the Statistical Theory of Turbulent Diffusion." Geophys. Res. Paper No. 19. Cambridge, Massachusetts.

Gibbs, J. W. (1948). "The Collected Works," Vol. 1. Yale Univ. Press, New Haven, Connecticut.

Grant, F. C. (1971). "Proposed Technique for Launching Instrumented Balloons into Tornadoes. *NASA-TN*-D6503, Washington D.C.

Gustafson, T., and Kullenberg, B. (1936). Untersuchungen von Trägheitsströmungen in der Ostsee, *Sv. Hydrograf Biol. Komm. Skrifter, Ny Ser: Hydrograf* **13**.

Haltiner, G. J., and Martin, F. L. (1957). "Dynamical and Physical Meteorology." McGraw-Hill, New York.

"Handbook of Meteorology" (1945). McGraw-Hill, New York.

Harleman, D. R. F., and Ippen, A. T. (1967). "Two-Dimensional Aspects of Salinity Intrusion in Estuaries," Tech. Bull. 13. US Corps of Engineers, Vicksburg, Mississippi.

Haurwitz, B. (1940). *J. Mar. Res.* **3**, 254.

Haurwitz, B. (1941). "Dynamic Meteorology." McGraw-Hill, New York.

Hewett, T. A., Fay, J. A., and Hoult, D. P. (1971). "Atmospheric Environment." Pergamon Press, Vol. 5, 767.

Hinze, J. O. (1959). "Turbulence." McGraw-Hill, New York.

Holland, J. Z. (1953). "A Meteorological Survey of the Oak Ridge Area," USAEC Rep. ORO-99. Oak Ridge National Laboratory, Oak Ridge, Tennessee.

Howard, J. E., Matheson, J. E., and North, D. W. (1972). The Decision to Seed Hurricanes, *Science* **176**, 4040.
Kamotani; Y., and Greber, H. (1972). Experiments on a Turbulent Jet in a Cross Flow, *J. A.I.A.A.*, **10**, 1425.
Knudsen, M. (1901). "Hydrografical Tables." Bianco Luno, Copenhagen.
Kampé de Fériet, J. (1939). *Ann. Soc. Sci. Bruxelles, Ser. I* **59**, 145.
Keenan, J. H. and Keyes, F. G. (1936). "Thermodynamic Properties of Steam," Wiley, New York.
Kochanski, A. (1964). *J. Geophys. Res.* **69**.
Lacombe, H. (1965). "Cours d'Oceanographie Physique." Gauthier-Villars, Paris.
Laikhtman, D. L. (1961). Physics of the Boundary Layer of the Atmosphere, GIMIZ Gidrometeorologicheskoe Izdatel'stvo, Leningrad.
Laplace, P. S. (1812). Recherches sur Plusieurs Points du Systéme du Monde, *Mem. Acad. Roy. Soc.* **88**, 75.
Lettau, H. (1950). A Re-examination of the Leipzig Wind Profile, Considering Some Relations between Wind and Turbulence in the Friction Layer, *Tellus* **2**, 125–129.
Lucas, D. H. (1958). The Atmospheric Pollution of Cities, *Int. J. Air Pollution* **1**, 71.
McBean, G. A., Stewart, R. W., and Miyake, M. (1971). The Turbulent Energy Budget near the Surface, *J. Geophys. Res.* **76**, 6540–6549.
McDonald, J. E. (1964). Exclusion Limits for Clouds of Water and Other Substances, *J. Geophys. Res.* **69**, 3669–3672.
"Meteorological Tables" (1951). 6th ed. Smithsonian Institute, Washington, D.C.
Monin, A. S., and Obukhov, A. M. (1954). Basic Laws of Turbulent Mixing in the Ground Layer of the Atmosphere, *Tr. Geofiz. Inst. Akad. Nauk SSSR* **151**, 163.
Munk, W. H. (1963). "On the Wind-Driven Ocean Circulation" (A. R. Robinson, ed.). Ginn (Blaisdell), Boston, Massachusetts.
Nan'niti, T., Akamatsu, H., and Nakai, T. A. (1964). A Further Observation of a Deep Current in the East-North-East Sea of Torishima, *Oceanogr. Mag.* **16**, 11–19.
Neumann, G. (1968). "Ocean Currents." Elsevier, Amsterdam.
Newton, Sir I. (1686). On the Attrition of Liquids. "Principia," Book II, Sect. IX. S. Pepys, London.
Newton, Sir I. (1946). "Principia in Modern English" (Motte's translation, revised by Cajori). Univ. of California Press, Berkeley.
Panchev, S. (1971). "Random Functions and Turbulence." Pergamon Press, Oxford.
Pasquill, F. (1968). "Atmospheric Diffusion." Van Nostrand-Reinhold, Princeton, New Jersey.
Petterssen, S. (1956). "Weather Analysis and Forecasting," Vol. 1, 2nd ed. McGraw-Hill, New York.
Phillips, O. M. (1969). "The Dynamics of the Upper Ocean." Cambridge Univ. Press, London and New York.
Pochapsky, T. E. (1966). Measurements of Deep Water Movements with Instrumented Neutrally Buoyant Floats, *J. Geophys. Res.* **71**, 2491–2504.
Poisson, S. D. (1837). Mémoire sur le Mouvement des Projectiles dans l'Air en ayant égard á la Rotation de la Terre, *J. Ecole Roy. Polytech.* **16**, 1.
Prandtl, L. (1932). Meteorologische Anwendung der Strömunglehre, *Beitr. Physik der freien Atmosphäre, Bjerknes Festschrift* **188**.
Pratte, B. D., and Baines, W. D. (1967). Profiles of the Round Turbulent Jet is a Cross-flow, *J. Hydraulics Div. ASCE*, HY6.
Priestley, C. H. B. (1956). A Working Theory of the Bent-Over Plume of Hot Gas, *Quart. J. Meteorol. Soc.* **82**, 165.
Randhawa, J. S. (1969). "Stratospheric Circulation" (W. L. Webb, ed.). Academic Press, New York.

Richardson, L. F. (1920). The Supply of Energy from and to Atmospheric Eddies, *Proc. Roy. Soc. London, Ser. A* **97**, 354.
Roberts, W. O. (1972). We're Doing Something about the Weather, *Nat. Geogr.* **141** (4), 519.
Rossby, C. G. (1941). "Climate and Man, Yearbook of Agriculture for 1941." US Govt. Printing Office, Washington, D.C.
Schilling, G. F. (1964). Forbidden Regions for the Formation of Clouds in a Planetary Atmosphere, *J. Geophys. Res.* **19**, 3663–3667.
Schmidt, E. (1969). "Properties of Water and Steam in SI-Units." Springer-Verlag, Berlin and New York.
Scorer, R. S. (1958). "Natural Aerodynamics." Pergamon, Oxford.
Shuleikin, V. V. (1953). "Physics of the Sea." Izd-vo. Akad. Nauk SSSR.
Silverman, R. (1973). Shoaling and Erosion due to Baroclinic Conditions in the Flow of an Estuary. M. S. Thesis. Dept of Mech. and Aerospace, Syracuse University, Syracuse, New York.
Simpson, G. C. (1928). Further Studies in Terrestrial Radiation, *Mem. Roy. Meteorol. Soc.* **3**, 21.
Stewart, H. J. (1945). Kinematics and Dynamics of Fluid Flow. "Handbook of Meteorology." McGraw-Hill, New York.
Sutton, O. G. (1949a). A Theory of Eddy Diffusion in the Atmosphere, *Proc. Roy. Soc., Ser. A* **A146**, 701.
Sutton, O. G. (1949b). Atmospheric Turbulence. "Methuen's Monographs on Physical Subjects." Methuen, London.
Sverdrup, H. U. (1942). "Oceanography for Meteorologists." Prentice-Hall, Englewood Cliffs, New Jersey.
Swinbank, W. C. (1964). The Exponential Wind Profile, *Quart. J. Roy. Meteorol. Soc.* **384**.
Taylor, G. I. (1921). Diffusion by Continuous Movements, *Proc. London Math. Soc. Ser. 2*, **20**, 196.
Taylor, G. I. (1931). *Proc. Roy. Soc. Ser. A* **132**.
Tennekes, H., and Lumley, J. L. (1972). "A First Course in Turbulence." MIT Press, Cambridge, Massachusetts.
Townsend, A. A. (1956). "The Structure of Turbulent Shear Flows." Cambridge Univ. Press, London and New York.
Turner, J. S. (1969). Buoyant Plumes and Thermals, *Ann. Rev. Fluid Mechanics*.
US Standard Atmosphere (1962). US Govt. Printing Office, Washington, D.C.
US Standard Atmosphere (1966). US Govt. Printing Office, Washington, D.C. Supplement.
Wood, B. D. (1969). "Applications of Thermodynamics." Addison-Wesley, Reading, Massachusetts.
Zemansky, M. W., and VanNess, H. C. (1966). "Basic Engineering Thermodynamics." McGraw-Hill, New York.

Index

C

H

I

J

K

L

M

T

U

V

W

Z

A
B 5
C 6
D 7
E 8
F 9
G 0
H 1
I 2
J 3